COLLECTED SCIENTOMETRICS

Myron Evans

Published by New Generation Publishing in 2015

Copyright © Myron Evans 2015

First Edition

The author asserts the moral right under the Copyright, Designs and Patents Act 1988 to be identified as the author of this work.

All Rights reserved. No part of this publication may be reproduced, stored in a retrieval system or transmitted, in any form or by any means without the prior consent of the author, nor be otherwise circulated in any form of binding or cover other than that which it is published and without a similar condition being imposed on the subsequent purchaser.

www.newgeneration-publishing.com

New Generation Publishing

COLLECTED SCIENTOMETRICS

These collected scientometrics are divided in to four parts. Part one shows the interest from the top twenty nine universities in the world, part two gives the daily scientometrics for www.aias.us   back to April 30th 2004, part three contains combined scientometrics from May 2002 to date, and part four is a survey carried out from Dec. 1 - 15 2014 that shows that essentially all the UFT papers are being read. The scientometrics are probably the most detailed ever carried out for a new paradigm shift in science, the "Post Einstein Paradigm Shift" of Alwyn van der Merwe. It shows intense and sustained interest from all the best universities, institutes and similar in the world, signalling the end of the Einsteinian Era in natural philosophy. This books records about 2% of the vast total interest in ECE theory and the three AIAS sites www.aias.us, www.atomicprecision.com and www.upitec.org. We begin with a short history of the ECE unified field theory

The Einstein Cartan Evans (ECE) unified field theory was initiated in March 2003 following upon my discovery at Cornell University in November 1991 of the well known B(3) field of electromagnetic radiation. It was pointed out by Vigier in early 1993 that the B(3) field is definitive evidence of identically non zero photon mass, the first such evidence since the idea of photon mass was put forward by Poincare in 1905. At about the same time Einstein proposed a theory in which photon mass is identically zero. In the early twenties de Broglie developed his famous wave particle dualism on the idea of finite photon mass, so the photon is a relativistic particle like any other. This led de Broglie to think that all particles can be particle like and wave like, leading to the famous equation of wave particle dualism. The wave and particle nature of the electron was verified by Compton in what has become known as Compton scattering. Both Compton and de Broglie were awarded Nobel Prizes for their work and Einstein famously described de Broglie's work as lifting a corner of the veil. Thereafter de Broglie continued to develop the theory of photon mass in many ways at the Institut Henri Poincare in Paris.

The existence of photon mass was brought to my attention by Vigier in a letter which arrived in January 1993 and which is in the archives of www.aias.us. I had sent Vigier a manuscript on the development of B(3) field to "Physics Letters A" of which he was an editor. He wrote back inviting more papers and suggesting a review paper in van der Merwe's eminent journal, "Foundations of Physics". The review was published, but Vigier's assistant Malenfant blocked the former's invitations for several more papers to "Physics Letters A". It is unknown how an assistant could block the invitation of such a famous scientist and statesman in such an arbitrary way, effectively dismissing Vigier's life work on the photon with mass while still remaining his assistant. I had run into a kind of war between the protagonists of photon mass and the protagonists of zero photon mass, the so called "standard model", now entirely obsolete as the book of scientometrics shows conclusively. One only has to flip through the following pages to become quickly aware of the overwhelming impact of ECE theory among the best in the world.

At the 1927 Solvay Conference de Broglie had run into trouble with Heisenberg and in Vigier's words was "shouted down" by Heisenberg. It is not clear why de Broglie allowed himself to be shouted down in an irrational way, but evidently Heisenberg's tirade had no effect on him. Vigier mentioned this to me in private at the first Vigier Symposium of 1995 which I had helped to organize. The Solvay Conference split physics permanently because of differing interpretations of quantum mechanics. Einstein, Schroedinger, de Broglie and many others maintained that quantum mechanics is causal, meaning that all observable phenomena have a cause, and that quantum mechanics is an extension of classical mechanics, both being examples of Baconian philosophy. This point of view has been proven overwhelmingly by ECE theory, which successfully bases physics on eminently knowable geometry. The latter is the most precise branch of mathematics and was equated by the ancient Greeks to beauty. The following scientometrics show that this point of view has been overwhelmingly accepted in a quiet revolution described by van der Merwe as "The Post Einsteinian Paradigm Shift".

The problem with Heisenberg and Bohr is that they developed a hugely overelaborate interpretation of a basically simple thing, a commutator. That is always a sign of lack of understanding. The thing about science is that it must be simple and understandable and must always be tested against data. If a theory cannot be tested against data it is not a scientific theory, it is a mere fantasy or idol of the Baconian cave. This was the long refuted "Heisenberg Uncertainty Principle" which should be described more accurately as the principle of indeterminacy or the principle of the absolutely unknowable. It has been refuted theoretically in UFT175 on www.aias.us, now a heavily studied paper. It has been refuted experimentally in many ways, notably by the Croca group in the University of Lisbon. It is trouble from the start, and blatantly anti Baconian in the sense that things are asserted by the human mind to be absolutely unknowable. This means that they cannot be measured experimentally and leads to all kinds of tortuous nonsense. ECE sought to dismiss the Heisenberg indeterminacy from the outset and constructed an interpretation based on geometry. The huge discontent with all that is shown in the following pages. Every UFT paper is linked to the others, so acceptance of one paper means acceptance of all papers.

Einstein immediately rejected Heisenberg's indeterminacy and continued to reject it scathingly all his life. After the second war Vigier was invited to work with Joliot-Curie but soon resigned in protest at the French atomic bomb. He was also invited by Einstein to become his assistant but was refused a visa despite being a member of the French Legion of Honour (Legion d'Honneur), France's highest honour. Such a crass act of bigotry could well have resulted in sharp words between the United States and France, but they had faced difficult times together and now had to face the stalinist threat. Joseph McCarthy had begun to throw his weight around against foreign nationals, even scientists like Vigier who were of the highest calibre, and even scientists like Einstein who had warned Roosevelt of what the atomic bomb could do in the wrong hands, in fact Heisenberg's own hands because he tried to develop an atomic bomb for the Nazi regime as is well known. He need not have done that. In contrast Schroedinger, an Austrian, left for Ireland at the invitation of de Valera. A brutal, anti human, regime like this could have crept its ugly way into power in any European country, for example Moseley in Britain then, Farage in Britain now, or the Vichy Gestapo who captured Vigier in France when he was a member of the General Staff of the French Resistance. For this exceedingly dangerous work, Vigier was also the recipient of the de Gaulle Resistance Medal (Medaille de Resistance). The problem for McCarthy is that Vigier was an increasingly prominent member of the French Communist Party, a democratic party similar to the Bevanite arm of the Labour Party. That was enough to prevent Vigier working with Einstein, a great loss to science because Vigier could have persuaded Einstein to begin work on photon mass. Dirac was also essentially a Communist but by the time Dirac moved to the United States McCarthy had lost influence and the U. S. Government had come to its senses. People of my generation have maintained the peace in Europe for seventy years. We have had to because the ever present alternative is nuclear annihilation. Those foolish enough to break this peace will probably bring an end to Humankind, and there are hideous bigots who are constantly trying to achieve this dubious aim. That is the insane, always present, side of human nature and it only takes good men to remain silent in the words of Edmund Burke. Vigier himself became an enlightened humanitarian and a prominent French Statesman, and had no problem working with people of any nationality. If he had been able to persuade Klaus Barbie of photon mass, he would probably have been delighted to have guided him away from the irons. He was due for an appointment with the less than delightful Barbie in Lyon, but his train was bombed and he escaped. Galileo, of course, was also shown the irons, but now he has been decreed to have been right after all. This is exactly what has happened in the following pages. It is more or less true that the people does not cause the problem, it is the ugly regime that they have allowed to creep in to power. Once this happens the people suffers most of all. Things are no different in physics. It is not the logic that causes the problem, it is the dogma that has been allowed to creep into power. With the nuclear capability hovering over all of us, they must not be allowed to creep any more.

In the enlightened and relieved atmosphere of post war Paris, Prince Louis de Broglie has no difficulty in inviting the Communist Vigier to work with him in the Institut Henri Poincare. They were both citizens of a French Republic. Vigier mentioned to me that de Broglie was descended from the dei Medici of another enlightened era, renaissance Florence. I remember in discussions with Vigier that the latter had a reverential attitude towards de Broglie, but at once distant from him. After all, Vigier represented the Republic, de Broglie's family was that of the Kings of France. They worked as colleagues for about forty years until de Broglie died in his early nineties, still researching actively. As is well known, Vigier persuaded de Broglie to come out openly against the then all powerful and ruthless Copenhagen interpretation. They could and can be as ruthless as any party organization gone to the bad, or totalitarian regime. By doing this de Broglie faced ostracization and mockery, bigotry and hatred, but he was sublimely immune from all of that human weakness. Vigier was himself ruthlessly suppressed so despite being a great scientist he was awarded only one honorary degree. I was fortunate enough to be present at the degree ceremony in York University, Toronto , Canada, in 1995. As in any war, honorary degrees are awarded only to the friends of those who have transient influence.

This is the background of a physics war, a Peloponnesian war, that I ran in to unknowingly in January 1993. Like Thucydides I could not fiddle but could accidentally build Athens, a little village, into a great city state. This is what the following pages show. Before that letter from Vigier I had not even heard of photon mass. My well known discovery of B(3) was based on experimental data in nonlinear optics. It was based on the ability of circularly polarized radiation of any frequency to produce magnetization in any material matter. I simply and innocently reinterpreted the received opinion in terms of a longitudinally directed and fundamental field B(3). I simply used vector analysis to infer that the conjugate product of nonlinear optics must produce a longitudinal field in free space. This inference was accepted and published in three papers of 1992 and 1993 in "Physica B". The originals are in the Omnia Opera of www.aias.us. Unknown to me, B(3) was anathema to the then standard model of physics because its dogma resulted in blatant nonsense, the assertion that free space electromagnetic fields are transverse. If one takes the vector cross product of a transverse magnetic field plane wave and denotes it as B(1), and then denote its conjugate product by B(2), the result in vector algebra is iB(0)B(3)* where B(0) is the magnitude of B(3) and the asterisk denotes complex conjugate. This was so trivially obvious to me in November 1991 that I had no hesitation in submitting the idea to Physica B, which published it.

I was appointed a full professor of physics with tenure for my discovery of B(3) and began work at University of North Carolina, Charlotte in August 1992 from Cornell University's Theory Center. The Copenhagen establishment began to wake up to the fact that its dogma had just disintegrated. There was a full professor on the loose who had innocently reduced its logic to rubble. In fact this is easy to do because the dogma is so ludicrous, but under the Copenhagen regime, no one dared to do it. In my innocence, I did not even know that the regime existed. The following pages show that this ancient mariner dogma has been entirely rejected by a new generation of some of the brightest scientists and engineers in the world. The dogma is still taught, but it falls on deaf ears. The best scientists in the world silently study ECE theory, carefully keeping themselves hidden so as not to get their tenure or funding chopped, so as not to fail exams. This is what happens under any dictatorship, the propaganda is studiously ignored to coin an awful pun. In the time of Copernicus, sailors knew that the world was not flat, but they kept carefully silent.

We do not have to go in to what happened at UNCC, it was a vile farce described in the UNCC Saga on www.aias.us, but it does illustrate the lengths to which people will go to protect themselves. Under any regime, the people enthusiastically talk nonsense and survive. Not any more, as the following pages show. So the following pages are therefore of great historical importance. They narrate the Peloponnesian war in the lucid manner of Thycidides, without bias, without bigotry. Computer chips do not lie. Back in 1993 Vigier showed his character when he snubbed Barron, who had made an underhand attempt to turn Vigier against me. So the Oxford / Cambridge standard establishment failed to destroy the idea of B(3) and became infuriated. A massive attack was launched on B(3) and the doors of all journals were slammed shut without argument. There was one exception, van der Merwe's journals and book series. These continued to resist like the Spartans at Thermopylae so that every trivial argument against B(3) was refuted.

I was finally chopped by UNCC on laughable contrivances but continued to develop B(3) theory in almost complete isolation here in Craig Cefn Parc. Both Vigier and van der Merwe were very helpful and fortunately the e mail system was available. There was international outrage against UNCC, but those responsible were never brought to justice. The following pages, however, record overwhelming poetic justice. Having failed to crush the idea of B(3) the establishment resorted to blank censorship, pressure was brought on van der Merwe to try to prevent the appearance of inconvenient books such as "The Enigmatic Photon". Some of the most unpleasant elements resorted to outright abuse in the form of hate mail and so on. This was directed at my first wife, an entirely innocent party. One such abuser was caught and severely reprimanded at York University Canada and effectively forced to retire. In this atmosphere of the mid and late nineties the subject of O(3) electrodynamics was developed in many papers and two review articles in "Advances in Chemical Physics".

After a short stay in Ithaca, New York from 1998 to 2001, I returned to this house, after being naturalized a U. S. Citizen at Cornell University in 2000. The theory of O(3) electrodynamics continued to be developed, and in May 2002 key help was given in the form of a website, organized by Bob Gray of Biophan Inc. in New York State. The following pages record the earliest impact of that first website, in May 2002, when there were 859 files downloaded or "hits" from 155 distinct visits or study sessions, 0.02 gigabytes downloaded. I thought that this was delightful, I had never experienced the pleasure of seeing interest in my work recorded objectively and in real time by computer feedback programs. In the latest month of recorded scientometrics, March 2015, there were 133,452 files downloaded from 22,902 distinct visits or study sessions, and a thirteen year record high of 37.78 gigabytes downloaded. In the following pages this interest is very carefully filtered and selected, a quality assurance exercise. It is clear that the interest comes from the best in the world, and has been sustained for thirteen years. In these thirteen years there have been at least fifty million readings off the AIAS websites: www.aias.us, www.upitec.org and www.atomicprecision.com. I own the first of these and the second two are owned by Sean MacLachlan who works for Hewlett Packard in Boise, Idaho, U. S. A.

The first website of May 2002 records the material that I was able to send over by hand using primitive equipment. In scientific terms it records the gradual transition from O(3) electrodynamics to a theory which included gravitation. At that time the Omnia Opera had not yet been constructed and I was living in the most difficult of circumstances after the chaos caused quite deliberately by UNCC. This was a Peloponnesian war after all. My only lifeline to the outside world was the enlightenment of van der Merwe as editor of "Foundations of Physics Letters" and the new website, www.aias.us. It was prepared and constructed in an entirely voluntary way by Bob Gray. The establishment tactics were to close off every avenue of publication. This was done by harassing editors with a massive barrage of e mail, and frightening them in to breach of contract. World Scientific breached contract twice and Kluwer once. It is quite easy for a few ruthless individuals to frighten off publishers, because the editors of large journals rarely read the papers they are publishing. After a while they turned their unwelcome attention on van der Merwe, and destroyed or took over his journals: "Foundations of Physics" and "Foundations of Physics Letters". That is probably an all time low in physics. The complete intellectual dishonesty of that era is exposed in the following pages in a cool, rational and completely objective manner. It is now beyond all doubt that the best scientists and engineers in the world have rejected not only the standard model but the entire, rotten system of physics propaganda and censorship. One cannot stop the march of ideas, and why try to censor imagination? The answer is that such censorship is needed in order to obtain money from the taxpayer. That is not a scientific argument, or one that should be tolerated by society in an era of increasing problems for humankind.

The ECE theory crystallized after the very tough winter of 2002 to 2003 when I came across the well known book by Sean Carroll: "Spacetime and Geometry: an Introduction to General Relativity", particularly the end of chapter three which is a brief synopsis of a type of geometry developed by Elie Cartan in the nineteen twenties. Elie Cartan was probably one of the greatest of mathematicians of any era, and his work was carried on for many years by his son Henri Cartan. The Cartan geometry is elegant but abstract, so I have reduced it in the past thirteen years in many different ways to a format that is understandable to non specialists. The key inference by Cartan is to recognize the existence of a fundamental property of all geometry called torsion. In the era before Cartan, torsion was unknown, so the geometry used by Einstein from 1905 to 1915 is incorrect and the entire twentieth century era has been transformed and corrected by ECE theory. Torsion is defined in the first structure equation of Maurer and Cartan in terms of the tetrad and spin connection. I noticed that the first structure equation, written in terms of differential forms, has the same format as some vector equations of O(3) electrodynamics. That brought to mind the words of Johannes Kepler "Ubi materia, ibi geometria" - "Where there is matter there is geometry". Arthur Koestler in his brilliant book "The Sleepwalkers" (open source online) describes Kepler and his contemporaries as they brought in the post Aristotelian enlightenment: Copernicus, Galileo, Brahe, Kepler and Newton.

In March 2003 I inferred the first ECE hypothesis, that the electromagnetic potential is a Cartan tetrad within a scalar proportionality A(0). It followed that the electromagnetic field is the Cartan torsion within the same proportionality constant A(0). It follows that the gravitational potential is also a Cartan tetrad within another scalar proportionality, and that the gravitational field is also a Cartan torsion. The elegant geometry of

Cartan meant that the B(3) field could be incorporated naturally into classical electrodynamics. There is also a counterpart to B(3) in gravitational theory. This work of March 2003 produced the first successful unified field theory in the history of physics. It has been overwhelmingly accepted by the enlightened part of the intellectual world as the following pages show. So I posted UFT1 on the www.aias.us website with the help of Bob Gray and it was published in "Foundations of Physics Letters" by Alwyn van der Merwe, among the best editors of the late twentieth century.

All of van der Merwe's papers were edited by him, so he read and copy edited every paper himself. Each paper was refereed at least twice, and if there was disagreement, a third referee was consulted. The first fifteen UFT papers were all published in this way, using two or three referees, and eventually all posted. Sean MacLachlan took over from Bob Gray as postmaster of www.aias.us, which is hosted by Dave Burleigh's Annexa Inc. of Arizona. Sean redesigned the www.aias.us site and constructed a new website www.atomicprecision.com. As the following pages show these began to attract a vast readership and from inception, ECE theory was studied by the best in the world. All this is known with complete objectivity. The study takes place in the manner of Francis Bacon's Invisible College. It was dangerous then to question dogma too openly and in fact Bacon was subjected to a conspiracy and banished from Court, being imprisoned for a short time. Galileo was also placed under house arrest, intimidated and threatened by the trolls of his day.

These days things are no different. It is dangerous to find the truth in an innocent way. It is obvious I developed ECE theory in internal exile, very similar to that of Sakharov. So there is really no difference between dogmatic totalitarianism to the east or west of the ex iron curtain. So the study of ECE theory takes place quietly and carefully. The vast amount of interest recorded in the following pages is only the peak of an immense mountain of invisible readership. It is dangerous these days to question long held dogma, especially profitable dogma. The ECE theory is Cartan geometry itself, with a little bit of imagination. For example the tetrad postulate of Cartan geometry can be developed into a wave equation of geometry. It soon became clear from March 2003 onwards that all the equations of physics can be derived quite easily from geometry with the exception of the Heisenberg indeterminacy. There is nothing unknowable about geometry, which is physics.

I was delighted and astonished by the huge tide of international interest in my work. In those days I was still working on the theory more or less on my own, with the key help of the new websites. I still thought of websites as the poor relation of book and journal publishing, a method I had used from early graduate days to May 2002. Gradually I began to realize that the website method was fast, efficient and immensely powerful. Everything stood or fell on pure merit. The profession and readership discard poor material and keep on reading good material. As the following pages show they keep on reading ECE theory in droves.

The value of the computer feedback sites became clear. At first I had recorded the feedback in a few scattered notes but from April 30th 2004 I began to make a systematic record of it in notebooks. By now the scientometrics are used to measure the impact of ECE theory in many different ways. May 2002 coincided, more or less, with the start of the knowledge revolution on the internet. The www.aias.us website was way ahead of its time in using open source methods.

The Peloponnesian war heated up about this time into a series of battles between the protagonists of ECE and a few fringe zealots of the standard model. By now no one really takes the standard model seriously any more, as the following pages show. The censorship system of the old physics has also been brushed aside. Nothing now stands between good ideas and their readership. In the year 2004 van der Merwe began to come under pressure as an editor because of his enlightened support of ECE theory. Eventually his journals were cynically destroyed, but they are regarded by the vast majority as models of scientific enlightenment and editorial integrity. I became aware of a deep dishonesty among the fringe elements of the old standard model, and I also became aware for the first time of something called "wikipedia".

PART 1: SURVEY OF INTEREST IN ECE FROM THE BEST UNIVERSITIES IN THE WORLD

This survey compiled from the Book of Scientometrics shows that the ECE theory has been read regularly in the top ranking universities in the world since 30th April 2004 when the Book of Scientometrics was started. The top ranking universities are defined using the top twenty in Webometrics and the top twenty in the Times Higher Education World Reputation rankings. In order of ranking they are as follows. The asterisk denotes repeat, often numerous, downloads. "Wisconsin" denotes Wisconsin Madison, and "Minnesota" denotes Minnesota Twin Cities. There are twenty nine universities in the two lists in all. The numbers for each month give the total number of the top 29 universities from which visits were received in a given month.

Webometrics

Harvard, MIT, Stanford, Cornell, Michigan, Berkeley, Columbia, Washington, Minnesota, Pennsylvania, Texas Austin, Wisconsin Madison, Penn State, UCLA, Toronto, Yale, Oxford, Cambridge, Purdue, Texas A and M.

Times Higher Education World Reputation Rankings

Harvard, MIT, Stanford, Cambridge, Oxford, Berkeley, Princeton, Yale, Caltech, UCLA, Tokyo, Columbia, Imperial, Chicago, Michigan, ETH, Cornell, Johns Hopkins, Kyoto, Toronto.

2015

January: Cambridge*, Imperial, Caltech, MIT, Princeton, U Penn, Texas, Washington*, Wisconsin, ETH, Toronto*, Tokyo 12

February: Berkeley*, Columbia*, Cornell, MIT*, Washington*, Yale*, ETH, Oxford*, Cambridge*, Imperial*, Toronto*, Caltech, Johns Hopkins, Texas A and M, Chicago, Wisconsin Madison 16

March: Berkeley, Cornell, Illinois, MIT, Princeton*, Stanford, Texas A and M, UCLA, Michigan*, U Penn, Penn State, Washington, Wisconsin, Yale, Oxford*, Cambridge*, Imperial*, ETH, Toronto 19

2014

January: MIT, Princeton*, Stanford, Minnesota*, Texas A and M, Chicago, Wisconsin Madison, ETH, Oxford*, Cambridge*, UCLA. 12

February: Berkeley*, Caltech*, Michigan, Harvard, MIT, Penn State, Texas A and M, Toronto*, Washington, ETH*, Cambridge*, Imperial*, Oxford, Purdue. 14.

March: Berkeley, Caltech, Columbia, Cornell*, Harvard, Pennsylvania*, Princeton*, Penn State, Washington*, Wisconsin, Stanford*, Texas A and M, Cambridge*, Oxford. 14.

April: Berkeley, Caltech*, Michigan*, Princeton, UCLA*, Oxford*, Cambridge*, Imperial, Harvard, Penn State, UCLA, Pennsylvania, Wisconsin, Yale*, ETH.15.

May: Berkeley, Caltech*, Columbia*, Cornell*, Harvard, Michigan, Stanford*, Minnesota, Wisconsin Madison, Imperial*, Oxford*, Toronto, Imperial*. 13.

June: Columbia, Cornell, Cambridge*, Imperial, Tokyo*, ETH, Imperial. 7

July: Princeton, Michigan, ETH, Cambridge*, Imperial*, Columbia, Cornell*, Harvard, Oxford, Toronto, Tokyo, Johns Hopkins, Purdue. 13.

August: Caltech, MIT*, Princeton, Michigan, Texas, ETH*, Cambridge, Oxford, Berkeley, Columbia, Purdue*. 11

September: Cornell*, Stanford, Chicago, Minnesota, Pennsylvania, ETH, Cambridge*, Imperial, Harvard, Princeton, Penn State, Chicago, Minnesota, Pennsylvania, Wisconsin Madison, Oxford*, Cambridge, Toronto, Tokyo, Purdue. 20.

October: Caltech, Princeton, Penn State, Stanford, Texas A and M, Wisconsin Madison, Yale, Cambridge*, Oxford*, Toronto*, Tokyo, Berkeley*, Caltech, Columbia, Harvard*, Penn State, Stanford, Texas A and M*, UCLA, Washington, Imperial, Tokyo, Purdue*. 23.

November: Berkeley, Caltech*, Cornell, Princeton, Penn State, Texas A and M*, Wisconsin Madison*, ETH*, Cambridge*, Oxford*, Toronto*, UCLA, Pennsylvania, Kyoto, Purdue. 16.

December: Caltech*, Columbia, Cornell, Chicago*, UCLA, Pennsylvania, Washington*, Cambridge*, Tokyo*, Michigan, Purdue, Johns Hopkins. 12.

2013

January: Caltech*, Cornell*, Harvard, Michigan*, Chicago*, Cambridge*, Kyoto, Tokyo*, Berkeley, Columbia*, Johns Hopkins, MIT*, Purdue, Stanford, Texas A and M, Minnesota, Texas, Wisconsin Madison, Oxford*, Toronto*, ETH, UCLA*. 23

February: Berkeley*, Caltech*, Columbia*, Cornell*, MIT*, Purdue, Chicago*, Michigan*, Minnesota, Pennsylvania*, Texas*, Washington*, Wisconsin Madison*, Cambridge*, Imperial*, Oxford*, Toronto, Tokyo, Harvard*, Johns Hopkins, MIT*, Princeton, Penn State, Kyoto, ETH. 25.

March: Berkeley*, Caltech*, Michigan*, Columbia, Cornell, Harvard, MIT*, Princeton*, Stanford, Chicago*, Washington, Madison Wisconsin*, Yale, Imperial, Cambridge*,Toronto*, Tokyo, Purdue, ETH*, UCLA. 20.

April: Berkeley*, Caltech*, Harvard, Johns Hopkins, MIT, Princeton*, Penn State, Purdue, Chicago*, Texas*, Imperial*, Oxford*, Toronto*, Columbia*, Cornell*, Michigan, Texas A and M, Wisconsin Madison, Cambridge*. Tokyo*, ETH. 21.

May: Berkeley, Caltech, MIT*, Minnesota*, Stanford, Michigan*, Wisconsin Madison*, Texas*, Washington*, Cambridge*, Oxford*, Texas A and M, Imperial*, Kyoto, Tokyo, UCLA. 16.

June: Columbia*, MIT*, Chicago, Cambridge*, Imperial*, Oxford*, Toronto* Kyoto, Tokyo, Cornell, Michigan, Stanford*, Madison Wisconsin, ETH, UCLA*. 15

July: Chicago, Columbia*, Harvard, Michigan*, Penn State, Texas A and M, Wisconsin Madison*, Oxford, Berkeley, Cornell*, MIT, Princeton, Penn State, Yale, Cambridge*, Tokyo, ETH*. 17.

August: Cornell*, Harvard*, Princeton, Texas A and M, Washington*, Cambridge*, Oxford*, Tokyo, Texas, Minnesota, ETH, Yale. 12

September: Columbia*, Princeton*, Stanford, Texas A and M*, Michigan*, Minnesota*, Texas*, Wisconsin Madison*, Washington*, Imperial*, Harvard, MIT*, Penn State, Chicago, Texas, Oxford*, UCLA. 17.

October: Caltech, Columbia*, Cornell, MIT, Princeton*, Penn State*, Purdue, Texas A and M, Chicago, Yale, Imperial*, Oxford*, Caltech, Purdue, Harvard, ETH, Cambridge, UCLA* 18.

November: Berkeley, Caltech*, Columbia, Cornell, Princeton*, Stanford*, Texas A and M, Chicago, Michigan*, Pennsylvania, Washington, Cambridge*, Oxford*, Imperial*, Harvard, MIT*, Purdue, ETH, Toronto, UCLA. 19

December: Berkeley, Caltech, Harvard, MIT, Michigan*, Princeton*, Texas, Yale, Cambridge*, Oxford*, Columbia, Texas A and M, Imperial, Chicago. 14

2012

January: Toronto, Berkeley, MIT, Texas A and M*, Cambridge*, Kyoto, Caltech*, MIT*, Princeton, Purdue*, Minnesota, Texas, ETH, Imperial*, Oxford*, Tokyo. 16

February: Berkeley*, Caltech*, Harvard*, Stanford*, Texas A and M, UCLA*, Minnesota*, Washington, Cambridge*, Oxford*, Cornell, Johns Hopkins, Michigan*, Pennsylvania, Yale*, Purdue, ETH, Toronto*, Tokyo. 18

March: Caltech*, Michigan*, Cornell*, Harvard*, MIT*, Chicago, Washington*, Wisconsin Madison*, Oxford*, Imperial*, Berkeley, Minnesota, Penn State, Texas A and M. 14

April: Berkeley*, Columbia, Cornell*, Harvard*, MIT*, Princeton, Penn State*, Purdue, Texas A and M*, Chicago*, UCLA, Michigan*, Cambridge*, Imperial*, Oxford*, Tokyo, Minnesota, Pennsylvania, Washington, Yale, Texas. 21

May: Caltech*, Cornell*, Purdue, MIT*, Princeton*, Penn State, Texas A and M, Chicago*, Texas, ETH*, Imperial, Cambridge*, Oxford*, Tokyo, Harvard, Minnesota*, Washington. 17.

June: Cornell*, Chicago*, UCLA, Michigan, Washington*, Cambridge*, Oxford*, Caltech, Johns Hopkins, Michigan, Stanford, Pennsylvania, Wisconsin, ETH, Imperial*, Toronto, Tokyo*. 17

July: Berkeley*, Caltech*, Columbia*, Purdue, Stanford, Minnesota, Texas*, Washington, Imperial, Toronto, Cornell, Harvard, MIT, Michigan*, ETH. 15

August: Caltech, Michigan*, Penn State, Purdue, Columbia, Stanford, Texas, Cambridge, Imperial. 9

September: Berkeley*, Caltech*, Columbia*, Cornell*, Harvard*, MIT, Princeton, Penn State*, Purdue*, Texas A and M, Toronto*, UCLA, Chicago, Michigan*, Washington, Wisconsin Madison, ETH, Cambridge, Oxford*, Johns Hopkins, Minnesota, Texas, Kyoto. 23.

October: Berkeley, Caltech, Columbia*, Cornell*, Harvard, Johns Hopkins, MIT*, Princeton*, Penn State, Purdue*, Texas A and M, UCLA, Pennsylvania, Texas*, Yale*, Cambridge*, Imperial*, Oxford*, Toronto, Michigan*, UCLA*. 21

November: Caltech*, MIT*, Penn State*, Purdue, Stanford*, Texas A and M*, Chicago, UCLA*, Texas*, Wisconsin Madison, Yale*, ETH*, Cambridge*, Imperial*, Oxford*, Toronto, Columbia, Cornell, Harvard, Michigan, Minnesota, Kyoto. 22

December: Berkeley, Caltech, Columbia*, Cornell, Harvard, MIT, Stanford, Texas A and M*, Chicago, UCLA, Pennsylvania*, Texas, Washington*, Cambridge*, Imperial, Kyoto, Tokyo*, Johns Hopkins, Oxford. 19

2011

January: Caltech*, Columbia, Johns Hopkins*, MIT*, Penn State*, Purdue, Stanford*, Texas, Oxford*, Tokyo, Berkeley, Harvard, Michigan, Pennsylvania, Washington, Cambridge*, Toronto. 16.

February: Berkeley*, Caltech*, Harvard, Johns Hopkins*, MIT*, Penn State, Stanford*, Texas A and M*, Washington, Imperial*, Oxford*, Tokyo*, Columbia, Cornell, Princeton, Purdue, Chicago, Minnesota, Pennsylvania, ETH, Cambridge*, Tokyo. 22.

March: Caltech*, Cornell*, Harvard, MIT*, Princeton, Penn State, Purdue, Stanford*, Washington, Wisconsin, ETH*, Cambridge*, Oxford*, Kyoto*, Yale. 15.

April: Berkeley, Caltech*, Harvard*, MIT*, Princeton, Penn State*, Texas A and M, Minnesota, Washington, Wisconsin*, Yale*, Oxford*, Cambridge, Imperial*, Toronto*, Columbia*, Cornell, Purdue, Chicago*, Pennsylvania, Texas, Toronto, Tokyo. 23.

May: Johns Hopkins, MIT, Princeton, Purdue*, Stanford, Michigan*, Minnesota*, Washington, Imperial*, Cambridge*, Oxford*, Berkeley, Kyoto, Caltech*, Harvard, Penn State, UCLA. 17.

June: Columbia, Cornell*, UCLA, Texas, Cambridge*, Imperial, Oxford*, Berkeley, Purdue, Texas A and M, Michigan, ETH, Tokyo. 13.

July: Columbia*, Cornell*, MIT, Princeton, UCLA, Michigan*, Texas*, Wisconsin, Cambridge*, Oxford, Caltech, Johns Hopkins, Texas A and M, Chicago, Washington, Imperial, Kyoto*, Tokyo. 18.

August: Harvard, Stanford, Texas A and M, Michigan, Minnesota, Texas*, Wisconsin, Princeton, Cambridge*, Cornell*, MIT, Princeton, Penn State, Purdue*, Texas A and M, Oxford*, Imperial*. 17.

September: Berkeley*, Cornell*, Harvard*, Texas A and M*, Wisconsin*, Imperial, Oxford*, Columbia, MIT*, Princeton, Penn State, Purdue, Stanford, Michigan, Texas*, Washington*, Cambridge*. 17.
October: Caltech, Columbia*, Harvard*, Purdue, MIT*, Texas A and M, Chicago*, Texas*, Wisconsin, Yale*, ETH, Cambridge*, Toronto, Cornell, Princeton*, Purdue, UCLA, Oxford, Imperial*, Toronto*. 20.

November: Berkeley, Caltech*, Columbia, MIT*, Princeton, Penn State*, Purdue*, Chicago, Michigan, Minnesota, Texas, Washington*, Wisconsin, Yale, Cambridge*, Imperial*, Oxford*, Toronto*, Tokyo, Columbia, Cornell*, Harvard*, ETH. 23.

December: Caltech*, Columbia*, Cornell*, MIT*, Princeton*, Penn State, Purdue*, Stanford*, Texas A and M, Texas*, Michigan*, Washington, Wisconsin*, UCLA, ETH*, Cambridge*, Oxford*, Toronto, Imperial, Kyoto*, Tokyo, Harvard. 22.

2010

January: Berkeley*, Caltech*, MIT*, Princeton*, Stanford*, Chicago, Kyoto, Michigan*, Johns Hopkins, Princeton, Penn State*, Purdue*, U Penn, Yale, Cambridge, Oxford, Toronto, Tokyo. 18.

February: Berkeley*, Caltech, Columbia*, Michigan*, Harvard, Johns Hopkins, MIT*, Penn State*, Texas A and M*, Chicago, UCLA*, Minnesota, U Penn*, Washington, Cambridge*, Oxford*, Imperial, Cornell*, Texas*, Toronto. 20.

March: Caltech, Chicago, Cornell, Harvard*, MIT*, Texas A and M, Minnesota*, Imperial*, Oxford*, Michigan, Penn State, Purdue, U Penn. 13.

April: Berkeley, Caltech*, Michigan*, Harvard, Princeton*, Penn State, U Penn*, Washington, ETH, Cambridge*, Oxford*, Toronto, Stanford, UCLA, Minnesota, Texas, Wisconsin, Imperial, Kyoto. 19.

May: Cornell, Harvard, Purdue, MIT*, Princeton*, Penn State*, Minnesota, U Penn, Texas*, Washington, Cambridge*, Oxford*, Tokyo*, Harvard, Imperial*, Tokyo*. 16

June: Caltech*, Stanford, Texas A and M, Minnesota, U Penn, Washington, Wisconsin, Yale, Cambridge*, Oxford*, Tokyo, Columbia*, Cornell. 13.

July: Caltech, Harvard*, Johns Hopkins, MIT*, Princeton*, Penn State, Michigan*, Imperial*, Oxford*, Toronto, Stanford, UCLA, Washington, ETH, Cambridge*, Tokyo. 16.

August: Berkeley, MIT, Texas, Yale, Cambridge*, Tokyo, Cornell*, Harvard, MIT*, Texas A and M, Michigan, Imperial, Oxford. 13.

September: Caltech, Columbia*, Cornell*, Harvard*, MIT*, Purdue, Stanford, Texas A and M*, Wisconsin*, Cambridge*, Imperial, Oxford, Berkeley*, Princeton, Purdue, Michigan, Minnesota, Texas*, Yale*. 19.

October: Harvard*, MIT, Princeton, Penn State*, Purdue, Stanford*, U Penn*, Texas*, Yale*, Cambridge*, Berkeley, Columbia, Cornell*, Johns Hopkins*, MIT*, Princeton*, Penn State, Chicago*, Washington, ETH*, Imperial*, Oxford*, Tokyo* 23. out of over 700 visits from universities, institutes and similar during the month.

November: Berkeley*, Caltech*, Columbia*, Cornell*, Purdue, Johns Hopkins*, MIT, Penn State*, Stanford, Texas A and M, Chicago*, UCLA, Michigan*, Minnesota*, U Penn, Washington, Wisconsin*, Yale, ETH*, Cambridge*, Imperial, Oxford*, Toronto*, Tokyo*, Harvard*, Texas*. 26.

December: Berkeley, Caltech, Colorado, Cornell*, Harvard*, Johns Hopkins, Stanford, Chicago*, UCLA*, Michigan*, Minnesota*, U Penn*, Texas, Cambridge, Imperial. Oxford*, Toronto, Tokyo*, Kyoto. 19. 2009

January: Harvard*, Princeton, Purdue, Texas, Michigan, U. Penn, Washington, Wisconsin, Cambridge*, Oxford*, Stanford*, Toronto*, Chicago, UCLA, Yale, ETH, Tokyo. 17.

February: Caltech, Columbia, Cornell*, Harvard*, MIT*, Penn State*, Purdue, UCLA*, Minnesota, Texas*, Washington*, ETH, Cambridge*, Imperial*, Oxford*, Toronto*, Tokyo, Berkeley*, Wisconsin. 19.

March: MIT*, Princeton*, Stanford, Texas A and M, Minnesota, Wisconsin, Cambridge*, Oxford*, Imperial*, Toronto, Penn State*, Chicago, Michigan*, Yale. 14.

April: Caltech, Columbia, Harvard, Johns Hopkins*. MIT*, Penn State*, Stanford*, Texas A and M*, Toronto, Minnesota, U Penn, Washington*, Oxford, Purdue, UCLA, Michigan*, Texas, Wisconsin, Cambridge*, Imperial. 20.

May: Berkeley, Columbia, MIT*, Princeton*, Stanford*, Michigan*, Texas, Cambridge*, Oxford*, Toronto, Caltech, Cornell, Harvard, Johns Hopkins, Chicago*, Minnesota, U Penn, Washington. 18.

June: Caltech, Harvard*, Purdue, Texas A and M*, Washington, Oxford*, Imperial, Toronto, Purdue, ETH. 10.

July: Columbia, Purdue, Michigan, Texas, Princeton*, Kyoto, Cornell*, Harvard, Johns Hopkins, Penn State*, UCLA, Washington, Wisconsin, Imperial. 14.

August: MIT*, Penn State*, Stanford, Texas A and M, Chicago*. Michigan*, U Penn*, Texas, Washington, ETH, Oxford*, Toronto, Berkeley, Caltech, Columbia, Harvard, Princeton*, Purdue*, Minnesota, Cambridge*, Imperial. 20.

September: Columbia, Cornell, Harvard, Johns Hopkins*, MIT*, Penn State, Purdue, Stanford*, Texas A and M*, Minnesota, Texas, Washington, Wisconsin*, Cambridge*, Imperial*, Oxford*, ETH. 17.

October: Columbia, Cornell, Harvard, MIT*, Princeton*, Penn State*, Purdue*, Texas A and M, Chicago*, U Penn, Washington*, Wisconsin,

Cambridge*, Imperial*, Oxford*, Tokyo*, Caltech*, Cornell*, Johns Hopkins, Stanford, Toronto, Texas*, Yale. 23.

November: Caltech, Columbia*, Cornell*, MIT*, Penn State*, Purdue*, Chicago*, Texas*, Washington, Wisconsin*, Yale, ETH*, Cambridge*, Oxford*, Kyoto, Johns Hopkins, Princeton, Michigan, Minnesota, Imperial, Toronto, Tokyo*. 22.
December: Caltech, Columbia, Harvard*, Johns Hopkins, MIT*, Purdue, Texas A and M, Washington*, Yale, Cambridge, Imperial*, Oxford*, Tokyo, Michigan*. 14.

2008.

January: Johns Hopkins, ETH*, Cambridge*, Imperial*, Toronto*, Caltech, Michigan, Harvard, MIT, Minnesota, Wisconsin, Oxford*, Kyoto, Tokyo. 14.

February: Harvard*, Purdue*, Stanford, Minnesota, Washington, Cambridge, Michigan*, Johns Hopkins, U Penn, Cambridge*, Imperial, Oxford. 12.

March: Caltech, Johns Hopkins, MIT*, Princeton, Stanford, Texas A and M, UCLA*, Michigan, Wisconsin, Washington, ETH, Toronto, Berkeley, Harvard, Purdue. 15.

April: Caltech, Cornell, Princeton*, Stanford, ETH, Berkeley, Harvard, Texas A and M, ETH, Cambridge*, Imperial*, Toronto. 12.

May: Columbia, Cornell, MIT, Michigan, Purdue, Washington*, Imperial, Caltech, MIT, Texas, Toronto. 11.

June: Berkeley, Johns Hopkins, MIT, Princeton*, Washington, Cambridge*, Imperial*, Cornell, Minnesota, 9.

July: Berkeley, Caltech, Johns Hopkins, MIT, ETH*, Cambridge, Cornell, Harvard, Princeton*, Purdue, Stanford*, Minnesota, Toronto. 13.

August: Cornell, MIT, ETH*, Cambridge, Oxford, Texas A and M, Washington, 7.

September: Columbia, Michigan, Washington, Cambridge, Imperial*, Berkeley, Princeton, Purdue, Texas A and M, Washington, Oxford*, Toronto, 12.

October: Berkeley, Stanford, Chicago, Cambridge*, Imperial*, Oxford*, Toronto, Columbia*, Harvard, MIT, Penn State 11.

November: Caltech*, Columbia*, Johns Hopkins, MIT*, Texas A and M, Toronto*, Minnesota, Washington, Yale, Cambridge*, Imperial*, Oxford*, Cornell*, Princeton*, UCLA, Texas, Berkeley. 17.

December: Berkeley, Cornell*, MIT, UCLA, ETH, Cambridge, 6.

2007

January: Berkeley, Harvard, Johns Hopkins*, MIT, Princeton*, Penn State*, UCLA, U Penn, Washington, Yale, Cambridge*, Toronto, Purdue, Texas A and M, Texas*, Imperial, Tokyo. 17.

February: Harvard, Purdue, Stanford*, Chicago, Minnesota, Cambridge, Imperial*, Oxford*, Toronto, Tokyo, Berkeley*, Caltech, Harvard, MIT, Princeton, Penn State, UCLA, Washington*, Purdue, Michigan. 20.

March: Princeton, MIT, Penn State, Stanford, Texas A and M, UCLA, Michigan, Minnesota, Washington*, Wisconsin*, Cambridge*, Imperial*, Oxford*, Toronto, Tokyo, Berkeley, Johns Hopkins, Purdue, Yale. 19.

April: Caltech*, Columbia, Harvard*, Johns Hopkins, MIT*, Penn State*, Purdue*, Texas A and M, Washington, Oxford*, Berkeley*, Cornell, Stanford*, Cambridge*, Imperial, Tokyo. 16.

May: Berkeley, Columbia, Harvard, Johns Hopkins*, MIT, Princeton, Cambridge, Imperial, Tokyo, Caltech, Stanford, Oxford*. 12.

June: Stanford*, Texas A and M, Chicago, Wisconsin, Imperial, Toronto*, Caltech, Cornell, Johns Hopkins, MIT, Purdue*, Minnesota, ETH, Tokyo. 14.

July: Cornell*, Texas A and M*, UCLA, U Penn, Washington*, ETH, Imperial, Oxford*, Johns Hopkins*, MIT. 10.

August: Berkeley, Cornell, MIT*, Cambridge*, Oxford, Imperial*, Columbia, Johns Hopkins, Penn State, Stanford, Minnesota, Tokyo. 12.

September: Texas A and M, Michigan, Washington, Wisconsin, Oxford, Berkeley. 6.

October: Caltech*, Johns Hopkins, Purdue, U Penn*, Cambridge, Imperial, Oxford, Harvard* Penn State, Stanford, Imperial*, Kyoto. 12.

November: MIT, Princeton, Texas A and M*, Cambridge*, Imperial*, Oxford, Toronto, Kyoto, Tokyo, Johns Hopkins, Purdue, U Penn, Texas, Washington, Yale. 15.

December: MIT, Penn State, Purdue*, Texas, Washington, Cambridge, Imperial*, Oxford, Princeton. 9.

2006

January: Berkeley, Caltech*, Michigan, Columbia, Cornell, Harvard, Johns Hopkins, MIT, Princeton, Purdue, Stanford*, Texas A and M, Chicago, UCLA*, Michigan, U Penn*, Minnesota, Texas, Washington*, Yale, Cambridge. 21.

February: Caltech, Columbia*, Cornell*, Johns Hopkins, MIT, Princeton*, Texas A and M*, UCLA*, Michigan*, Minnesota*, U Penn, Texas, Wisconsin, Cambridge*, Oxford*, Toronto, Tokyo, Berkeley, Johns Hopkins*, MIT, Penn State, Purdue, Stanford*, Washington*, Yale, ETH, Imperial*. 27.

March: Berkeley*, Caltech*, Columbia*, Cornell*, Harvard*, Johns Hopkins*, MIT*, Princeton*, Purdue*, Stanford*, Texas A and M*, UCLA*, Michigan*, Minnesota, Washington*, Yale, ETH, Cambridge*, Oxford*, Imperial*, Toronto*, Kyoto, Tokyo*, Penn State*, U Penn, Texas*, Wisconsin*, ETH*. 28.

April: Berkeley*, Caltech*, Columbia, Harvard*, Johns Hopkins*, MIT*, Michigan*, Princeton*, Penn State*, Purdue*, Stanford*, Chicago, UCLA, Minnesota*, U Penn, Washington*, Wisconsin*, Yale*, ETH*, Cambridge*, Oxford*, Toronto*, Tokyo*, Cornell*, Texas A and M, Texas* 26.

May: Berkeley*, Caltech, Johns Hopkins, MIT*, Penn State*, Texas A and M, Michigan, Texas, Washington, Yale, Cambridge, Oxford*, Toronto, Tokyo*, Harvard, MIT*, Princeton*, Purdue*, Stanford, Washington*, 19.

June: Berkeley, Caltech, Johns Hopkins*, MIT, Princeton*, Stanford*, Texas A and M*, Chicago*, Michigan*, U Penn*, Washington*,

Wisconsin, ETH, Oxford*, Toronto*, Kyoto*, MIT, Purdue, UCLA, Texas, Cambridge*, Oxford*. 22.

July: Berkeley, Johns Hopkins*, MIT, Princeton, Purdue, Stanford, Texas A and M, UCLA, Michigan, U Penn, Texas, Washington*, Wisconsin, ETH, Oxford*, Toronto*, Kyoto, MIT, Purdue, Stanford, UCLA, U Penn, Texas, Cambridge* 24.

August: Berkeley*, Johns Hopkins*, MIT, Princeton, Purdue, Stanford, Texas A and M, Texas, Washington, Wisconsin, Cambridge, Oxford*, Toronto, Tokyo. 14

September: Berkeley, Caltech, Columbia, Harvard*, MIT, Princeton, Penn State*, Purdue*, Texas A and M*, Chicago*, Washington*, ETH, Cambridge, Oxford*, Cornell, Johns Hopkins, MIT, Princeton, UCLA*, Michigan*, Toronto, Kyoto, Tokyo. 23.

October: Berkeley, Caltech*, Harvard, MIT*, Penn State*, Purdue*, Texas A and M*, Chicago*, UCLA*, Michigan*, Minnesota*, U Penn*, Washington*, Wisconsin*, Yale*, Imperial*, Oxford*, Toronto*, Kyoto, Tokyo, Stanford, Texas, Cambridge. 23.

November: Harvard*, MIT, Princeton*, Penn State*, Stanford*, UCLA, Michigan*, Minnesota*, Yale*, Cambridge*, Oxford*, Toronto, Berkeley, Cornell*, Purdue, Texas A and M, Tokyo. 17.

December: Caltech, Columbia, MIT, Penn State*, Texas A and M, Michigan, Wisconsin, Washington, Cambridge*, Tokyo*, Johns Hopkins, ETH, Imperial*. 13.

2005

January: Yale*, Kyoto*, Berkeley*, Princeton*, Imperial*, ETH*, UCLA, Michigan, Harvard*, Chicago*, Washington*, Stanford*, Oxford*, U Penn*, Toronto*, MIT*,Caltech*, Purdue, Cambridge*, Penn State*, Texas, Cornell, Imperial*, Wisconsin*, Washington. 25.

February: MIT*, Chicago*, Minnesota*, Yale*, Oxford*, Michigan*, Cornell*, Texas*, Texas A and M, Tokyo*, UCLA, Toronto*, Stanford*, Columbia*, Penn State*, Cambridge*, Havard*, Caltech*, Washington*, Imperial*, Princeton*, Johns Hopkins*, Purdue*, U Penn, MIT, Wisconsin. 25.

March: Berkeley*, Washington*, Kyoto*, Cambridge*, ETH*, Michigan*, U Penn, Texas*, Wisconsin*, Oxford*, MIT*, Princeton*, Toronto*, Harvard*, Purdue*, Stanford*, Caltech*, Cornell*, Penn State*, UCLA*, Wisconsin*, Columbia*, Cornell*, Columbia*, Imperial*, Johns Hopkins*, Texas A and M*, Chicago 27.

April: Toronto*, ETH*, Caltech*, Princeton*, Michigan*, Wisconsin*, Oxford*, Cambridge*, Imperial*, Johns Hopkins*, MIT*, Yale*, Chicago*, Purdue*, Washington*, Cornell*, Stanford*, UCLA*, Berkeley*, Minnesota*, Columbia*, Texas A and M*, Texas*, Tokyo, Harvard*, U Penn*, Penn State*. 26.

May: MIT*, Cambridge*, Toronto, ETH*, Caltech*, Michigan*, Berkeley*, Columbia*, UCLA*, Cornell*, Harvard*, Yale*, Penn State, Imperial, Stanford*, Purdue*, Kyoto*, Washington*, Oxford*, Texas, Chicago, Minnesota, U Penn, Johns Hopkins, Tokyo, Wisconsin*, Texas A and M*, Toronto 28.

June: Berkeley*, Cornell*, MIT*, Texas A and M*, Washington*, Yale*, Imperial*, Toronto*, Princeton*, Stanford*, Michigan*, Texas, MIT*, Oxford*, Cambridge*, Columbia, Kyoto*, Caltech*, Johns Hopkins*, Tokyo*, UCLA, ETH, Michigan*, U Penn*, Texas, Harvard*, Penn State*, Purdue*, Minnesota. 29.

July: Berkeley*, Cornell*, MIT*, Texas A and M*, Washington*, Yale*, Imperial*, Toronto*, Princeton*, Stanford*, Michigan*, Texas*, MIT*, Oxford*, Cambridge*, Columbia, Kyoto*, Caltech*, Johns Hopkins, Washington*, Tokyo*, UCLA, ETH, Michigan*, U Penn*, Toronto*, Harvard*, Penn State*, Purdue*, Minnesota, 29.

August: Caltech*, U Penn, Imperial*, Columbia, Johns Hopkins*, Stanford*, Minnesota*, Texas, Washington, ETH, Wisconsin, Cornell, Harvard, Berkeley, Penn State, Texas A and M, Tokyo, Cambridge*. 18.

September: Columbia, Harvard*, MIT*, Penn State*, Purdue, Stanford, U Penn*, Texas*, Washington*, Cambridge*, Oxford*, Cornell, Michigan, Minnesota*, Imperial, Caltech, Princeton, Texas A and M, Michigan, Yale, ETH, Tokyo. 22.

October: Cambridge, Oxford*, Imperial, Berkeley, Caltech, Michigan*, Columbia*, Harvard, MIT, Princeton, Penn State, Stanford, Texas A and M, Chicago, U Penn, Texas*, Washington, Wisconsin, Yale, Tokyo, Toronto. 21.

November: Cambridge*, Imperial, Oxford*, Toronto, Berkeley, Caltech, Michigan*, Columbia, Cornell, Harvard, Johns Hopkins, MIT, Princeton, Penn State, Stanford, Texas A and M, Texas, Chicago, UCLA, Minnesota, U Penn, Washington, Wisconsin, Yale, ETH. 24.

December: Berkeley*, Caltech, Columbia*, Cornell*, Harvard*, Johns Hopkins, MIT, Princeton*, Texas A and M, UCLA*, Michigan*, Minnesota*, U Penn, Cambridge*, Oxford*, Imperial, Kyoto, Stanford, Washington, Yale, ETH. 21.

2004:

April 30th: Purdue, Cambridge.

May: Michigan*, U Penn, Wisconsin*, Penn State, Texas A and M, Chicago, Texas*, Cambridge*, Berkeley*, Caltech*, MIT*, Columbia, Cornell*, Harvard*, Yale, UCLA, Kyoto, Washington, Tokyo*, Oxford*, Penn State, Stanford, Princeton. 23.

June: UCLA*, Washington*, Cambridge*, Imperial*, Stanford*, Oxford*, Chicago, Berkeley*, Columbia, Johns Hopkins, Purdue*, Tokyo* , Penn State, Kyoto, ETH, Harvard, MIT*, Caltech, Minnesota. 19. French presidential staff on 29/6/2004.

July: Cambridge*, Imperial, Oxford*, Purdue, Washington*, Tokyo*, ETH*, Princeton*, Yale*, Michigan*, Stanford*, Harvard*, Berkeley, Wisconsin*, MIT*, Chicago, Minnesota, Texas A and M, Yale, Penn State, Tokyo. 21.

August: Michigan*, Harvard, Columbia*, Yale, Purdue*, Stanford*, MIT, Washington, Princeton, Chicago, Imperial*, Cambridge*, Caltech, ETH, Penn State, UCLA, Berkeley, Texas, Kyoto, Oxford. 20.

September: Johns Hopkins*, Michigan*, Purdue, Texas A and M*, Stanford*, Columbia*, Wisconsin*, Cornell*, Berkeley, MIT, Toronto*, Purdue*, U Penn, Princeton, MIT, Kyoto, Caltech*, Penn State, Chicago, Harvard. 20.

October: Harvard*, Michigan*, U Penn*, Caltech*, MIT, UCLA, Toronto*, ETH*, Stanford*, Cambridge*, Oxford*, Imperial*, Columbia*, Texas, Berkeley*, Johns Hopkins, UCLA, Purdue*, Yale*, Washington, Kyoto*, Texas A and M, Minnesota. 23.

November: Columbia*, Harvard*, Michigan*, Chicago*, Texas*, Caltech, Stanford*, Tokyo*, Cambridge*, Oxford*, MIT*, Minnesota, Princeton*, Purdue, U Penn, Johns Hopkins, Berkeley, Imperial, Yale, ETH*, Penn State. 21.

December: Columbia, Johns Hopkins*, MIT*, Chicago, Michigan, U Penn, Imperial*, Oxford*, Harvard*, Penn State*, Texas*, Toronto, Cornell, Princeton, Yale, Wisconsin, Tokyo, ETH, Purdue. 19.

PART 2: DAILY SCIENTOMETRICS for <u>www.aias.us</u> SINCE 30th APRIL 2004

This overview should be updated with reference to review papers UFT100, UFT200 and UFT281 to UFT288. The 3111 source papers to date of Einstein Cartan Evans (ECE) theory are posted on <u>www.aias.us,</u> <u>www.atomicprecision.com</u>, www.upitec.org and <u>www.et3m.net.</u> They are also published in seven volumes to date of M. W. Evans, AGenerally Covariant Unified Field Theory@ (Abramis 2005 to 2009) and in journals. The appendix shows that the theory has developed into a new school of thought in physics, and is used routinely in all sectors. The <u>www.aias.us</u> site also contains many articles and similar by colleagues who have applied and developed the theory in several directions. This is a brief overview of the papers, showing how the standard model of physics is rendered obsolete. The ECE theory can be applied in physics, mathematics, chemistry and electrical engineering, notably in the search for new energy. During the course of its development, nearly all the main concepts of the Astandard model@ have been shown to be obsolete. The demonstrations of its obsolescence are straightforward and based on the application of the fundamental commutator method. Each source paper is accompanied by several background notes which give comprehensive detail. There are about a thousand of these notes posted on <u>www.aias.us.</u> All this material is regularly studied by the international community in all its sectors: academic, corporate, institutional, military, governmental and individual scholars. The Appendix is about 1 - 5 % of the unprecedented total interest for the <u>www.aias.us</u> site, recorded from 30th April 2004 to present. These data have been chosen to represent the highest quality interest, which is assumed to be that from universities, institutes, government departments, military facilities, large household name corporations, and organizations and similar. The complete interest is archived on computer from mid 2006 onwards for <u>www.aias.us.</u> The total interest in ECE theory can only be described as astounding, and for the three ECE sites mentioned already runs into many millions of files downloaded (Ahits@). The monthly summary of interest is recorded back to the year 2002 on the file monthly.doc attached to this overview file. The relative impact of the theory has been measured against about fifty other commensurable sites and is given in the comparative impact table on <u>www.aias.us.</u> This shows clearly that ECE theory is the one that is making most impact by far at present. The impact is recorded by feedback activity software and is measured in many ways. The feedback archives are available on request. They are analysed daily and the Appendix is the overall result of five and a half years of daily analysis.

It is convenient to give a brief overview of the ECE theory in reverse chronological order. In later papers methods were strengthened and simplified to the point where the fundamental incorrectness and obsolescence of the standard model becomes clear to those with a minimum of mathematical training. A key concept in this development is that of the commutator of covariant derivatives. This is an operator which acts on a vector or any tensor in general to produce the fundamentals of geometry. The ECE theory is based directly on geometry, specifically on the concept of spacetime torsion, ignored incorrectly in the standard model. This basic error persisted throughout the twentieth century, meaning that a great amount of gravitational and cosmological work is incorrect, a major disaster for the old physics. This resulted in an unacceptable waste of public funding, it is estimated that the order of a hundred million dollars has been wasted on incorrect theory.

The overview of the contents of the ECE source papers is given first. Papers 122 and 131 to 134, and the five proofs on the homepage of www.aias.us introduce the antisymmetry theorem of ECE, and apply it to gravitation and electromagnetism. Both the gravitational and electromagnetic sectors of the old physics (standard model) are shown to be incorrect by straightforward use of the commutator operator in these papers and their background notes. The main result in gravitation is that the connection is antisymmetric in its lower two indices, not symmetric as in the old physics. This alone is enough to show that the whole development of twentieth century cosmology is meaningless. In ECE theory it has been replaced entirely by a development based on the correct use of torsion. In the electromagnetic sector papers 131 to 134 show that the antisymmetry theorem is enough on its own to render the Maxwell Heaviside theory meaningless, and again, this has been replaced entirely by a theory and engineering model based on the correct geometry, and correct use of the commutator. The two sectors are unified naturally by the use of geometry and by a simple fundamental hypothesis. Both sectors are rendered generally covariant as required by the philosophy of relativity. In these papers, the concept of spin connection resonance is shown to be central to energy from spacetime, as is the antisymmetry theorem itself.

Papers 129, 130 and 135 give the ECE equations of the fermion and the SU(2) equations of the electromagnetic field. The main philosophical advance of these papers is the final rejection in ECE theory of unobservables in physics. These are non-Baconian entities that cannot be observed experimentally. In the old physics unobservables proliferated, so that the entire subject became vastly over complicated and almost meaningless to science. Two examples rejected in these papers are the Dirac sea and negative energy. They are replaced by Baconian concepts based on geometry. The Dirac equation is shown to be based on geometry, and is simplified to an equation based on 2 x 2 matrices. The use of the 4 x 4 matrices of Dirac is shown to be over complicated. The anti-fermion is described in a much simpler way than that of Dirac, in accordance with Okham's Razor of philosophy, the simpler theory is preferred.

Papers 123 to 128 develop new cosmologies based on torsion, and without use of the incorrect Einstein equation. To dedicated scholars, the latter has been known to be incorrect almost since it was first inferred in 1915. For example Schroedinger and Bauer independently pointed out some problems with it as early as 1918. Since then many others have criticised it, notably Eddington, Dirac and Levi-Civita, one of the pioneers of tensors. The equation is based on the incorrect use of a symmetric connection, and the subsequent omission of torsion. In consequence, papers 93, 95, 96, 100, 117 and 120 of ECE theory show through the use of computer algebra that the equation is incorrect, sometimes wildly so. These computer algebra calculations will be collected shortly in a monograph, ACriticism of the Einstein Field Equation@. The Appendix shows that all these papers have been intensely studied since they were posted. The logic behind them is relatively simple and based on the Cartan Bianchi identity of standard differential geometry. The identity has been proven in all detail in papers such as 112, 109, 104, 99 - 103, 88 and 15. Again all these papers have been intensively studied worldwide, some for several years. The proofs are self checking and irrefutable.

Paper 121 gives the conservation theorems of ECE and paper 116 the continuity theorems. The most important result is that spin connection resonance obeys the theorems, so electric power from spacetime can be obtained without violation of any basic conservation or continuity theorem. Spin connection resonance (SCR) is a Bernoulli Euler resonance which does not violate any basic theorem.

Papers such as 49, 76, 77, 105, 106, 108, 110, 111, 114, 117, 118 and 120 deal with cosmology based on torsion and correct geometry, and no longer use the incorrect Einstein field equation. Paper 88 shows that the so called Asecond Bianchi identity@ used by Einstein is incorrect, again because of its use of an erroneous symmetric connection and erroneous neglect of torsion. Paper 88 gives the correct identity from the Cartan Bianchi identity. The Appendix shows that these key papers are studied routinely throughout the world, so funding of incorrect cosmology and of experiments that are badly designed should be abandoned. The Gravity Probe B experiment was shown in paper 117 for example to have produced no result at the cost of the order of several millions of dollars. The ECE theory gives the order of magnitude of the result without use of the Einstein field equation and without any expenditure of public funds. Any experiment designed to Aprove@ an incorrect theory is of course obsolete and should not be funded. Again, the antisymmetry theorem of ECE is enough to show that the theory behind CERN is incorrect. The CERN theory is based on electroweak theory which is in turn based on a U(1) sector symmetry for electromagnetism. This sector symmetry corresponds to the thoroughly obsolete Maxwell Heaviside theory. This has been heavily criticised throughout the twentieth century by many dedicated scholars. It was replaced in 1992 by B(3) theory and similar theories due for example to Horwitz et al, Amoroso et al., Lehnert et al., Barrett, and Harmuth et al. The B(3) field is observed in the inverse Faraday effect and is the precursor concept of ECE theory. The Maxwell Heaviside theory does not explain the inverse Faraday effect without the ad hoc introduction of the conjugate product of non-linear optics. All this is well known, and billions are wasted on CERN because it is based on a theory that is not only incorrect but one which is full of unobservables such as the Higgs mechanism and boson, and full of adjustable parameters. There are currently fears that CERN may not reach its specified energy level, and it has been seriously delayed. The Higgs boson was not observed in the LEP cooperation that preceded CERN. The reason made clear by ECE theory and by many other scholars, is that the Higgs boson does not exist.

Papers 63, 94 and 107 are papers in electrical engineering which are among the most read of the ECE papers as the Appendix shows. They use the concept of spin connection resonance introduced in papers 52, 53, 59 - 65, 61, 68 and 74 in response to well developed devices manufactured and sold by the Alex Hill company in Mexico City. The US Navy asked me to come up with a plausible explanation for the unprecedented spikes of electric power observed when it tested these devices, invented by Aureliano Horta and developed by Alex Hill using advanced robotics and microcircuits. It is significant in this context that all the major computer and circuit corporations have been studying ECE theory for nearly six years. For example IBM, Intel, Motorola, Microsoft, Sun, Hewlett Packard and many others. Again the Appendix shows this clearly. A range of devices based on SCR have now been developed and code developed by Lindstrom and Eckardt to simulate them.

Papers 85 to 87 give the ECE theory of the radiative corrections. Here ECE is preferred to QED by Ockham's Razor. As is typical of the old physics, QED is full of unobservables such as virtual particles, electrons that can go backwards in time, indeterminacy, and adjustable parameters such as dimensional regularization and renormalization. It is not known whether the series used in QED converges even after thousands of terms are used. To any objective scientist, QED is unworkable and QCD more so. During the course of development of paper 85 the disturbing discovery was made that the data of the NIST standards laboratory were wildy self-inconsistent, so the claim to accuracy of QED is false in several ways, In ECE the vacuum is filled with a primordial voltage, which causes the various radiative corrections. This primordial voltage can be amplifed by spin connection resonance, and the Tesla resonance is a manifestation of spin connection resonance.

Papers 80 - 84 develop the theory of the inverse Faraday effect (IFE) and radiatively induced fermion resonance (RFR) for one electron. If developed, RFR leads to magnetless ESR, NMR and MRI at ultra high resolution. This would be of immense benefit to chemistry and medicine. Again we see in the Appendix that the medical community studies ECE papers and articles regularly.

Papers such as 71 replace the obsolete gauge principle by one based on geometry, the invariance of the tetrad postulate under the general coordinate transformation. In the old physics, gauge theory was developed from the Heaviside potentials and elevated incorrectly into a principle of physics. The antisymmetry principle of the latest ECE papers shows that gauge theory is incorrect fundamentally. It is also incompatible with the Proca equation and photon mass theory as is well known to dedicated scholars.

Papers such as 40, 41, 54 to 57 deal with basics such as coordinate invariance in classical and quantum mechanics, geodesic analysis and second quantization in quantum field theory. It is shown in these papers that everything can be developed from geometry, a basic need of relativity. Paper 50 is the first review paper, and paper 48 describes straightforwardly an simply how the interaction of gravitation with electromagnetism can produce a change of polarization, as indeed observed experimentally when light grazes a massive object. This is one of the many experimental confirmations of ECE theory given in a table on www.aias.us and in the books by Kerry Pendergast on www.aias.us. Paper 47 introduces the dielectric version of ECE theory, a simplified version, and papers 45 and 46 give the ECE theory of the Sagnac effect. Papers 43, 44 and 107 give theories of the Faraday disk generator, paper 107 incorporating SCR into the theory to describe experimental reports of surges of electric power observed recently in a variable frequency generator. As can be seen in the Appendix these are always well read papers along with papers 63 and 94, and articles to do with new energy. This area is badly in need of public funding.

Paper 39 replaces the concept of dark matter with torsion based on the correct geometry and papers such as 37 and 38 with aspects of intrinsic spin, strong field theory and similar. Papers 33 - 36 prove the tetrad postulate in about ten different proofs, and prove the ECE Lemma in several ways. The Lemma is the geometrical basis of the wave equations of physics, and of quantum mechanics without indeterminacy. Papers 13 and 32 give a generally covariant commutator equation of wave mechanics which replaces the now obsolete Heisenberg indeterminacy principle. The latter is well known to dedicated scholars to have been thoroughly refuted using microscopy. By advanced optical microscopy in the Croca group, (Lisbon University), and electron microscopy in which atoms are observed directly. The latter was first achieved by Sir John Thomas and his group in the seventies. Kerry Pendergast gives a vivid description of how this was done, he himself being an observer of atoms. According to Heisenberg, Kerry, the measurer, made the atoms exist by invocation of the occult and by supplication to the idols of the cave. This illustrates vividly the grotesque decadence and intellectual dishonesty of the old physics, kept alive as it is by sinecures for the public purse and by attempted suppression of new thought. Mediaeval methods.

Papers 27 and 28 deal with the Aharonov Bohm and inverse Faraday effects, and papers 19 and 20 with electroweak theory and neutrino oscillation theory, both based on geometry. Paper 15 is always well studied because of its detailed and self checking proofs of the mainstays of Cartan geometry.

The first paper of ECE theory was written in March 2003, giving the field equations, and was followed shortly by the ECE wave equation in paper 2. The latter has been studied intensely for nearly six years. Paper 3 gave the basics of the unified field theory, and paper 4 gave the Dirac equation as a limit of the ECE equation. Finally paper 5 gave the theory of the strong and gravitational fields.

ARTICLES BY COLLEAGUES.

Over the past few years several articles, overviews and lectures have been prepared by colleagues on ECE theory. As can be seen from feedback these articles are all read each month, and are a valuable contribution to the progress of science. In July 2009 for example the most read articles were as follows.

1) ADevices for Spacetime Resonance Based on ECE Theory@ by Dr. Horst Eckardt, AIAS Director and 2009 TGA Gold Medallist. This is an article that has been studied intensively for nearly six years and gives an overview of how devices taking energy from spacetime may be constructed using the principles of spin connection resonance.

2) AEinstein Cartan Evans - a New View on Nature@ by Horst Eckardt and Lar Felker. This is a valuable overview of ECE theory which has been translated into several languages, and is always intensively studied each month. (Referred to as ECE).

3) AThe ECE Theory, History and Key Persons@, this is a valuable public lecture by Dr. Horst Eckardt, written on a level that the general public can comprehend.

4) AHow do Space Energy Devices Work?@ by Dr. Horst Eckardt, Spanish translation by Alex Hill. This is an introductory article which along with several other articles, has been translated into Spanish by Alex Hill.

5) AECE Engineering Model, the Basis for   Electromagnetic and Mechanical Applications@, by Dr. Horst Eckardt. These slides give the basic ECE equations of electrodynamics and dynamics.

6) AHow do Space Energy Devices Work, An Explanation by ECE Theory@, English language version of article 4.

7) AThe Life of Myron Evans@, a biography by Kerry Pendergast, an excellent book, always well read, and scheduled to be made into a film directed by Ken Russell.

8) ASpace Time Resonances in the Ampere Maxwell Law@ by Dr. Douglas Lindstrom, one of several articles on spin connection resonance with original input by Dr Lindstrom, sometime of the Alberta Research Council.

9) AComprehensive Advanced Aerospace Program@, by Charles Kellum and Associates, giving a description of a device family for new energy and counter gravitational applications.

10) ACrystal Spheres@, an original and always well read book by Kerry Pendergast which includes a history of the British Civil List scientists and many other things of interest.

11) Spanish translation of article 1.

12) AOn the Development of ASpacetime@ and a Unified Theory of Physics@ by Dr. Gareth J. Evans, an article which gives an excellent overview of ECE theory, pure and applied.

13) Spanish translation by Alex Hill of AThe Evans Equations of Unified Field Theory@ by Laurence Felker, whose original book was published by Abramis in 2007 and which was a Avirtual best seller@ on www.aias.us. (Referred to as F4).

14) Spanish translation of ECE.

15) Severe and definitive criticisms of black hole theory by Stephen Crothers, 2008 TGA Gold Medallist. These are given in several of the ECE papers and in an article, ABlack Holes, Unicorns and All that Jazz@.

16) AGalaxy Structures described by ECE Theory@ by Horst Eckardt, always a well read article in both English and Spanish, giving a popular account of galactic dynamics in ECE theory.

17) F13 in Spanish.

18) Spanish translation of article (16).

There are about forty of these articles and presentations in all, and every month every article is read worldwide.

SUMMARY ANALYSIS OF FEEDBACK PATTERNS.

The feedback patterns in the appendix have been studied over nearly six years and show a steadily increasing and continuous interest in ECE theory, both papers and articles. It can be seen that over this span of time, essentially all institutions to do with theoretical physics have begun to study the theory continuously. This implies an overwhelming rejection of the obsolete standard model among staff and students of several disciplines. The great amount of material downloaded from the site indicates that this is not casual interest, it is deep study. A table of the world's best universities and institutes is available on the internet, so it is possible to deduce objectively that they all study ECE theory routinely. A feature of the feedback is that there is considerable student interest, because of the periodicity of the visits. During semester or term time the number of university visits is greater than during vacation time. Although important, the university and institute sector interest is small compared with the industrial sector. Literally thousands of corporations study ECE theory routinely by now. Sometimes the appendix records major or household name corporations. In the computing sector for example there are repeated visits from IBM, Intel, Motorola, Hewlett Packard, Microsoft and so forth. In the aerospace sector there are repeated visits from Lockheed Martin and Northrop Grumman for example. Again, it is known that these are not casual visits because of the number of gigabytes downloaded and the number of files downloaded. Over nearly six years the latter amount to several million, and the former to several hundred gigabytes. The number of individual visitors per month peaked in June 2009 at over twenty five thousand during the month for all ECE sites: www.aias.us, www.atomicprecision.com and www.unifiedfieldtheory.info.

A significant feature of the interest is military and governmental. As can be seen form the appendix the US military sector continuously studies ECE theory. All four services have repeatedly visited www.aias.us: Army, Navy, Airforce and Marines. These visits include ones from the Pentagon and Tactical Commands. Many US Departments also study the site continuously, notably Departments of State, Transport, Health, NASA, NOAA, Government Laboratories, Government science agencies such as NSF and NIH, several sections of NASA, and so forth. There have been visits from Head of State staffs of several countries, and from the US Congress. A similar pattern of visits occurs from Europe.

Visits have occurred monthly from up to 116 countries. The leading interest occurs from the major industrialized countries as may be expected. The main areas of research work in theoretical physics may be identified clearly from the Appendix as North and Central America (USA, Canada and Mexico) and Europe. However there is also considerable interest from the Far East: (Australia, New Zealand, Japan, Taiwan and South Korea). The university and institute visits are uniformly distributed in each region, for example they cover the entire United States and the entire area of Western and Central Europe. Visits from the Russian Federation and China occur more rarely but they do occur continuously. There are also visits from the near East, notably from India, Singapore, Thailand, Indonesia and Malaysia. Visits also occur routinely from the entire continent of South America, notably from Argentina and Brazil.

There is an overwhelming number of visits from individuals. These are not recorded in the Appendix, but are totalled up in the appended file of monthly visits from 2002 to present. The complete feedback data for www.aias.us is backed up on computer from 2006 to present and is available on request. There are also many automated search engine visits, and ECE is picked up by every major search engine.

A significant feature of the feedback is that every month, over 1,500 documents are read off the www.aias.us site, including all the 135 ECE source papers to date, and all the forty or so books and articles written by other ECE scholars. It is also significant that the background notes to each paper are read every month. There are several hundred of these, and give more much more detail than the 135 source papers. This means that the theory is continuously studied in depth.

It may be concluded from every measure of feedback therefore that ECE is a new school of thought by international usage. Conversely the standard model of physics has been rejected as obsolete and unworkable. Government funding should therefore be switched away from expensive white elephants such as CERN to urgent matters of sources of new energy. Quality controlled website publishing has made journal publishing obsolete. Journal citations have been overtaken by measure of actual usage, a far better indicator of relevance, immediacy, timeliness, acceptance and usage. This is especially true when an obsolete elite restricts journal publication to the standard theory. It becomes known over time that this is happening, so confidence in these journals is eroded to almost nothing. Accordingly the readership has turned to the ECE websites, literally in their hundreds of thousands, if not millions. This is an unprecedented event in the history of science.

Evans of Glyneithrym,
British Civil List Scientist,
August 2009.

SELECTED FEEDBACK INTEREST IN ECE,
APRIL 30th 2004 TO PRESENT

* Denotes Multiple User Visits from Same URL.
Each URL may denote many visits from an individual user.

April 30th 2004
   Teledynefluid, Ruhr Bochum University, Northern Illinois University, Purdue, Utah, Elysee Palace France, Polytechnic University Hong Kong, University College Cork Ireland, Chubu Univ. Japan, American Physical Society, KBSU Russia, University of Cambridge.

May 1st 2004 (137 1574)
   Bochum, Univ Texas, Trinity College Dublin, Univ Pisa, Navy (NMCI).

May 2nd 2004 (144 1808)
   KU Leuwen Belgium, Teledynefluid, Colorado, Duke, Louisville, Louisiana State, CCF France, NASA (Lerc), NIST, Gov. Virginia, Univ. Pisa.

May 3rd 2004 (113 607)
   HWCDSB Ontario, Univ British Columbia, Univ Calgary, York Univ. Toronto, Motorola Corp., Colorado, Georgia State, Princeton IAS, Indiana, Univ. Michigan, U. Penn., Univ San Francisco, Wisconsin, Univ Pisa, Sandia, HQ European Space Organization.

May 4th 2004 (145 811)
   Gov. Alberta, CERN, Motorola*, Teledynefluid Corp., Ruhr Univ. Bochum, Denver, Oklahoma State, Penn State, Rowan, San Jose State, Texas A and M, Univ. Chicago, Univ Texas, CNRS Grenoble, INPG France, INSA Lyon, Lorraine, Toulouse, CNR (RM) Italy, Pisa, Navy (NMCI), Maastricht Univ., American Physical Society, Trinity College Cambridge, Heriot Watt, Luton, St Andrews.

May 5th 2004 (162 1321)
   DCC Gov. New Zealand, EPF Lausanne, Microsoft, Motorola*, Ruhr Univ Bochum, Georgia Tech., UCLA, ENSC Rennes, SNEF France, ENEA Italy, Army (ARL), KAU Sweden, Open University.

May 6th 2004 (140 892)
   Australian National Univ., FPMS Belgium, UNESP Brazil, EPF Lausanne, ETH Zurich, Algonquin College, Microsoft, Teledynefluid,

34

KFA Juelich, TU Cottbus*, Berkeley*, Caltech*, Central Michigan, Michigan State, Stonybrook, Utah, Wisconsin*, WUSTL, Xavier, CERMEP France, NASA (JPL), Army Medical Corps, Navy (NCSC), Sandia, NU Singapore.

May 7th 2004 (140 798)
    Motorola*, Philips, HU Berlin, Univ Frankfurt, Haverford, MIT, Illinois Chicago, Texas, William and Mary, BIPM France, Paris Sud, NASA (LERC), Trieste, Pisa, VSU Russia, British Gov.*, BIPM France.

May 8th and 9th 2004 (221 1536)
    Latrobe Univ Australia, TU Cottbus*, St Andrews, Caltech, Zaragoza, Leiden, LTH Sweden, Bristol, Melbourne, Merck, Mainz, Union College, TEE Greece, Latrobe Univ Australia, Teledynefluid.

May 10th 2004 (165 1631)
    William and Mary, Liege, Valle Univ. Colombia, Motorola, Teledynefluid, TU Cottbus, Heidelberg, Bowdoin, Brandeis, Cornell, Harvard, UWM, Williams, Wisconsin, Yale, Valencia, Southampton.

May 11th 2004 (150 980)
    Citicorp, Motorola, Hanover, Heidelberg, Stuttgart, Denver Univ., Temple, UCLA, BIPM France, UJF Grenoble, US Bureau of Land Management, Univ Patras, Army EPG, Liverpool, Southampton, Swansea, British Government.

May 12th 2004 (149 1000)
    Kuyshu Univ., UNP Argentina, York Univ. Canada, Boeing, Shell, Teledynefluid, Cornell, Newhaven, Ohio State, U. Mass., Univ Texas, UJF Grenoble, Univ Poitiers, Sandia, Univ Kyoto, NTNU Norway, NTHU Taiwan, Southampton, BBC, British Gov.*.

May 13th 2004 (138 417)
    Univ Western Ontario, Gov. Canada, Kodak*, Motorola, Rockwell, FGAN Germany, Univ California system, Georgia State, MIT, New York Univ., Rockefeller Univ., Univ Nevada Reno, UM Spain, Oak Ridge, Pisa, US Airforce Pik., US Army EPG, UIO Norway, British Gov.

May 14th 2004 (96 451)
    Dalhousie Univ., Northrop Grumman, Raytheon, Jena, Washington, INRA Bordeaux, INSA TLSE France, Tohoku Univ., US Army EPG, Gov. Taiwan.

May 15th 2004 (86 320)
   Teledynefluid, Iowa State, Christ's College Cambridge.

May 16th 2004
   Hamburg, Weizmann, Pisa, Tohoku, Waikato Univ., Bristol, BOR Tech. South Africa.

May 17th 2004 (145 1100)
   UIBK Austria, ARC Alberta, Motorola, Teledynefluid, LRZ Munich, Munster, Monroe, Washington, CNAM France, Poitiers, Toulouse, ICTP Trieste, Univ. Tokyo, US Airforce AFIT, US Edwards Airforce Base, Airforce RL, Army EPG, Leiden Univ., TU Delft, USU Russia, Bristol.

May 18th 2004 (153 932)
   Buffalo, Melbourne, NRC Canada, Calgary, CERN, Motorola, Pfizer, Ruhr Univ Bochum, TU Darmstadt, Tuebingen, CTC Univ., GASOU Univ., Kansas State, Univ. California Riverside, Univ Iowa, Univ Maryland, Toulouse, Univ Patras, Pisa, KNCT Japan, TMD Japan, Miyasaki, Tokyo, Ibaraki, US Army EPG, KUN Netherlands, TU Delft, Lancaster, Physics Oxford, MRC South Africa.

May 19th 2004 (139 584)
   Woolongong Univ Australia, Philips, Teledynefluid, Univ. Bonn, Indian River CC, LLUMC, Ohio State, Genoa, Pisa, US Army EPG, US Army Stewart, Univ Durham.

May 20th 2004 (109 734)
   Georgia Tech., Penn State, Stanford, UV Spain, US Army EPG, Univ. Luton

May 21st 2004 (105 331)
   Teledynefluid, Brigham Young, Colorado State, CSU Santa Barbara, Harvard, ISI, New York Univ., Ohio State, Delaware, UJF Grenoble, Lawrence Livermore, NIH, TMD Japan, Massey Univ., New Zealand, Southampton.

May 22nd 2004 (74 965)
   Murdoch Univ., Australia, US Army EPG.

May 23rd 2004 (82 543)
   Australian National Univ., CERN.

May 24th 2004 (123 389)

Australian National Univ., Univ Sidney, Lockheed Martin, Motorola, Teledynefluid, Heidelberg, Mississipi, UCHSC, Michigan, Texas, USAL Spain, Caen, Technion Israel, US Army EPG.

May 25th / 26th 2004

Lockheed Martin, Motorola*, Teledynefluid, TU Cottbus, CSUN, Florida State, Indiana, Temple, UC San Diego, U Texas, NASA (Lerc), Weizmann Institute, Univ Turin, UIO Norway, American Physical Society, Metu Univ. Turkey, Oxford, Hampshire County Council, Australian National Univesity, Lockheed Martin, Motorola, Teledynefluid, Heidelberg, Mississipi, UCHSC, Michigan, U Texas, CSIC Spain, USUAL Spain, Caen, Technion, US Army (EPG), ARIS Romania, Cankaya Univ Turkey.

27th May 2004 (118 1316)

ESI Austria, Latrobe Univ Australia, Univ Freiburg, Teledynefluid, Wuppertal, KB Denmark, Florida Tech., Harvard, UMBC, Toulouse, KTH Sweden, Sheffield.

28th May 2004 (116 711)

Motorola, Raytheon, AGI Germany, Hamburg, Leipzig, Caltech, CWSL, Central Washington Univ., BIUS France, Lawrence Berkeley National Laboratory, NASA (LERC), Pope US Airforce Base, Bangor.

29th May 2004 (77 332)

Max Planck Dresden, Princeton, Chapel Hill, TEE Greece.

30th May 2004 (172 974)

Teledynefluid, Duke, OTE Greece, KTH Sweden, NTHU Taiwan.

31st May 2004 (117 358)

TPL Toronto, Cornell, Harvard, City Univ Hong Kong, Pavia, Kaiyodai Univ. Japan, NU Singapore, CU Turkey.

1st June 2004 (138 1464)

Latrobe Unv. Australia, Teledynefluid, Ruhr Univ Bochum, Paderborn, ESF, UCLA, Maryland, Kentucky, Washington, JYU Finland, Bordeaux 2, US PLK Airforce Base, Bristol, Cambridge.

2nd June 2004 (119 714)

UC Riverside, Lockheed Martin, Cedar Crest, Ohio, Stanford, Valdosta, JYU Finland, INSA Lyon, US Randolph Airforce Base, US

Military DARPA, American Physical Society, KOU Turkey, Imperial College.

3rd June 2004 (139 784)

FUNDP Belgium, Motorola, FU Berlin, CSU Ohio, De Paul, Denver, Fullerton, NEIU, UCLA, Indian Statistical Institute, US Wright Patterson Airforce Base, UJ Poland, Imperial, Lancaster, Oxford.

4th June 2004 (118 771)

Australian National University, EPF Lausanne, Motorola, Max Planck Mainz, US Dept of Energy, NASA (Lerc), TEE Greece, US Pope Airforce Base, NU Singapore, London School of Economics, Hampshire County Council.

5th June 2004 (117 1016)

Univ Waterloo Canada, IBM Almaden, FU Berlin, Georgia Tech., TEE Greece, UNN Russia, NTU Singapore.

6th June 2004 (149 738)

City University Hong Kong, Indiana Tech., U Chicago, NTUA Greece, Tel Aviv, Kobe Univ Japan, Univ York, ULA Venezuela, National Tech. Univ. Athens.

7th June 2004 (192 635)

Defence Department Canada, Univ. Calgary, EPF Lausanne, Hewlett Packard, Lockheed Martin*, Motorola, NKTRC Denmark*, Berkeley, Columbia, CSU Fresno, Florida Tech., Stanford, SUNY Stonybrook, UV Spain, IAP France, French Particle / High Energy Lab., TEE Greece, GARR Italy*, Pavia, Kyushu, Sandia, Victoria Univ New Zealand, UJ Poland.

8th June 2004 (160 672)

Oregon State, Vienna, UNICAMP Brazil, Motorola, Philips, TUBS Germany, Karlsruhe, Johns Hopkins, Urbana Champaign, HUT Finland, Univ Patras, Kyushu, UJ Poland, Oxford (physics), Southampton, British Government, PUK South Africa.

9th June 2004 (174 984)

Queensland, NS Canada*, Motorola, Polaroid, Munich, Arizona State, Indiana, North Carolina State Univ., Purdue, Rutgers, Univ South Carolina, Williams, UPM Spain, AMI Finland, TREAS, GARR Italy*, INFN Italy, Postech South Korea, Nottingham University.

10th June 2004 (168 880)

Queensland, Intel, Institute of Physics, Motorola, Sun, Munich, Duke, Maryland, Vermont, NASA (GSFC), DUTH Greece, CU Hong Kong, Weizmann, Victoria Univ NZ, American Physical Society, KTH Sween.

11th June 2004 (130 493)

McGill, Motorola, Maricopa County California, US Gov. PNL, EWHA South Korea, HIS Iceland, NTU Singapore, OCMS Oxford.

12th June 2004 (116 309)

TU Vienna, Jena, Ege Turkey, Cranfield, Exeter College Oxford.

13th June 2004 (102 653)

Phelps Dodge, Georgia Tech., U Connecticut, Saunalahti Finland, Univ Tokyo, Ege Turkey, Gantep Turkey.

14th June 2004 (150 620)

Univ Ottawa, Motorola, Univ Trier Germany, Australian National Univ., Berkeley, Penn State, Purdue, Urbana Champaign, Vermont*, UV Spain, KFKI Hungary, European Space Agency, ICTP Trieste, Kyoto Univ., US Military DCAA, US Naval Research Laboratory, American Physical Society, NTU Singapore, Heriot Watt, Southampton.

15th June 2004 (136 693)

FH Aargau, Motorola, GWDG Germany, Hannover, Leipzig, Paderborn, Middlebury, Ohio State, UCLA, UV Spain, BIPM France*, Patras, KUN Netherlands, Imperial College.

16th June 2004 (146 574)

Brigham Young, Univ Western Australia, ETH Zurich, Teledynefluid, Paderborn, Columbia, Harvard, Hawaii, New York Univ., University College Cork, Pisa, Leeds, New College Oxford.

17th June 2004 (153 989)

Hewlett Packard, TU Cottbus, Colby, Case Western Reserve, Urbana Champaign, UV Spain, NASA (GSFC), US Airforce DM, PUK South Africa.

18th June 2004 (146 1485)

HCRC Taiwan, Freiburg, Motorola, Teledynefluid, NKTRC Denmark, LeHigh, AFKI Hungary, INFN Italy, Victoria Univ New Zealand, TU Kielce Poland.

19th June 2004 (106 1003)

Leipzig, Univ Swabia, Valdosta, US Gov PTO.

20th June 2004 (125 806)

NTU Singapore, Honeywell, Teledynefluid, CLU, Univ Tokyo, Oxford.

21st June 2004 (154 829)

Govt. Australia, Freiburg, Motorola, Washington Univ., UPV Spain, ENSTA France, Paris Psud, UJF Grenoble, CEC European Govt., Sandia, Norwegian Military, PUB Romania, Univ. York.

22nd June 2004 (157 487)

Gov. Australia, Univ of Zurich, Motorola, Teledynefluid, Jena, SUNY Stonybrook, UC Santa Barbara, UPC Spain, INLN France, NASA (Lerc), CEC European Govt., INFN Milan, Univ Venice, Navy (NSWC), BUAP Mexico, MSUIIT Philipines, Univ York.

23rd June 2004 (138 1139)

Western Ontario, Heidelberg, Motorola, Deutsches Museum, CSU Ohio, MIT, Fort Smith, Vermont, Xavier, UV Spain, ONERA France, Perpignan, Fermilab, Jerusalem, INIT Latvia, American Physical Society, Iyute Turkey, BBK Britain.

24th June 2004 (148 670)

Defence Dept. Australia, Saskatchewan, Marquardt, Motorola, Bonn, Ulm, Illinois Tech., Vermont, AENA Spain, UV Spain*, CNES France, INFN Italy, Postech S. Korea, Kluwer, Cambridge, Oxford.

25th June 2004 (109 376)

Lausanne, Arizona, Caltech, FHDA, Stanford, Unican Spain, ENAC France, ONERA France, Trinity College Dublin, US Army (EPG), Metu Turkey, Aberdeen, Imperial, Italian Academic and Research Network.

26th June 2004 (83 353)

Teledynefluid, Charles Univ. Prague, INLN France, PUK South Africa.

27th June 2004 (92 477)

Caltech, Sydney, General Electric, Essen, Auburn, Purdue, HUT Finland, NTU Taiwan, Oxford.

28th June 2004 (230 707)

NU Singapore, Siemens Austria, Queen's Univ. Canada, Boeing, Disney, General Electric, Lockheed Martin, Raytheon, Teledynefluid,

Charles Univ. Prague, Columbia, ECU, Humboldt, Nevada Reno, Utah, Kentucky, William and Mary, Washington, UGR Spain, UV Spain, Italian Academic and Research Network, Messina, YLC Taiwan, Derby College (Britain).

29th June 2004 (276 853)

Gov. Canada, General Electric, Motorola, Perimeter, MUNI Czechia, Bonn, IUPUI, MIT, IC Irvine, Urbana Champaign, Nevada Reno, Xavier, UCM Spain, US Spain, UC Spain, Elysee Palace (French Presidential Staff), IAP France, Montpellier 2, US Navy (NSWC), KTH Sweden, Oxford, QMW Britain.

30th June 2004 (152 1273)

Sydney, Motorola, Teledynefluid, Heidelberg, Florida Atlantic, PUC, Rutgers, SWT, Minnesota, UV Spain, CEA Saclay France, ESPCI France, KYU Tech Japan, KTH Sweden, Mahidol Thailand, KOU Turkey, SCU Taiwan.

1st July 2004 (172 1167)

IBM Almaden*, Motorola, Volvo, MIT, STTHOM, Texas Tech, Valdosta*, UV Spain, HUT Finland, Argonne, Milan, RUG Netherlands, American Physical Society, ICI Romania, NU Singapore, NCTU Taiwan, NTU Taiwan, Cambridge, Imperial, Zululand South Africa.

2nd July 2004 (154 627)

TU Vienna, New Brunswick, Motorola, Maryland, UV Spain, P and M Curie, Nice, INFN Italy, UTL Portugal, Oxford.

3rd July 2004 (96 497)

TIE Chile, NTUA Greece, Technion, National Technical Univ. Of Athens.

4th July 2004 (86 250)

Royal Military College of Canada, NJUST China, Trieste, DLSU Philippines.

5th July 2004 (127 556)

CERN, Basel, TIE Chile, General Electric, Max Planck Heidelberg, NKTRC Denmark, Purdue, Valdosta*, Washington, UPC Spain, UPM Spain, UV Spain, HUT Finland, NTUA Greece, INFN Italy, Osaka, Tokyo, NU Singapore, Oxford, Warwick.

6th July 2004 (135 633)

Australian National Univ., Boeing*, German Synchrotron Facility, Konstanz, Southern Ilinois, UPS, UV Spain, US DoE, NUI Galway, INFN Italy, Utrecht, Bydgoszcs, Anadolu, Cardiff. SUN South Africa.

7th July 2004 (152 456)

ETH Zurich, CLU, Pittsburgh, Renselaer, SDSC, Arkansas, Xavier, UV Spain, INSA Lyon, French Polyechnique System , US Gov BLM, NASA (Larc), Technion, Milan Tech., Rome 1, IJS Slovenia, KH Taiwan, Southampton.

8th July 2004 (142 545)

CERN, Basel, ETH Zurich, Lockheed Martin, Bremen, Heidelberg, Indian River CC, UAF*, Illinois Chicago, UMB, New Mexico, Valdosta, US Gov FAA, US Gov NREL, Oak Ridge, INFN Italy, Trieste, Sandia, Porto, BBC.

9th July 2004 (191 566)

TIE Chile, Nissan, Shell US, Max Planck Heidelberg, TU Berlin, Swabia, Georgia Tech., Princeton IAS, Minnesota, Xavier, Yale, INPL Nancy, US Gov DOEAL, Honk Kong Tech., Univ College Cork, RUG Netherlands, DLSU Philippines, TCRC Taiwan, Oxford, Warwick.

10th July 2004 (128 533)

Max Planck MPIS, Florida State, Univ. F. Comte, Strathclyde.

11th July 2004 (93 602)

City Univ Hong Kong, Simon Frazer, TU Ilmenau, SUNNY Stonybrook, NTUA Greece, TMN Russia, Univ Ljubliana Slovenia.

12th July 2004 (129 715)

City Univ Hong Kong, Defence Dept. Canada, Gov. Canada Space Agency, Alberta, CERN, Motorola*, Nokia, Florida, UMSL, Western Washington, Yale, BIPM France, NOOS France, Metropolitan Univ Japan, US Airforce Wright Patterson, US Navy NWC, TU Eindhoven, Hull.

13th July 2004 (143 1074)

Memorial Health, Motorola, Swabia, Central Michigan, UNI, UVA Spain, TEE Greece, Technion, INFN Italy, Pavia, US Robins Airforce Base, Utrecht, Massey Univ NZ, Cambridge. British Gov.

14th July 2004 (187 553)
   Alberta, Quebec, Motorola, Cologne, Arizona, North Carolina State, North Dakota, Ohio Univ., UV Spain, US Gov BLM, Brookhaven, Lawrence Berkeley, Surrey, BBC, British Gov., PUK South Africa.

15th July 2004 (167 753)
   Brigham Young, Dow, IBM Almaden*, Kodak, Motorola, Stuttgart, BGSU, Iowa State, KSBE*, Stanford*, TCU, UAF, Vanderbilt, Xavier, UPM Spain, US Gov BLM., TEE Greece, ADSI Hungary, European Space Agency, INFN Italy, US Nellis AF base, Eindhoven, Free Univ. Amsterdam, FUW Poland, GU Sweden, GTSI Slovakia, Bristol, Lauder, British Gov.*

16th July 2004 (151 422)
   Univ Victoria Canada, Basel, Motorola, IFH Germany, Erlangen, Florida State, Johns Hopkins, KSBE, Princeton, URV Spain, Fermilab, NASA (Lerc), UST Hong Kong, INFN Italy, Milan Tech., ICTP Trieste, Ibaraki Univ. Japan, Metropolitan Univ. Japan, American Physical Society, Jefferson National Laboratory, Univ. Anglia, Univ Surrey, British Govt.

17th July 2004 (121 365)
   Univ. New Brunswick Canada, General Electric, Univ. Florida, ICTP Trieste, UNIBA Slovakia.

18th July 2004 (109 868)
   Florida State, Lousiana State.

19th July 2004 (172 1513)
   CERN, Motorola, TU Chemnitz, Berkeley, Harvard, SUNY Stonybrook, TMC, Huston, Wisconsin, YSU, UV Spain*, HUT Finland, American Physical Society, US Govt. BLM, City Univ. Hong Kong, US Navy Pacific Northwest, KUN Netherlands, SP Singapore, UNIBA Slovakia, Sheffield, Stafford, British Govt.

20th July 2004 (211 775)
   Brigham Young, Australian Gov. Defence, Sao Paolo, Victoria, AT and T, IBM Almaden, Motorola, Paderborn, American Univ Cairo, Central Michigan, Florida Atlantic, LCC, MIT, Chicago, Minnesota, Washington, Wayne State, UV Spain, HTV Finland, US Gov BLM, Sandia, US Gov USBR, US Army MSL, Otago NZ, ORU Sweden, Cambridge, Queen Mary, Surrey.

21st July 2004 (190 576)

UMR, Massey Univ NZ, Vienna, UFAL Brazil, Motorola, Northrop Grummond, Teledynefluid*, German Bundestag, Max Planck MPPMU, STW Bonn, Karlsruhe, Regensburg, KSBE, North Carolina State, Northwestern, Stanford, Michigan, Washington Univ St Louis, UC Spain, EDCSM France, OLEANE France, Yamagachi Univ Japan.

22nd July 2004 (154 681)

Harvard, Motorola, Erlangen, Harvard, UC Irvine, INLN France, NASA (Gsfc), UTH Greece, INFN Florence, US Pope Airforce Base, Univ York.

23rd July 2004 (148 498)

Manitoba, Motorola, Stanford, Texas A and M, Dallas, UNED, UNICAN Spain, UV Spain, BIPM France, ENS CACHAN France, INFN Bologna, US Navy NRL, Imperial.

24th July 2004 (125 494)

Manitoba, Teledynefluid*, Paderborn, UCLA, Yale, UTH Greece, ICM Poland.

25th July 2004 (129 1379)

IBM Almaden*, MIT, Bordeaux 1, NTUA Greece, Massey Univ. NZ, NTU Singapore.

26th July 2004 (152 789)

Defence Gov Australia, ISI, MSUIIT Philipines, IIHE Belgium, CERN, Motorola, Teledynefluid, Thyssen, German Synchrotron Facility, Mainz, Penn State, Kentucky, SCS Spain, UV Spain, US Dept of State, European Govt. CEC, Trieste, NTU Singapore, NCL Britain.

27th July 2004 (154 644)

ISI, National Research Council Canada, AFU Canada, Motorola, HU Berlin, Max Planck Heidelberg, MPIPKS Dresden, Arizona, Central Michigan, Case Western Reserve, ISI*, MIT, Princeton, UV Spain, HTV Finland, ONERA France, BARC India, KEK Japan, US Airforce Pope, US Airforce Wright Patterson, MOH Govt. NZ, BC Yugoslavia.

28th July 2004 (162 1149)

ISI*, Defence Gov Australia, CERN*, Motorola*, Teledynefluid, Springer, Stuttgart, KUMC, Michigan, Vermont, WHOI, UPC Spain, UV Spain, US Gov. Courts, Kyoto, Univ Tokyo, Utrecht, Bristol.

29th July 2004 (153 1001)

Australian National Univ., Gov British Columbia, Motorola, Indian River CC, ISI, Urbana Champaign, Wisconsin, UV Spain, Lawrence Livermore, Technion, Univ Trieste, Univ Tokyo, NTU Japan, FSA British Govt.

30th July 2004 (167 785)

Linz, Simon Frazer, Boeing, Hewlett Packard, Motorola, Teledynefluid, Thyssen, Pittsburgh, American Physical Society, NTHU Taiwan, Univ College London, Romanian Educational Network.

31st July 2004 (135 684)

Sandia, Bath.

1st August 2004 (154 748)

Teledynefluid, Wuppertal*, Michigan, Sheffield.

2nd August 2004 (164 1056)

CERN*, Motorola*, Goettingen, Harvard, Rochester, IAP France, Univ. Trieste, LIU Sweden, Ljubliana, Swansea, UMIST.

3rd August 2004 (180 832)

ECU Australia, CERN, Motorola, Teledynefluid, Stuttgart, IPFW*, Najah, Rockefeller, Delaware, Univ. College Dublin, American Physical Society, KTH Sweden, Lancaster, Lawrence Berkeley.

4th August 2004 (157 601)

KU Leuwen, CERN, Lockheed Martin, Motorola, Columbia, Renselaer, Stanford SLAC, SUNY Stonybrook, North Carolia Charlotte, UNL, CEA France, SNPE France, Rennes 1, Argonne, Los Alamos, Trieste, US Military NIPR, Massey Univ NZ, Confederation of British Industry, NTU Singapore, British Gov DSTL.

5th August 2004 (158 722)

Adelaide, Motorola, Teledynefluid, Heidelberg, AUS Denmark, Boston, UC Santa Cruz, USC, William and Mary, Yale, NOOS France, US Dept of Energy, NASA (JPL), US Govt. NREL, Massey Univ NZ, The Childrens' Hospital, Edinburgh.

6th August 2004 (169 563)

US Airforce Academy, Motorola, Max Planck Potsdam, TU Darmstadt, Arizona State, UNL, Amiedu Finland, US Army Stewart, Kluwer, MSU Russia, Umea Sweden.

7th August 2004 (153 486)
Sydney, Motorola, Teledynefluid*, Texas Lutheran.

8th August 2004 (135 500)
CERN, Texas Arlington.

9th August 2004 (145 435)
Petro Canada, Teledynefluid*, IT Volvo, UJEP Czechia, Max Planck MPQ, Bremen, Ohio State, SWRI, BIPM France, ESPCI France, IHES France, Strasbourg, US Gov BLM, Fermilab, NASA (ARC), US Airforce McGuire, NTT Russia*.

10th August 2004 (179 1443)
UNS Argentina, Manitoba, Heidelberg, Intel, Lockheed Martin, Motorola, Dartmouth, OU, Purdue, Stanford, Chapel Hill, ICTP Trieste, Ameican Physical Society, Lkubliana, NTCB Taiwan.

11th August 2004 (168 477)
Chiba Univ Japan, Tsinghua China, Heidelberg, Colorado, Georgia Tech., Purdue, Stanford, Florida, Michigan, Pasteur Univ France, City Hong Kong, Yamaguchi Univ Japan, US Military USCG, Massey Univ. NZ, Jefferson National Lab., NTU Taiwan, Doncaster, Manchester, St Andrews.

12th August 2004 (167 511)
CERN, Motorola, Dortmund, Schwabia, Tuebingen, Columbia, MIT, Nevada Reno, Washington, Xavier, INM Spain, Helsinki, Fermilab, ELTE Hungary, Mathematical Institute Hungarian Academy, Tel Aviv, Technion, KYU Tech Japan, Victoria Univ NZ, American Physical Society.

13[tH] August 2004 (176 700)
NIST, UNAL Colombia, Hewlett Packard, Philips, Tuebingen, Ulm, CORD, Princeton, Chicago, BIPM France, Lawrence Berkeley, NASA (NAS), NIST, City Honk Kong, Warsaw, AIT Thailand, Imperial.

14th August 2004 (140 450)
Gov Wisconsin, Motorola, Fermilab, ICTP Trieste, CCU Taiwan.

15th August 2004 (135 980)
Motorola, Munich, OU, Washington, CCU Taiwan, PUK South Africa.

16th August 2004 (163 1256)

UNICAMP Brazil, Western Ontario, EPF Lausanne, Motorola, Michigan, UOPHX, Ericsson Sweden, Cambridge Univ. Press, Sheffield.

17th August 2004 (236 1096)

Feedback Site down.

18th August 2004 (209 889)

Western Australia, Montreal Tech., IBM Yamato, Motorola, Teledynefluid, Regensburg, Caltech, Rochester, Cincinatti, Michigan*, Southern Florida, Utah, Washington, Ecole Normale Superieure Lyon, Lawrence Berkeley, Lawrence Livermore, Technion, Trieste, Osaka Univ Japan, US Airforce Wright Patterson, Sandia, Hedmark Norway, Durham, Sheffield, Science Applications International Corp.,

19th August 2004 (219 822)

Australian National Univ., INPE Brazil, Motorola, IFN Magdeburg, TU Chemnitz, Tuebingen, North Carolina State, Michigan*, Washington, French Particle / High Energy Lab., Polytechnique, Lawrence Livermore, Dublin Institute of Advanced Studies, Hokudai Univ., US Army Research Laboratory, NTNU Norway, Umea Sweden, Warwick*.

20th August 2004 (225 1052)

Gov Wallonie Belgium, Gov. British Columbia, Geneva, Motorola*, Perimeter, Teledynefluid, Zlin, TU Darmstadt, Giessen, Stanford, Michigan*, Kentucky, Washington, UPC Spain, IIAP India, Trieste, US Navy NSWC, SDU Turkey.

21st August 2004 (180 644)

Lockheed Martin, Buffalo, CSU Hayward, Cambridge particle physics.

22nd August 2004 (160 537)

Australian National Univ., Daily Herald, Teledynefluid, Columbia, Leiden.

23rd August 2004 (206 1743)

Michigan*, Sydney, Liege, ETH Zurich, IBM Almaden, IBM (USF), Motorola, Max Planck Bonn, Stuttgart, SUNY Stonybrook, Florida, Kentucky, MAP Spain, Amiedu Finland, Oak Ridge, SWU Thailand, Manchester, Sheffield.

24th August 2004 (211 876)

Tasmania, UNLP Argentina, UNE Australia, Zurich, Intel, Motorola, Philips, Teledynefluid, Leipzig, New Mexico State, Michigan, MAP Spain, IMAG France, NOOS France, US Gov BLM., Fermilab, NIH, US Treasury, Massey Univ NZ, Linkoping Sweden, Imperial.

25th August 2004 (248 1028)

Motorola, Harvard, New Mexico State, Penn State, UCAR, UC Irvine, UCLA, Michigan, Kentucky, Washington, HUT Finland, NASA Ames, Sheffield.

26th August 2004 (265 1246) - New record number of visits per month.

Deakin Australia, UNISA Australia, Alberta, EPF Lausanne, Motorola, Teledynefluid, TU Chemnitz, TU Darmstadt, Tuebingen, AUB Denmark*, Berkeley, North Carolina State, Purdue, Stanford, Stanford (SLAC), Texas Arlington, Washington, US Gov NOAA, US Pope Airforce Base, Exeter.

27th August 2004 (213 988)

Washington, USP Brazil, UTA Chile, Motorola, AUN Denmark, New Mexico State, Purdue, UC Santa Barbara, Michigan, UCM Spain, UJF Grenoble, NASA (GSFC), US Gov USDA, Dublin City, NGI Italy, Trieste, KYU Tech Japan, Kluwer, Norwegian Military, Anadolu Univ, Turkey.

28th August 2004 (183 988)

IBM Almaden, Teledynefluid, Rice Univ., Stanford, Washington, Xavier, HTK Finland, Hong Kon Tech., Sandia, GDA Poland, NU Singapore.

29th August 2004 (182 839), New record gigabytes downloaded.

RMIT Australia, Emory, Illinois Chicago, Washington, Kyoto, TU Delft, Manchester, Swansea.

30th August 2004 (244 1646)

Sandia, New South Wales, Boeing, Motorola, Teledynefluid, FH Regensburg, Hannover*, Colorado, JMU, North Dakota, New Mexico State, Northwestern, UIS, Michigan*, Vermont, Trieste, Waseda, US Military USCG, Utrecht, KTH Sweden, Uppsala Sweden, CCU Taiwan, Oxford.

31st August 2004 (221 942)

UIBK Austria, Motorola, Hannover, Georgia Tech., MIT, Purdue, SELU, Michigan*, Washington State, USC Spain, CEA France, Rennes 1,

Los Alamos, Dublin IAS, CNR ITIA Italy, Massey Univ NZ, CCU Taiwan, Swansea.

1st September 2004 (210 1479)
    Australian National Univ., Gov. Australia, UNICAMP Brazil, Motorola, Teledynefluid, Zlin, Johns Hopkins, SUNY New Paltz, Connecticut, UC Santa Barbara, Urbana Champaign, USU, UTSI, Xavier, UV Spain, ASTRO Italy, INFN Italy, Sandia, Kluwer, KTH Sweden, Umea Sweden, NU Singapore, Strathclyde, Swansea.

2nd September 2004 (244 1176)
    ASN-KTN Austria, Linz, USP Brazil, EPF Lausanne, Buffalo, Florida State, Rutgers, SELU, Arkansas, UC Santa Barbara, Urbana Champaign, Michigan, UC3M Spain, UPM Spain, Finn Tech., Paris Psud, Marseilles, Fermilab, US Gov NREL, Sandia, Toshiba, Kluwer, YLC Taiwan, Bristol.

3rd September 2004 (162 556)
    UNESP Brazil, EPF Lausanne, Intel, Northrop Grummon, Teledynefluid, KU Denmark, Howard, Indiana, Purdue, Texas A and M, Cincinnati, UML, Wheaton, MAP Spain, ONERA Farnce, LeHavre, IRB Hungary, Technion, Trieste, UNICT Italy, Groningen, Jefferson National Laboratory, GU Sweden.

4th September 2004 (164 1677)
    KMUTT Thailand, Columbia, Stanford, UC Santa Barbara, William and Mary, ACN Greece.

5th September 2004 (151 676)
    Teledynefluid, Duesseldorf, Johns Hopkins, Renselaer, Idaho, Vermont, Washington, Leiden.

6th September 2004 (217 1790)
    Frankfurt, Heidelberg, Regensburg, Florida State, Kansas State, SUNYIT, Conecticut, Delaware, South Carolina, Texas, Washington, Zaragoza, UC Spain, Technion, LEARN Sri Lanka, RCS engineering Britain, Bor Tech South Africa.

7th September 2004 (205 2378)
    TU Graz, Sydney, Newfoundland, Quebec, UNAL Colombia, IBM Almaden, IT and T, Motorola, Philips, Teledynefluid, FHG Germany, Arizona, LTU, Michigan State, mew Mexico State, Oklahoma, UAB, Connecticut, Omaha, South carolina, Wisconsin, Xavier, APIC France, US

Gov. BLM, NASA (JPL), INFN MIB Italy, TU Szczecin Poland, KTH Sweden, KMUTT Thailand, Cambridge.

8th September 2004 (205 648)
    Chubu Univ. Japan, MIT Australia, Yukon College, Lockheed Martin, Motorola, Wuerzburg, Berkeley, Florida International, Grace, Santa Rosa, Syracuse, Connecticut, South Florida, Vermont, UV, UTH Greece, City Univ Hong Kong, INFN LNGS Italy, US Military DISA, Us Navy NUWC, Linkoping, Umea, Edinburgh, Keele.

9th September 2004 (214 1393)
    Cornell, Sydney, Alberta, Motorola, Teledynefluid, UO Cuba, Berkeley, Florida State, Gettysburg, JHMI, MIT, Oklahoma, Connecticut, Urbana Champaign, Michigan, South Carolina, US Naval Academy, Valdosta, Wisconsin, MAP Spain, NIH, INFN Bologna, US Army AMRDEC, Twente, DSLU Philipines, Lviv, Liverpool.

10th September 2004 (181 807)
    Curtin Univ Australia, Gov. Brazil, Alberta, Toronto, SLUB Dresden, Bonn, Clark, North Dakota, Ohio State, South Carolina, YSU, UV Spain, Oak Ridge, City Univ Honk Kong, Indian Statistical Institute, Bologna, TNRC Taiwan, Bristol.

11th September 2004 (158 926)
    OEAW Austria, Motorola, Teledynefluid, Niels Bohr Institute, Colby, Oklahoma State, Hebrew Univ Jerusalem, UWC South Africa.

12^{tH} September 2004 (158 691)
    Sao Paolo, CSU Australia, UGDSB Ontario, Shell Canada, CERN, Berkeley, North Dakota, Stanford, Syracuse, Florida, Vermont, ICTP Trieste, Postech S. Korea, Bath.

13th September 2004 (164 527)
    Berkeley, Dalhousie, McGill, Raytheon, Teledynefluid, IFH Germany, Frankfurt, Potsdam, Albany, Berkeley, Georgia Tech., Michigan State, North Dakota, St Cloud State, Tufts, UV Spain, French Particle High Energy National Laboratory, Univ F. Comte, US Gov, BLM, US Dept of Energy, Los Alamos, NIST, TIFR India, CNR ITIA Italy, INFN, Waseda, KAIST S. Korea, SUN South Africa.

14th September 2004 (208 1258)
    Latrobe, Frankfurt, Gov Australia, Gov. Alberta, Motorola, Zlin, Charles Univ. Prague, Schwabia, Berkeley, MIT, New York Univ.,

Syracuse, UC Santa Barbara, New Mexico, UC Spain, Paris Psud, Rennes 1, Patras, CNS ITIA, Genoa, KAIST S. Korea, US Airforce Arnold, US Navy NUWC, US Military OSD, UIO Norway, American Physical Society, GU Sweden, Winchester College.

15th September 2004 (194 1269)

   BGSU Argentina, NU Singapore, Melbourne, USP Brazil, Toronto, CERN, IBM Almaden*, Motorola, Volvo, TU Cottbus, Frankfurt*, Arizona, Florida International, North Dakota, UMSMED, New Hampshire, Utah, JYU Finland, US Gov BLM, Los Alamos, NASA (GSFC), Groningen, Utrecht, KOU Turkey, NCTU Taiwan, Exeter.

16th September 2004 (223 615)

   Alberta, Montreal, Kodak, Motorola, Teledynefluid, Max Planck MPIS, Max Planck Berlin, Erlangen, Arizona, Cornell, Illinois Tech., Johns Hopkins, Lousiana State,Ohio State, Stanford SLAC, Michigan*, New Hampshire, Rouen, US Gov BNL, City Hong Kong, Calabria, KAIST S. Korea, US Military NIPR, FFI Norway, Amerian Physical Society, Jagiellonian Krakow, KTH Sweden.

17th September 2004 (244 1099)

   Princeton IAS, ADFA Australia, UNICAMP, Motorola, Charles Univ Prague, Boston, Pittsburgh, Central Florida, Delaware, Maryland, Indian Statistical institute, Chubu, US Army Research Laboratory, TU Delft, Massey NZ, NPIC Taiwan, National Health Service.

18th September 2004 (175 993)

   Schwabia, Oregon State, Urbana Champaign*, Pennsylvania, South Florida, Worcester Polytech. Institute, Hong Kong Tech, Indian Statistical Institute, Tritek South Africa.

19TH September 2004 (195 1403)

   Texas, Teledynefluid, Princeton, Kentucky, Vermont, Wisconsin, UCM Spain, IRFU Sweden.

20th September 2004 (279 2981)

   Princeton, National Research Council Canada, Calgary, Zlin, TU Cottbus*, TU Darmstadt, Leipzig, Wuerzburg, Columbia, Embry Riddle, Northwestern, Rochester, texas A and M, Michigan, UC Spain, rennes 1, US Gov Brookhaven, US Dept of Energy, CNR ITA Italy, Trieste, US Airforce AORCENTAF, US Airforce Wright Patterson, AERO, Methodist Health, KTH Sweden, NU Singapore, NDHU Taiwan, Brookes, Leeds, UP South Africa.

21st September 2004 (246 854), New Monthly Record of Gigabytes downloaded.

Kodak, Northrop Grummon, Leipzig, Schwabia, Tuebingen, Columbia, Florida Atlantic, Illinois Tech., Missouri, MIT, Montana, SUNY Stonybrook, Connecticut, Urbana Champaign, Maryland, Union, Washington St Louis, UPV Spain, BARC India, CNRITA Italy, INFN Padua, Kyoto Tech., Groningen, TU Delft, Massey NZ, Newman Foundation, FUW Poland, NTU Taiwan, Bristol.

22nd September 2004 (209 1052), New Monthly Record of Previews

Monash, Melbourne, Newfoundland, Belfort, Northrop Grummond, Max Planck MPIKG-GOLM, Leipzig, Mainz, Boston, Caltech, Clark, Cornell, North Carolina State, Northwestern, Iowa, Michigan*, New Mexico, Los Alamos, CNRITIA Italy, TU Delft, Massey NZ, Jefferson National, CCU Taiwan, NTU Taiwan, NCL, Southampton.

23rd September 2004 (233 721), New Monthly Record of Hits.

TU Graz, Australian Gov. Defence, McMaster, Alberta, CERN, TIJMU China, Lockheed Martin, Motorola, Springer, Leipzig, Schwabia, Arizona, Caltech, Columbia, Michigan State, NIU, Stanford, Tufts, Michigan, Union, UV Spain, NASA (Lerc), Hong Kong Tech., MKB Hungary, CNR ITIA Italy, INFN Turin, Osaka, US Airforce AORCENTAF, US Army AMRDEC.

24th September 2004 (233 933)

UNIMEP Brazil, Motorola, TU Chemnitz, Kassel, Leipzig, Brockport, ECU, Oregon State, Penn State, Chicago, UNLV, ESA Italy, CNR ITIA Italy, Trieste, American Physical Society, NU Singapore, Aberystwyth.

25th September 2004 (206 898) New Daily Record of Mbytes Downloaded.

Regensburg, MIT, Teledynefluid, Leipzig, Rostock, Grayson, MIT, Rice, Swarthmore, Xavier, Lawrene Berkeley, Thessalonika, Yamamashi.

26th September 2004 (179 1610)

Georgia Tech., RMIT Australia, McMaster, Leipzig, Regensburg, Georgia Tech., Rutgers, Michigan, Washington, CNRS INLN, Thesalonnika, UST Hong Kong, Technion.

27th September 2004 (251 2137)

Berkeley, Purdue, UNESP Brazil, USP Brazil, Quebec, Calgary, York Univ Toronto, IBM Almaden, Motorola, Charles Univ Prague, TU Berlin, Heidelberg, Leipzig, Regensburg, Lousiana State, MIT, North Carolina

State, NIU, New York Univ, Renselaer, Rutgers, Houston, Michigan*, UV Spain, F Comte, Indian Statistical Institute, ENEA Italy, Trieste*, Strathclyde.

28TH September 2004 (225 1300)

Gov. New South Wales, EPF Lausanne, IBM Almaden, Motorola, Teledynefluid, Siemens Germany, Bonn, Erlangen, Leipzig, Boston, Denver, Indiana, Maine, Pace, South Carolina, UCSC, WSFU, Worcester Polyetchnic Institute, US Gov GSA, US Navy NMCI, US Navy NRIMRY, UTL Portugal, KOU Turkey.

29th September 2004 (231 1187). New Monthly Record of Individual Visits.

Toronto, Memorial Newfoundland, Boeing, IBM Almaden*, IBM Thomas J Watson, Motorola, TU Cottbus, Frankfurt, Leipzig, Columbia, Lehigh, Montana, Syracuse, Houston, Michigan, Wayne State, UC Spain, IAP France, US Gov BLM, Patras, Physics Institute Croatia, Kyushu, US Navy NSWC, Neijmegen, Ccu taiwan*, NTHU Taiwan, Belgrade.

30th September 2004 (285 925)

NCU Taiwan, IBM Almaden, IBM Thomas J. Watson, Motorola, Zlin, Leipzig, Cornell, Harvard, New Mexico State, Ohio State, Pittsburgh, Rose Hulman, Stanford, U Mass., Michigan*, Thessalonika Greece, BARC India, IUCAA India, European Space Agency KAIST S. Korea. US Military NIPR, TU Delft, Twente, Methodist Health, NU Singapore, CCU Taiwan, British Parliament.

1st October 2004 (235 1291)

Linz, IBM Almaden*, IBM Thomas J. Watson, TU Chemnitz, Leipzig, Regensburg, Arizona, Harvard, NAU, UC Santa Barbara, UI, Michigan, U Pennsylvania, UV Spain, Marseilles, Thessalonika, IUCCA India, CNR RTIA, US Navy NMCI, Hacettepe Turkey, UP South Africa.

2nd October 2004 (183 853)

Massey NZ, HTLWRN Austria, IBM Almaden*, Teledynefluid, Leipzig, Regensburg, Harvard, Maine, South Carolina, Thessalonika, Milan Tech., US Airforce AORCENTAF, LUMS Pakistan, NTU Singapore.

3rd October 2004 (194 1999)

Leipzig, Caltech, Duke, Florida State, Georgia Tech., Hawaii, Iowa State, MIT, Florida, UNC Asheville, TCRC Taiwan, UCT South Africa.

4th October 2004 (232 1593)

California Lutheran, North Dakota, UCLA, Vienna, Adelaide, Ghent, Toronto, ETH Zurich, Motorola, TU Cottbus, Leipzig, Schwabia, Alaska, IUN, Louisiana State, MIT, Michigan State, US Gov BLM, Milan Tech., Pisa, Perugia, Rome 2, Utrecht, Jefferson National Laboratory, Tubitak Turkey.

5th October 2004 (228 960)

Australian National University, Glechenberg Austria, MQ Australia, UNICAMP, Monsanto, Teledynefluid, Zlin, Charles Univ. Prague, Leipzig, Regensburg, Appalachian State, Arizona, California Tech., Colorado, Harvard, Longwood, Louisiana, Michigan State, Stanford SLAC, SUNY Institute of Technology, Michigan, French particle high energy national lab., OBSPM France, Brookhaven, INFN Turin, US Airforce Aorcentaf, Linkoping Sweden, Cambridge, British National Physical Laboratory.

6th October 2004 (246 1389)

Leipzig, Regensburg, Toronto, Boeing, Philips, Toyota, Heidelberg, Leipzig, DTU Denmark, Caltech, Iowa State, MU, SFSU, Michigan, US Airforce Academy, Vermont, HTV Finland, CNRS INLN, SISSA Italy, US Airforce Lackland, US Army Reserve, TU Delft, BTH Sweden, Linkoping, NU Singapore, SWU Thailand, Metu Turkey, TYRC Taiwan, Sheffield.

7th October 2004 (308 2102)

Toronto, Dolby, Springer, TU Chemnitz, Leipzig, DTU Denmark, American, Colorado, Mayo Clinic, UC Santa Cruz, Michigan, HTV Finland, CEA France, Strasbourg, Lemans, US Gov BLM, Los Alamos, TEE Greece*, Trieste*, Yamagata, US Airforce Sheppard, KU Thailand, Imperial, Oxford, Queen's College London, York, BBC.

8th October 2004 (241 1768)

Leipzig, Free Univ. Brussels, Boeing, Teledynefluid, Max Planck Dresden, Leipzig, Northwestern, Connecticut, Michigan, Xavier, UV Spain, BIPM France, ENS Lyon, Strasbourg, NASA (ARC), UST Hong Kong, JNCASR India, Pisa, Leiden, Bath, Lancaster, London Business School.

9th October 2004 (222 3076)

Leipzig, Texas Arlington, UC Santa Cruz, Michigan, UST Hong Kong, St Andrews, Arellano Philipines.

10th October 2004 (200 1640)
   Toronto, Biola, Defence Gov Australia*, UPC Czechia, Leipzig, Columbia, Illinois Tech., Heriot Watt, Imperial, Sussex.

11th October 2004 (265 1507)
   Leipzig, Mississippi, CSIRO Australia, UNICAMP, Memorial Newfoundland, British Columbia, Waterloo, IHEP China, Teledynefluid, LRZ Munich, TU Chemnitz, Regensburg, Arizona, Arizona State, Boston, Duke, Harvard, Northwestern, Michigan, Texas, UAM Spain, CHU Lyon, SPECTRO Jussieu France, Nantes, Rennes 1, IASBS Iran, INFN Rome 1, Nagoya, Institute Electric Engineering, Kiev, Oxford, RL Britain.

12th October 2004 (264 1098)
   Kentucky, UNLP Argentina, Monash Australia, Waterloo, Lockeed Martin, Zlin, MUNI Czechia, Max Planck CIPS, STW Bonn*, Leipzig, Duke, Lasalle, MIT, MNSU, Michigan State, SUNY Rockland, Chicago, UMBC Maryland, Michigan, LUT Finland, ENSC rennes, CNR Naples, TUS Japan, Postech S Korea, American Physical Society, Linkoping, NCTU Taiwan.

13th October 2004 (245 939)
   Melbourne, Virginia, Nasda, Monash, Gov. Australia, UNICAMP, Brandon, Memorial Newfoundland, Victoria, Honeywell, Zlin, HU Berlin, TU Cottbus, Leipzig, ACUSD, Berkeley, Colorado, Columbia, CUIS, Duke, F Marion, Michigan State, ORST, OU, Michigan, Alcatel France, BNF France, ONERA France, Strasbourg, Los Alamos, Trieste, Florence, US Navy NMCI, Uppsala, Strathclyde, York, Belgrade.

14th October 2004 (223 784)
   TU Graz, Gov. Alberta, Simon Frazer, Calgary, Daiwa Tokyo, TU Dresden, Karlsruhe, Stuttgart, AUC Denmark, Arizona State, Indiana,Missouri, Michigan State, ORST, Illinois Urbana Champaign, Williams, UAB Spain, UV Spain, IMAG France, OBS Nice, RAIN France, NIST, Sandia, Jerusalem. PRL India, European Space Agency HQ, TU Delft, WAT Poland,. Warsaw, UTL Portugal, Edinburgh, Goldsmith, Oxford.

15th October 2004 (238 844)
   Monash, Defence Australian Gov., UNB Brazil, UNICAMP, NRC Canada, Simon Frazer, Gov. Quebec, Calgary, Toronto, Dupont, Lockheed Martin, Leipzig, AU Denmark, Florida International, Johns Hopkins, Ohio State, SUNY Stonybrook, UCLA, UC Santa Barbara, U Iowa, UNH,

Williams, BIPM France, Pisa, Tohoku, US Military DIA, Nijmegen, TU Delft, ADMU Phillipines, Ninho, Durham, Nescot.

16th October 2004 (266 1064)
ASN-KTN Austria, USP Brazil, Lockheed Martin, Charles Univ. Prague, Schwabia, Berkeley, Michigan, U Penn., JNCASR India, KTH Sweden, UTFORS Sweden.

17th October 2004 (181 875)
Kentucky, Memorial Newfoundland, Colorado State, F Marion, Wright, IBCP France, Waseda, US Navy NMCI, Brookes, Swansea.

18th October 2004 (261 1044)
Hong Kong Tech., UNICAMP, Geneva, Boeing, Springer, Fmarion, Florida State, Michigan State, Purdue, Florida, Michigan, U Penn., Kentucky, Yale, UAM Spain, UCA Spain, UV Spain, BIPM France, US Gov BLM, Los Alamos, Tel Aviv, Tohoku, US Andrews AF Base, US Airforce AORCentaf, UIB Norway, Metu Turkey, CMAT Uruguay. Belgrade.

19th October 2004 (279 1199)
QUT Australia, ETH Zurich, Mircrosoft, Northrop Grummon, KFA Juelich, Ulm, Purdue, Stanford, UC San Diego, Texas Arlington, UAM Spain, LPTHE France, Lyon 1, UTH Greece, US Military NIPR, Kluwer, Metu Turkey, Academica Sinica, Durham, Univ. College London.

20th October 2004 (259 1422)
UNS Australia, UA Belgium, Geneva, Shell, TU Cottbus, Ulm, Carleton, Embry Riddle, Fmarion, Johns Hopkins, NortH Dakota, New York Univ., Ohio State, Purdue, UAF, Yale, French national particle lab., UTC France, US Dept. of Energy, INFN Naples, US Airforce Centaf, US Military NIPR, KTH Sweden, Cankaya Turkey, Ege Turkey, Metu, Oxford*.

21st October 2004 (255 1956)
UNICAMP, NRC Canada, Toronto, Lockheed Martin, TU Cottbus, Ulm, Buffalo, Embry Riddle, Fmarion, Hamilton, Louisiana State, Rutgers, Stonybrook, UC Davis, Houston, Michigan*, UV Spain, INFN LE Italy, US Navy DT, US Navy NIPR, AUT NZ, Massey NZ*, American Physical Society, Columbia, Washington, JINR Russia, KTH Sweden, Cranfield, Oxford, AUB Britain, RHUL Britain.

22nd October 2004 (218 10667) New Hits Daily Record.

IBM Thomas J. Watson, Lockheed Martin, TU Berlin, KU Denmark, Albany, Berkeley, Harvard, Northwestern, SP College, U Mass Dartmouth, IAC Spain, US Gov BLM, Fermilab, Trieste, US Military NIPR, Methodist Health.

23rd October 2004 (183 604)

Ghent, Alberta, Florida International, Renselaer, SJCSF, USU, Toledo, Yale, CPL Industries Britain.

24TH October 2004 (229 1228)

Adelaide, Simon Frazer, Cornell, Johns Hopkins, North Carolina State, Northwestern, Victoria NZ, Aberystwyth.

25th October 2004 (316 1612)

Louisina Tech., Vienna, EPF Lausanne, UMAB Chile, FU Berlin, Goettingen, Mainz, Arizona, Colorado, Embry Riddle, Georgia Tech., Maine, Montana, North Carolina State, Northwestern, SIU, SUNY Stonybrook, UC Santa Cruz, Michigan, New Hampshire, UNL, Washington, Xavier, UPV Spain, UV Spain, French national particle lab., US Gov. BLM, INFN Aquila, Trieste, Turin, US Airforce Luken, Massey NZ, KTH Sweden, Metu, Heriot Watt, Leeds, St. Andrews.

26th October 2004 (268 1756) New Monthly Records Set.

KYU Tech Japan, US Military NIPR, Honeywell, Hewlett Packard, Lockheed Martin, Zlin, TU Cottbus, Duisberg, DTU Denmark, Caltech, Georgia Tech, Harvard, Purdue, Stanford, Florida, Winthrop, Washington St Louis, Los Alamos, Milan Tech*, Turin, Massey NZ, AMU Poland, NU Singapore, NTHU Taiwan, Bristol, Warwick, Belgrade.

27th October 2004 (328 693)

Virginia, Massey, Australian Defence Dept., Sao Paolo, Gov. British Columbia, McMaster, Memorial Newfoundland, Montreal, Lockheed Martin, Siegen, Arizona State, Drexel, Harvard, Northwestern, Penn State, UC Riverside, UC Santa Barbara, U. Mass., Michigan, UPV Spain, US Govt. LOC, Pisa, Kyoto, Chungbak S. Korea, US Military NIPR, American Physical Society, PW Poland, MB Slovenia, Hacettepe Turkey, BBSRC Britain, Sussex, Univ. College London, York.

28th October 2004 (261 1020)

Newcastle Australia, Gov British Columbia, British Columbia, Montreal, Honeywell, Teledynefluid, JBC, MTU, Ohio State, Rutgers, SIUE, Texas A and M, Michigan, Minnesota, Kyoto, UIO Norway,

American Physical Society, Methodist Health, Bristol, Queen Mary College, St Andrews, Parliament.

29th October 2004 (270 773)
IMEC Belgium, UNICAMP, McMaster, Montreal, Max Planck MPIWG Berlin, Caltech, Sheperd, Temple, TU Varna, Michigan, Minnesota, Williams, Xavier, Brookhaven, NASA (JPL)*, Oak Ridge, US Gov SSA, IISC India, INFN Bari, US Military NIPR, Methodist Health, NU Singapore, Bristol, Paisley, Sussex.

30th October 2004 (191 638)
Teledynefluid, Georgia Tech., New Hampshire, Porto, Cambridge.

31st October 2004 (272 1395) Five Monthly Records Set.
MSVU Canada, Ohio State, Stanford, NTUA Greece, Jerusalem, University College London.

1st November 2004 (309 1122)
Drexel, NIU, ASN KTN Austria, Montreal, CERN, Hewlett Packard, Microsoft, TU Cottbus, Bonn, Magdeburg, Buffalo, Columbia, Harvard, MU, North Dakota, NEIU, Plymouth, Chicago, Houston, Michigan, Texas, Mulhouse, BME Hungary, Jerusalem, US Navy NMCI, VU Netherlands, NTNU Norway, Massey NZ, Linkoping, LTH Sweden, Swansea.

2nd November 2004 (271 937)
Honeywell, Lockheed Martin, Teledynefluid, Zlin, TU Cottbus, Albany, BC, Caltech., Colby, Colorado State, Davidson, Najah, Oregon State, Stanford, U California system, UC Santa Cruz, Illinois Chicago, Urbana Champaign, South Florida, USU, French national particle laboratory, ONERA, SUPELEC, Mulhouse, Fermilab, CNR ITIA Italy, Trieste, Krakow, NU Singapore, Bristol, Oxford, SBU, Sussex.

3rd November 2004 (212 521)
Queensland, Tokyo, Valle Colombia, Intel, Lockheed Martin, Random House, TU Cottbus, Leipzig, Renselaer, Washington St Louis, ONERA, UST Hong Hong, Jerusalem, CNR ITIA Italy, Rome 3, UNAM Mexico, KTH Sweden, Bilkent Turkey, Bangor, Bath, Cambridge, Oxford.

4th November 2004 (245 1413)
Australian National University, Ghent, Gov. Alberta, McMaster, Calgary, Saskatchewan, TU Darmstadt, Augsburg, Colby, Harvard, MIT, Ohio, Oregon State, SFUSD, UC Santa Barbara, Maryland, Helsinki, UTU Finland, RAIN France, Paris Psud, BARC India, CNR CBF Italy, KAIST

S. Korea, US Military NIPR, American Physical Society, ADMU Philippines, NU Singapore, TCRC Taiwan, Cambridge, Napier, Queen Mary College.

5th November 2004 (176 9804)

Siemens, UNICAMP, CERN, UNAL Colombia, Boeing, Random House, Teledynefluid, Duke, Florida Atlantic, Michigan State, Urbana Champaign, Michigan, UV Spain, ONERA France, UPS-TLSE France, Lawrence Berkeley, Trinity College Dublin, INFN Padua, ICTB Trieste, Rome 1, US Navy NMCI, KUN Netherlands, Linkoping, NU Singapore, Cardiff.

6th November 2004 (160 639)

ARCS Austria, McMaster, UNBC Canada, OU, SUNY Stonybrook, Williams, Jagielonski, Croydon.

7^{tH} November 2004 (196 2033)

TU Cottbus*, CLU, Hawaii, JMU, MIT, MU, Renselaer, St Mary's, UCOK, Minnesota, Metu Turkey.

8th November 2004 (238 914)

ARCS Austria, Vienna, UTS Australia, Waterloo, EPF Lausanne, Lockheed Martin, Teledynefluid, MUNI Czechia, IFW Dresden, TU Cottbus, American, Drexel, Penn State, St Thomas, U California system, Michigan, Minnesota, Wyoming, Vermont, UV Spain, Tellas, Trieste, US Edwards AF Base, Bristol, UMIST.

9th November 2004 (200 1418)

Princeton, Curtin, Liege, UNICAMP, Alberta, Ottawa, Hewlett Packard, Microsoft, Springer, TU Cottbus, Boston, Colorado, Florida Atlantic, Georgia Tech., Indiana, LCC, Oklahoma State, Rochester, San Diego, SJSU, Nevada Reno, Vanderbilt, NOOS France, Poitiers, IITKGP India, US Navy NMCI, US Military NIPR, Massey, SIT NZ, UDSM Tanzania, Queen's Univ. Belfast, Sheffield, Warwick, USB Venezuela.

10th November 2004 (241 822)

Vienna, UERJ Brazil, Calgary, Montreal, Tsinghua China, Hewlett Packard, TU Cottbus, Leipzig, BGSU, Duke, GCSU, Indiana, Northwestern, Princeton, Urbana Champaign, US Spain, CNES France, Pasteur Lille, INFN Padua, INFN Pavia, SNU S. Korea, US Navy NUWC, US Military SOCOM, Methodist Health, SHU Britain, Surrey.

11th November 2004 (206 829)

Tulane, Gov. Australia Defence, Sao Paolo, York Univ. Canada, National Univ Colombia, Motorola, Teledynefluid, Zlin, MBI Berlin, TU Cottbus, Dortmund, Frankfurt, Tuebingen, Boston, Central Michigan, Georgia State, MSOE, Ohio State, SIU, SUNY Stonybrook, Syracuse, Delaware, Michigan, UNL, UV Spain, Lawrence Livermore, US Airforce Tinker, Methodist Health, PK Poland, Bilkent, BBSRC Britain, Bristol, Swansea, York.

12th November 2004 (140 643)

UERJ Brazil, UNICAMP, British Columbia, Calgary, CERN, Valle Colombia, Kodak, Lockheed Martin, Merck, Duisburg, Indiana, Missouri, New Mexico Tech., Michigan, U Penn., UCM, UV Spain, Poitiers*, Trieste, HVA Netherlands, Jefferson National Laboratory, NU Singapore, Gov Slovakia, TNRC Taiwan, BBSCR Britain, Cardiff, St Andrews, Univ. College London, BBC.

13th November 2004 (180 944) Feedback Website Down.

14th November 2004 (192 1453)

Rice, Geneva, Teledynefluid, HU Berlin, Mainz, Duke, Texas, Fermilab., US Gov. JCCBI, CEU Hungary, Kitami Japan, York.

15th November (249 1373) Feedback Website Down.

16th November 2004 (260 1670)

Tuaisi Romania, Vienna, Siemens Austria, Gov Canada, Toronto, Valle Colombia, Lockheed Martin, Northrop Grummond, German Synchrotron Facility, Springer, TU Cottbus, Frankfurt, Mainz, Alfred State, American, Berkeley, Caltech., CMSU, Colorado State, Columbia, CWRU, Hawaii, Johns Hopkins, Kent State, Kutztown, MIT, OIT, Oregon State, Rice, Syracuse, Tulane, UC San Diego, U Mass., Michigan, UNLV, New Mexico, US Airforce Academy, Kentucky, UTM, Toledo, UPF, ENSTA France, French Particle Lab., Lawrence Berkeley, US Gov USPS, AUTH Greece, Technion, European Space Agency, CNR ITIA, INFN Bologna, Univ. Tokyo, KUNIV Taiwan, US Airforce AORCENTAF, US Army Houston, US Military NIPR, Massey NZ, UA Portugal, NU Singapore, Inonu Turkey, NTHU Taiwan, Cambridge, City, Manchester, RHUL, Southampton, Stoke College, Univ College London, York.

17th November 2004 (255 1555)

British Columbia, Kochi Tech Japan, Royal Military College Canada, EPF Lausanne, Lockheed Martin, Northrop Grummond, TU Darmstadt,

Frankfurt, DTU Denmark, Berkeley, Embry Riddle, JSU, Missouri, Ohio State, OU, Syracuse, Dallas, Nevada Reno, Oregon, Helsinki, US Dept. of Energy, NASA (Larc), Technion, US Airforce Eglin, US Naval Research Laboratory, Massey NZ, UTL Portugal, Imperial, Manchester, Napier, Southampton, Eastern England, Belgrade.

18th November 2004 (238 1347)

Harvard, Graz, Mount Royal Alberta, McGill, Regina, Western Ontario, Intel, Kodak, Raytheon, Leipzig, Wuerzburg, BGSU, Louisville, OU, Rutgers, Stanford, U Conn., Urbana Champaign, Michigan*, Washington, UV Spain, Strasbourg, Poitiers*, US Gov. BLM, Fermilab, Trieste, Pisa, Shonon Tech. Japan, Waseda, EEA Ottawa, Jefferson National, DYU Taiwan, Southampton, York.

19th November 2004 (221 1013)

UNICAMP, McMaster, McGraw Hill, Rockwell Collins, German Synchrotron Facility, TU Cottbus, Colorado State, Drexel, Lousiana, Urbana Champaign, Michigan, Kentucky, Yale, French national particle lab., Poitiers*, CNR ITIA, INFN Genoa, US Navy IH, AERO, American Physical Society, A-Star Singapore, NU Singapore, Southampton, York.

20th November 2004 (184 1890)

MQ Australia, Laval Canada, Frankfurt, Mainz, CSU Hayward, Ohio, Illinois Chicago, UTU Finland, US Gov. NOAA, HVA Netherlands, Cambridge, RDG Britain.

21st. November 2004 (208 1089)

New Mexico, Linz, ETH Zurich, EMU, Cornell, Georgia Tech., LeHigh, Purdue, Michigan, Williams, Aberdeen, Oxford.

22nd November 2004 (245 2262)

Stanford, Linz, Gov Brazil, Toronto, Geneva, Intel, Northrop Grummond, Max Planck CPIFS, Max Planck Duesseldorf, AUB Denmark, Clemson, Case Western Reserve, MIT, MU Ohio, Princeton, UCLA, South Florida, Kentucky, Zaragoza, Poitiers*, US Gov Maricopa County, ACN Greece, Weizmann, Indian Statistical Institute, HQ of the European Space Agency, US Hill Airforce Base, US Army USACE, Linkoping, Bristol, Durham, Imperial, Lancaster, RDG Britain.

23rd November 2004 (297 5077)

Melbourne, McGill, British Columbia, Toronto, CERN, Northrop Grummond, Shell, TU Cottbus, Erlangen, Heidelberg, AUC Denmark, BGSU, UC Davis, UC Santa Barbara, Florida, Texas, UTSI, UV Spain,

IAP France, KLTE Hungary, University College Dublin, ESA HQ, INFN Florence, ICTP Trieste, Univ Trieste, Titech Japan, Yamanashi Medical, Glasgow, Oxford.

24th November 2004 (259 2618)

Gov. Australia, Linz, Vienna, Concordia, Saskatchewan, Toronto, Boeing, Lockheed Martin, KFA Juelich, Ruhr Bochum, Springer, Regensburg, CSUS, Ohio State, UCLA, UI, Maryland, UM Spain, UV Spain, Helsinki, INPG France, Bourgogne, Poitiers, NTUA Greece, TEE Greece, Parma, Hiroshima, Univ Tokyo, UEC Japan, UIB Norway, Aberystwyth, Bristol, Cambridge, Durham, Liverpool, Hertford, Queen's University Belfast, Strathclyde, UCL.

25th November 2004 (311 1365)

Columbia, Delaware, Curtin, IMEC Belgium, Alberta, Calgary, Toronto, York Univ. Toronto, Neufchatel, Max Planck MPIS, TU Cottbus, Leipzig, Delaware, Kentucky, UNED, TEE Greece, Trinity College Dublin, Tel Aviv, NTU Taiwan, Cambridge, Cardiff, Exeter, Glamorgan, Portsmouth.

26th November 2004 (219 1261)

U. Connecticut, GSI Germany, Ruhr Univ. Bochum, Fullerton, Iowa State, UC Davis, IISC India, CNR ITIA, FUW Poland, Linkoping, MDH Sweden, Cambridge*, Lancaster, Queen's Univ. Belfast, RL Britain, Royal Society, Southampton.

27th November 2004 (167 1408)

ESI Austria, Charles Univ. Prague, Egmont Denmark, Iowa State*, Kentucky, Chapel Hill, William and Mary, Wellesley, US Navy Pacific South West, Bath, Oxford.

28th November 2004 (199 2337)

MIT, OU, Linz, UWS Australia, TU Cottbus*, Konstanz, BGSU, Colorado, Memphis, SUNY Stonybrook, Syracuse, Washington State, LAAS France, City Hong Kong, IASBS Iran, HIG Sweden, IRFU Sweden, Cambridge, Derby, Leeds, Southampton.

29th November 2004 (257 1445)

RAIN France, ARCS Austria, Linz, Alberta, Guelph, ETH Zurich, Boeing, Lockheed Martin, Microsoft, Solvay, Mainz, Drexel, Embry Riddle, Iowa State, MSCD, Missouri State, North Dakota, Purdue, Delaware, Florida, U Mass., Michigan, Kentucky, US Gov. BOP, US Dept of Energy, LN Hong Kong, BIU Israel, Indian Statistical Institute, ENEA

Frascati national lab, SNU S. Korea, US Brookes Airforce Base, Massey Univ. NZ, Bristol, Cambridge, Heriot Watt, King's College London, Manchester, Oxford, Sussex.

30th November 2004 (265 1896)

Liege, Ghent, Gov. British Columbia, Laval, York Univ. Toronto, TU Cottbus, Cologne, Paderborn, Stuttgart, Arizona, Boston College, Grace, Penn State, Reed, Stanford, UCLA, Dallas, U Mass., Virginia, Zaragoza, Helsinki, Rennes 1, Univ. College Cork, Jerusalem, IASBS Iran, Milan Tech., LEARN Sri Lanka, US Army Arlington, American Physical Society, NCKU Taiwan, Bristol, Cambridge, King's College London, Lancaster, Leeds, Manchester, UKC.

1st December 2004 (234 2159)

Univ. Sao Paolo, Gov. British Columbia, Calgary, Lockheed Martin, Northrop Grummond, Shell, Max Planck Tuebingen, TU Chemnitz, Heidelberg, Leipzig, Univ. Tuebingen, DTU Denmark, Drexel, Harvard, MIT, MSCD, MU Ohio, Stanford. UCLA, U Mass., Maryland, UNL, North Texas, Washington, URJC Spain, BIPM France, Lille 1, Los Alamos, Lawrence Livermore, US Dept. of State, CNR ITIA, INFN Calabria, Gov. NZ, Aberdeen, Napier, St Andrews, Strathclyde*.

2nd December 2004 (258 1377)

Columbia, JCU, Ghent, Alberta, British Columbia, Calgary, Manitoba, Montreal, Sherbrooke, Boeing, Heidelberg, Lockheed Martin, Freiburg, Heidelberg, Colby, Colorado State, Gallaudet, IUP, Johns Hopkins, MIT, Northwestern, Chicago, Urbana Champaign, Michigan, U Penn, URI, Virginia, HUT Finland, BIPM France, US Govt BLON, NASA (Jpl), US Gov. USITC, US Army Arlington, US Gov Beaumont School, Linkoping, NTHU Taiwan, Bangor, Imperial, Oxford.

3rd December 2004 (233 1612)

York Univ. Canada, TU Cottbus, Karlsruhe, AUC Denmark, Gettysburg, Hawaii, Lacoe, Louisiana State, MIT, Ohio State, Chapel Hill, UNCO, Texas Dallas, AC Bordeaux, Bourgogne, Strasbourg, NASA (gsfc), TEE Greece, Trieste, LEARN Sri Lanka, US Robins AF Base, US Navy CNET, TU Delft, Twente, UTL Portugal, UVT Romania, Linkoping, Bristol, Sussex.

4th December 2004 (183 1024)

Sydney, UFPE Brazil, WH Stuttgart, Auburn, Berkeley, Brigham Young, CSU Pomona, Swarthmore, UAF, UC Riverside, Minnesota, Kentucky, TUT Finland, Turin Tech., Linkoping.

5ᵗʰ December 2004 (213 887)

CSUMB, Michigan State, New York Univ., UTK, Australian Gov BOM, Saskatchewan, Potsdam, Ruhr Univ. Bochum, BGSU, Caltech, Colby, Drexel, Harvard, Missouri State, New York Univ., Penn. State, Stanford, Minnesota, South Florida, Texas Arlington, IASBS Iran, Hosei Japan, US Navy NUWC, ICM Poland, Linkoping, Cambridge.

6ᵗʰ December 2004 (371 1676)

Rowan, Linz, UNICAMP, McGill, Simon Frazer, Toronto, Hewlett Packard, IBM Almaden, LRZ Munich, AU Denmark, NIU, SWRI, UCCS, UCLA, UC San Diego, U Mass., Minnesota, Texas, Wiliam and Mary, Western Washington, French national particle laboratory, INSA Lyon, Lemans, NTUA Greece, FSB Hungary, UNFN Florence, American Physical Society, Opole Poland, Chalmers Sweden, NTOU Taiwan, TCRC Taiwan, Manchester, NTU, Oxford, Swansea, Univ College London, York.

7ᵗʰ December 2004 (305 1626)

Vienna, Australian Gov. BOM, UNESP Brazil, UNICAMP, AEI Canada, Queen's Canada, Toronto, CERN, Basel, Jena, Columbia, Cornell, CSUS, Harvard, JMU, Louisiana State, Montana, North Carolina State, NIU, New York Univ., Princeton, Stanford, Texas Tech., ITCS, Michigan, UNO, Wellesley, HUT Finland, INPL Nancy, Poitiers, Sandia, US Treasury, Technion, Trieste, Genoa, US Eglin AF Base, US Wright Patterson AF Base, US Army PICA, US Military Central Command, Utrecht*, Massey NZ, LUMS Pakistan, NTHU Taiwan, Aberystwyth, Durham, SHU, York.

8ᵗʰ December 2004 (270 1574)

Fermilab, Ghent, Winnipeg, McMaster, British Columbia, Montreal, EPF Lausanne, Motorola, UO Cuba, Bosch, TU Chemnitz, Frankfurt, WH Stuttgart, DDRE Denmark, Brigham Young, Central Michigan, Iowa State, Kent State, Kansas State, NAU, PDX, Pittsburgh, Princeton, Urbana Champaign, Michigan, U. Penn., South Florida, Texas, Wellesley, Argonne, US BLM, US Dept. of Energy, NASA (Jpl), Szeged, Jerusalem, INFN LNF, INFN Rome 1, Univ. Tokyo, US Airforce AORCENTAF, Chalmers Sweden, Physto Sweden, Heriot Watt, Imperial, King's College London, Liverpool, Oxford*, Queen's Univ. Belfast.

9ᵗʰ December 2004 (238 981)

BARC India, WAU Netherlands, Graz, Gov. Australia ANSTO, Alberta Research Council, Montreal Tech. Guelph, Boeing, Siemens, Leipzig, Mainz, SDU Denmark, Columbia, New Mexico State, Ohio State,

Salem, Texas A and M, Chicago Illinois, Chapel Hill, South Florida, William and Mary, Xavier, UAM Spain, Helsinki, Finnish Military, CNES France, French national particle lab., Poitiers, NASA (gsfc), Catania, Groningen, TU Delft, NTNU Norway, KTH Sweden, Cambridge, Durham, Imperial.

10th December 2004 (406 1441) New daily visitors record.

UAF, ASN-KTN Austria, TIE Chile, Lockheed Martin, Charles Univ. Prague, TU Cottbus, Karlsruhe, Niels Bohr Institute, Amherst, Berkeley, Colorado State, CSUN, Johns Hopkins, Montana, Pace, Rutgers, Texas Lutheran, Chicago, Minnesota, U Penn., Yale, CSIC Spain, UV Spain, NIH, Hong Kong Tech., Technion, INFN LNGS, Rome 1, Physto Sweden, TCRC Taiwan, Durham, Imperial, Oxford, Queen's Mary London, Swansea.

11th December 2004 (236 1624)

Winnipeg, Teledynefluid, TU Cottbus, IOT Denmark, UC Davis, Connecticut, UCRA, Urbana Champaign, UKY, Washington State, Genoa, UCV Romania, NTHU Taiwan, Manchester, St Andrews.

12TH December 2004 (214 1328)

LRZ Munich, Studentwhonheim, HMC, Missouri State, NCSU, OU, SUNY Stonybrook, UMBC, URI, Massey NZ, UMIST Britain, Warwick.

13th December 2004 (273 956)

Adelaide, UNICAMP, Winnipeg, Toronto, Shell, IFW Dresden, Berkeley, Colorado, CSU Ohio, Drexel, Harvard, Michigan State, Purdue, Florida, Urbana Champaign, Wisconsin, Washington State, Yale, French Gov. Civil Aviation, Weizmann, INFN Turin, Pisa, Rome 1, Aoyama, EWHA S Korea, LEARN Sri Lanka, US Airforce Wright Patterson, Naval Research Laboratory Washington DC, US Navy USCG, TU Eindhoven, Kluwer, UNIBA Slovakia, CHULA Thailand, Academica Sinica, Manchester, Oxford, Strathclyde.

14th December 2004 (273 1604)

KF Graz, Winnipeg, Calgary, EPF Lausanne, Nokia, RWTH Aachen, STW Bonn, Heidelberg, Albany, Buffalo, Cornell, Digipen, Harvard, Lousiana Tech., Michigan State, Delaware, Virginia, William and Mary, Helsinki, ENST Brittany, ETCA France, French national particle lab., Weizmann, Turin, Trieste, Tsukaba, US Army Research Lab., LUTH Sweden, Cambridge, British Gov. Daresbury Laboratory, UMIST.

15th December 2004 (231 1092)

UIBK Austria, Linz, Melbourne, Australian Gov. BOM, KU Leuven, Liege, Winnipeg, Geneva, TU Cottbus, Tuebingen, Berkeley, Florida Tech. MIT, Princeton, Rutgers, SMCM, UC San Diego, Michigan, UMS Medical, Western Michigan, UPF Spain, Rennes 1, NIH, OTE Greece, HQ European Space Agency, CNR ITIA, ENEC Brindisi, Turin, Utrecht, Kluwer, UIB Norway, Krakow, KTH Sweden, Swansea.

16th December 2004 (409 1383) New Record of Daily Visitors.

ULB Belgium, Sao Paolo, Canadian National Research Council, Queen's Canada, BMW Group, Intel, Lockheed Martin, IU Bremen, TU Darmstadt, Heidelberg, Leipzig, BU, Harvard, UNL, New Mexico, U Penn., South Florida, Helsinki, Aerospaciale France, INSA Lyon, US Gov. Dept. of Energy, US Gov NREL, IASBC Iran, ENEA Pisa, Univ Tokyo, AERO, UMU Sweden.

17th December 2004 (185 937)

ATI Austria, UNB Brazil, CERN, ETH Zurich, Freiburg, Honeywell, Max Planck Heidelberg, TU Cottbus, MIT, North Carolina State, New Mexico State, Penn State, UC San Diego, Virginia, Vermont, BIPM France, French particle lab., Poitiers, Reims, UTC France, CNR IPCF, TUT Japan, US Army SMDC, KU Nijmegen, NTNU Norway, Rutherford Appleton.

18th December 2004 (156 965)

Leipzig, U. Mass. Dartmouth, US Airforce AORCENTAF, SU Sweden.

19th Dec. 2004 (193 1521)

UWS Australia, TU Hamburg, Frankfurt, BU, Harvard, Minnesota, US Gov TVA.

20th December 2004 (260 1275)

NTNU Norway, Curtin, UNICAMP, Winnipeg, Queen's Canada, Neufchatel, Raytheon, Charles Univ. Prague, Bosch, Siemens, Springer, TU Cottbus*, Duisburg, AUB Denmark, Colorado, Purdue, Rice, Georgia, Washington, NASA (Arc), UST Hong Kong, HQ European Space Agency, CNR ITIA, Trieste, Wroclaw, IRFU Sweden, Oxford, York.

21st December 2004 (339 1174)

IBM Almaden, Massey NZ, ASN KTN Austria, Queensland, Sydney, Gov Australia MUS, Sao Paolo, TIE Chile, CUJAE Cuba, Springer, Potsdam, Georgia Tech., Oregon State, Pittsburgh, Purdue, Arkansas, St Etienne, Gov Maryland Dept of Energy, INFN Turin, Rome 1, Kyoto, US

Airforce Headquarters, US Army PICCA, Leiden, LUMC Netherlands, Kluwer.

22nd December 2004 (232 1079)
   Leipzig, Indian River CC*, Johns Hopkins Applied Physics, Rutgers, Tufts, Michigan, US Dept. of Energy, Fermilab, NASA (gsfc), US Gov. OSTI, Patras, Jerusalem, Weizmann, Mitre, IRFU Sweden, Physto Sweden, IJS Slovenia, Anadolu Turkey, BBK Britain, Central England.

23rd December 2004 (237 1753)
   ANU Australia, BAS Belgium, Canadian NRC, Calgary, Waterloo, Pfizer, TU Cottbus, Stuttgart, Michigan, USU, Washington, US Airforce OLN-AFMC, US Navy NUWC, Kluwer, Chula Thailand, Metu Turkey, Strathclyde, Warwick, British Gov. GSI.

24TH December 2004 (240 1759)
   DUTH Greece, NTUA Greece, NCTU Taiwan.

25th December 2004 (146 962)
   Weizmann, US Army SILL, Chalmers Sweden.

26TH December 2004 (165 1056)
   Leipzig, Weizmann, KN Poland, NTU Taiwan, Oxford.

27th December 2004 (294 1164)
   McGill, Lockheed Martin, TU Cottbus, Leipzig, Berkeley, Michigan, Zaragoza, Technion, PRL India, LU Sweden.

28th December 2004 (217 1799)
   Jersey City, CTT Czechia, Leipzig, MCG, North Carolina Charlotte, Technion, Naples, US Airforce Andrews, RUP Slovakia, Ankara.

29th December 2004 (252 1683)
   Winnipeg, RWTH Aachen, Leipzig, NEIU, Texas Medical Center, Medical Michigan, Xavier, Poitiers, US Gov BLM, NASA (Lerc), Jerusalem, PRL India, Postech S Korea, Methodist health, Hacettepe.

30th December 2004 (209 1029)
   TU Vienna, IBM (USF), Microsoft, Creighton, Harvard, New Mexico State, Urbana Champaign, French national particle lab., Univ College Cork, US Airforce Kirtland, Porto, IRFU Sweden, Hermes Slovenia, NTHU Taiwan, NERC-BAS British Gov., Queen's Univ. Belfast.

31st December 2004 (163 689) New record number of monthly visitors.
   Colorado State, Medical Michigan, Tartu Estonia, EDF France, NIH, Jerusalem, National Health Service.

1st January 2005 (223 843)
   NRC Canada, TIE Chile, Arizona, Boston Univ., Yale.

2nd January 2005 (213 1276)
   Leipzig, Kyoto, STW Bonn, Univ. Bonn, Berkeley, Princeton, Yale, Imperial.

3rd January 2005 (331 1861)
   South Australia, ETH Zurich, Bosch, Max Planck Halle, TU Cottbus, TU Darmstadt, Duke, Maine, Mississippi, UC Irvine, UCLA, Michigan, Poitiers, Fermilab, Us Gov USITE, HKBU Hong Kong, Jerusalem, Tel Aviv, CNR ITIA, Kyoto, Uppsala, Metu Turkey, TCRC Taiwan.

4th January 2005 (235 1849)
   Springer Austria, City of Winnipeg, General Electric Power, IT and T, Leipzig, Berkeley, Drexel, Florida International, Harvard, Mew Mexico State, Ohio State, UC Santa Barbara, Michigan, UWM, Washington, Tartu, Helsinki, Poitiers, Fermilab, Technion, JNCASR India, CAEN Italy, Pisa, US Army PICA, US Military NIMA, Maastricht, FOI Sweden, Glasgow.

5th January 2005 (250 2194)
   QUT Tasmania, UA Belgium, UNICAMP Brazil, Winnipeg, Ontario Power, Max Planck Heidelberg, KU Denmark, Elmhurst, Harvard, North Dakota, Sulross, TCU, Chicago, Michigan, Texas Arlington, Washington, ESPCI France, US Dept. of Energy, Us Edwards Airforce Base, US Navy NRL, Mahidol Thailand, Ogu Turkey, NTNU Taiwan, Bristol, Imperial, Jenner.

6th January 2005 (270 3201)
   Michigan, Melbourne, Australian Govt. BOM, UNICAMP, ETH Zurich, General Electric, Frankfurt, Central Michigan, Harvard, Maine, Sabanci, Stanford, Delaware, Michigan Medical, Washington St. Louis, Perugia, MIE-C Japan, Us Navy NUWC, US Military OSD, Groningen, US Gov Central Bureau of Intelligence, Birmingham, Napier, Oxford, Strathclyde, UCL.

7th January 2005 (270 2794)

Regina, Corning, IT and T, Lockheed Martin, Northrop Grummond, Shell, Harvard, Michigan Medical, Michigan, U. Penn., Washington, HTV Finland, UTU Finland, Picardie, European Parliament, Trieste, Ritsumei, US Navy NUWC, Groningen, TU Delft, US Gov. CBI, Unibuc Romania, Chula Thailand, Ege Turkey, Metu Turkey, Tubitak, TNRC Taiwan, British Gov. BBSRC, Nottingham, National Health Service.

8th January 2005 (177 1104)

STW Bonn, LAPL Organization, WAT Poland.

9th January 2005 (419 2194) New Record of Daily Visitors.

TIE Chile, AUB Denmark, CSU Pomona, Princeton, Chicago, USC, Gov. Estonia, UAM Spain, Gov Hong Kong, Sienna, US Army Medical Corps, Metu Turkey, TCRC Taiwan, BBC.

10th January 2005 (441 3481) New Record of Daily Visitors

Renselaer, Sydney, UNICAMP, McMaster, NRC Canada, Wellington CSSb Ontario, Toronto, GWDG Germany, TU Cottbus, Dusseldorf, Hamburg, Schwabia, Tuebingen, Harvard, OGI, Ohio State, Renselaer, SHIP, Delaware, Michigan, UNO, Vermont, WHOI, UV Spain, HUT Finland, ENST France, Paris Psud, Poitiers, US Gov. BNL, Patras, Jerusalem, Tel Aviv, JNCASR India, INFN Naples, INFN Turin, Pavia, Osaka FU, UAM Mexico, TU Delft, Free Univ Amsterdam, US Gov. CBI, Gdynia, Lodz, LTH Sweden, IATP Ukraine, Queen Mary London, UKC, WYKE.

11th January 2005 (468 3960) New Record of Daily Visitors.

Sydney, UA Belgium, Memorial Newfoundland, British Aerospace, Honeywell, Hughes, Philips, Heidelberg, Regensburg, Arizona, Boston College, Florida Atlantic, MIT, North Dakota, UC Davis, Maryland, Michigan, Washington, UV Spain, Descartes ADC France, Poitiers, Gov. Virginia, CNR IPCF, US Navy USNO, Warsaw, Chalmers Sweden, Bristol, Exeter, Imperial, Liverpool, Oxford, Mansfield, Queen Mary, RHUL Britain, Belgrade.

12th. January 2005 (404 2234)

Graz, CSIRO Australia, Australian Gov. Defence, KU Leuven, Dalhousie, Saksatchewan, Regina, Microsoft, Max Planck Heidelberg, DTU Denmark, KU Denmark, Berkeley, Caltech, Creighton, Dartmouth, Elmhurst, Purdue, Stanford, UCLA, UC Santa Cruz, Michigan, UV Spain, ENS France, Nantes, NASA (dfrc), US Gov UCIA, BA Tech. Italy, US

Army Research Laboratory, US Gov CBI, Ankara Turkey, Cambridge, Essex, Napier, St Andrews.

13th January 2005 (429 2087)
UNICAMP. Winnipeg, McMaster, York Univ. Canada, RWTH Aachen, Kiel, Evergreen, Georgia Tech., Harvard students, Maine, MIT, OU, UCLA, South Carolina, Washington State, UAB Spain, UV Spain, Helsinki, ESPCI France, DRP Jussieu France, Paris Psud, Poitiers, US Gov. EPA, NIST, US Gov. USGS, Univ College Cork, Technion, CNR ITIA, Pisa, Nagoya, Wakayama Medical, US Army CJTF180, Nijmegen, US Gov. CBI, Metu Turkey, TNRC Turkey, Imperial, Oxford.

14th January 2005 (377 1452)
TU Vienna, British Columbia, Gillette, IBM Almaden, Northrop Grummond, UPC Czechia, TU Cottbus, Leipzig, WH Germany, AGRSKI Denmark, Danish Tech (DTU), DVC, Georgia Tech., GCSU, Maine, Pittsburgh, Stanford, Florida, Michigan, Vermont, Wayne State, UAB Spain, UV Spain, IIT India, IABS Iran, US Airforce AFIT, US Airforce Wright Patterson, US Army Research Laboratory, Siemens Portugal, ATOM Slovakia, THU Taiwan, TNRC Taiwan, Exeter, Imperial, Leeds, Oxford BNC, Oxford physics, Warwick.

15th January 2005 (311 3834)
Toronto, WH Stuttgart, Oklahoma State, Princeton, Penn State, U Penn., Texas, UTU Finland, IISC India, Kanazawa, NKI Norway, Cambridge.

16th January 2005 (361 2104)
Michigan, IFH Germany, Hohenheim, Colorado, Cornell, Princeton, Princeton students, UC San Diego, Michigan Medical, USU, NIH, Weizmann, Meijo, US Military DISA, US Navy NUWC, VU Netherlands, Imperial.

17th January 2005 (406 1357)
NU Singapore, Linz, Latrobe, British Columbia, ETH Zurich, Boeing, Lockheed Martin, TU Cottbus, Leipzig, DTU Denmark, Georgia Tech., Penn State, Rutgers, Michigan, U Penn., UAB Spain, Vigo, JYU Finland, UTU Finland, Poitiers, CEU Hungary, Technion, CNR ITIA, Tsukuba, Groningen, PAP Poland, Aberystwyth, Birmingham, Durham, Nottingham, Swansea, Belgrade.

18th January 2005 (393 1947)

British Columbia, Calgary, Saskatchewan, Hewlett Packard, RWTH Aachen, TU Dresden, Frankfurt, Bradley, Georgia Tech., Hawaii, Louisville, Northwestern, Richmond, Roanoke, Rose Hulman, UCLA, UC Santa Cruz, South Carolina, Washington, Wisconsin, UV Spain, UTU Finland, CEA France, CICRP Jussieu, Nantes, Paris 8, NASA (JPL), Samos, Technion, CNR ITIA, Kyoto, US Airforce Malmstrom, US Airforce Wright Patterson, US Army, US Navy NUWC, Groningen, BKKB Norway, BBSCR Britain, Cambridge, Durham, Oxford.

19th January 2005 (367 1897)

Australian Gov. BOM, SMU, Fujitsu, ARCS Austria, Adelaide, Queensland, UIA, Maplesoft Ontario, Manitoba, Waterloo, CERN, Intel, Max Planck MPIS, Leipzig, Marburg, Tuebingen, AU Denmark, Colorado, Cornell, Miami, Oklahoma State, Princeton, RIT, Renselaer, Stanford, Michigan, Southern Florida, Texas, YU, UV Spain, UPS Toulouse, Oak Ridge, Us Gov. TVA, Jerusalem, Weizmann, CNR ITIA, ENEA Frascati, SISSA Italy, Rome 1, US Army Medical, UIO Norway, US Gov. CBI, US Poland, Lisbon, UPJS Slovakia.

20th January 2005 (381 3028)

ADMU Philipines, Linz, ULB Belgium, McGill, Canadian NRC, Ottawa, CERN, IBM Almaden, IT and T, McGraw-Hill, Albert Einstein Institute, KU Denmark, CN, Columbia, Drexel, Evergreen, Iowa State, Maine, Nevada, Renselaer, SIU, U Penn., US Naval Academy, Utah, Washington State, UC Spain, Helsinki, Marseilles 3, Fermilab, Sandia, CNR ITIA, Titech Japan, Tohoku, TU Eindhoven, Otago NZ, NU Singapore, Birmingham, Cambridge, Heriot Watt, Oxford, RL Britain.

21st January 2005 (370 1741)

Erlangen, UTK, UNISA Australia, Disney, Teledynefluid, TU Cottbus, Kassel, Arizona State, Caltech, Colorado, Georgia Tech., Kansas, Northwestern, Pittsburgh, UC Santa Cruz, Chapel Hill, Washington, WPI, UV Spain, French national particle lab., Lawrence Livermore, NASA (Lerc), NIH, CNR ITIA, SISSA Italy,Kyoto, UNAM Mexico, Leiden, US Gov CBI, Torun Poland, Physto Sweden, Bilkent, TNRS Taiwan, Bristol, Cranfield, Durham, Oxford, Sussex.

22nd. January 2005 (393 1758)

UNLP Argentina, Zurich, Leipzig, Aizona, Columbia, Rutgers, California system, Connecticut, UPC Spain, Kyoto, UEC Japan, LTH Sweden, Utfors Sweden, Edinburgh.

23rd January 2005 (369 1312). New Record of Monthly Visitors.

Memorial Newfoundland, Queen's Canada, CERN, Caltech, Marietta, ODU, Michigan, Wyoming, WCUPA, ENS France, IITB India, NAU Ukraine, St. Andrews.

24th. January 2005 (510 2929), New Record of Daily Visitors.

Linz, Gov. Brazil, UNICAMP, McGill, British Columbia, EPF Lausanne, Philips, Teledynefluid, Max Planck Bonn, Hannover, Karlruhe, Tuebingen, Arizona, CLU, Central Michigan, Duke, Florida International, Florida State, Iowa State, Lousiville, North Carolina State, Northwestern, Penn State, Rochester, UC Irvine, Michigan, U Penn., WCUPA, WFU, Western Washington, UV Spain, French national particle lab., NASA (Jpl), CNR ITIA, Kyoto, Tohoku, KAIST S. Korea, US Army Research Laboratory, Groningen, TU Eindhoven, US Gov CBI, BTH Sweden, NTNU Taiwan, TRC Taiwan, BCUC Britain, OCC Britain, Oxford, RDG Britain, Strathclyde, Univ Wales Institute Cardiff.

25th January 2005 (401 2751) New Record of Monthly Hits.

KU Leuven, UNB Brazil, King's Canada, Alberta, British Columbia, Manitoba, York Univ. Canada, Intel, Northrop Grummond, MUNI Czechia, Albert Einstein Institute, Max Planck Heidelberg, Arizona, Boston Univ., Buffalo, Clemson, Columbia, CSU Hayward, Northwestern, Ohio State, SMU, Stanford, UC Santa Cruz, Oregon, Sioux Falls, Wisconsin, UV Spain, ENSAE France, EAP Greece, Jerusalem, IUST Iran, Milan Tech., UNAM Mexico, US Gov. CBI, Methodist Health, Shihlin Taiwan, TTRC Taiwan, Bristol, Glasgow, Heriot Watt, Oxford, Univ College London.

26th January 2005 (499 2122)

AN-KTN Austria, Lockheed Martin, McGraw-Hill, German Synchrotron System (DESY), Max Planck Halle, Max Planck MPIS, Karlsruhe, Konstanz, Appalachian State, Georgia Tech., GCSU, Harvard, Haverford, North Mexico State, Ohio, Penn State, Renselaer, UC Irvine, UCV, Florida, Michigan Medical, UNL, URI, Vermont, UV Spain, ANN Jussieu, SNV Jussieu, UHP Nancy, US Gov Dept of Transport, Weizmann, UEC Japan, US Army Monmouth, Gov. of Mexico, TU Eindhoven, US Gov. CBI, Chula Thailand, MOE Taiwan, TTRC Taiwan, Birmingham, Durham, Luton, Manchester, Salford.

27^{tH} January 2005 (527 2320) New Record of Daily Visitors.

Arizona State, St Vincent, Starman, ASN-KTN Austria, UNICAMP, Manitoba, New Brunswick, Northrop Grummond, Teledynefluid, DTU Denmark, Colorado, Iowa State, Oklahoma State, Chicago, UC Irvine,

Connecticut, Houston, Michigan Medical, Minnesota, North Carolina Charlotte, Vanderbilt, URV Spain, UV Spain, UTU Finland, UPS Toulouse, Los Alamos, NCMR Greece, Weizmann, CNR ITB Italy, INFN Pisa, INFN Turin, Trieste, US Military Fusion, US Naval Research Laboratory, Gov. Mexico, KUN Netherlands, Lodz, Chalmers Sweden, Physto Sweden, Ljubljiana Sweden, MOE Taiwan, TTRE Taiwan, Kyiv, Imperial, Kent Canterbury.

28th January 2005 (637 1837) New Record of Daily Visitors.
    TU Berlin, Maryland, IATP Ukraine, ASN-KTN Austria, MQ Australia, Acad. Canada, Alberta, Lockheed Martin, Wolfram, Xerox, Charles Univ. Prague, Max Planck Heidelberg, TU Cottbus*, Freiburg, Heidelberg, Rostock, Central Michigan, Columbia, Duke, MIT, Ohio State, Rutgers, UC Davis, UCLA, U Conn., UC Riverside, Delaware, Florida, Michigan, Minnesota, UNLV, Utah, West Virginia, EHU Spain, UM Spain, UV Spain, INSA Lyon, US Gov BLM, NTUA Greece, CNR ITIA Italy, INFN Genoa, US Army Monmouth, Porto, Bton Britain, Cambridge, Sheffield, Macmillan.

29th January 2005 (541 3436) New Record of Monthly Gigabytes downloaded.
    British Columbia, ASN-KTN Austria, Winnipeg, Max Planck MPPMU, Boston University, Buffalo, Rochester, SISSA Italy, UKIM, IIMCB Gov Poland, Oxford*, St Andrews.

30th January 2005 (374 1520) New Record of Monthly Previews, Ten new records set in Jan.
    Teledynefluid, Stuttgart, UUC Denmark, Boston Univ., JSUMS, Rutgers, U Conn., UC Riverside, Florida, NIH, Tohoku, Newman Foundation, SPBU Russia, MOE Taiwan, Nottingham.

31st January 2005 (476 1542)
    Texas, Univ. Tokyo, KF Graz, UIA Belgium, Alberta, Toronto, IBM Watson, HU Berlin, Leipzig, Oldenburg, Drexel, Georgia Tech., New Mexico State, Pittsburgh, Southern, SUNY OCC, Syracuse, Texas A and M, Temple, Chicago, UC Irvine, Michigan, WSU, EHU Spain, Zaragoza, UV Spain, Helsinki, INERIS France, LAAS France, OBSPM France, NASA (Gsfc), NASA (lerc), UOA Greece, Indian Statistical Institute, Florence, UIO Norway, Bath, Cambridge.

1st February 2005 (518 2221).
    ASN-KTN Austria, UNICAMP, Alberta, Calgary Disney, IBM Thomas J. Watson, TU Cottbus, Erlangen, Leipzig, Rostock, Berkeley, City Univ

New York, Dartmouth, Georgia Tech., MIT, Michigan State, PISD, Rochester, Santa Fe, Chicago, Michigan Medical, Minnesota, Chapel Hill, Oregon, Yale, UV Spain, INRIA France, Strasbourg, Poitiers, Rennes 1, Thessalonika, TIN Italy, Milan, Unimo Italy, Nagoya, KNU S Korea, US Navy NUWC, Gov. Mexico, UANL Mexico, NU Singapore, NCTU Taiwan, TTRC Taiwan, Bath, Essex, Manchester, Oxford*, RL Britain, St Andrews, Univ College London.

2nd February 2005 (500 3533)
   Texas Dallas, Canadian NRC, Toronto Medical, Compaq, IBM Thomas J. Watson, Teledynefluid, LRZ Munich, TU Berlin, TU Cottbus, Leipzig, Potsdam, Stuttgart, Arizona, Arizona State, Boston College, Berkeley, Boston University, SDSMT, Chicago, Michigan, Michigan Medical, UNL, Oregon, Wednet, Wesleyan, USC Spain, UV Spain, USP Fiji, Sandia*, UOA Greece, CNR ITIA, US Army Medical, US Army NVL, Otago, ABC Programs, Jefferson National Lab., UJ Poland, US Poland, LTU Sweden, UMU Sweden, TUKE Slovakia, MOE Taiwan.

3rd February 2005 (428 2997)
   Nagoya, ASN-KTN Austria, UIA Belgium, UFPE Brazil, UQAM Canada, Raytheon, STW Bonn, Cornell, Embry Riddle, Georgetown, North Dakota, New Mexico Tech., Oregon State, Rutgers, Samford, SUNY Stonybrook, Florida, Urbana Champaign Michigan Medical, U.S. Naval Academy, USU, Texas, UWM, WVSC, Xavier, UPM Spain, UV Spain, BIPM France, Polytechnique, Poitiers, Thessalonika, Tel Aviv, CNR IPCF, US Army Research Laboratory, US Army cjtf180, US Army Polk, Twente, MTU net Russia, Aberystwyth, Edinburgh, Manchester, Westminster.

4th February 2005 (316 1401)
   Ghent, UNICAMP, TU Cottbus, Frankfurt, Paderborn, Arizona, Bucknell, Cerritos, Fort Lewis, North Carolina State, Rutgers, Texas A and M, Connecticut, Urbana Champaign, UC3M Spain, UV Spain, ESPRI France, Los Alamos, Army Research Laboratory,
   US Military DISA, TU Delft, PAP Poland, Aberystwyth, Edinburgh, St Hughs Oxford, Queen Mary London, Swansea, Macmillan.

5th February 2005 (407 1896)
   ASN KTN Austria, Teledynefluid, Colorado, Indiana, SP College, U Mass., UAM Spain, Perugia, Yokohama CU, NTNU Norway, Lisbon, NAU, Bristol, Jesus Colleeg Oxford.

6th February 2005 (314 2591)

Missouri, UC San Francisco, Stuttgart, Minnesota, Vanderbilt, Marseilles, Technion, Tsukuba, Univ Tokyo, LUMS Pakistan, NAU Ukraine, Huddersfield, BNC Oxford.

7th February 2005 (394 2178)

Southern Michigan, Central Florida, UCLA, New South Wales, Liege, Montreal Tech., British Columbia, Toronto, Glaxo Wellcome, Intel, IU Bremen, LRZ Munich, SBS, Hannover, IHK Denmark, Albany, Biola, Florida State, Haverford, MUM, North Carolina State, Renselaer, Rutgers, Stanford, Texas A and M, Temple, Illinois Chicago, Michigan Medical, UV Spain, UTU Finland, CEA France, Montpellier 2, NASA (Lerc), US Gov USDA, Trinity College Dublin, CNR ITIA Italy, Hiroshima, TUS Japan, US Military WAC, US Navy SPAWAR, Groningen, Twente, LUTH Sweden, NTU Singapore, Ljubljana Slovenia, SWU Thailand, BBSRC British Gov., Masnsfield Oxford, York.

8th February 2005 (643 2203) New record of visits per day.

SIT NZ, ASN - KTN Austria, Free Univ Brussels, UIA Belgium, McGill, McMaster, Laval, UQAM Canada, Western Ontario, Boeing, Nokia, SUN, Teledynefluid, Charles Univ. Prague, Max Planck, TU Cottbus, Central Michigan, Columbia, City Univ. Ne York, GAC, Michigan State, Najah, Pittsburg, Penn State, SUNY Stonybrook, UMDNJ, UAM Spain, UV Spain, ENAC France, Bordeaux 1, Poitiers, NIST, Technion, IASBS Iran, CNR ITIA, INFN Rome 1, Rome 1, US Army Research Laboratory, HIN Norway, UIO Norway, US Gov CBI, MOE Taiwan, TTRC Taiwan, Bristol, Cambridge, Durham, Heriot Watt, Kingston, Manchester, MAGD, Swansea College.

9th February 2005 (367 1731)

Harvard, Tsukuba, Loeben Austria, SBS Germany, TU Cottbus, Munich, IOT Denmark, Arizona, Brigham Young, Caltech, Indian River CC, Michigan State, OU, Pomona, SIU, Urbana Champaign, Michigan, Washington, CSIC Spain, UPC Spain, UV Spain, Helsinki, Gov of New York City, Trinity College Dublin, NIC India, Univ Tokyo, US Army Research Laboratory, US Army SILL, Twente, Otago NZ, LUMS Pakistan, NU Singapore, Ljubljiana, SWU Thailand, NAU Ukraine, Manchester, Warwick.

10th February 2005 (288 2092)

ULB Belgium, UNICAMP, Nove Brazil, McGill, McMaster, Memorial Newfoundland, York Univ. Canada, Intel, Lockheed Martin, FHG Germany, Max Planck Heidelberg, TU Cottbus, Dortmund, AU Denmark,

Arizona, Cal Polytech, Columbia, ESU, Lousville, STUY, Idaho, Michigan, UTC, Texas, UC Spain, ISEP France, Poitiers, CNR Rimini Italy, Pisa, US Airforce AFIT, US Army Bragg, Twente, KTH Sweden, Cambridge, Nottingham, Oxford, Univ College London, BBC.

11th February 2005 (268 2180)

Alberta, British Columbia, Boeing, Merck, Teledynefluid, LRZ Munich, Cottbus, Essen, Goettingen, Colorado, Cornell, Georgia Tech., Penn State, UC Santa Cruz, Florida, Michigan, Washington, UV Spain, ENSAM France, Fermilab, CNR ITIA Italy, CSI Italy, INFN Milan, ICTP Trieste, Utrecht, HIN Norway, UMU Sweden, Aberdeen, Birmingham, City.

12th February 2005 (329 1180)

ASN-KTN Austria, Goettingen, Boston University, Colorado, Cornell, Johns Hopkins Medical Institute, New Mexico State, Texas A and M, UC Santa Cruz, U Mass., New Hampshire, US Navy Pacific South West, LUTH Sweden, Metu Turkey, NCTU Taiwan, Univ College London.

13th February 2005 (331 1578)

Caltech, US Navy ONR, Montreal, FU Berlin, Central Michigan, Colorado, MIT, Penn State, UC Santa Cruz, Minnesota, Utah, Texas, Szeged, Jerusalem, Linkoping, Bilkent, Cambridge.

14th February 2005 (473 2246)

ASN-KTN Austria, Adelaide, Western Australia, Gov Brazil Camara, Boeing, Philips, Siemens, Palacky, TU Chemnitz, Giessen, Kiel, Mannheim, ACCD, Arizona State, Columbia, ENMU, Florida International, HMC, Iowa State, JWU, MIT, RIT, Renselaer, Rutgers, South Carolina, UC Davis, Chicago, UIC, Michigan, Chapel Hill, UNL, Texas, Unican Spain, UV Spain, HTV Finland, Cergy France, Paris Psud, Szeged, Trieste, US Airforce Gunter, US Navy NMCI, US Navy SSP, TU Delft, Twente, Torun, Metu Turkey, MOE Taiwan, BBSCR British Gov., Edinburgh, Huddersfield, Heriot Watt, Leeds, Oxford, Strathclyde, Trinity-CM.

15th February 2005 (523 3351)

ASN-KTN Austria, TU Vienna, IMEC Belgium, UNICAMP, Memorial Newfoundland, Montreal Tech., Victoria, Mobile Gas, Northrop Grummond, Sun, Palacky, FGAN Germany, FHG, Hannover, Jena, Arizona, ATU, Brown, Columbia, CSU Ohio, CUP, Georgia Tech., Iowa State, JMU, MS State, North Dakota, Northwestern, New York Univ., Penn State, Rutgers, Stolaf, Stonybrook, UNCG, Villanova, Washington,

WFU, UV Spain, HTV Finland, HUT Finland, ENS France, Jerusalem, INFN Pavia, US Airforce Edwards, US Airforce Keesler, US Airforce Maxwell, US Army AMRDEC, US Army Research Laboratory, Leiden, Twente, Kluwer, LUMS Pakistan, Utrecht, NU Singapore, MOE Taiwan, Edinburgh, Oxford, Surrey.

16th. February 2005 (505 2231)

Maricopa, UNICAMP, Brandon, OCDSB Ontario, Saskatchewan, Laval, Montreal, York Univ. Toronto, New Brunswick, DIPF Gemany, Max Planck Dresden, Arizona State, Caltech, Colorado, Harvard, Hawaii, Louisiana State, Missouri, MIT, NIU, Ohio State, Tulane, Chicago, UGA, Michigan, Oregon, US Naval Academy, Utah, Western Michigan, UAM Spain, UV Spain, Polytechnique, Lemans, BIU Israel, CNR ITIA Italy, SNS Italy, US Airforce Edwards, Gov. Mexico, Groningen, Twente, US Gov. CBI, TPC Taiwan, Bath, London School of Economics, Nottingham, Oxford*, Queen's Univ. Belfast.

17th. February 2005 (498 1947)

New South Wales, Bucknell, ULB Belgium, McGill, Western Ontario, IHEP China, Tsinghua China, Mobile Gas, Charles Univ. Prague, Max Planck CPFS, SBS Germany, Boston University, Central Michigan, CTC, CWU, GMU, Lousiana Tech., MIT, North Carolina State, Renselaer, Rutgers, SUNY Stonybrook, Texas A and M, UC Davis, UI, Kentucky, Michigan Medical, U Penn., Washington, Wayne State, UV Spain, Polytechnique, Paris Psud, US Gov BLM, Lawrence Livermore, ACN Greece, OTE Greece, TEE Greece, CNR Rimini, PIT Nagano, UEC Japan, KAIST South Korea, US Army Amedd, US Army Research Laboratory, US Army Wiesbaden, US Navy NMCI, UNAM Mexico, TU Delft, Eindhoven, Twente, Kluwer, NTNU Norway, NTHU Taiwan, Imperial, Warwick.

18th February 2005 (475 1848)

UPC Czechia, Rowan, ASN-KTN Austria, New South Wales, UQAM Canada, Waterloo, Basel, Apache, BMW, IBM (USF), Lockheed Martin, Charles Univ. Prague, German Synchrotron Facility, FGAN, Max Planck IPP, TU Darmstadt, Greifswald, Kassel, Kiel, DTU Denmark, Arizona State, Baylor, Emporia, Florida International, Georgia Tech., ILSTU, Kent State, Maricopa, Northwestern, New York Univ., Penn State, Toronto, TTU, U. Mass. Michigan Medical, UB Spain, UC Spain, CEA France, Mulhouse, Paris 1, US BLM, US Gov.LOC, TEE Greece, PMN Italy, US Airforce Los Angeles, Twente, UTL Portugal, LUTH Sweden, Umea, KSUT Taiwan, Oxford, York.

19th February 2005 (453 2533)

ASN-KTN Austria, KU Leuwen, Stuttgart, Berkeley, Boston University, Colorado State, Princeton students, Penn State, UH, Maryland, Utah, Washington, Paris Psud., Paris 1, US Army Leavenworth, UTL Prtugal.

20th February 2005 (359 1137)

Toronto, Charles Univ. Prague, AU Denmark, Brown, Harvard, HMC, Hope, Missouri, MSOE, MS State, MU, MWSU, Princeton, Toronto, Tufts, Chicago, Delaware, Vemont, NIPS Japan, Porto, ITTI NZ, FUW, Umea, MOE Taiwan, Cambridge, Hull.

21st February 2005 (481 2691)

Johns Hopkins, ASN-KTN Austria, McGill, CERN, Merck, Bosch, Dortmund, Marburg, Biola, Columbia, CSULB, CWRU, Georgia Tech., Harvard, HSC, Lousiana, North Carolina State, Pittsburgh, Stanford, Toronto, Florida, Urbana Champaign, Maryland, Michigan Medical, South Florida, Texas Arlington, Utah, William and Mary, UM Spain, UV Spain, ENSAM France, Nice, Poitiers, Fermilab, INFN Milan, INFN TS Italy, Perugia, KNU S. Korea, TU Delft, Twente, Kluwer, IFJ Poland, MOE Taiwan, TTRE Taiwan, BBSCR British Gov., Bristol, Cambridge, Durham, Harrow School, Luton, Keble, Oxford, St. Andrews.

22nd February 2005 (700 1854) New Record of Daily Visits.

Chicago, Vienna, UNICAMP, Gov Alberta, Quebec, Tsinghua China, Microsoft, Max Planck MPQ, Arizona, Caltech, Colorado, Cornell, CSU Pomona, Iowa State, IWU, Louisville, MIT, Michigan State, North Carolina State, Florida, Stanford, Georgia, U Mass., Maryland, Michigan Medical, New Mexico, Utah, Washington St Louis, UV Spain, HUT Finland, EDF France, UJF Grenoble, Marseilles, Poitiers, US Gov ANL, Gov California, Fermilab, NASA (KSC), Sandia, AUTH Greece, City Hong Kong, Univ College Dublin, CNR ITIA Italy, ICTP Trieste, Pavia, US Military Pentagon, Gov. Mexico, TU Eindhoven, Massey NZ, Jefferson National Laboratory, IFJ Poland, TTRC Taiwan, Birmingham, Bristol, Canterbury, Hull.

23rd February 2005 (410 1552)

TNRC Taiwan, UIBK Austria, McMaster, Alberta, Windsor, Raytheon, Mac Planck MPI-SB, TU Cottbus, Berkeley, Caltech, Central Michigan, CSU, CSU Ohio, Georgia Tech., Indiana, Johns Hopkins, Lousiana State, North Carolina State, North Dakota, ONU, Princeton, Penn State, Purdue, Sabanci, St Cloud, SVSU, Tufts, Florida, Michigan Medical, WOU, WVU, Andalucia Gov., USC Spain, UV Spain, OULU Finland, CNRS Director

France, Poitiers, NASA (GSFC), Oak Ridge, UOA Greece, Dublin City, CNR ITIA Italy, US Army Research Laboratory, UNAM Mexico, NTNU Norway, FUW Poland, UTL Portugal, CMB Romania, Saratov Russia, Cambridge, Glasgow, Imperial. LEA Vale of Glamorgan.

24th February 2005 (378 2211)

RWTH Aachen, Tulane, Siemens Austria, Curtin, Sydney, AEI Canada, York Univ Toronto, Intel, Lockheed Martin, Siemens PSC, FHG Germany, Max Planck MPIS, Ruhr Univ. Bochum, SBS, Bayreuth, Niels Bohr Denmark, Berkeley, Cornell, CUW, Duke, Harvard, Illinois Tech., JWU, Northwestern, Purdue, SIU, California University System, U. Mass Dartmouth, Michigan Med., New Mexico, Texas, Wyoming, Yale Medical, UV Spain, UTU Finland, Marseilles, US Gov Brookhaven National Laboratory, Sandia, US Gov USGS, AUTH Greece, Dublin City, PRL India, INFN LNF, US Navy NASKW, US Navy NMC, Twente, Kluwer, NTNU Norway, UIO Norway, Edison Poland, ITN Portugal, Porto, ASTRAL Romania, CMB Romania, NIPNE Romania, LTH Sweden, LUTH Sweden, Birmingham, Cambridge, Durham, Edinburgh Leeds, Southampton, Warwick, British Gov DERA, Kings Ely School Cambridge.

25th February 2005 (302 1344)

UTAS Austria, KF Graz, Flinders, UFU Brazil, Alberta, York Univ Toronto, UNAL Colombia, Max Planck Mainz, TU Darmstadt, Berkeley, Caltech, Florida International, Indiana, IWU, Louisville, MSU, OU, UC Santa Barbara, UMBC, UMR, U Penn, UV Spain, US Gov Federal Aviation Authority, Trinity College Dublin, NTNU Norway, KTU Turkey, MSU Turkey, MOE Taiwan, NETU Taiwan, Cambridge*, King's London, Oxford, RDG.

26th February 2005 (276 1012)

City of Winnipeg, IBM Thomas J. Watson, Berkeley, Evergreen, Rice, Texas A and M, Delaware, UST Hong Kong, NTNU Romania, PK Poland, Cambridge, Imperial, ULA Venezuela.

27th February 2005 (438 1183)

ASN-KTN Austria, Memorial Newfoundland, Alberta, Toronto, Leipzig, Oldenburg, Buffalo, Cornell, MIT, MS State, New Mexico State, Urbana Champaign, Wisonsin, WPI, NASA (LARC), Cambridge*, Edinburgh.

28th February 2005 (362 1359) Appointed to the British Civil List
Unesp Brazil, Brock, Bayreuth, Brigham Young, Caltech, Colorado, Connecticut College, Drexel, MU Ohio, South Florida State, Stanford, Stonybrook, Texas A and M, Michigan Medical, Virginia, UV Spain, Poitiers, US Gov BLM, AUTH Greece, TEE Greece, UST Hong Kong, Gov Latvia, US Army Research Laboratory, Twente, LUTH Sweden, Chula Thailand, Aberystwyth, Cambridge, Edinburgh, Oxford, York.

1st March 2005 (448 1979)
Brigham Young, Australian National, Queensland, Laurentian, McMaster, British Columbia, Ford, Motorola, TU Cottbus, Greifswald, Berkeley, Brown, Boston Univ., Columbia, EOU, Indian River CC, Montana, MS State, Northwestern, Ohio State, ONU, UA, UC Santa Barbara, Michigan Medical, Washington State, USAL Spain, UV Spain, ONERA France, Poitiers, Argonne, UOC Greece, Dublin City, TIN Italy, Pisa, Kyoto Univ., UNAM Mexico, Groningen, Twente, Kluwer, Jefferson National Laboratory, Bristol, Cambridge.

2nd March 2005 (498 2656)
Defence Gov. Australia, New Brunswick, NS, Simon Frazer, ETH Zurich, Canon, IU Bremen, TU Darmstadt, Frankfurt, Muenster, Stuttgart, Amherst, Brandeis, Boston Univ., Caltech, CWRU, Emporia, Hertford, Indiana, Maine, Missouri, Michigan State, Northwestern, Ohio State, Stanford, TMC, UALR, Chicago, California University System, Houston, Michigan, U Penn, Texas, Vermont, Washington, Wisconsin, UCM Spain, Lawrence Livermore, NASA JPL, Oak Ridge, Thessalonika, Rome 2, Kanazawa, Metropolitan Univ., Torun, IJS Slovenia, MOE Taiwan, Bristol, Cambridge, Cambridge, Edinburgh, Oxford.

3rd March 2005 (479 2896)
Australian NLA, Waterloo, ETH Zurich, Ford, Palacky, Max Planck Immunobiology, Mainz, Brandeis, Buffalo, Embry Riddle, MIT, Morehouse, Mississippi, Princeton, SMSU, Stillman, Stonybrook, TCU, California System, UC Santa Cruz, Delaware, Michigan, Michigan Medical, Chapel Hill, Texas Dallas, Texas, Xavier, HTK Finland, French national particle lab., ISEN France, Lille 1, ELTE Hungary, ICRA Italy, TIN Italy, Trieste, JAIST Japan, Tohoku, KAIST S. Korea, NTNU Norway, Massey NZ, Lisbon, KTH Sweden, UDSM Tanzania, Aberystwyth, Bristol, Cambridge, Imperial, Leeds, Manchester, Napier, Oxford.

4th March 2005 (420 2217)

Vienna, JCU Australia, Leuven, UNICAMP, Sao Paolo, Memorial Newfoundland, UCV Chile, Berkeley, Bridgew, BSU, Georgia Tech., Lousville, MIT, NEU, OU, Stanford, Stevens, Toronto, Texas Tech., UC Santa Barbara, UC San Diego, Urbana Champaign, Maryland, Chapel Hill, U Penn., Vanderbilt, Wayne State, EBSC Rennes, ICTP Trieste, Pisa, NTNU Norway, Massey NZ, LTH Sweden, LUTH Sweden, Ljubljiana, Bristol, Essex, Lancaster, Leeds, Manchester, Oxford*.

5th March 2005 (372 2349)

Unilever, RWTH Aachen, Amherst, DU, MIT, Najah, Renselaer, Stanford, Stonybrook, KAIST S. Korea, UTCLUJ Romania, Novosibirsk, Manchester.

6th March 2005 (274 1643)

Stonybrook, Melbourne, McGill, Microsoft, Charles Univ Prague, IU Bremen, Brown, HMC, Northwestern, Renselaer, Washington, Manchester, Oxford, Univ. College London.

7th March 2005 (382 1464)

UC Riverside, McGill, Waterloo, CERN, TU Cottbus, Erlangen, Berkeley, Brown, Columbia, Graceland, Harvard, Purdue, Stanford, Delaware, Michigan, UPMC, South Florida, Utah, UV Spain, BIPM France, Marseilles, INFN LNF, Rome 1, Chungbuk S. Korea, US Airforce Wright Patterson, VU Netherlands, NERSC Norway, UIO Norway, IPB Portugal, NIPNE Romania, RIT Thailand, Anadolu, MOE Taiwan, BBK Britain, Durham, Imperial, King's London, Linst., Oxford, Queen Mary London, RL, Salford.

8th March 2005 (348 1903)

Calvin, KF Graz, Univ Queensland, UNICAMP, New Brunswick, Victoria, York Univ. Canada, TU Cottbus, Tuebingen, Auburn, Boston Univ., Columbia, CSU, Florida International, Galileo, Georgia Tech., MIT, QC, SHU, U Conn., UC San Diego, Florida, Maryland, Medical Michigan, Nevada Reno, South Florida, Washington, IAC Spain, UJI Spain, UMA Spain, UV Spain, UTU Finland, CEA France, Poitiers, Argonne, NASA JPL, NASA LERC, Univ. College Cork, Weizmann, Twente, NTNU Norway, FUW Poland, Exeter, Manchester, RDG, Warwick.

9th March 2005 (324 1554)

Gov. Australia Defence, Toronto, Waterloo, ETH Zurich, Intel, Microsoft, Siemens, TU Berlin, Jena, Arizona State, Bowdoin, Colorado State, City Univ New York, IPFW, Ohio State, Purdue, Stanford, Stanford

SLAC, UC Davis, UC Santa Barbara, Delaware, Florida, UH, Michigan, South Florida, Utah, UWM, Wisconsin, WSU, YSU, UV Spain, Polytechnique, Poitiers, US Gov JCCBI, US Gov. NIH, DUTH Greece, NTUA Greece, Miskolc Hungary, Weizmann, IIAP India, Trieste, Sienna, Turin, Free Univ Amsterdam, US Gov. CBI, Jefferson National Laboratory, Aberdeen, Aberystwyth, Bristol, Edinburgh, Glasgow, Manchester, Warwick County Council.

10th March 2005 (340 1681)

MIT, Delaware, Curtin, UNICAMP, York Univ Toronto, ETH Zurich, Boeing, Raytheon, DFG Gemany, FU Berlin, LRZ Munich, TU Cottbus, Buffalo, Caltech, Colorado, Columbia, Cornell, CUNY, Maine, MIT, Ohio State, Penn State, Rowan, Akron, UCLA, Delaware, Florida, Urbana Champaign, U Mass., Maryland, Michigan Medical, South Florida, Webster, Wright, West Virginia, IFAE Spain, UV Spain, Oulu Finland, OAMP France, Polytechnique, Poitiers, US Gov. Ames Lab., NASA ARC, TUC Greece, City Hong Kong, Weizmann, IIAP India, HI Iceland, Trieste, Kyoto, US Army PICA, US Naval Research Laboratory, Twente, US Gov CBI, FUW Poland, KTH Sweden, Ankara, KU Turkey, Aberdeen, Bath, Birmingham, Cambridge, Lancaster, York, Medway County Council, Kings Ely School Cambridge.

11th March 2005 (356 1425)

Ohio State, UNR Argentina, KF Graz, TU Graz, Dalhousie, Laval, ETH Zurich, LRZ Munich, Bonn, Hamburg, Stuttgart, California Polytech., Columbia, Harvard, Howard CC, Maine, Missouri, New Mexico State, Northwest Missouri, New York Univ., Pace, Stanford, Stanford SLAC, SUNY Stonybrook, UCLA, UH, Michigan, New Hampshire, Oregon, South Florida, Angers, St. Etienne, TUC Greece, Colaisteide Ireland, CNR ITIA, INFN Florence, TIN Italy, Trieste, US Airforce Pope, US Navy NUWC, UAQ Mexico, UITM Malaysia, TU Delft, Twente, Torun, NU Singapore, Swedish Institute of High Energy Physics, Imperial, Milton Keynes Council.

12th March 2005 (307 2529)

Linz, Shell, Berkeley, CSU Pomona, Miami, RIT, St Olaf, TCU, UIC, Michigan Medical, NTNU Norway, MOE Taiwan, Durham, Imperial, Nottingham.

13th March 2005 (366 3032)

QUT Australia, Institute of High Energy Physics China, Hewlett Packard, Wells Fargo, RBS Denmark, Caltech, Kent State, Lehigh, North

Carolina State, Stanford, TUT Finland, CNRS INLN, Jerusalem, KAIST S. Korea, UNAM Mexico, Twente, NTNU Norway, Edinburgh.

14th March 2005 (428 2645)

Brigham Young, Monash, UNICAMP, McGill, Simon Frazer, Western Ontario, Gov. of Yukon, Andes, Hewlett Packard, IBM Thomas J. Watson, Lockheed Martin, TU Ilmenau, Berkeley, Brandeis, Buffalo, Central Michigan, City Univ. of New York, Dartmouth, FHDA, Florida State, Georgia State, Johns Hopkins Applied Physics, Macalester, Michigan State, Ohio, Princeton, Penn State, Connecticut, Maryland, Michigan, UMR, Wisconsin, Unican Spain, Zaragoza, UV Spain, Poitiers, Fermilab, NASA (JPL), Univerisy College Dublin, Weizmann, PRL India, TIFR India, ICTP Trieste, Univ Trieste, Catania, Rome 1, US AF Base Tyndall, US Navy NPS, Twente, Kluwer, NTNU Norway, UIO Norway, Methodist Health, Royal Society of Chemistry, CMB Norway, NIPNE Norway, LUTH Sweden, Bristol, Edinburgh, Glasgow, Imperial, Leeds.

15th March 2005 (419 1770)

Adelaide, US Australian Gov. Defence, McGill, Queen's, Sherbourne, ETH Zurich, SBS, TU Chemnitz, TU Cottbus, Essen, Jena, Regensburg, CSULB, Depaul, Harvard, Haystack, Hope, MU Ohio, Ohio State, Penn State, SHU, Stanford, Stanford SLAC, SVSU, UCOP, UMBC, Maryland, Texas Dallas, Willamette. UV Spain, OAMK Finland, Poitiers, NTUA Greece, Thessalonika, Technion, INFN Perugia, Tohoku, KAIST, US Airforce Pope, UAEM Mexico, Twente, Free Univ Amsterdam, VU Netherlands, NTNU Norway, Massey NZ, Lisbon, LLH Sweden, LU Sweden, NCTU Taiwan, Bangor, Birmingham, Exeter, Imperial, Leicester, Manchester, NCL, Nottingham.

16th March 2005 (353 2393)

Korean Army, Postech Korea, US Army Pica, US Military Disa, Twente, NTNU Norway, LU Sweden, Upsala, City Univ. London, Heriot Watt, ROE, Swansea, Medway Council.

17th. March 2005 (384 1433)

Urbana Champaign, URBK, UIA Belgium, TCHE Brazil, Alberta, Gov. British Columbia, Gov Canada. CERN, Glaxo Wellcome, Nokia, FHG Germany, KFA Juelich, TU Berlin, TU Cottbus, Cornell, Georgia State, Harvard students, Lesley, Princeton, Penn Sate, Richmond, Stanford, Michigan Medical, Virginia, Vermont, Wright, Helsinki, Brest, Poitiers, UPS TLSE France, NTUA Greece, City Hong Kong, ELTE Hungary, Haifa, Technion, INFN Perugia, Bari Tech., Milan Tech., Dendai, US Military DISA, US Navy NMCI, NTNU Norway, Massey NZ, AERO,

Warsaw, Wroclaw, LTH Sweden, MOE Taiwan, Imperial, Strathclyde, York, BL Britain.

18th March 2005 (372 1573)

NTNU Norway, AGR Canada, McMaster, CERN, UTA Chile, Lockheed Martin, Albert Einstein Institute, Augsburg, Bayreuth, Buffalo, Georgia Tech., Harvard, South Florida State, Stanford, Tufts, Xavier, Jussieu ifr924, UPS Toulouse, Los Alamos, Univ. College Dublin, CNR IPCF Italy, SISSA Italy, Postech Korea, PDN Sri Lanka, US Army PICA, US Army Sill, US Military Fusion, Twente, Lisbon Tech., NTU Singapore, KOU Turkey, Rutherford Appleton, British Gov. DERA.

19th March 2005 (292 1590)

Oregon, NTNU Norway, UIB Norway, BBK Britain, Bristol, Cambridge, Winchester College.

20th March 2005 (351 3438)

Haverford, ASN KTN Austria, Monash, Melbourne, Central Michigan, MIT, South Florida, Hanyang Korea, NTNY Norway, Jefferson National, NSYSU Taiwan.

21st March 2005 (434 2892)

National Univ. Singapore, Linz, Swinburne, VU Australia, UNICAMP, Sao Paolo, Gov. Quebec, Queen's Canada, Disney, Rockwell Collins, TU Cottbus, Ulm, Brandeis, Central Michigan, Colorado, Emory, Florida State, Lehigh, NOAO, St Cloud State, Stonybrook, SUNY Stonybrook, Dayton, Michigan, UV Spain, F. Comte, Poitiers, US Gov JCCBI, City Hong Kong, IED Hong Kong, Tel Aviv, MRI India, HQ European Space Agency, Hanyang, US Military JCSE, Twente, NTNU Norway, Massey NZ, Jefferson National, Royal Society of Chemistry, PIT Poland, Torun, Physto Sweden, NTU Singapore, SU Thailand, NDHU Taiwan, Bangor, Bristol, Cambridge.

22nd March 2005 (383 2128)

KF Graz, Alberta, Western Ontario, Boeing, Glaxo Wellcome, Lockheed Martin, Siemens, Leipzig, DODEA, GRCC, Harvard, MTU, North Carolina State, Ohio, PNC, Princeton, Texas A and M, UC Davis, UCLA, Iowa, Michigan, Wright, UB Spain, UV Spain, Spanish Parliament, Fermilab, Los Alamos, ACN Greece, BME Hungary, Univ College Dublin, Hiroshima CU, KAIST, Twente, NTNU Norway, AMU Poland, Metu Turkey, Queen's Univ. Belfast.

23rd March 2005 (352 1390)

Linz, McGill, Alberta, Waterloo, Western Ontario, Hughes, Monsanto, Rockwell Collins, Cyprus, Ulm, Florida International, Harvard students, Kent State, Manhattan, Northwestern, Purdue, Texas A and M, Chapel Hill, Nevada Reno, Vanderbilt, MLV France, INFN Florence, Turin Tech, Pisa, Peruvia, KAIST, Postech Korea, US Army Pica, Groningen, Twente, Utrecht, NTNU Norway, LTH Sweden, Umea, NU Singapore, Bilkent, Boun Turkey, Cankaya Turkey, Cambridge, Imperial, Oxford.

24th March 2005 (507 2113)

NTNU Norway, ASN KTN Austria, GU Australia, ULB Belgium, Dow, FH Kiel, MIT, Michigan State, Santa Fe, SIU, SPSU, UC Irvine, Utah, CNES France, Fermilab, NASA (Lerc), TEE Greece, Perugia, US Army MSL, KTH Sweden, NU Singapore, Kiev, Cardiff, Cranfield, King's College London.

25th March 2005 (469 2047)

ASN KLN Austria, Siemens, McGill, Motorola, Caltech, Swarthmore, Michigan Medical, Utah, West Virginia, Poitiers, Hiroshima CU, US Navy NUWC, Twente, NTNU Norway, Chalmers Sweden, LTH Sweden, KMITNB Thailand, Cambridge, Oxford.

26th March 2005 (285 1320)

Waterloo, City of Winnipeg, Boston Univ., Johns Hopkins, Louisiana Tech., New Jersey Tech., TEE Greece, Technion, US Army Korea, Twente, NTNU Norway.

27th March 2005 (308 2925)

ADFA Australia, Gov. Canada, Toronto, Hawaii, U Mass., William and Mary, BIU Israel, Gov. Jordan, NTU Singapore, Exeter.

28th March 2005 (278 1731)

Buffalo, UNICAMP, Charles Univ.Prague, Danish Tech., Bellamine, Iowa State, IWU, Colorado School of Mines, Ohio State, Renselaer, SUNY Stonybrook, Michigan, NASA JPL, Hong Kong Univ., Hanyang.

29th March 2005 (310 1807)

Stanford, HWCDSB Ontario, CERN, Intel, TU Cottbus, Erlangen, Carleton, Colorado, Embry Riddle, Galileo, Johns Hopkins, Purdue, Syracuse, Akron, Florida, UWStout, Washington, Xavier, UCM Spain, UV Spain, Poitiers, Rennes 1, Argonne, Los Alamos, US Gov PPPL, US Gov. SSA, City Hong Kong, Trieste, US Army Research Laboratory, US Navy

MUGU, US Navy ONR, Twente, Massey NZ, Lisbon, Neva Russia, NU Singapore, York.

30th March 2005 (289 1553)

Western Australia, Leuven, Northrop Grummond, Arizona State, Berkeley, Boston Univ., Columbia, Florida State, JMU, Kennesaw, MU Ohio, New York Univ., Ohio State, Chicago, U Penn., Texas Arlington, Utah, Washington, Wisconsin, UV Spain, Jussieu, F Comte, Lawrence Livermore, Univ. of the Aegean, IITKGP India, Sony, US Airforce Dover, US Naval Research Lab., Twente, NSK Russia, Birmingham, Cambridge, Oxford.

31st March 2005 (338 1672)

Western Ontario, North Carolina State, ASN KTN Austria, TU Vienna, Centennial College Canada, Waterloo, Motorola, Philips, Giessen, Kaiserslautern, Mainz, Berkeley, Caltech, Cornell, CSUN, Duke, Indian River CC, Louisville, Nebo, Ohio State, Oklahoma, Penn State, Purdue, Renselaer, Rutgers, SINTE, Stonybrook, Michigan Medical, Western Florida, Wisconsin, Xavier, Jussieu LPTL, TVT Finland, Rennes 1, UST Hong Kong, UIO Norway, Massey NZ, Methodist Health, Ljubljiana, Academica Sinica, Cambridge, NERC Monkswood, Royal Society, Southampton, UMIST, Wolverhampton.

1st April 2005 (292 1235)

Toronto, ETH Zurich, Microsoft, Wells Fargo, LRZ Munich, TU Darmstadt, Erlangen, Karlsruhe, Leipzig, Caltech, Clarkson, Central Michigan, Galileo, Indian River CC, Montana, Montgomery College, Ohio State, Pomona, Pratt, Princeton, Rowan, UC Davis, UC San Diego, UC San Francisco, Michigan Medical, South Florida, Utah, Wisconsin, BRGM France, ENSICA France, Maths Jussieu, Poitiers, NIST, Sandia, IUCAA India, Tohoku, Postech, US Army Roberts, NTU Singapore, Oxford, Univ College London, National Physical Laboratory.

2nd April 2005 (311 2369)

ASN KTN Austria, Saskatchewan, IBM Thomas J. Watson, FU Berlin, IU Bremen, Galileo, North Carolina State, Rutgers, Dublin City Univ., KNU Korea.

3rd April 2005 (209 1349)

Calgary, Curtin, Conordia, Alberta, Toronto, Caltech, Embry Riddle, Nwhealth, ONU, ICU Korea, Twente, NU Singapore, Institute of High Energy Physics Russia, Cambridge, Imperial

4th April 2005 (302 1440)

ASN KTN Austria, Flinders, Tasmania, UNICAMP, Sao Paolo, Hydro Quebec, Laval, Amgen, Heidelberg, Sony, FH Kiel, FU Berlin, Max Planck Bonn, Hannover, Karlsruhe, Kaiserslautern, Leipzig, Boston College, CU Denver, Indian River CC, Colorado School of Mines, Pomona, IC Irvine, California system, U Mass, Michigan Medical, UNO, US Airforce Academy, Utah, California system, U Mass, Utah, Washington, UAM Spain, UHP Nancy, Poitiers, Hong Kong, Twente, FUW Poland, Evora Portugal, CANDI Britain, Imperial, Oxford, Sheffield, Swansea College.

5TH April 2005 (246 1082)

Latrobe, Curtin, KU Leuven, Calgary, Nine Chile, Northrop Grummond, Bosch, FH Jena, Leipzig, Arizona State, CSU Ohio, Denison, HMC, Johns Hopkins, Louisiana State, MIT, Princeton, SWRI, Urbana Champaign, Maryland, Western Michigan, Vermont, Yale, UCM Spain, UV Spain, Hong Kong Univ., IPN Mexico, UNAM Mexico, KUN Netherlands, Massey NZ, NTU Singapore, UPJS Slovakia, Cambridge, UMIST.

6th April 2005 (263 1448)

Toronto, Gov Australia BOM, UNICAMP, Canadian Defence Dept., Lockheed Martin, Lucent, UO Cuba, Heidelberg, Buffalo, Cooper, CWRU, Drexel, Hawaii, ILSTU, Indiana, Knox, Ohio Univ., Chicago, UC Irvine, Houston, Urbana Champaign, Michigan, Washington, UV Spain, Univ College Cork, Trieste, NTNU Norway, AERO, UA Portugal, Cambridge.

7th April 2005 (282 2225)

Xavier, KF Graz, Canadian Dept. of Defence, Nokia, Pfizer, IU Bremen, Leipzig, Stuttgart, National Library of Denmark, Cornell OPAC, Duke, Millikin, NCAT, North Carolina State, UC Irvine, Michigan, Chapel Hill, South Carolina, Xavier, UV Spain, CEA France, UJF Grenoble, Los Alamos, US Treasury, Hong Kong Tech., US Airforce Tinker, US Army Research Laboratory, NTNU Norway, UIO Norway, IFJ Poland, KTH Sweden, Cambridge, Sussex.

8th April 2005 (234 1940)

TU Vienna, Australian Gov. BOM, Shell, TU Cottbus, Leipzig, Boston University, Galileo, Princeton, SDSC, Florida, Maryland, Michigan Medical, Yale, UAM Spain, UPM Spain, CEA France, CNRS Grenoble, Paris Psud, UHB France, Glasgow, Nottingham, Oxford, Queen Mary College, Sheffield, Sussex, Swansea, USB Venezuela.

9th April 2005 (198 1915)

NTNU Norway, UCL Belgium, EPF Lausanne, TU Cottbus, Leipzig, Boston University, Paris Psud, Argonne, Hiroshima CU.

10th April 2005 (209 809)

Merck, British Columbia, Calgary, Leipzig, Danish Tech., Caltech, Clerkson, Colorado State, Indiana, Maine, McKenna, Rutgers, Stanford, UCLA, Wayne State, Jerusalem, NTNU Norway, Sheffield, Sheffield Hallam.

11[tH] April 2005 (335 1440)

KF Graz, Swinburne, Sydney, USP Brazil, Gov. British Columbia, Carleton, Manitoba, McMaster, Gov. Quebec. Ottawa, Cisco, Lockheed Martin, Qinetiq, Raytheon, Dortmund, Leipzig, Albany, Berkeley, Buffalo, COA, Johns Hopkins, Maine, North Dakota, Northwestern, Purdue, RIT, Sheperd, Stanford, SUN Stonybrook, Temple, Chicago, UC Santa Barbara, UC Santa Cruz, Iowa, Michigan Medical, Union, Catalonia, UCM Spain, UV Spain, HUT Finland, Lille 1, Argonne, IUST Iran, Rome 3, US Navy NMCI, US Naval Research Laboratory, Twente, Ericsson Sweden, Ljubjliana, Metu Turkey, FJU Taiwan, Kiev, Bangor, Bristol, North East Wales Tech., Oxford, St Andrews, York College.

12th April 2005 (279 1127)

Acadian, McGill, ETH Zurich, Algonquin College, Motorola, Northrop Grummond, MUNI Czechia, KFA Juelich, SBS, TU Cottbus, Giessen, Leipzig, Arizona State, Berkeley, Cornell, Iowa State, Michigan State, Stanford SLAC, Tufts, UC Davis, Chicago, Urbana Champaign, Michigan, Michigan Medical, Minnesota, Vanderbilt, Washington, Polytechnique, Oak Ridge, Univ College Dublin, Weizmann, PRL India, US Naval Research Laboratory, NTNU Norway, Bangor, Oxford, Winchester College.

13th April 2005 (301 1132)

Vienna, Latrobe, Sydney, Western Australia, Centennial College, Dalhousie, Boeing, Arizona State, Boston University, Colorado, Columbia, Cornell, Missouri, Montana, North Dakota, Purdue, Rice, Santa Fe, South Carolina, St Olaf, SUNY Stonybrook, Texas A and M, UC Riverside, Iowa, Michigan Medical, UNL, UNLV, US Airforce Academy, Texas, Virginia, Washington, Wellesley, Wisconsin, UV Spain, UTU Finland, Jussieu, Tel Aviv, Technion, Osaka, US Army Research Laboratory, US Navy NUWC, US Military OSD, HVU Netherlands, Groningen, TU Delft, Twente, HIST Norway, Umea, Heriot Watt, KIAD, NTU, Warwick.

14th April 2005 (349 1572)

Adelaide, AEI Canada, Toronto, Lockheed Martin, Merck, Northrop Grummond, German Synchrotron Facility, IKS Jena, Max Planck MPL, Freiburg, Caltech, CSU Chico, Georgia Southern, Hawaii, Lehigh, Louisiana State, MIT, Michigan State, Ohio State, Santa Fe, SUNY Stonybrook, Houston, Minnesota, NC Charlotte, Chapel Hill, US Naval Academy, Texas, Yale, TVT France, Weizmann, SISSA Italy, Univ. Tokyo, KUN Netherlands, Twente, NTNU Norway, UIO Norway, AGH Poland, Bangor, Cambridge, Edinburgh, Imperial, Queen's Univ Belfast, Southampton, Sussex, Warwick, Westminster, BK Yugoslavia.

15th April 2005 (299 1720)

Free Univ. Brussels, NRC Canada, Laval, Victoria, CERN, Max Planck MPIS, Bonn, Hannover, Munich, Albany, Arizona, A RTIC, Colorado State, Columbia, City Univ. New York, FIT, Harvard, Johns Hopkins, Maine, North Dakota, Rochester, Santa Fe, TMC, Michigan Medical, UNO, Texas, Wisconsin, NASA (gsfc), NIH, Patras Greece, Hong Kong Univ., PHY Croatia, Weizmann, INFN Turin, Umea, NU Singapore, Bolton, Oxford, Queen's Univ. Belfast, Southwark, Surrey, Univ, College London.

16th April 2005 (264 1157)

Berkeley, U Penn, Washington State, ADFA Australia, IAS, Michigan, UML, USF, NTNU Norway, Aberystwyth.

17th April 2005 (301 1450)

Florida Atlantic, KF Graz, Monash, Melbourne, Australian Gov. BOM, IU Bremen, Leipzig, Embry Riddle, Holy Cross, Maryland, Omaha, Willamette, NTNU Norway.

18th April 2005 (390 1887)

ESI Austria, Monash, Queensland, UFMG Brazil, British Columbia, CERN, Basel, Lockheed Martin, Max Planck Berlin, TU Cottbus, TU Darmstadt, TU Dresden, Kassel, Caltech, Central Michigan, Columbia, Duke, Georgia Tech., Indiana Tech., MIT, Monroe County CC, NAZ, Pittsburgh, Simmons, St Cloud State, Stonybrook, TTU, Iowa, Michigan Medical, Michigan, Texas, Washington, Williams, HUT Finland, GSFC NASA France, Hiroshima CU, Groningen, Twente, Auckland, US Gov. CBI, Chalmers Sweden, UWCM Britain.

19th April 2005 (409 1681)

MTU, Washington, DYU Korea, Curtin, Simon Frazer, Calgary, UNAL Colombia, canon, Microsoft, LRZ Munich, Max Planck Garching, Jena, Kaiserslautern, Lueneburg, IOT Denmark, Arizona, Central Michigan, Columbia, Fort Lewis, Georgia Tech., Montana, MS State, New Mexico State, Princeton, Penn State, Rutgers, TCNJ, UC Davis, Iowa, Maryland, Michigan Medical, U Penn., Washington, Williams, Yale, UV Spain, HUT Finland, INRIA France, Fermilab, Oak Ridge, AUTH Greece, Titech Japan, Tohoku Japan, US Army Pica, US Navy NSWC, Twente, Kluwer, US Gov, CBI, Wroclaw Poland, Anadolu Turkey, Metu, Bangor, Birmingham, Glamorgan, Liverpool, Southampton, Belgrade.

20th April 2005 (553 1580)

Stanford SLAC, ASN KTN Austria, SCC Australia, Sydney, Sao Paolo, Geneva, IU Bremen, TU Darmstadt, Brandeis, Caltech, Choate, Georgia Tech., Illinois Tech., New Mexico State, Princeton, Rice, UC Davis, Maryland, Michigan Medical, Montana, UMR, Chapel Hill, Utah, Wyoming, Vanderbilt, Washington, UCM Spain, UV Spain, Montpellier 2, NASA (Larc), University College Dublin, IUCAA India, Pisa, Tohoku, US Army TACOM, Groningen, Twente, US Gov CBI, Royal Society of Chemistry, Lisbon, Chalmers Sweden, Metu Turkey, Aberystwyth, Liverpool, UEL.

21st. April 2005 (423 1269)

Vienna, Australian Gov. BOM, British Columbia Hydro, Dehavilland, Boeing, Lockheed Martin, Charles Univ. Prague, Leipzig, Amherst, Angelo, Indiana Tech., LETU, Missouri, MNSCU, Penn State, STSCI, U. Conn., US Naval Academy, Virginia, Washington Univ St. Louis, UCM Spain, UPM Spain, UV Spain, Fermilab, US Gov INEL, Los Alamos, NASA JPL, NIST, Sandia, Jerusalem, Florence, Titech Japan, US Navy UAR, TU Delft, Utrecht, Umea, AIC Thailand, Aberdeen, Oxford, SHU, York, NHS.

22nd April 2005 (434 1588)

ASN KTN Austria, Linz, Calgary, Max Planck IPP HGW, AUC Denmark, Central Michigan, Georgetown, Haystack, Indian River CC, Lousiana Tech., Northwestern, New York Univ.,Oregon State, Rochester, Trinity College, Michigan Medical, UPM Spain, French particle laboratory, NASA (gsfc), IRB Croatia, Univ. College Cork, US Army Pica, Twente, Kluwer, FUW Poland, Wroclaw, Bilkent, Bangor, Imperial, Nottingham, Swansea, Univ. College London,Warwick, York.

23rd April 2005 (272 1585)

Clarkson, Mount Holyoke, New York Univ., Penn State, Purdue, Rochester, Kentucky, UV Spain, US Army PAC, Sterling Students, Twente, Liverpool.

24th April 2005 (224 967)

Cornell, Georgia Tech., Vanderbilt, Tel Aviv, RSSI Russia, Bilkent Turkey, Cambridge, Cranfield, Herts, Liverpool.

25th April 2005 (349 2574)

UCLA, CSIRO Australia, KU Leuven, UNICAMP, Canadian Govt. UCDSB Ontario, West Ontario, CERN, BAO China, Albert Einstein Institute, Max Planck IPP HGW, Siemens, Cornell, CTC, Duke, Indiana, IWU, KUSD, Lehigh, Connecticut, Kentucky, UN Charlotte, Texas Arlington, Virginia, Wednet, Wisconsin, UV Spain, CEA France, ELF Atochem France, Nantes, Argonne, Univ College Cork, Hosei Japan, US Navy NMCI, US Military NIMA, Twente, Utrecht, NTNU Norway, Poznan, RSSI Russia, NU Singapore, Cambridge, Durham, Gold, Oxford, Queen's Univ. Belfast, UMIST.

26th April 2005 (386 1867)

Melbourne, VC Graz, Adelaide, ANU, McMaster, Alberta, Zurich, Amgen, Intel, DKFZ Heidelberg, IU Bremen, Hannover, Heidelberg, AUC Denmark, Arizona State, Brigham Young, Cornell, Harvard, Haystack, MIT, North Dakota, Pittsburgh, RPSLMC, Swarthmore, Texas A and M, Michigan Medical, Minnesota, Nevada Reno, Washington St. Louis, UMA Spain, UPC Spain, UV Spain, NASA JPL, US Gov. OSIS, Univ. College Cork, Tel Aviv, CNR ITIA Italy, INFN Bari, Sardegna, Medina, Milan, Baika Japan, TUS Japan, US Army Monmouth, Leiden, Twente, Massey Univ. NZ, BUU Thailand, Bangor, Imperial, Keele, Liverpool, Luton, Swansea.

27th April 2005 (395 1416)

KF Graz, ANU, Melbourne, New South Wales, Australian Gov. BOM, Lockheed Martin, FU Berlin, HS Bremerhaven, Albert Einstein Institute, TU Cottbus, Bielefeld, Regensburg, Berkeley, EOU, Iowa State, MIT, MU, New York Univ., Simons Rock, Urbana Champaign, Oregon, WFU, Wisconsin, UHP Nancy, Marseilles, UPS Toulouse, Fermilab, CNR IMM Italy, INFN Bari, Trieste, Catania, European Parliament Luxembourg, Utrecht, UVT Netherlands, US Poland, UW Poland, NU Singapore, SWH Slovakia, ITU Turkey, Metu Turkey, Bolton.

28th April 2005 (337 1102)

US Gov CBI, Belgrade, Curtin, New South Wales, Saskatchewan, National Univ. Colombia, Prentice Hall, German Synchrotron Facility (DESY), Freiburg, Leipzig, Binghampton, Boston Univ., Buffalo, Brigham Young Caltech, Central Michigan, Geogia Tech., North Dakota, Pittsburgh, Stanford, Tufts, Urbana Champaign, Helsinki, Jussieu, IS Gov. FAA, BARC India, US Airforce Schriever, Twente, US Poland, Krasnoyarsk, Aberystwyth, Cambridge, Glasgow, Southampton Institute, Strathclyde.

29th April 2005 (277 2141)

Linz, British Columbia, Freiburg, Neufchatel, Hewlett Packard, MUNI Czechia, Palacky, FHG, Kaiserslautern, American, Arizona, Berkeley, Boston Univ., Central Michigan, Colby, Colorado, Columbia, Cornell LSU, LUC, Maine, Morehouse, Rice, Robert Morris, Akron, Delaware, Minnesota, Utah, Texas, Wisconsin, IFAE Spain, UB Spain, UC Spain, IMAG France, USF Grenoble, Poitiers, US Gov. SFWMD, SISSA Italy, US Airforce Wright Patterson, Twente, Umea, Aberystwyth, Birmingham, Glasgow, Oxofrd, British Gov. DERA.

30th April 2005 (233 1919) : ONE YEAR of feedback

ANU, New South Wales, Embry Riddle, MIT, Purdue, UC San Diego, ENS France, City Univ Hong Kong, NTNU Norway, Oxford.

1st May 2005 (240 795)

FHG Germany, Leipzig, Arizona, Colorado, Columbia, Miami, MIT, U Mass., Maryland, WSU, HTV Finland, Trinity College Dublin, INFN LNF, NTNU Norway, Confederation of British Industries, LIU Sweden, Cambridge.

2nd May 2005 (336 1567)

ASN KTN Austria, CSIRO Australia, UNICAMP Brazil, Toronto, ETH Zurich, Boeing, Bosch, Oldenburg, Boston University, Caltech, Kansas, Loyno, Milligan, UMBC, Michigan, Williams, Western Kentucky, Yale, French national particle lab., FDA, US Gov. Llb, NASA (Larc), Perugia, US Army Lid, Twente, NTNY Norway, Porto, Tuiasi, KTH Sweden, Physto.

3rd. May 2005 (335 4070)

Delaware, Liege, Sao Paolo, Concordia, Gov. Quebec, CERN, General Electric, Intel, Lockheed Martin, TU Munich, Bayreuth, Karlsruhe,

Leipzig, Wuerzburg, Arizona, Arizona State, Berkeley, Bryn Mawr, Caltech, Columbia, Emory, Maine, Potsdam, Rutgers, UCLA, UWM, WPI, UV Spain, CNRS Grenoble, F Comte, NASA (gsfc), Sandia, NC Ireland, Univ College Dublin, Technion, INFN Milan, US Airforce Scott, TU Delft, Twente, NTNU Norway, Massey NZ, KTH Sweden, NU Singapore, Bradford, Cardiff, St Andrews.

4th May 2005 (385 1287)

Latrobe, UNISA Australia, Australian Gov. BOM, FPMS Belgium, YRBE Ontario, Alberta, Ford, MUNI Czechia, Oldenburg, WH Stuttgart, Cornell, Drexel, Harvard, Iowa State, Indiana, MIT, MU Ohio, Syracuse, Chicago, UC Irvine, Connecticut, UC San Diego, UGA, U Mass Dartmouth, UMR, New Hampshire, UNL, South Carolina, William and Mary, Yale, UC Spain, Mulhouse, NIST, US Gov USPS, IED Hong Kong, Szeged, ISAC India, Perugia, UNIUD Italy, Nihon, US Army AMRDEC, US Army PICA, US Navy NUWC, Groningen, KTH Sweden, Bradford, Cambridge*, Leicester, Nottingham, East Anglia.

5th May 2005 (335 2361)

Hewlett Packard, Northrop Grummond, Columbia, Cornell, Georgia Tech., MIT, New Mexico Tech., Oklahoma, Penn State, SPSU, TMC, Chapel Hill, Connecticut, UC Riverside, Urbana Champaign, Maryland, UWM, William and Mary, UV Spain, Lawrence Berkeley, INFN Bologna, Modena Gov., Catania, US Naval Research Laboratory, Twente, Durham, Imperial, St Andrews.

6th May 2005 (306 1475)

Stanford, FPMS Belgium, Free Univ. Brussels, British Columbia, Siemens, Tatra, Bayreuth, Niels Bohr Institute, Berkeley, Montgomery College, Oklahoma, Purdue, UC San Diego, UMBC, Maryland, Texas Arlington, Washington, UAB Spain, UPC Spain, OULU Finland, LODYC Jussieu, Lawrence Berkeley, NASA (gsfc), NTUA Greece, EDU Hong Kong, Perugia, Kyoto, Us Airforce Pope, US Navy NMCI, Massey NZ, UMEA Sweden, Ogu Turkey, Cambridge, Oxford, Medway Council.

7th May 2005 (220 1921)

Newcastle, CERN, Columbia, Davidson, MIT, UC Santa Cruz, Florida, U Mass., JNCASR India, Sterling Students, Oxford.

8th May 2005 (268 2517)

Australian National University, ASN KTN Australia, Caltech, COD, New York Univ., Chicago, Texas, Rennes 1, Jerusalem, Tel Aviv,

Technion, KAIST, Twente, BI Norway, Torun, Ljubjliana, Bristol, Cambridge, Imperial, York.

9th May 2005 (399 2526)

Central Michigan, ASN KTN Austria, Gov. Australia, ULB Belgium, RCT SC Brazil, UNICAMP, CERN, Belfort, Freiburg, Erlangen, Arizona, Caltech, CLU, Columbia, Cornell, Harvard, Indian River CC, Purdue, St Cloud State, UC Davis, Chicago, UCLA, UTK, Xavier, UGR Spain, UV Spain, Marseilles, Poitiers, Los Alamos, IUST Iran, INFN Florence, Titech, US Army NVL, US Military USCG, UIO Norway, LIU Sweden, LUTH Sweden, Edinburgh, HWLC, Portsmouth, Surrey.

10th May 2005 (342 1809)

ANU, CRC Canada, ETH Zurich, BMW Group, Hewlett Packard, Thyssen, Niels Bohr Institute, Arizona, Armstrong, Berkeley, Northwestern, Oklahoma, SJSU, Stanford, U Mass., Minnesota, UCM Spain, UV Spain, Trinity College Dublin, Turin Tech., TIM Italy, Perugia, Leiden, Physto Sweden, Edinburgh, Leeds, NCL.

11th May 2005 (335 3568)

Leuven, UNICAMP, Calgary, CERN, Hewlett Packard, Xerox, HU Berlin, TU Dresden, Arizona, Berkeley, Caltech, Colorado, ECW, GWU, Princeton Institute for Advanced Study, Illinois Tech., MIT, Purdue, Stanford, Tenet, New Mexico, U Penn., Washington, Williams, Velenciennes, NIST, US Gov WA, NUI Galway, Trinity College Dublin, UPC, NTNU Norway, Physto Sweden, Edinburgh, Open.

12[tH] May 2005 (371 3130)

ASN KTN Austria, CSIRO Australia, Monash, KU Leuven, IT and T, Charles Univ. Prague, FH Aachen, HU Berlin, TU Cottbus, Magdeburg, IHK Denmark, Cornell, CTC, Drexel, Johns Hopkins, Northwestern, PDX, SJSU, UC Irvine, UC Santa Barbara, Michigan, UV Spain, Poitiers, ELTE Hungary, Indian Statistical Institute, JNCASR India, CNR IFAC Italy, INFN Pavia, Trieste, Ocha Japan, Tohoku, Univ Tokyo, Aizawa, US Airforce Wright Patterson, NTNU Norway, Tuiasi Romania, Ljubljiana, Bath, Coleg Sir Ga^r, Imperial, Strathclyde, UWCM.

13th May 2005 (305 3489)

Gov. New South Wales, Sao Paolo, Hewlett Packard, Canadian MDA, Amgen, Hewlett Packard, Charles Univ. Prague, TU Cottbus, Schwabia, Central Michigan, Miami, MIT, Rutgers, TNTech, UC Davis, Washington,

Wisconsin, UV Spain, Poitiers, AUTH Greece, Hiroshima, US Army USMA, Massey NZ, Bath, Cambridge, ICR, Manchester, Imperial, NTU, Strathclyde.

14th May 2005 (295 1925)
Lawrence Berkeley, BC Hydro Canada, Concordia, McGill, Calgary, CSUS, MIT, Maryland, NTNU Norway, Tuaisi, Cambridge.

15th May 2005 (255 2369)
Leipzig, Concordia, Tsinghua China, Texas A and M, Maryland, UST Hong Kong, NTNU Norway, Chalmers, Durham.

16TH May 2005 (361 2402)
Northwestern, ASN KTN Austria, KU Leuven, Gov. Alberta, Alberta, IHEP China, Boeing, IBM Thomas J. Watson, Wells Fargo, Schwabia, Arizona, Coleman, Johns Hopkins, MIT, UNL, UAL Spain, ENS Cachan, U Cergy France, Strasbourg, NASA (Lerc), CNR IRA Italy, INFN Naples, UITM Malaysia, NTNU Norway, Massey Univ. NZ, Confederation of British Industries, Porto, Birmingham, Cambridge*, Edinburgh, Manchester, Sheffield, Sheffield College, Southampton.

17th May 2005 (340 4788)
New South Wales, Alberta Research Council, ETH Zurich, Lockheed Martin, Dortmund, Frankfurt, Buffalo, Caltech, Georgia Tech., Haystack, North Carolina State, Ohio State, Rutgers, Salk, Stanford, TMC, U Connecticut, UC Santa Barbara, Florida, Idaho, Maryland, Oregon, HTV Finland, TUT Finland, ESPCI France, French national particle laboratory, Sandia, CNR ISOF Italy, IDS SPA Italy, INFN Milan, INFN Naples, Wright Patterson Airforce Base, Groningen, Utrecht, VU Netherlands, NTNU Norway, Confederation of British Industries, KMITNB Thailand, Cambridge, Edinburgh, Heriot Watt, Imperial, Southampton, British Gov. DERA.

18th May 2005 (334 3378)
ETH Zurich, Andes, Dow, IBM Thomas J. Watson, IBM Yamato, UO Cuba, FH Aachen, Max Planck MPPMU, Frankfurt, Niels Bohr Institute, Albany, Arizona State, Cornell, Renselaer, Chicago, Western Michigan, Washington, Oak Ridge, Patras, Hong Kong Tech., Univ College Cork, ISAC India, Trieste, Kyoto, Titech Japan, US Airforce Elmendorf, US Navy NMCI, TU Eindhoven, Twente, NTNU Norway, URR Romania, Chalmers Sweden, LU Sweden, Boun Turkey, NCTU Taiwan, Birmingham, Oxford*, SUN South Africa.

19TH May 2005 (299 1628)

Caltech, Linz, Monash, EPF Lausanne, Zurich, Honeywell, Hewlett Packard, Northrop Grummond, Raytheon, Freiburg, Kaiserslautern, OU Denmark, Berkeley, Columbia, DVC, Emory, Georgia Tech., Mayo Clinic, OGI, UC San Francisco, North Carolina Charlotte, Utah, UHU Spain, HTV Finland, BARC India, INFN Bari, ICTP Trieste, Turin, US Airforce AFIT, US Airforce AFSPC, US Navy NMCI, TU Delft, Twente, Gov. NZ MOH, RISM Thailand, KPPRC Taiwan, Edinburgh, Glasgow, London Business School, Oxford, Univ College London.

20th May 2005 (335 2708)

KF Graz, Australian National Univ., Monash, Toronto, Boeing, Cisco, Max Planck Heidelberg, Jena, Stuttgart, F and M, Harvard students, Haystack, Princeton Institute for Advanced Studies, Maine, Pittsburgh, Texas A and M, UPS, South Carolina, Wisconsin, UCM Spain, French national particle lab., F Comte, Modena, Pisa, Titech Japan, US Airforce Edwards, US Navy IOR, Eindhoven, Twente, NTNU Norway, Chula Thailand, Cambridge, Durham, Edinburgh, Imperial, North East Wales Tech (Owain Glyndw^r), York.

21st May 2005 (303 1728)

Princeton Institute for Advanced Studies, Chicago, OTE Net, US Airforce Salem.

22nd May 2005 (316 2845)

Stuttgart, Wuerzburg, Berkeley, Drexel, Texas A and M, Bath, Birmingham, Southampton.

23rd. May 2005 (338 3437)

Vienna, Clarkson, CSU Santa Barbara, Duke, Maine, MSU, Stanford, UC San Diego, Delaware, Urbana Champaign, Utah, Washington, UCM Spain, Bordeaux 1, Lille 1, Trieste, Himeji Tech Japan, US Navy NMCI, US Navy NUMC, TU Delft, Gov. Poland IMPAN, KTH Sweden, NTHU Taiwan, BBK, Leeds, Oxford, Swansea.

24th May 2005 (368 3103)

JKU Austria, Adelaide, Tasmania, Sao Paolo, Philips, KFA Juelich, Max Planck IPP, RWTH Aachen, TU Chemnitz, Duisburg, Berkeley, Caltech., Indian River CC, Oregon State, TTU, Michigan Medical, Michigan, William and Mary, Zaragoza Spain, UV Spain, Poitiers, NASA (Lerc), Sandia, Jerusalem, Tel Aviv, INFN Naples, Shinshu Japan, US Airforce Schriever, TU Delft, Twente, Kluwer, LUMS Pakistan, UTT

Romania, LIU Sweden, Aberystwyth, Southampton, British Gov DSTL, National Health Service.

25th May 2005 (338 2799)
  New South Wales, Simon Frazer Canada, Boeing, IBM Yamato, Lockheed Martin, Motorola, Siemens, Ruhr Bochum, Carleton, TMC, Tartu Tech Estonia, Canaries, UV Spain, CNRS Grenoble, US Gov UCIA, Jerusalem, TU Eindhoven, Free Univ. Amsterdam, Mitre, IGF Poland, Lublin, Lisbon Tech., Edinburgh, Imperial, Oxford, Southampton.

26th May 2005 (334 3022)
  Compaq, RMIT Australia, Australian Gov BOM, KU Leuven, McGill, UQAM Canada, Sherbourne, Disney, Max Planck Halle, Colgate, Colorado, JMU, Maine, Marygrove, MIT, Stanford, UC Riverside, Delaware, UNL, Vanderbilt, Washington, JYU Finland, French Gov. Defence, US Gov. FDA, Weizmann, KAIST, UNAM Mexico, Gantep Turkey, Oxford, British Gov. DERA.

27th May 2005 (265 3045) New Record of Hits per Month
  UNICAMP, EPF Lausanne, Palacky, LRZ Munich, Leipzig, Caltech, Harvard, Oklahoma, Pittsburgh, UC Davis, UC Irvine, UB Spain, UJF Grenoble, F. Comte, Poitiers, Mathematics Institute Croatia, European Space Agency HQ, INFN Pisa, US Military DCMA, US Navy NMCI, US Navy NUWC, USON Mexico, Twente, Royal Society of Chemistry, Gov. Poland IMPAN, RMT Russia, NTU Singapore, Edinburgh, York.

28th May 2005 (268 3605)
  Bonn, Berkeley, Caltech, Cedarsville, Ithaca College, MIT, Northwestern, Minnesota, Yale Medical, ENS France, IITB India, US Airforce Wright Patterson, NTNU Norway.

29th May 2005 (125 1791) Feedback site partially down
  ENAC France, CSIRO Australia, HU Berlin, Northwestern, ESA Spain, UCM Spain, UV Spain, HUT Finland, Lorraine, NAO Japan, Twente, Lisbon, Imperial.

30th May 2005 (307 3571)
  Australian Gov BOM, Montreal Tech., Manitoba, UQAM, Chile, Max Planck Bonn, Erlangen, Muenster, US Gov, NOAO, Pomona, Yale Medical, UAL Spain, Montpellier 2, INFN LNF Italy, NTNU Norway, Strathclyde.

31st May 2005 (321 5118)

New South Wales, UCL Belgium, Sao Paolo, Toronto, LN China, Boeing, Perkin Elmer, Philips, Raytheon, Sun, Titan, TU Chemnitz, Konstanz, Paderborn, Caltech, Columbia, Cornell, MIT, PDX, Penn State, Delaware, North Carolina Charlotte, Chapel Hill, U. Penn., Washington, UCO Spain, URJC Spain, LUT Finland, US Gov, INEL, US Dept. of State, Univ College Dublin, Trieste, Hitachi, US Army Beruning, US Army Polk, US Navy NUWC, Utrecht, NTNU Norway, Auckland NZ, Cambridge*, Heriot Watt, Sheffield, Wakefield College, Acadinfo Britain.

1st June 2005 (284 4520)

Flinders, UFAM Brazil, Gov. British Columbia, NRC Canada, Laval, Basel, Botevgrad, Northrop Grummond, Raytheon, UO Cuba, Bonn, Berkeley, Cornell, MIT, Rutgers, Texas A and M, UC Davis, Maryland, South Carolina, Washington, Yale Medical, UV Spain, HUT Finland, ENS Cachan, French Defence Dept., Polytechnique, IITM India, INFN Genoa, Pisa, Tokyo, US Army Pica, Utrecht, NTNU Norway, KTH Sweden, Edinburgh, Imperial, Luton, NERC Wallingford.

2nd June 2005 (277 4324)

OMA Belgium, UQAM Canada, Toronto, Delta Airlines, Lockheed Martin, Drexel, MIT, Princeton, Rice, Stanford, Connecticut, UC San Diego, Michigan, Texas, Washington, ECP France, ENS Lyon, INSA Lyon, Paris Psud, US Gov. Dept. Of Transport, Lawrence Livermoore, NASA (gsfc), US Army Monmouth, NTNU Norway, Edinburgh, Strathclyde.

3rd June 2005 (242 2449)

WYD Australia, Queensland Australia, Memorial Newfoundland, UQAM Canada, Usherbrooke, Toronto, Boeing, Raytheon, Shell, Albert Einstein Institute, Frankfurt, BW Munich, Embry Riddle, Hawaii, MIT, SIU, UB Spain, Canadian Gov., Lawrence Berkeley, NASA (Gsfc), Postech Korea, US Army Monmouth, NTNU Norway, Imperial, Birkenhead, Wrexham, NS Yugoslavia, SUN South Africa.

4th June 2005 (217 4093)

Sydney, RWTH Aachen, Bonn, Lehigh, Northwestern, Ohio State, Florida, Maryland, Union, Yale Medical, Ibaraki, NTNU Norway, Joint Institutes of Nuclear Research Russia, Ege Turkey, NTU Taiwan, Oxford.

5th June 2005 (267 3654)

British Columbia, Botevgrad, Microsoft, Berkeley, MTU, Yale Medical, Nagoya, NTNU Norway, Massey NZ, PK Poland, Cambridge*, Luton.

6th June 2005 (331 3711)

KF Graz, Canning College Australia, Australian Gov. BOM, UNICAMP, McMaster, Montreal Tech., EPF Lausanne, Charles Univ Prague, Max Planck MPIM Bonn, TU Berlin, Hannover, Jena, Columbia, Georgia Tech., Ohio State, Yale Medical, French national particle lab., Nancy, Canadian Dept. of Transport, Trieste, Kyoto, Tsukuba, Postech, US Army Research Laboratory, Twente, NTNU Norway, Royal Society of Chemistry, MSUIIT Philipines, Porto, Lisbon, Bristol, Sheffield.

7th June 2005 (308 3000)

Vienna, Siemens Austria, New South Wales Dept of Transport, UFAM Brazil, UNICAMP, Alberta Research Council, Gov. Quebec, Boeing, Duke Energy, Lockheed Martin, Wuerzburg, Caltech, Cornell, Georgia Tech., Johns Hopkins Applied Physics, Stanford SLAC, Washington, UGR Spain, IRCAM France, Argonne, Canadian Dept. of Transport, US Gov. EPA, NASA (Gsfc), US Gov USGS, Patras, Jerusalem, Rome 2, Ichinoseki, Tohoku, Univ. Tokyo, SNU Korea, Twente, NTNU Norway, Wroclaw, NU Singapore, UPJS Slovakia, Bangor, Oxford.

8th June 2005 (335 3497)

UNICAMP, City of Vancouver, Alberta, Univ Chile, Boeing, Duke Energy, Northrop Grummond, UO Cuba, Palacky, Erlangen, Stuttgart, Wuerzburg, Georgia Tech., Illinois Tech., Indian River CC, Maine, MIT, North Carolina State, Ohio State, Pacific U, Houston, UNL, UWM, UV Spain, French Dept. Defence, NASA (gsfc), Tel Aviv, INFN LNF, INFN LNGS, Saga Japan, Shizuoka, US Army Tobyhanna, US Military DISA, Groningen, Twente, NTNU Norway, Oslo, Victoria Wellington, Imperial, Oxford, Sheffield.

9th June 2005 (295 3039)

Queensland, Liege, UNICAMP, Canadian Dept. of Defence, Toronto, Chile, Honeywell, Northrop Grummond, Philips, Kaiserslautern, Brigham Young, Maine, Texas A and M, Urbana Champaign, IUCAA India, CNR IFAC Italy, Kyushu, NTNU Norway, Massey NZ, Royal Society of Chemistry, Lisbon, KTH*, HWLC Britain, Imperial.

10TH June 2005 (323 3344)

Alberta Research Council, EPF Lausanne, Univ. Chile, Boeing, Botevgrad, Browning, Duke Energy, SBS Germany, Heidelberg, Bates, Brown, Ohio State, Chicago, Urbana Champaign, YU, UAB Spain, UPC Spain, UPS Toulouse, INFN Milan, Milan Tech., Pisa, Pavia, Groningen, Twente, NTNU Norway, ITT Romania, Imperial.

11th June 2005 (219 3061)

Max Planck RZ Berlin, Leipzig, UAH, Tel Aviv, IMSC India, Imperial. British Gov., Wrexham Gov.

12th June 2005 (237 2650)

FH Aachen, Florida, Vermont, NU Singapore, NTHU Taiwan, SUN South Africa.

13TH June 2005 (269 2753)

UNICAMP, Calgary, Botevgrad, Pfizer, FHG Germany, Marshall, New York Univ., UCLA, Michigan, HTV Finland, CEA Saclay, INFN TS Italy, US Airforce PLH, US Army Research Laboratory, Massey NZ, JINR Russia, FOI Sweden, KTH Sweden, Ljubljiana, KKU Thailand, NCTU Taiwan, Cambridge, Cardiff, Edinburgh, Pembrokeshire, PWV Gov South Africa.

14TH June 2005 (332 4417)

Queensland, UNICAMP, Toronto, Albert Einstein Institute, Siemens, Munich, MIT, Texas A and M, UCLA, Maryland, Minnesota, INPG France, US Gov DC, Chiba, Hiroshima, Osaka, Univ Tokyo, US Airforce Brooks, Free Univ. Amsterdam, Torun, KTH, LU Sweden, Cambridge, Oxford, Warwick.

15th June 2005 (333 2294)

Adelaide, Flinders, UFSC Brazil, ETH Zurich, Univ Chile, Browning, IBM Thomas J. Watson, McGraw Hill, Max Planck Heidelberg, TUBS Germany, Leipzig, Magdeburg, Tuebingen, Columbia, North Dakota, New Jersey Institute of Technology, Princeton students, Rochester Tech., UC Irvine, Michigan Medical, UNL, U Penn, Texas, Catania, Iwate Japan, Kanazawa, US Airforce AFRE, US Army Sill, US Military Fusion, Cambridge, Edinburgh, Leeds, SHU, York.

16th June 2005 (290 2892)

Ghent, CERN, EPF Lausanne, Browning, Hewlett Packard, Intel, FGAN Germany, STW Bonn, TU Berlin, Duisburg, Rostock, Columbia, Embry Riddle, Georgia Tech., US Naval Academy, Washington Univ St.

Louis, Yale, RAIN France, US Gov BLM, Lawrence Livermore, Sandia, INFN Italy, ICTP Trieste, Samsung, US Mil. DCMA, US Naval Research Laboratory, US Navy NUWC, Chula Thailand, NCTU Taiwan, Bangor, Lancaster, UMIST.

17th June 2005 (294 4220)

Vienna, FundP Belgium, EPF Lausanne, IBM Thomas J. Watson, Raytheon, TU Dresden, Erlangen, Hamburg, Fullerton, Washington, UAM Spain, ICTP Trieste, Pavia, Univ Tokyo, Korea, OU Sri Lanka, Jefferson National Laboratory, Lisbon, YM Taiwan, Sheffield, Anchorage Alaska.

18th June 2005 (199 1240)

British Columbia, Max Planck RZ Berlin, RWTH Aachen, Marshall, Urbana Champaign UML, UNO, Technion, Campagna, Edinburgh.

19th June 2005 (212 2795)

Latrobe, Queensland, Alberta, Magdeburg, Schwabia, Michigan State, UC Davis, Urbana Champaign, Unideb Hungary, Titech Japan, NTHU Taiwan, Cambridge.

20th June 2005 (241 3909)

Adelaide, Sydney, UNICAMP, Calgary, Toronto, Zurich, Botevgrad, Browning, Sun, HU Berlin, Max Planck MPIFR Bonn, Buffalo, Columbia, CWRU, Duke, Galileo, Harvard, ODU, Washington, ENSAE France, Argonne, Los Alamos, NIH, BME Hungary, Jerusalem, Modena, Pavia, Yamanashi Univ., Fuji Electric, Chonnam Korea, US Airforce Rhein Main, Royal Society of Chemistry, JINR Russia, KTH Sweden, Ege Turkey, NTU Taiwan.

21st June 2005 (257 4791)

UNICAMP, Alberta, Saskatchewan, CERN, Browning, Hewlett Packard, TU Darmstadt, Frankfurt, Danish Tech., Harvard, Indian River CC, Oberlin, UC Davis, Central Florida, UC San Diego, Illinois Chicago, Michigan Medical, Oregon, Washington, OBSPM France, Fermilab, INFN LNF Italy, UNIBG Italy, Naples, Hitachi, US Naval Research Laboratory, Groningen, Twente, Lublin, Szczecin, Gliwice, Chula Thailand, NCU Taiwan, Cambridge, Durham, Southampton, York.

22nd June 2005 (276 2822)

ULB Belgium, Calgary, CERN, Browning, Charles Univ. Prague, Erlangen, Magdeburg, MUM, Penn State, U Penn., INSA Rennes, ONERA France, IITB India, ICTP Trieste, Kyoto Univ., NHS.

23rd June 2005 (266 3205)

ETH Zurich, Freiburg, Air Products, Boeing, Browning, Dow Corning, IT and T, Lockheed Martin, Solvay, Charles Univ Prague, NRI Czechia, Bosch, IU Bremen, Stuttgart, Berkeley, Georgia Tech., North Carolina State, Rochester, Houston, Urbana Champaign, UMR, Chapel Hill, U Penn., South Florida, Helsinki, Onera France, IRB Croatia, IISC India, Tohoku, US Navy Naskw, Twente, LTH Sweden, NU Singapore, Bath, Imperial.

24th June 2005 (226 1896)

Alberta, Waterloo, York Univ Toronto, Boeing, Browning, RWTH Aachen, Goettingen, Colorado State, LCC, MU, New York Univ., Purdue, UC Davis, Onera France, Poitiers, NASA (Grc), NASA (Jpl), NASA (Larc), NIST, Oak Ridge, Gov. Virginia, ENEA Frascati National Lab., INFN Milan, ICTP Trieste, Calabria, Tohoku, Royal Society of Chemistry, Edinburgh, NCL.

25TH June 2005 (174 1530)

UNICAMP, Caltech, Raytheon, Central Michigan, Mercy, ORST, Urbana Champaign, OU Hong Kong, IMSC India, KAIST Korea, Postech Korea.

26th June 2005 (244 2095), New record number of hits per month.

South Australia, Sydney, Defence Gov. Australia, Botevgrad, Kaiserslautern, Berkeley, Houston, Urbana Champaign, Virginia, Paris Psud, Los Alamos, Weizmann, Free Univ. Amsterdam.

27th June 2005 (263 3324)

Laval, Toronto, UTA Chile, Tsinghua China, Browning, Calltech, Duke Energy, IBM Almaden, Arizona, Berkeley, CSUDH, Drexel, Harvard, Rutgers, UC Los Angeles, UMDNJ, Minnesota, Uni2 Spain, BIPM France, NASA, Oak Ridge, Twente, Lisbon, PSU Thailand, KOU Turkey, Menai, Guardian.

28th June 2005 (253 4662) New record of gigabytes downloaded.

TU Graz, JKU Austria, UNICAMP, Memorial Univ. Newfoundland, Yukon Gov., Canada, Browning, Duke Energy, IBM Almaden, Lucent, Cornell, CSUDH, Drexel, Harvard Medical, IUPUI, Urbana Champaign, Wayne, Canaries, CEA France, Poitiers, Fermilab, Sandia, Messina, US Airforce Vandenberg, US Navy NMCI, US Naval Research Laboratory, Lisbon, NU Singapore, NTHU Taiwan, Edinburgh, Surrey.

29th June 2005 (280 4477)

Cobham, UNICAMP, Gov Canada IC, CERN, Amgen, Botevgrad, Browning, Duke Energy, IBM Almaden, Intel, KFA Juelich, LRZ Munich, Cornell, ETSU, MIT, Wyoming, Vanderbilt, UPC Spain, URJC Spain, JYU Finland, Paris Psud, Strasbourg, US Gov FDA, NIST, UOI Greece, BARC India, Trieste, US Naval Research Laboratory, Massey NZ, Gov. NZ MOH, IJS Slovenia, Birmingham, Nottingham.

30th June 2005 (274 4486) Two New Records Set for the Month.

TU Graz, Queensland, UFMG, Memorial Newfoundland, Queen's, BMW Group, Botevgrad, Browning, Raytheon, Albert Einstein Institute, Colorado, Iowa State, TVI, Maryland, UV Spain, Lawrence Livermore, Hokudai, US Military Difas, US Navy NUWC, Groningen, Utrecht, UNAC Peru, Lisbon, NTUST Taiwan, Bangor, Cambridge, Southampton.

1st July 2005 (258 1871)

MRSE Argentina, FUNDP Belgium, CERN, Botevgrad, CIS Hollywood, FU Berlin, KFA Juelich, Leipzig, Georgia Tech., Iowa State, IUPUI, New York Univ., Penn State, Washington, West Virginia, Poitiers, US Gov. BNL, Ikaruga Japan, KAIST Korea, US Navy NUWC, Krakow, Kiev, Sussex.

2nd July 2005, stats site down.

3rd July 2005 (211 2237)

Queensland, Canadian NRC, IBM Almaden, TU Berlin, Iowa State, Johns Hopkins, New York Univ., Renselaer, IITM India, Okayama, Univ Tokyo, US Army, ITRI Taiwan.

4tH July 2005 (326 1947)

Australian National Univ., Gov. Bhutan, Gov British Columbia, Simon Frazer, Gov. Chile, Calltech, Qinetiq, Shell, MUNI Czechia, FH Lausitz, Munich, North Dakota, UCLA, Caen, Poitiers, TIFR India, Pisa, Titech Japan, Groningen, Twente, Metu Turkey, Leeds.

5tH July 2005 (302 3302)

Sao Paolo, Simon Frazer, ETH Zurich, Botevgrad, Browning, Duke Energy, Max Planck MPI Heidelberg, Wuerzburg, Oklahoma, Minnesota, Frenc national particle lab., Paris Psud, Los Alamos, PRL India, CNR Rimini, Milan Tech., Massey NZ, Avonside School NZ, Peru, KTH Sweden, NU Singapore, PSU Thailand, Liverpool, Warwick.

6TH July 2005 (312 1773)

UMH Belgium, UGRJ Brazil, Hewlett Packard, Lockheed Martin, Max Planck MPIS, Regensburg, Stuttgart, Colorado, Cornell, Johns Hopkins Applied Physics, KSBE, Mercer, TVI, UMSMed, UPCT Spain, Marseilles 3, ELTE Hungary, US Airforce Hurlburt, US Airforce RL, US Navy CNET, US Military OSD, Marshall Hopsital, Marshal Medical, STV Slovakia, Cambridge, British Gov. Ministry of Defence.

7TH July 2005 (338 3568)

CERN, Botevgrad, Browning, Calltech, Hewlett Packard, Lockheed Martin, Albert Einstein Institute, Heidelberg, GPC, Marshall, Purdue, SUNY Stonybrook, UC Santa Barbara, Iowa, Urbana Champaign, Wright, Strasbourg, Trinity College Dublin, PRL India, ENEA Casaccia, Tohoku, Postech Korea, Samsung, US Navy NUMC, Massey NZ, Mahidol Thailand, Imperial.

8th July 2005 (244 1897)

McGill, Browning, Duke Energy, Cyprus, Ruhr Bochum, TU Cottbus, Arizona, Colorado, Georgia Tech., Iowa State, Oklahoma State, Central Florida, Minnesota, BNE Spain, Valencia, French national particle lab., Lorraine, Jerusalem, PRL India, Perugia, pavia, Kluwer, Oak Ridge, Philipines, Porto, Lisbon, Chula Thailand, DL Britain, NS Yugoslavia.

9th July 2005 (263 1819)

Botevgrad, TU Berlin, Northwestern, Ohio State, BNE Spain.

10th July 2005 (244 3309)

TU Vienna, Alberta, Botevgrad, IBM Almaden, Stonybrook, South Carolina, Szeged, INFN LE Italy.

11th July 2005 (279 1800)

UIBK Austria, Australian Gov. BOM, UNICAMP, CERN, Botevgrad, Browning, Microsoft, Max Planck MPI-SB, TLS Tautenberg, Columbia, Duke, ECU, Georgia Tech., North Carolina State, Penn State, Texas Tech., Connecticut, North Carolina, Nevada Reno, US Airforce Academy, NIST, Princeton Plasma Lab., Tokushima, Yokohama, TU Eindhoven, Twente, JINR Russia, NTU Singapore, AHRB Britain.

12TH July 2005 (287 3395)

Leoben Austria, Sofia Bulgaria, Toronto, Wetern Ontario, Geneva, Zurich, Boeing, Botevgrad, Browning, IBM Almaden, Lincoln Electric, Qinetiq, FU Berlin, Max Planck CNS, Stuttgart, AFIT, Drexel, Indian River CC, MIT, Illinois Chicago, Maryland, UV Sapin, IAP France, LPTH

Jussieu, Polytechnique, Paris Psud, Rennes 1, Lawrence Berkeley, NASA (Grc), NIH, US Gov.Nrel, Sandia, Calabria, US Army Amrdec, US Naval Research Laboratory, Joseph Stefan Institute Slovenia, Chula Thailand, NTU Taiwan, Napier, Southampton, Swansea.

13th July 2005 (308 2518)

UNICAMP, CERN, Zurich, Calltech, Duke Energy, IBM North Carolina, Ontario Power Generation, Leipzig, Georgia Tech., Johns Hopkins Medical Institute, Chicago, UCHSC, Washington St Louis, Yale Medical, US Gov BLM, Fermilab, Lawrence Berkeley, NTUA Greece, Indian Statistical Institute, Univ Tokyo, US Army Sill, Groningen, TU Delft, Porto, NTHU Taiwan, Huddersfield, King's College London, Open.

14th July 2005 (323 2604)

Melbourne, UFPB Brazil, Alberta Research Council, Canadian Defence Dept., Alberta, Saskatchewan, CERN, Valle Colombia, Botevgrad, Browning, Auburn, Brigham Young, Fisk, Florida State, MU, Ohio State, Santa Fe, Chicago, Union, Wisconsin, UB Spain, UV Spain, Los Alamos, Aok Ridge, Jerusalem, Titech, US Army Sill, UITM Malaysia, Royal Society of Chemistry, Iyte Turkey, King's College London, Liverpool.

15TH July 2005 (227 1800)

USQ Australia, UFAL Brazil, Alberta Research Council, McGill, CERN, Botevgrad, Browning, General Electric, IBM Almaden, Berkeley, Bridgew, Cornell, Princeton, Rutgers, Illinois Chicago, Virginia, Washington, UTU Finland, CNR Bologna, SISSA Italy, Titech, SNU Korea, Groningen, National Health Service,

16th July 2005 (274 4153)

Botevgrad, IBM Almaden, HU Berlin, Hawaii, Yale Medical, French national particle lab., US Gov. NOAA, Okayama, Titech.

17TH July 2005 (335 2226)

Western Australia, Gov. Bhutan, IBM Almaden, Lockheed Martin, Columbia, Indiana, ENS France, Jerusalem, Weizmann, Kyoto Univ., Okayama, Titech.

18th July 2005 (333 4513)

Alberta Research Council, Toronto, ETH Zurich, Amgen, Botevgrad, Browning, Northrop Grummond, Sysco, Brandeis, Georgia Tech., MIT, Northwestern, Texas A and M, Toronto, Chicago, Chapel Hill, Nevada Reno, Wisconsin, OBSPM France, Lawrence Berkeley, INFN Bari, INFN

Milan, Postech, Massey NZ, PWR Wroclaw, Bristol, King's College London, London School of Economics.

19TH July 2005 (309 2823)

Australian National Univ., ETH Zurich, Botevgrad, IBM Almaden, IBM Yamato, Lockheed Martin, Max Planck MPIP Mainz, Hamburg, Kaiserslautern, TDC Denmark, Buffalo Medical, Georgia Tech., Kennesaw, Mercer, Chicago, Michigan, Nevada Reno, UNT, INERIS France, LORIA France, Lawrence Berkeley, DUTH Greece, Univ College Cork, CAT India, Turin Tech., Kyutech Japan, US Army NVL, US Military DLA, Massey NZ, IRFU Sweden, Essex, Heriot Watt, London School of Economics, Rutherford Appleton, UMIST.

20th July 2005 (323 3330)

Loeben Austria, Australian Defence Dept., Montreal, Botevgrad, Northrop Grummond, HU Berlin, Max Planck IPP, Heidelberg, Leipzig, Danish Tech., Niels Bohr, Berkeley, Columbia, GMU, Montana, NSO, Princeton, UNL, Omaha, Washington, UJF Grenoble, US Gov KAPL, Indian Statistical Institute, Turin, Kyoto Univ., US Navy Medical, Twente, Cranfield.

21st July 2005 (269 2204)

New South Wales, IBM Almaden, Raytheon, Bosch, IU Bremen, Augsburg, Niels Bohr, A RTIC, Florida Atlantic, Kutztown, North Dakota, Oklahoma, Stevens Tech, Toronto, TVI, UC Santa Barbara, Florida, Maryland, Virginia, INSA Lyon, US Army Research Laboratory, US Naval Research Laboratory, TU Delft, TCU Taiwan, London School of Economics.

22nd July 2005 (247 1707)

TU Graz, KABSI Austria, Queen's, Basel, Duke Energy, Qinetiq, Rockwell Collins, UO Cuba, HU Berlin, Max Planck MPIS, Cottbus, Munich, Georgia Tech., Purdue, Stanford, Michigan, U Penn., TSAI Spain, URJS Spain, French Gov. Ministry of Labour, ONERA France, UPS Toulouse, US Gov USPTO, Athens, IKM Norway, Massey NZ, Cambridge, Sheffield.

23rd July 2005 (214 1727)

Botevgrad, IBM Almaden, Lepzig, TVI, ILL Grenoble, Univ Tokyo, TCU Taiwan, Lees, Oxford.

24th July 2005 (188 1082)

Gov British Columbia, Botevgrad, IBM Yamato, Central Florida, UC Santa Barbara, Dublin Institute for Advanced Studies, Fujitsu, SUN South Africa.

25[tH] July 2005 (290 4031)
CNA Austria, Canberra, UNICAMP, Memorial Newfoundland, IBM Thomas J. Watson, IT and T, Kodak, Microsoft, Nokia, Munich, Harvard, UC Davis, New Hampshire, UPV Spain, UV, AGNS France, Lyon 1, Fairfax County Gov., Los Alamos, NASA (Jpl), Indian Staitsical Institute, BARC India, ICTP Trieste, Univ Trieste, Univ Tokyo, US Naval Research Laboratory, Govt of Mexico, Groningen, VU Netherlands, AERO, TCU Taiwan, Bangor, Cranfield, Swansea.

26th July 2005 (289 3055)
Hewlett Packard, IBM Almaden, Microsoft, Siemens, Heidelberg, AFIT, MIT, Stanford, UC San Diego, Michigan, UMR, TVO Finland, USP Fiji, ACN Greece, GAU Hungary, Jerusalem, Keio Univ Japan, NIFS Japan, US Army NAMRL, Gov. Mexico, TCU Taiwan, Bolton.

27[tH] July 2005 (300 2745)
Gov. Western Australia, Siemens Canada, Browning, IBM Almaden, IBM Yamato, Turin Tech., FU Berlin, Max Planck MPIS, Albany, Caltech., Georgia Tech., Haystack, Indian River CC, UPV, OTE Greece, Univ College Dublin, CNR ITIA, US Navy NMCI, Medu Turkey, Glasgow.

28th July 2005 (266 1815)
Gov Argentina, ACU Australia, Australian Gov BOM, UNICAMP, Zurich, AT and T, Browning, Calltech Company, IBM Almaden, Shell, RWTH Aachen, Leipzig, Colorado, Cornell, Georgia Tech., North Carolina State, Princeton, Minnesota, Washington, UV Spain, LORIA France, US Gov BLM, US Gov NOAA, Bar Ilan Univ., ENEA Frascati, Rome 1, Sony, US Army DLA, Kluwer, Warsaw, King's College London, Napier, Strathclyde.

29[tH] July 2005 (288 2970)
Latrobe, CERN, LN China, Browning, Calltech Corporation, IBM Almaden, IBM Yamato, Bielefeld, Wuppertal, Georgia Tech., MIT, Pittsburgh, Purdue, Minnesota, Marseilles, US Gov BPA, Kobe Univ. Japan, US Airforce Headquarters, Lisbon, Imperial, HCMLAW Venezuela, RAU South Africa.

30th July 2005 (320 1978)

IBM Almaden, Microsoft, Chicago Illinois, Minnesota, Oner France, Gov. Mexico, TCU Taiwan, King's College London, Manchester.

31st July 2005 (208 1781)

Calltech, IBM Almaden, IBM Yamato, Florida Atlantic, Ohio State, Ohio, Rowan, Western Michigan, IITM India, Fuji Electric, Belgrade.

1st August 2005 (316 2644)

ECU Australia, UNICAMP, Boeing, Browning, Calltech, Duke Energy, Max Planck Garching, Konstanz, AU Denmark, Brigham Young, Carleton, CSU Chico, Florida Atlantic, SMCM, U Penn, Texas Arlington, French national particle lab., OBSPM France, US Gov Loc., Gov Virginia, Technion, Indian Statistical Institute, ICTP Trieste, Kyushu Univ., Tsukuba, US Brooks Airforce Base, US Wright Patterson Airforce Base, US Military OSD, Twente, Otago NZ, Uniba Slovakia, Imperial, leeds, Southampton.

2nd August 2005 (322 4370)

Northern Queensland, UIBK Austria, Australian National Univ., Monash, Australian Defence Dept., Government of Canada, Browning, IBM Almaden, Charles Univ. Prague, HU Berlin, Leipzig, Columbia, Drexel, DUQ, Indiana, Johns Hopkins Applied Physics, Mercer, MU Ohio, Stanford, Chapel Hill, Minnesota, Texas, Washington, French national particle lab., NASA (Jpl), US Gov NOAA, Titech, KAIST S korea, US Army Kuwait, US Navy NRL, Univ College London, National Health Service.

3rd August 2005 (303 2908)

Australian Gov. Defence, KU Leuven, UFAL Brazil, CERN, ETH Zurich, Browning, Calltech., IBM Almaden, Qinetiq, Leipzig, Georgia Tech., Johns Hopkins Applied Physics, U Mass., UWP, Wisconsin, CNES France, US Gov Customs, Oak Ridge, Trinity College Dublin, CNR ITIA Italy, US Army Amedd, US Army Rucker, Ubiv Philipines, LIU Sweden, Chula Thailand, Kiev.

4th August 2005 (350 2324)

UTA Austria, Queen's, Waterloo, EPF Lausanne, Calltech., Hewlett Packard, IBM Almaden, Shell, Max Planck MPIM Bonn, Erlangen, AUC Denmark, Cornell, Duke, Harvard, Northwestern, Ohio State, Stanford, SVSU, UC San Diego, Minnesota, WSU, UPM Spain, LPTL Jussieux, CAT India, Messina, Univ Tokyo, US Navy NMCI, Gov. Mexico, Twente, Kluwer, Bath, Warwickshire County Council.

5th August 2005 (273 1943)

Neufchatel, Calltech, IBM Almaden, IBM Colorado, KFA Juelich, Indiana, Rochester Institute of Technology, Los Alamos, Oak Ridge, Unibs Italy, US Naval Research Laboratory, Lisbon, National Univ Singapore.

The method of recording feedback is changed at this point and is eventually systematized to half monthly for the USA, Europe and the Rest of the World.

August 12 - 24,2005

Universities and Similar.

SBG Austria, Australian National University, Sydney, Royal Military Academy of Belgium, Liege, Alberta Research Council, St. John's Alberta, Brandon, McGill, UFRJ Brazil, Sao Paolo, Saskatchewan, CERN, Norte Colombia, Max Planck Halle, Heidelberg, Leipzig, Magdeburg, Regensburg, Auburn, Berkeley, Caltech, CSU Chico, Elon, NAIC, NCAT, Penn State, Stanford SLAC, Texas A and M, TNTech, UC Riverside, UC San Diego, UC San Francisco, Illinois Chicago, Urbana Champaign, EHU Spain, UJI Spain, UV Spain, HUT Finland, CIUP France, US Gov. DHS, US Gov. Dept of Energy, Los Alamos, Lawrence Berkeley, US Gov PNL, Gov. Virginia, Univ College Cork, Trieste, Venice, Hiroshima, Hokudai, Shizuoka, Univ. Tokyo, US Airforce Eglin, US Airforce Wright Patterson, US Military ARO, US Army Lewis, US Military DISA, US Navy NMCI, Gov. Of Mexico, NTNU Norway, Auckland, NU Singapore, Agricultural Univ. Slovakia, YLC Taiwan, Cambridge Univ. Press, Cambridge, Cranfield, Edinburgh, Imperial, Keele, London School of Economics, Sheffield, British Gov. DERA.

August 2005, Small Selection of Household Name Corporations.

Boeing, Botevgrad, Browning, Duke Energy, Hewlett Packard, IBM Almaden, IBM Yamato, Intel, Lockheed Martin, Motorola, Philips, Qinetiq, Raytheon, Yahoo, Siemens, Hitachi.

First Two Weeks of September 2005

US Universities.

Arizona, Boston, Buffalo, Columbia, Drexel, Georgia Tech., Harvard, HMC, Indiana, Kansas, MIT, Montgomery College, Mount St. Vincent, MTU, North Carolina State, North Dakota, Oakton, Penn State, Purdue, Radford, Rice, Saddleback, Stanford, Stonybrook, SUNY Stonybrook, TMCC, UC Irvine, UC San Diego, UI Urbana Champaign, Kentucky, U Mass., UMBC, U Penn, South Florida, USU, Texas, UTHSCSA, VSU, Vermont, Washington State.

British Universities.

Bristol, Cambridge, Cranfield, Glasgow, Napier, Oxford, Queen Mary College London, Univ. College London.

German Universities and Institutes

Max Planck MPI-SB, Max Planck Stuttgart, Bochum, Leipzig, Mainz, Ulm, Wuerzburg.

French Universities and Institutes

ENS, ENSMA, OBSPM, F. Comte, Poitiers.

September 2005,

Universities and Similar Worldwide.

TU Graz, McQuarrie Australia, QUT Australia, Melbourne, New South Wales, Sydney, Belgian Royal Military Academy, Ghent, UFC Brazil, UNICAMP, Alberta Research Council, McGill, NCF Canada, National Research Council Canada, Montreal Tech., Saskatchewan, Zurich, UO Cuba, German Synchrotron Facility, IU Bremen, Albert Einstein Institute, Max Planck MPI, Max Planck Stuttgart, Ruhr Bochum, Siemens Company, TU Cottbus, TU Darmstadt, Leipzig, Mainz, Manheim, Saarland, Ulm, Wuerzburg, CVT Denmark, Arizona, Boston Univ., SUNY Buffalo, Colorado State, Columbia, Cornell, Fresno, Drexel, Emporia, Georgia Tech., Georgetown, Harvard, Harvey Mudd, Indiana, Kansas State, Kansas, Lane, Maine, MIT, Montgomery College, Mount St. Vincent, MTU, North Carolina State, North Dakota, New York Institute of Technology, Oakton, Oklahoma State, Penn State, Purdue, Radford, Rice, Saddleback, San Jose State, Stanford, SUNY Stonybrook, Stonybrook, TMCC, Texas Tech., UAH, Central Florida, UC San Diego, Urbana Champaign, Kentucky, U Mass., UMBC, Maryland, Michigan, Minnesota, UNMC, Oregon, U Penn, South Florida, USU, Texas, UTHSCSA, Kentucky, VSU, Vermont, Washington, Washington State, Xavier, Yale, UCM Spain, UPV Spain, LUT Finland, UTU Finland, AGNS France, CEA France, CNES France, ENS France, ENS France, ENSMA France,

OBSPM France, F Comte, Lemans, Poitiers, US Gov BLM, Brookhaven, US Dept. of Energy, Federal Aviation Authority, Fermilab, Lawrence Berkeley, NASA (Larc), NIST, US Gov. NOAA, Oak Ridge, US Dept. of State, AUTH Greece, NTUA Greece, Patras, ELTE Hungary, Bose India, CNR ITIA Italy, INFN LNF Italy, INFN Perugia, INFN Milan, Bari Tech., Trieste, Genoa, Padua, TUS Japan, SNU Korea, US Army AMRDEC, US Military DTIC, US Military OSD, Gov. Mexico, UNAM Mexico, HVU Netherlands, Nijmegen, Groningen, TU Eindhoven, Twente, Utrecht, Free Univ Amsterdam, UIO Norway, Philipines, Krakow, Poznan, Warsaw, UCV Romania, KSTU Russia, LU Sweden, Chula Thailand, KMITL Thailand, Mahidol Thailand, PSU Thailand, NTNU Taiwan, NTUST Taiwan, Bristo , Cambridge, Cranfield, DMU, Napier, NERC Wallingford, Oxford, Queen Mary London, Univ. College London, Hampshire County Council, Belgrade.

4-26 October 2005

British Universities.

Aberdeen, Bath, Birmingham, Bradford, Bristol, Cambridge, Cardiff, Dundee, Durham, Edinburgh, Glasgow, Greenwich, Imperial, Lancaster, London Business School, Liverpool, Llandrillo, Manchester, Menai, Nottingham, Balliol Oxford, Physics Oxford, Queen's Belfast, Sheffield, St. Andrews, Strathclyde, Swansea.

German Universities and Institutes.

FU Berlin, Albert Einstein Institute, Max Planck Stuttgart, Max Planck MPPMI, Cologne, Ruhr Bochum, TU Cottbus, Bonn, Dortmund, Frankfurt, Freiburg, Heidelberg, Leipzig, Magdeburg, Mainz, Munich, Oldenburg, Potsdam, Saarland, Siegen, Stuttgart, Tuebingen, Wuerzburg.

US Universities and Similar.

Arizona, Arizona State, Austin CC, Bellevue, Berkeley, Brandeis, Brown, Boston Univ., Brigham Young, Caltech, CCC, Central Michigan, Colorado, Colorado State, Columbia, California Tech. CP, CSUN, CSU Ohio, CUNY, Denver, Duke, Florida Atlantic, FHSU, Florida Institute, FRC, GCC, Harvard, Hawaii, HMC, Illinois Tech., JMU, McKenna, Missouri, MIT, Montana, MUM, North Carolina State, Newchurch, New Jersey Tech., New Mexico State, Mew Mexico Tech., Northwestern, New York University, ODU, Ohio, Oklahoma State, Mississippi, Pittsburgh, Pratt, Princeton, Penn State, Radford, Rice, Rochester, Rockefeller, Rutgers, South Carolina, SJCA, Stanford, Stonybrook, SWU, Texas A and M, Central Florida, Chicago, UC Irvine, UC Merced, UC Santa Cruz, Delaware, Florida, Idaho, Urbana Champaign, U Mass., Maryland,

Michigan, Minnesota, UMSL, North Carolina Charlotte, UPC, U Penn, South Carolina, South Florida, USM, US Naval Academy, USU, Texas Arlington, Utah, Texas, UTHSCSA, Kentucky, Washington, Wesleyan, Wisconsin, WPI, Yale; US Gov BLM, US Gov CDC, Fermilab, US Gov INEL, Lawrence Livermore, NASA, NOAA, Oak Ridge, US Gov. OSTI, Sandia, US Gov. Treasury, Gov. Virginia.

Other European Universities and Institutes.
ARCS Austria, OEAW Austria, Linz, Leuven, Royal Belgian Military Academy, CERN, EPF Lausanne, ETH Zurich, Lausanne, Zurich, Charles Univ. Prague, Palacky, CVT Denmark, Danish Tech., UAM Spain, UCM Spain, UDL Spain, Uni2 Spain, US Spain, UV Spain, UVA Spain, Vigo, Helsinki, Stat Finland, TUT Finland, ENS Cachan, CEA France, CNRS GIF France, ECP France, ENS Lyon, INPG France, INRIA France, INSA Lyon, AOMC Jussieu, LPTL Jussieu, NOOS France, ONERA, Bourgogne, Paris Psud, Strasbourg, Univ MLV France, Marseilles, Nantes, Poitiers, Rennes 1, Tours, AUTH Greece, NTUA Greece, TEE Greece, Patras, UTH Greece, IRB Croatia, ELTE Hungary, MIE Ireland, ENEA Italy, INFN Bari, INFN LNF, INFN TS, ICTP Trieste, Trieste, Bari, Bologna, Padua, Pisa, Pavia, Turin, Nijmegen, Leiden, Groningen, Delft, Eindhoven, Twente, Utrecht, UIB Norway, UIO Norway, ICM Poland, IFPAN Poland, UWM Poland, Wroclaw, Porto, UTGJIU Romania, Novosibirsk Academy of Sciences, Chalmers Sweden, KTH Sweden, LIU Sweden, LU Sweden, Space Institute Sweden, SU Sweden, UMU Sweden, UU Sweden, ARNES Slovenia, MB Slovenia, Anadolu Turkey, Bilkent, Ege, Itu, KOU Turkey, Gov. Turkey, TBMM Turkey, PTT Yugoslavia.

Australian, Canadian and Japanese Universities and Similar.
Hiroshima, Hue, Osaka, Tokyo Tech., TUS, Tokyo, Monash, Melbourne, New South Wales, Wollongong, AMSTO, Alberta Research Council, Canadian Dept. of Defence, British Columbia, Gov. Canada, McGill, National Research Council, Queen's, Simon Fraser, Alberta, Montreal, New Brunswick, Regina, Toronto, Wightman, York Univ. Toronto.

Central and South America.
UFC Brazil, UNESP Brazil, UNICAMP, Sao Paolo, Chile, UO Cuba, BUAP Mexico, Gov. Mexico, UAM Mexico, UNAM Mexico.

Asia Minor and Asia.
UGM Indonesia, Jerusalem, Indian Statistical Institute, BARC India, Bose India, PDN Sri Lanka, MMU Malaysia, KAIST Korea, AIT

Thailand, PSU Thailand, CCU Taiwan, NCTU Taiwan, NTHU Taiwan, ITP China.

November 2005

British and Commonwealth Universities and Institutes.
   Aberystwyth, Bath, BGS, Bradford, Catherine Cambridge, Jesus Cambridge, Sidney Sussex Cambridge, Particle Cambridge, Cardiff, Durham, Edinburgh, Glasgow, Imperial, Leeds, Leicester, Liverpool, Manchester, Menai, MMU, Napier, Nottingham, Open, Balliol Oxford, BNC Oxford, Oriel Oxford, Phys. Oxford, Queen Mary London, QMU, RDG, Rutherford Appleton, Sheffield, Southampton, Strathclyde, Stoke, Surrey, Sussex, Swansea, Central England, Univ. College London, Leicester, Warwick, WKAC, Wolverhampton, York, York SJ, BBC, ACT Australia, Adelaide, ANU, Curtin, Griffith, Monash, Melbourne, New South Wales, Wollongong, Tasmania, Australian Gov. BOM, Australian Gov, NT, Queensland Gov., Canadian Gov., GVPL, Laurentian, McGill, MTA, Quebec, Simon Frazer, British Columbia, Calgary, Montreal, New Brunswick, Saskatchewan, Toronto, Western Ontario, York Univ. Toronto, Perimeter Institute, IITB India, IUCAA India, MRI India, Pune, IMSC India, NU Singapore, Auckland, CPIT New Zealand, Waikato.

US Universities and Institutes.
   ACCD, Adelphi, Albany, Andrews, Arizona, Arkansas State, Austin College, Arizona Western, Boston College, Berkeley, Brandeis, Brown, BSU, Boston Univ., BW, Brigham Young, Caltech, Central Michigan, Colorado, Columbia, Cornell, CSU East Bay, CSU Ohio, City Univ. New York, Dartmouth, Drexel, Duke, Elon, Embry Riddle, Evergreen, Florida International, Florida Institute of Technology, Florida State, GCC, Hamilton, Hartford, Harvard, Hawaii, Iowa State, IMSA, Indiana, Indian River CC, Johns Hopkins, JMU, Kent State, Kansas State, Louisiana State, Miami, MIT, Montana, Michigan State, MTU, NCA&T, North Carolina State, ODU, Ohio State, Ohio University, OU, Plymouth, Princeton, Penn State, Rice, Rocky, Renselaer, Rutgers, Sabanci, South Dakota, San Francisco State, San Jose State Stanford, Stonybrook, Texas A and M, Texas Tech., Tufts, Texas State, Univ California System, UC Davis, UC Fresno, Chicago, UC Irvine, UCLA, U Conn., UC Santa Barbara, UC Santa Cruz, UC San Diego, Delaware, UD Mercy, Florida, Houston, Idaho, Urbana Champaign, U Mass., UMBC, Maryland, Michigan, Minnesota, Chapel Hill, UNCA, UNL, U Penn, South Carolina, South Florida, US Naval Academy, USU, Texas Arlington, Utah, UTEP, Texas, Kentucky, Vanderbilt, Virginia, Vermont, Washington, Wesleyan, WFU,

Wisconsin, WSU, Washington St Louis, West Virginia, Western Washington, Yale.

US Government and Military

Bureau of Land Management, Brookhaven National Laboratory, Dept. Of Energy, Federal Aviation Authority, Fermilab, KAPL, Lawrence Berkeley, Lawrence Livermore, NASA (JSC), NASA (LARC), City of New York, Oak Ridge, PNL, Treasury, USGS, USPS, Gov. of Utah, US Airforce AFSPC, USAFA, USAAF Wright Patterson, US Army AMRDEC, US Army Monmouth, US Army USMA, US Military DISA, US Navy IH, US Navy Medical, US Navy MURGU, US Navair IDTE, US Navy NMCI, Naval Research Laboratory, US Navy NUWC, US Navy Pacific Northwest Command, US Navy Spawar.

Continental Europe, Ireland and Turkey

Vienna, FUNDP Belgium, KU Leuven, Belgian Royal Military Academy, ULB Belgium, Liege, Ghent, CERN, EPF Lausanne, ETH Zurich, Scherrer, Geneva, Zurich, Cyprus, Czech Army, Charles Prague, German Synchrotron Facility, IU Bremen, MPI Stuttgart, MPI Berlin, Cologne, TU Darmstadt, TU Ilmenau, TU Munich, Bielefeld, Bonn, Frankfurt, Heidelberg, Karlsruhe, Kaiserslautern, Leipzig, Saarland, Schwabia, Tuebingen, Ulm, AU Denmark, Danish Tech., UT Estonia, CNM Spain, EHU, ESA, UAM, Barcelona, UGR, Jaen, CSIC, EHU, ESA, Madrid Autonomous, Itreia, Zaragoza, UPM, UPV, US, USAL, USC, UV, Vigo, Academica Finland, Helsinki, HUT Finland, Acreunion France, CEA, EC Nantes, ENS, National Particle Lab., AOMC Jussieu, Lorca, Onera, Paris Psud, Strasbourg, Poitiers, Rennes, IRB Croatia, Chemres Hungary, Hungarian Academy Institute of Mathematics, Miskolcz, Debrecen, NUI Galway, Trinity College Dublin, CNTR IASF, CSI Italy, ENEA Bologna, INFN Catania, INFN Milan, INFN MIB, INFN Naples, INFN Padua, INFN Perugia, INFN Rome 1 and Rome 2, Milan Tech., Trieste, Bari, Bologna, Calabria, Florence, Subria, Psa, Pavia, Rome 3, Sardinia, Turin, BTV Latvia, Nijmegen, Groningen, TU Delft, Twente, Utrecht, Free Univ. Amsterdam, NTNU Norway, AGH Poland, IFJ Poland, Gov. Poland, Katowice, Krakow, Torun, Warsaw, UA Portugal, Lisbon, Porto, Lisbon Tech., BH Romania, JIU Romania, Krasnogorsk, RMT Komsomol, Troitsk, HIG Sweden, HIK, KTH, LUTH, UMU, UU Sweden, Josef Stefan Slovenia, Itu Turkey, Iyte, Metu, ACN Greece, AUTH, HOL, NTUA Greece, TEE, UOC, Patras.

At this point the feedback is recorded systematically every two weeks up to present (August 2009). From mid 2006 it is posted on the blog of www.aias.us. every two weeks. It is collected here.

1-15 December 2005

US Universities and Similar

Albany, ANNA, American Physical Society, Arizona State, Boston College, Berkeley, Berklee, Brown, Buffalo, Brigham Young, Caltech, Rhode Island CC, Colorado, Colorado State, Columbia, Cornell, CSU Chico, CSU East Bay, CSU Long Beach, City Univ New York, Dartmouth, De Vry, East Strousburg, Florida State, Georgia Tech., Georgetown, Harvard, Indianapolis, Ivy Tech CC, Johns Hopkins Applied Physics, Louisiana State, Maine, Missouri, MIT, North Carolina State, Northern Illinois, Northwestern, New York Univ., Oakton, Old Dominion, Phoenix, Pittsburgh, Plymouth, Pomona, Princeton, Radford, Rariton Valley, Rio Hondo, Rochester, Renselaer, Sabanci, Sanford, South Florida State, Sam Houston State, Seattle Pacific, Stanford, Stonybrook, Swarthmore, Texas A and M, Texas Tech, Tufts, Albuquerque, Telvoc Institute, Univ California System, Univ. Corporation for Atomic Research, UC Fresno, Chicago, UC Irvine, UCLA, UC Riverside, UC San Diego, Florida, Illinois Chicago, Iowa, Urbana Champaign, U Mass., Maryland, Michigan, Minnesota, U Penn, South Carolina, South Florida, US Naval Academy, Texas Arlington, Utah, Texas, Tennessee, Tulsa, Vermont, Virginia, Virginia Tech., Washington, Western Carolina, Wesleyan, Whitman, Wisconsin, Western Illinois, Western Kentucky, William and Mary, Western Michigan, Washington St Louis, Western Washington, Argonne, BLM, Brookhaven, Fermilab., US House of Representatives 109th Congress, Los Alamos, Lawrence Berkeley, Lawrence Livermore, NASA (JPL), NASA Johnson, NASA Langley, NIST, Oak Ridge, US Dep. of Justice, USAAF Wright Patterson, US Army Heidelberg, US Army PICA, US Defense Agency, US Navy MUGU, Naval Research Laboratory, US Navy NUWC, US Military OSD, US Military Command.

Britain and Commonwealth, Universities and Institutes.

Aberystwyth, Birkbeck, Birmingham, Bradford, Engineering Cambridge, Pembroke Cambridge, Trinity Cambridge, Particle Cambridge, Cardiff, Cranfield, Durham, Essex, Glasgow, Highbury, Heriot Watt, Imperial, Lancaster, Leeds, Liverpool, Luton, Manchester, Napier, Northampton, Balliol Oxford, Queen Mary London, Southampton, Strathclyde, UMIST, Warwick, BBC (four visits), Australian National Univ., Melbourne, Queensland, Sydney, Defence, Alberta Research Council, Gov. Alberta, Canadian National Water Research Institute,

Concordia, Catholic Schools Ontario, Canadian National Research
Council, Toronto, Perimeter, Dawson, Gov. Quebec, Hydro Quebec,
Queen's, Calgary, Manitoba, New Brunswick, Victoria, Waterloo, York
Univ. Toronto, IUCAA India, NSC India, UITM Malaysia, NU Singapore.

Continental Europe, Ireland and Asia Minor, Universities and Institutes..
    OEAW Sutria, Vienna, UIBK Austria, Belgian RMA, Leuven, CERN,
EPF Lausanne, Basel, Zurich, Charles Univ Prague, HU Berlin, IU
Bremen, Max Planck CNS, Max Planck IPP, Max Planck Garching, Max
Planck MPG, Max Planck MPPMU, Max Planck MPE, Max Planck MPG,
TU Hamburg, Bayreuth, Frankfurt, Heidelberg, Kaiserslautern, Konstanz,
Leipzig, Oldenburg, Schwabia, Siegen, CVT Denmark, ITU Denmark, UA
Madrid, Barcelona, Madrid 2, Salamanca, Valencia, LUT Finland, CNES
France, CNRS Grenoble, French national particle lab., Polytechnique
system, SHOM France, Montpelier, Marseilles, Nantes, Poitiers, NTUA
Greece, Thessalonika, Univ College Ireland, Technion, Weizmann, INFN
Milan, INFN Turin, ISI, Milan Tech., SISSA, Bari, Bologna, Florence,
Padua, Messina, Pavia, Sardinia, Turin, Leiden, Groningen, Eindhoven,
Twente, Utrecht, FFI Norway, HIN Norway, NTNU, UIB Norway, IFJ
Poland, US Poland, Krakow, Lisbon, Porto, Lisbon Tech., PU Russia,
Chalmers Sweden, KTH Sweden, Physto Sweden, UMU, UU Sweden,
Edus Slovenia, Ljubjiana, Truni Slavakia, Agric. Slovakia, Institute of
High Energy Physics Russia, Akeleniz Turkey, Gazi, Hacettepe, Metu
Turkey.

Far East, Universities and Institutes.
    Dendai, Kindai, Kobe, Kwanse, Kyoto, Nagoya, Tohokai, UEC Japan,
PCJ Japan, GIST Korea, Postech Korea, NCTU Taiwan, Academica Sinica
Taiwan, TNRC Taiwan, Philipines.

South American Universities.
    UNICAMP, Sao Paolo.

15-30 December 2005

US Universities and Similar.
    AUB, Berkeley, Brown, Columbia, Cornell, Cornell College, Georgia
Tech., Harvard, Hawaii, Illinois Tech., Lousville, MIT, Michigan State,
North Carolina State, New Mexico State, Oklahoma, Princeton, Stanford,
Tufts, Univ. California System, UC Irvine, UCLA, UIS, Urbana
Champaign, Michigan, Minnesota, Nevada Reno, South Florida, Utah,
UTC, Texas, UWEC, Western Michigan, Virginia, Washington,
Wisconsin, WITC, Washington St. Louis, Xavier, Yale, US Gov. BLM,

Brookhaven, Fermilab, US Gov. GSA, NASA (JSC and LARC), NIH, NSF, Oak Ridge, Sandia, USAAF Wright Patterson, USAAF Yokota, Army Research Laboratory, US Army Monmouth, US Navy NCSC, US Navy NMCI, US Navy NUWC.

15-30 December Europe and Asia Minor, Universities and Institutes.
Vienna Tech., CERN, ETH Zurich, Charles Univ. Prague, Max Planck Heidelberg, Max Planck MPPMU, TU Chemnitz, TU Ilmenau, Greifswald, Karlsruhe, Leipzig, Regensburg, Saarland, Siegen, Madrid 2, UPV, Valencia, CEA Saclay France, Polytechnique, SHON France, Bordeaux, Poitiers, NTUA Greece, NUIM Ireland, Univ College Dublin, Jerusalem, Weizmann, Reykjavik, INFN Genoa, INFN Naples, INFN Rome 3, SISSA Italy, SNS Italy, ICTP Trieste, Florence, Perugia, Pavia, Delft, Twente, Utrecht, HIST Norway, NTNU Norway, UIB Norway, Gdansk, Wroclaw, Lisbon, HTU Sweden, Gov. Slovakia, TRNI Slovakia, Agric. Slovakia, Ankara, Ege, KSU Turkey, Metu, Engineering Cambridge, Particle Cambridge, Durham, Glasgow, Heriot Watt, King's College London, Manchester, MMU, Oxford Isis, Southt, Univ College London, Warwick, British Gov., British Gov NICS, Belgrade.

Asian University
RUH Sri Lanka.

January 2006

U.S.A, Universities, Institutes and Similar.
SUNY Albany, Appelachian State, Arizona, Arizona State, Berkeley, BHC, Brooklaw, Boston, SUNY Buffalo, Butler, Brigham Young, Caltech*, Citrus College, Central Michigan, Columbia, Cornell, CTC, CWRU, Dartmouth, Denison, Drexel, Duke, EOU, ESU, ETSU, Evergreen, Florida Atlantic, Florida International, Florida State, Georgia Tech., Georgetown, Hamilton, Harvard, Howard, Indiana, Indian River CC, Princeton Institute for Advanced Studies, Indiana, Johns Hopkins, Lousiana, Lousiana State, LTU, JWU, MC3 McKendree, Memphis, Missouri, Missouri State, Michigan State, MIT, Michigan State, NAU, North Carolina State, New Mexico State, North Dakota, NOVA, New York Institute of Technology, NY University, Oberlin, Ohio State, Oswego, PDX, Phoenix, Pittsburgh, Princeton, Princeton Institute for Advanced Studies, Purdue, Rochester Tech., Rochester, Roosevelt, Renselaer, Rutgers, Sabanci, SCCNC, Stanford*, SUNY Stonybrook, U Mass., Maryland, Texas A and M, Texas Tech, UC Davis, Central Florida,

Chicago, U Conn., UC Santa Barbara, UC Santa Cruz, Michigan, UMKC, Minnesota, Chapel Hill, UAF, U California System, UC Davis, UCLA, UC Riverside, UC San Diego, UC Santa Cruz, Florida, UC Davis, UCLA, Urbana Champaign, Kentucky, UNL, Michigan, UML, Minnesota, UMR, UNCG, Union, New Mexico, Oregon, U Penn, Texas Arlington, UT Dallas, Texas, Kentucky, Wyoming, Virginia, Vermont, Washington*,William and Mary, Worcester Polytechnic Institute, Wright, WSU, Xavier, Yale, Argonne, UPMC, Vanderbilt, Virginia, Vermont. Washington, Wichita, Yale, Gov California, US Dept of Energy, Fermilab, NASA (JPL), NIST, NSF, Oak Ridge, US Gov, USGS, Gov. Virginia, USAAF Eglin, USAF Kirtland, USAAF Moron, US Military DISA, US BLM, US Dept. of Energy, USAAF Aorcentaf, U Military MDA, US Navy NMCI, US Navy NUWC, US Navy UAR.

1-14 January 2006

Europe and Asia Minor Universities, Institutes and Similar.

Eduhi Austria, Gov. Austria, RMA Beglium, Liege, Free Univ Brussels, Ghent, Sofia, EPF Lausanne, FU Berlin, Max Planck MPIBPC, HU Berlin, Max Planck MPE, Max Plank MPIA Heidelberg, Max Planck MPP, Ruhr Bochum, STW Bonn, UDK Berlin, Bremen, Heidelberg, Leipzig, Regensburg, Saarland, AUB Denmark, UCM Spain, Uni2 Spain, UV Spain, UTU Finland, CEA France, ENS France, ENS Cachan, ENS Lyon, French national particle lab., INSA Toulouse, Sciences-PO France, Bourgogne, Lemans, Lyon 1, Poitiers, EAP Greece, Patras, Koncar Croatia, KLTE Hungary, Dublin City, Bar Ilan, MACAM Israel, Weizmann, CDT Europe, CEC Europe, INFN Bari, INFN Genoa, INFN LNF, SNS Italy, Bologna, Catania, Messina, Perugia, Pisa, Pavia, European Parliament Luxembourg, Porto, Groningen, Twente, Utrecht, Free Univ. Amsterdam, NTNU Norway, UIO Norway, AGH Poland, FUW Poland, UJ Poland, WAT Poland, Krakow, Lodz, Lublin, Warsaw, Lisbon, Chalmers Sweden, UU, KTH Sweden, EDUS Slovenia, Joseph Stefan Institute, KPI Kharkov, Aberdeen, Aberystwyth, Bristol, Brunel, Churchill Cambridge, CSX Cambridge, Engineering Cambridge, Cardiff, Exeter, Glasgow, Imperial, Liverpool, Manchester, Napier, Oxford, Sheffield, Surrey, UWIC, Warwick, York SJ, Ceredigion, British Gov. DSTL, British Gov. GSI, National Health Service, Anadolu Turkey.

15 - 31 January 06

Europe and Asia Minor: Universities, Institutes and Similar.

ESI Austria, Graz, Vienna, WU Vienna, Leuven, UMH Belgium, Ghent, Sofia, EPF Lausanne, Belfort, Geneva, Max Planck IPP-HGW,

Max Planck Heidelberg, RWTH Aachen, Tautenberg, TU Berlin, TU Chemnitz, Freiburg, Karlsruhe, Leipzig*, Munich, Regensburg, Tuebingen, TDC Denmark, Barcelona, UDL Spain, UGR, Rioja, UPC, USC, JYU Finland, TUT Finland, UTU Finland, CNRS Grenoble, French Ministry of Culture, French national particle lab*, INSP Jussieu, NOOS France, Polytechnique system, SHOM France, Strasbourg, BP Clermont, Poitiers, AUTH Greece, DUTH Greece, Bar Ilan, Jerusalem, Technion, Weizmann, CNR ITIA, City of Modena, Milan Tech., Turin Tech., Trieste, Perugia, Pavia, Nijmegen, Leiden, KVI, Rotterdam, Groningen, Shell Netherlands, Delft, Eindhoven, Twente, Utrecht, Free Univ Amsterdam, NTNU Norway, UIO Norway, Porto, Minhoi, Lisbon Tech., Orenburg CCI Russia, Chalmers, IRFU, LTH Sweden, MB Slovenia, Agric. Slovakia, Omu Turkey, Aberdeen, Birmingham, Bolton, Bristol*, Cambridge*, Cardiff*, Dundee, Durham, Edinburgh, Glasgow Caledonian, Imperial, Lancs. NGFL, Leicester, Liverpool, Manchester, Nottingham, Oxford*, ROE, Sheffield*, St Andrews, St David's, Stirling, Surrey, Swansea, Univ College London, UKC, UMIST, Warwick, BBC, NHS.

1-14 January 2006

Rest of World Universities and Institutes

Australian National University, RMIT Australia, Melbourne, Gov. Victoria, Gov. of Canada, Laurentian, New Brunswick, Olds College, Maplesoft, Gov. of Quebec, Royal Military College, Calgary, Montreal, Victoria, York Univ. Toronto, Keio, Titech, Univ. Tokyo, KEK Japan, NAIST Japan, Dongeui Korea, KNU Korea, NU Singapore, NCTU Taiwan, NCU Taiwan, NTU, Academica Sinica, TYC Taiwan.

15 - 31 January 2006

Rest of World Universities and Institutes.

Melbourne, Queensland, Australian Research Council, Centennial College Canada, Canadian Dept of Defence, Gov. Quebec, Lakehead, McGill, McMaster, National Research Council of Canada, Perimeter, Queen's, British Columbia, Calgary, Laval, Toronto, Windsor, Western Ontario, York Univ Toronto, DU India, IITB India, IITHGP India, NU Singapore, Agric Sri Lanka, UITM Malaysia, Jaring Malaysia, Kyoto, Nagasaski, Univ Tokyo, Yamagata, CIT, Hitachi, Postech, Academica Sinica*, Gov Brazil*, UFPA, UFPB, Sao Paolo, UO Cuba, IPN Mexico, CIO Mexico, ULA Venezuela.

US Universities, Institutes and Similar.

Arizona State, Brown, Caltech., Central Michigan, Colorado, Columbia, Cornell, CSU Eastbay, CSULB, CSU Ohio, CSU Pomona, CWRU, Delhi, Elon, Emory, Florida Tech., Georgia State, Princeton Institute for Advanced Studies, Iowa State, Johns Hopkins Applied Physics, Lousiana State, Manhattan, MC3, McKenna, MIT, Michigan State, North Carolina State, ND, NEU, New Mexico State, North Dakota, Northwestern, Ohio State, Mississippi, OCU, Pittsburgh, Princeton, Rochester, SHIP, SJSU, Stanford, SUNY Stonybrook, Texas A and M, Univ. California System, UC Davis, Chicago, UC Irvine, UCLA, UC Santa Barbara, UC San Diego, Iowa, Urbana Champaign, U Mass., Maryland, Michigan, Minnesota, UMR, Chapel Hill, New Mexico, UNMC, UNT, U Penn., South Florida, USU, Texas, Vanderbilt, Washington, Wisconsin, WPI, Washington St Louis, IS Gov BLM, US Gov BPA, US Dept of Energy, US Gov ED, US Gov FDA, Fermilab, Lawrence Berkeley, NIH, Oak Ridge, US Gov WA, USAAF AFSPC, USAAF ANG, US Army MSL, US Army USMA, US Navy IH, US Navy NUWC.

Europe and Asia Minor: Universities and Institutes.

Vienna, Leuven, Royal Military Academy of Belgium, Liege, PU Academy Belgium, Sofia, German Synchrotron Facility, GFZ Potsdam, Albert Einstein Institute, Stuttgart, Max Planck MPIKG-GOLM, Max Planck MPIP Mainz, Bonn Students, Karlsruhe, Kassel, Kiel, Leipzig*, Rostock, Saarland, Ulm, Aarhus Denmark, AUC, TDC Denmark, EHU Spain, UCM Spain, UNi2 Spain, Ovi, JYU Finland, CEA France, CNRS LPN, ENSEA France, ENST Brittany, French national particle lab., INPL Nancy, INRIA, LISIF Jussieu, SHOM, Bordeaux, Picardy, Poitiers, Rennes 1, AUTH Greece, Patras, Univ College Cork, Jerusalem, CNR IGI, ENEA Frascati, INFN Milan, INFN Sardinia, Messina, Pisa, Pavia, Turin, Leiden, Groningen, Eindhoven, Twente, Utrecht, Amsterdam, UIB Norway, PAP Poland, Warsaw, Wroclaw, Lisbon, NIPNE Romania, IITP Russia, Chalmers, LUTH, MSI Sweden, Gantep Turkey, Omu Turkey, Aberdeen, Aston, Bangor, Bath, Birmingham, Bristol, Churchill Cambridge, Physics Cambridge, City, Durham, Edinburgh, Glamorgan, Hull, engineering Oxford, physics Oxford, Pursglove, Sutcol. Britain, Swansea, Univ College London, UKC, Univ Wales Institute Cardiff, Warwick, BBC*, National Physical Laboratory, British Gov. DSTL and Ministry of Defence.

Rest of World: Universities and Institutes

Australian National University, Queensland, Australian Gov. BOM, Univ. British Columbia, NCF Canada, Sheridan College Ontario, Perimeter, Queen's, Manitoba, Sherbrooke, Toronto, Victoria, Waterloo, York Univ. Toronto, Pune, IMSC India, City Hong Kong, UST Honk Kong, Massey NZ, National Univ. Singapore, KMITL Thailand, KMITNB Thailand, RU South Africa, NIFS Japan, Osaka, Tokyo Tech., Univ. Tokyo, ULIS Japan, Waseda, KAIST Korea, NTU Taiwan, UNLP Argentina, UNRC Argentina, UFF Brazil, UNICAMP Brazil.

14-26 February 2006

US Universities, Institutes and Similar.

Appalachian State, Arizona, Arizona State, Austin CC, Berkeley, Boise State, Boston, Caltech, Cedarville, Central Michigan, Colgate, Colorado State, Columbia*, Cornell, CSU Fresno, CSU Ohio, Denver, Duke, Georgia Tech., Goshen, Hartford, Hartwick, Hawaii, Princeton Institute for Advanced Studies, Iowa State*, Johns Hopkins, Johns Hopkins Applied Physics, Kansas State, Lewis, Lousiana, MCW, Missouri, MIT, Michigan State, Mount Holyoke, North Carolina State, New Mexico State, New York Univ., Ohio State*, Mississippi, Oregon State, PDX, Pittsburgh, Princeton, Penn State, Purdue, Richmond, Rooseveldt, Renselaer, Stanford*, Texas A and M, Temple, TJU, Univ. California system, UC Davis, Chicago, UC Irvine UCLA, UC Santa Barbara, UC San Diego, UC San Francisco, Delaware, Georgia, Michigan, Minnesota, New Mexico, Oregon, USU, Kentucky, Virginia, Vermont, Washington*, Washington State, Washington St Louis, Yale, US Gov BNL, US Gov DHS, US Gov. EPA, NASA (gsfc), NASA (jpl), USAAF Hanscom, USAAF Robins, US Army Medical, US Army NVL, US Army USMA, US Navy NMCI, US Navy NUWC.

Europe and Asia Minor, Universities and Institutes and Similar.

TU Vienna, KU Leuven, Belgian RMA, Free Univ. of Brussels, Ghent, CERN, ETH Zurich, Basel, Palacky, GFZ Potsdam, Max Planck MPI-SB, MPI-CBG, RWTH Aachen, TU Berlin, TU Munich, Ulm, Hamburg, Hannover, Heidelberg, Kaiserslautern, Konstanz, Leizpig*, Stuttgart, Tuebingen, AUC Denmark, CVT Denmark, Aena Spain, Madrid Autonomous, UDG, Madrid 2*, Leon, UPM, UV, Helsinki, JYU Finland, VTT Finland, CEA France, CNRS LPN, ENS System, ENS Cachan, ENS Lyon, Jussieu lcpp6, NOOS, OBSPM, Strasbourg, Lorraine, Poitiers, Tours, AUTH Greece, TEE Greece, Tellas, Patras, Trinity College Dublin, Univ. College Dublin, Bar Ilan, CNR IGI Italy, INFN Bari, INFN Bologna, INFN Naples, INFN Padua, SISSA Italy, Pisa*, Rome 3,

Sardinia, RSU Latvia, ULSE Netherlands, MJIT, Groningen*, TU Delft, TU Eindhoven, Twente*, Utrecht, Free Univ. Amsterdam, Norut Norway, Gdynia, IPB Partugal, Minho, NW Russia, HH Sweden, KTH, LUTH Sweden*, Agric. Univ. Slovakia, Institute of High Energy Physics Russia, Gyte Univ Turkey, Metu, ITL Ukraine, Aberystwyth*, Bristol*, Brunel, Cambridge, Durham, Glamorgan*, Imperial*, Lancaster, Leeds, Liverpool, Manchester, Nottingham, Hertford Oxford, Lincoln Oxford, Queen Mary, Reading, RHUL, Southampton*, Univ College London, UMIST, Warwick, York, Belgrade.

Rest of World, Universities, Institutes and Similar.

Newcastle Univ. Australia, Queensland, Sydney, Defence Gov. Australia, Gov. New South Wales, Alberta Canada, Edmonton, Gov. British Columbia, Dalhousie, McGill, Saskatchewan, Sherbrooke, Windsor, York Univ. Toronto, BARC India, JMI India, Fukuoka Univ. Japan, Hokuai, Nagoya, Tokyo Tech., Tsukuba, Ryukyu, UAM Mexico, UTM Mexico, UNS Argentina, UFU Brazil, UNAL Columbia, CCIT Taiwan, NCTU and NTHU Taiwan.

1 -14 March 2006

US Universities, Institutes and Similar.

Albany, Berkeley*, Berklee, Brandeis, Brown*, Caltech, Columbia*, Cornell*, FSCWV, Georgia Tech*, Hartford, Harvard, Herzing, HMC, Princeton Institute for Advanced Studies, Iowa State, Indiana, Inter, Indian River CC, Johns Hopkins Applied Physics*, MCCSC, MIT*, Michigan State, North Carolina State*, New Jersey Institute of Technology, New Mexico State, New Mexico Tech, Northwestern, Ohio State*, Oklahoma, Princeton*, Purdue*, Rhodes, Rice, Rochester Tech, Rutgers*, SDSC, SIU*, Stanford, St Cloud State, Stonybrook, Texas A and M*, Trinity College, Texas State, UAH, Univ California System, UC Davis, Chicago, UCLA*, UC Santa Barbara, UC Santa Cruz, UC San Diego, U Mass, UMBC Maryland, Maryland, Michigan*, Minnesota, UMR, Univ North Carolina Charlotte, UNCO, Dhaka, Oregon, South Florida*, Utah, Washington*, William and Mary, Yale*, US Gov BLM, US Gov BPA, US Gov Central Intelligence Agency, US Gov Dept. of Energy, US Gov IHS, Los Alamos, Lawrence Livermore, NASA (Jpl), Sandia, Gov Clayton State, USAAF AFMC, USAAF Los Angeles, US Army Medical, US Military CIFA, US Navy NMCI*, US Navy NSWC, US Navy Spawar, US Military NGA, US Military OSD.

European Universities, Institutes and Similar,

Graz, Vienna, Belgian RMA, Liege, Bulgarian Academy of Sciences, Swiss Federal Government Bern, EPF Lausanne, ETH Zurich, Belfort, Charles Univ. Prague, Max Planck Halle, MPG MPPMU, TU Clausthal, TU Cottbus, Bielefeld, Erlangen, Frankfurt, Giessen, Goettingen, Heidelberg, Karlsruhe, Leipzig*, Muenster, Regensburg, Stuttgart, WSS Stuttgart, DKU Denmark, Niels Bohr Institute, SDU Denmark, Inta Spain, Complutensian Madrid, Univ 2, UPV, Valencia*, Vigo, CIUP France, CNES France, Polytechnique System, Strasbourg, Marseilles, Poitiers, Tours, Patras, Dublin City Univ., UC Cork, Trinity College Dublin, Univ. College Dublin, Tel Aviv, INFN Milan, INFN Pisa*, INFN Pavia, INFN Turin*, Turin Tech, SNS Italy, ICTP Trieste, Bologna, Padua, Pisa*, Pavia, Turin, Groningen, Shell Netherlands, Eindhoven, Twente, Utrecht, Oslo, Royal Society of Chemistry, Torun, UVT Romania, Joint Institutes for Nuclear Research Russia, Bratislava, Ryazan Russia, Gyte Univ Turkey, Hacettepe, British Gov. BBSRC, Cambridge*, Cranfield, Durham, Glamorgan*, Imperial*, Leeds*, Liverpool*, Luton, Napier, Oxford*, Queen's Univ Belfast*, Southampton*, St David's College*, Swansea*, Univ London Computer Centre, UMIST, West Thames Authority, British Gov. Energis, British Ministry of Defence, National Health Service, AUB Lebanon.

Rest of World Universities, Institutes and Similar

UNLP Argentina, Australian Ministry of Defence, Gov Brazil, UFMG Brazil, UFPB, UNICAMP Brazil, Sao Paolo, AEI Canada, Dalhousie, Gov. Quebec, Hydro Manitoba, McGill*, Memorial Univ. Newfoundland, NS Health Canada, HPCDSB Ontario, Montreal Tech., Hydro Quebec, Queen's, Alberta, Laval, Saskatchewan, Usherbrooke, Toronto*, Waterloo, IUCCA India, IASBS Iran, Hokudai Univ Japan, Kanazawa, Kobe, Kyoto, Tokyo, Postech Korea, SLT Sri Lanka, Auckland NZ, AUT, Massey NZ, CCU Taiwan, NTU Taiwan.

15 - 29 March 2006

US Universities, Institutes and Similar

Alleghenny, Berkeley*, Adams, Binghampton, Brooks, Brown*, Brigham Young UH, Caltech*, Clemson, Carnegie Mellon, Cornell*, CSU Ohio, CSUS, City Univ of New York, Case Western Reserve, Delta, Evangel, Florida State, Georgia Tech*, GCC, Geneseo, Georgetown*, Harvard*, Indiana*, IUPUI, Ivy Tech, Johns Hopkins*, JJC, Kent State, Lousiana, Maine, MCCSC, Mercynet, MIT*, MNSU, Michigan State*, Muhlenberg, NIU, Northwestern*, New York Tech., Ohio State*, Pittsburgh, Princeton*, Penn State*, Purdue*, Radford, Richmond,

Rochester Tech.*, Rochester*, Rowan, Rutgers, South Dakota State, Stanford*, SUNY Stonybrook, Syracuse, TVI, UAH, Univ California system*, UCAR, UC Fresno, UC Irvine, UC Los Angeles*, U Conn., UC Santa Barbara*, UC San Diego*, Delaware, Florida, UIC, U Mass*, Michigan*, UML, UMR, Chapel Hill, Unic. North Carolina Charlotte, UNL*, UNLV, U Penn, South Carolina, Utah*,Texas*, Kentucky, UWM, Virginia, Vermont*, Washington*, Wayne State*, Wisconsin*, Western Michigan, Worcester Polytechnic Institute, Yale*, US Gov BNL, US Dept. of Energy, Fermilab*, US Gov GSA, Lawrence Livermore, Lawrence Berkeley, NASA (Gsfc), NASA (Jpl)*, NOAA, NSF, Oak Ridge, Gov Virginia, US Military AFIT, US Army Pica, US Army USMA, US Military DISA, US Navy NMCI*, US Navy NUWC*, US Military NCSC, US Military OSD, US Military USMC.

Europe and Asia Minor: Universities, Institutes and Similar

ASN Vienna, HTL Salzburg, KF Graz, WU Vienna, Salzburg Research, RMA Belgium, Free Univ Brussels, Province of Liege, Bulgarian Academy of Sciences, Swiss Federal Govt., CERN, EPF Lausanne, ETH Zurich*, Bern*, Palacky, Max Planck Stuttgart, TU Berlin, Bielefeld, Leipzig*, Wuerzburg, Aarhus Denmark, CVT, Niels Bohr Institute, TDC Denmark, Barcelona, Madrid 3, Madrid Complutensian*, UPF, UV*, Vigo, HTV Finland, HUT Finland*, AC Amiens, EC Nantes, ENS Cachan, French national particle lab., INPL Nancy, Polyechnique, Poitiers*, NTUA Greece, Miskolc Hungary, NUIM Ireland*, Bar Ilan, JCT Israel, Tel Aviv, INFN Bari, INFN Florence, INFN Genoa, INFN Milan, INFN Presidium, INFN Rome 1, Bari, Genoa, Pavia*, Turin*, Parma, VTU Latvia, KVI Netherlands, RU Netherlands, Twente*, Utrecht*, Free Univ. Amsterdam, HIB Norway, NTNU Norway, UIO Norway*, ANI Poland, IFPAN, Polish Congress, Poznan, Warsaw, Lisbon Tech., IBRAE Russia, Joint Institutes for Nuclear Research Russia*, Chalmers, LUTH, Physto Sweden, UMU Sweden, Agric Univ Slovakia, Ankara, Metu, Space Gate Ukraine, Kharkov, Odessa, Bath, BBSCR, Birmingham, Bradford, Brunel, Durham*, Edinburgh*, Essex, Heriot Watt*, Imperial*, Leeds*, Luton, Manchester, NCL, Open, Oxford*, QMW, RDG, ROE, Salford, Sheffield*, Southampton*, Sussex, Swansea*, Warwick, York*, BBC*, Ceredigion, NHS*, Eton College.

Rest of World: Universities, Institutes and Similar.

UNER Argentina, UNRC Argentina, Gov. Argentina, ARA Military Argentina, CSIRO Australia, Adelaide, ANU, Griffith, Sidney, Tasmania, Defence, Gov Brazil, INPE Brazil, RSI Brazil, UFRGS Brazil, UNICAMP, Unisul, Sao Paolo, Alberta Research Council Canada, Gov. British Columbia, Dalhousie*, McGill*, Nova Scotia, Ontario, Toronto*,

Gov. Quebec, Queen's, Simon Frazer*, Manitoba, Montreal, Guelph, Saskatchewan*, UIFSM Chile, Tsinghua China, IIIT India, IMSC India, IASBS Iran, Hirsohima, Kyoto, Tohoku, Tokyo, KAIST Korea, Postech*, Woosuk* Korea, UTBB Mexico, ITESM Mexico, UTM Mexico, Monash Malaysia, ISU Saudi Arabia, SU Thailand, NCTU Taiwan, YLC Taiwan, ULA Venezuela, Massey NZ, SQU Oman.

1 - 14 April 2005

US Universities, Institutes and Similar.

Austin CC*, Bennington, Berkeley*, Binghampton, Brown*, Brigham Young, Caltech*, Chungang, Carnegie Mellon*, Colorado, Colorado State, Columbia, Dartmouth*, Elon, GCC, Georgetown*, Harvard*, HMC, Johns Hopkins, Johns Hopkins Applied Physics, JMU*, Kansas, U Mass, MIT*, Michigan State*, North Dakota, Northwestern, New York Univ.*, Mississippi, Pittsburgh*, Princeton*, Penn State*, Purdue*, Radford, Rice*, Richmond, Rochester, Rutgers, Stanford*, Stevens Tech.*, SWAU, Syracuse, Tufts*, California System, UC Davis*, Central Florida, Chicago, UCHSC, UC Irvine, UCLA, UC Riverside, UC Santa Barbara, UC Santa Cruz, UC San Diego*, Urbana Champaign*, U Mass, Michigan, Minnesota, Chapel Hill, UNL*, U Penn, UPRM*, South Carolina, Utah*, UT Dallas, Texas*, Kentucky, UTMEM, Virginia*, Vermont*, Washington*, Wayne*, Wisconsin, Washington St Louis, Yale*, Argonne, Los Alamos*, Lawrence Livermore, NASA (Gsfc), NIH, Oak Ridge, USAAF AFSPC, US Navy NCSC, US Navy NMCI, US Military NRSC.

Europe and Asia Minor: Universities, Institutes and Similar.

Vienna*, WU Vienna, Graz, RMA Belgium, Liege, Free Univ Brussels, Ghent, CERN, EPF Lausanne, ETH Zurich, Charles Univ. Prague, UNA Cyprus, Ruhr Bochum, TU Munich, Duisburg, Freiburg, Heidelberg*, Karlsruhe, Kaiserslautern, Leipzig*, Ulm, Aarhus, Danish Tech., Barcelona, UPF, URV, Helsinki, JYU Finland, French national particle lab., INRIA France, Strasbourg, Poitiers*, Rennes 1, Athens, NTUA Greece, Olympic Greece, NUI Galway, Trinity College Dublin, Technion, MoD Israel, HI Iceland, Astro Italy, Bari Tech., ICTP Trieste*, Calabria, Padua, Turin, AUB Lebanon, Groningen, Delft, Twente, Utrecht, NTNU Norway, IFJ Poland, Gdansk, Warsaw, INFIM Romania, Tuaisi, UU Sweden, KOU Turkey, Aberystwyth*, BBSCR, Brunel, Cambridge*, Durham, Liverpool, Manchester, Oxford, Southampton, Surrey, Swansea, Univ College London, East Anglia, Univ. London Computer Centre, Wolverhampton, Environment Agency, Ministry of Defence.

Rest of World: Universities, Institutes and Similar

Gov. Argentina, Adelaide, Monash, Swinburne, Gov. Victoria, UFC Brazil, UFG, UNICAMP, Carleton, Concordia, McGill, McMaster, NRC Canada, Simon Frazer, Alberta, British Columbia, Montreal, Quebec, Toronto*, Waterloo, UNAL Colombia, Bose India, IIITB India, IITD India, Chuo Univ. Japan, Tohoku, Tokyo, UEC, Kobe, CAU Korea, SNU and UITM Malaysia, Massey NZ, AIT Thailand, NSRRC Taiwan, UDSM Tazmania, MOET Venezuela.

15 - 28 April 2006

US Universities, Institutes and Similar.

Berkeley*, SUNY Binghampton, Boise State, Brown*, Bucknell, SUNY Buffalo*, Caltech*, Carnegie Mellon, Colorado*, Colorado State, Cornell*, CSU Pomona, CWRU, Drexel, Duke, Emory, FEASD, Georgia Tech., Georgetown*, Harvard*, Hawaii, Immaculata, Indiana*, Indian River CC*, Kansas State, Maine, MNSCU, Montana, Mississippi State, NIU, New Mexico Tech., North Dakota*, Northwestern*, Missisippi*, Oregon State, Princeton*, Renselaer, Rutgers, Scranton, Seattle, SIU*, SJSU, Stanford, St Cloud State, Stevens Tech., Texas A and M, UC Davis*, Chicago, UC Irvine*, UCLA, U Conn., UC Santa Barbara*, UC San Diego*, Delaware, Florida, Houston*, Iowa, Urbana Champaign*, Kentucky, U Mass, Minnesota, North Carolina Charlotte, UNCW*, Oregon, Texas Arlington, Texas*, UTK*, Toledo, Vanderbilt, Vermont, Washington, Wisconsin*, Washington State*, Yale*, York Hospital, Brookhaven, NASA (Gsfc), Oak Ridge*, Pacific National Laboratory, USAAF Robins, US Army USAce, US Navy NUWC, US Military USCG.

Europe and Asia Minor: Universities, Institutes and Similar.

Rudolfinum Austria, Red Cross Austria, RMA Belgium*, Liege, FU Brussels, TU Sofia, CERN*, EPF Lausanne*, ETH Zurich, Basel, UNA Cyprus, Palacky, DAAD Germany, German Synchrotron Facility, FH Aachen, HU Berlin, Karlsruhe, Students Bonn, TU Dresden, Augsburg, Erlangen, Hamburg, Karlsruhe, Leipzig*, Mainz, Marburg, Ulm*, Wuerzburg, Danish Tech*, RUC, IHK Denmark, Barcelona, UC3M, Madrid 2*, UPC, USC*, Vigo, Valencia, HTV Finland*, LUT, UTA Finland, Rouen, CEA, ENS System France, ENST Brittany, French national particle lab., INRIA, INT Evry, Jussieu Lisif, Grenoble 3, Strasbourg, Marseilles, Poitiers*, HIVE Iceland*, NGI Italy, Turin Tech, ICTP Trieste, Unimib Italy, Sardinia, RSU Latvia, Eindhoven, Twente, Utrecht*, UIB Norway, Czest Poland, Lisbon, Porto, Lisbon Tech., Atom Romania, Kht Russia, Chalmers Sweden, Agric Univ Slovakia, Anadolu

Turkey, Bilkent, Hacettepe, Kiev Ukraine, Odessa, Birmingham, Bristol. Cambridge*, Cardiff, Durham*, Glasgow, King's College London, Student Halls London, NEWI* (Owain Glyndw^r), Reading, Swansea, Warwick, Ceredigion, Belgrade.

Rest of World: Universities, Institutes and Similar.
APRA Australia, ANU, Griffith, Swinburne, Queensland, Western Australia, Gov. Brazil, UFU Brazil, Alberta, AEI*, British Columbia, McGill*, Toronto*, Western Ontario, York Univ Toronto, TIY Chile*, Tsinghua China, UNAL Colombia, City Univ. Hong Kong, Hokudai Japan, Kyushu, Osaka, Shizuoka, Univ Tokyo*, Yamagata, KAIST Korea, UNAM Mexico, USON Mexico, UITM Malaysia*, Auckland NZ, Waikato, Indian Statistical Institute, JNCASR India, JNU India, NCTU Taiwan, ULA Venezuela.

THIS COMPLETES TWO YEARS OF FEEDBACK

1- 15 May 2006

US Universities, Institutes and Similar.
American, Arizona, Berkeley*, Brandeis, BVB, Caltech, Carnegie Mellon*, Colorado State, Columbia, CWRU*, Dartmouth, Denison, Hartford, IVCC, Johns Hopkins Applied Physics, JSU, Kent State*, Kansas State, Linfield, Louisiana State, Maine, MIT*, MTU, North Dakota, Northwestern, Oregon State, Pittsburgh, Pitzer, Penn State, Rice, SIUE, Texas A and M, Akron, California System, UC Fresno, UC Irvine, UC Santa Barbara, Iowa, Urbana Champaign, Maryland, Michigan, UMR, Chapel Hill, UNCW, New Hampshire, UPR, Texas, Kentucky, Washington*, Xavier, Yale, Brookhaven, US House of Representatives, NASA (Larc), NIH, Oak Ridge, St. Lucie Co., US Gov USAP, USAAF Edwards, US Navy NMCI.

Europe and Asia Minor: Universities , Institutes and Similar
TU Graz, TU Vienna*, RMA Belgium, Liege, Free Univ Brussels, CERN, EPF Lausanne, Basel, German Synchrotron Facility, IU Bremen, Mannheim, Max Planck Halle, Bochum, Bonn students, Duisburg, Erlangen, Leipzig*, Magdeburg, Muenster, Wuppertal, Aarhus, Barcelona, Uni2 Spain, USC Spain, UV Spain, AC Amiens, ENST Brittany, IHES, Jussieu AOMC, OBSPM, Lille 1, Poitiers, DUTH Greece, NTUA Greece, UOC greece*, Trinity College Dublin, Univ. College Dublin, Weizmann, AUB Lebanon, ITS Italy, Bari Tech., ICTP Trieste*, Calabria, Milan*, Insubria, Perugia, Rome 3, Twente, Utrecht, NTNU Norway*, UC

Portugal, Porto, Tuiasi, Karelia Russia, LTH Sweden*, UU Sweden, Agric. Univ Slovakia, Marmara Turkey, Kiev Ukraine, Cambridge, Durham, Edinburgh*, Hereford, Heriot Watt, Oxford graduates, Keble College Oxford, Magdalen College Oxford, Physics Oxford, Swansea*, Warwick.

Rest of the World: Universities, Institutes and Similar.
    Gov Brazil, UNICAMP, Murdoch, Swinburne, Queensland, Gov Victoria, Gov. British Columbia, Gov. Canada, Toronto, BIC Pune India, IUCAA India, SBU Iran, Nittai Japan, Univ Tokyo*, KAIST Korea, SLT Sri Lanka, UAM Mexico, UNAM Mexico, Manukau NZ, Otago, ADNU Philipines, NTHU Taiwan, TNRC Taiwan.

16 - 30 May 2006

US Universities, Institutes and Similar.
    Berkeley*, CSU Pomona, Dartmouth, Harvard, Indiana, Lanecc, MIT*, Ohio State*, Princeton*, Penn State*, Purdue*, Stanford, UC Santa Barbara, U Mass*, Utah, Kentucky, Washington*, Wayne State, NASA Jet Propulsion Lab., US Army Polk, US Military MDA, US Navy MUGU, US Navy NMCI.

Europe and Asia Minor: Universities, Institutes and Similar.
    Belgian RMA, CERN, Palacky, TU Berlin, TU Munich, Bonn*, Leipzig*, Magdeburg, WH Stuttgart, KU Denmark, AENA Spain, Uni2, UPM, UV, BNLL France, CEA France*, CH Soissons, Lille 1, Metz, Poitiers, Trinity College Dublin, Univ College Dublin, Jerusalem, Tel Aviv, Weizmann, ENEA Italy, Trieste, Bologna, Turin, KLM Netherlands, UPC Norway, Lisbon Tech, Agric. Univ. Slovakia, Rusian Academy of Sciences Novosibirsk, Aberystwyth, Durham, Edinburgh*, Swansea, Ynys Mo^n School, Belgrade.

Rest of World: Universities, Institutes and Similar.
    Australian National Univ., Latrobe, Alberta, Ontario, Shell Canada, Rio de Janeiro Brazil, Ocha Japan, KAIST Korea, UITM Malaysia, Massey NZ, ADNU Philipines.

June 2006

US Universities, Institutes and Similar.
    Berkeley, BGSU, Caltech, DUQ, Georgia Tech., Hawaii, IMSA, Johns Hopkins, Johns Hopkins Applied Physics, Kansas*, MIT, PDX, Princeton, Rutgers, Stanford, Texas A and M, Akron, Chicago*, UC Santa Cruz*,

Florida, Urbana Champaign, U Mass, Michigan, Chapel Hill, U Penn*, Kentucky, Vermont, Washington*, Wisconsin, Washington St Louis, US Gov FAA, NOAA, St Lucie County, USAAF AFSPC, US Military OSD, US Army Pentagon, Browning, IBM Almaden*, IBM Austin, IBM UK, IBM Thomas J, Watson, Intel, IT and T, Lockheed Martin, Microsoft, Northrop Grummond, Philips, Raytheon.

Europe and Asia Minor: Universities, Institutes and Similar.

KG Graz, Vienna, KU Leuven, RMA Belgium, Free Univ, Brussels, OMA Belgium, Bulgarian Academy of Sciences, ETH Zurich, Charles Univ. Prague, German Synchrotron Facility (DESY), Max Planck Garching, Max Planck Heidelberg, TU Darmstadt, Erlangen, Freiburg, Leipzig, Munich, Muenster, Aarhus Denmark, AUB Denmark, UAM Spain, Uni2, URV, Helsinki, TUT Finland, CEA France, EMN, NOOS, Polytechnique system, Poitiers, UTT Greece, Technion Israel, INFN Bologna, Turin Tech., Trieste, Calabria, Rome 2, Salerno, Uniud Italy, ITI Latvia, KLI Latvia, Anna Maria Netherlands, IPACT, NMI, ONS Net students, Groningen, Eindhoven, Twente, Utrecht, Free Univ Amsterdam, UIT Norway, UJ Poland, ISTS Poland, Wroclaw, ONI Portugal, UC Portugal, Porto, Lisbon Tech., Ministry fo Industry Romania, Lipetsk Russia, LTH Sweden, Agric Univ. Slovakia, Metu Turkey, Omu Turkey, Odessa Ukraine, VN Ukraine, Aberystwyth, Birmingham, Dundee, EPSRC, King's College London, London School of Economics, Magdalen College Oxford, Queen's College Oxford, Portsmouth, Sheffield, Southampton, York.

Rest of World : Universities, Institutes and Similar.

Southern Queensland, Australian Gov. BOM, New South Wales, UFPB Brazil, UNICAMP, Alberta Research Council, Simon Frazer, Toronto, Victoria, Windsor, IMSC India, IMS Japan, Keio Univ., Kobe, Kyoto, KAIST Korea, UAM Mexico, UITM Malaysia, Polytechnic Nepal, Massey NZ, UNAC Peru, ISU Saudi Arabia, NTHU Taiwan, Fisica Uruguay, MOET Venezuela.

1 - 26 July 2006

US and European Major Corporations

Adobe, Amazon, Anderson Global, Boeing, Botevgrad, Dow Corning, Fidelity, Fujitsu Siemens, General Electric Power, Honeywell, IBM Almaden, Intel, Layered Technologies*, Lockheed Martin, Microsoft, Mobile Gas, Motorola, MSN, New York Life, Northrop Grummond, Raytheon, Sun, TCH Machines*, Xerox, Intergenia, Siemens, Renault, Samsung.

US Universities, Institutes and Similar.

Arizona State, Berkeley, Clemson, CSUS, CWRU, Dartmouth, Drexel, Indiana, Johns Hopkins Medical Institute, Johns Hopkins Applied Physics, Kansas State, Kansas, Lousville, Mayo Clinic, MIT, North Carolina State, Northwestern, New York Univ., Ohio State, Pittsburgh, Princeton, Purdue, Rice, Rochester, Rutgers, South Dakota State, Stanford, Syracuse, Texas A and M, UAH, Chicago, UCLA, UC Santa Cruz, UC San Diego, U Mass, Maryland, Michigan, Chapel Hill, North Carolina Charlotte, UNCG, U Penn., Texas, UVM Net, Wyoming, Washington, WFU, WIU, Brookhaven, Gov California, US Dept of Health, US Federal Aviation Authority, Fermilab, NASA (Rcs), NASA (grc), NIH, NSF, US Gov PNL, US Gov WA, USAAF Patrick, USAAF Vandenburg, USAAF Wright Patterson, US Army Heidelberg, US Navy NMC, US Navy NPS, US Navy NUMC, US Military OSD.

Europe and Asia Minor: Universities, Institutes and Similar.

KF Graz, Vienna, Australian Univ. of Applied Sciences, Belgian RMA, CERN, Basel, Geneva, Neufchatel, Charles Univ. Prague, German Synchrotron Facility (DESY), FHT Stuttgart, IU Bremen, Max Planck Dortmund, Nord Akademie, RWTH Aachen, TU Ilmenau, Duesseldorf, Essen, Heidelberg, Jena, Kassel, Koeln, Leipzig, Oldenburg, Regensburg, Saarland, CVT Denmark, Elion Estonia, AENA Spain, ICIA Spain, Madrid Autonomous, Madrid 2, UPC, Salamanca, HTV Finland, UTU Finland, CEA France, NOOS France, SHOM France, Lille 1, Poitiers, DUTH Greece, Patras, Trinity College Dublin, Technion, Netheimur Iceland, ENEA Italy, SISSA Italy, Trieste, Bologna, Catania, Genoa, Milan, Turin, Parma, Utrecht, Free Univ. Amsterdam, FFI Norway, PT Lithuania, Szczecin Poland, Warsaw, Lisbon, ASTRAL Romania, HITV Russia, Chalmers Sweden, KTH, LTH, EUBA Slovakia, Siberian Academy of Sciences Novosibirsk, Anadolu Turkey, Gyte, Kiev Ukraine, Clare College Cambridge, Trinity College Cambridge, Particle Cambridge, Cardiff, EPSRC, Glasgow, Lancaster, Leeds Metropolitan, Manchester, Maths Oxford, Physics Oxford, STX Oxford, Temp Oxford, Surrey, Warwick.

Rest of World: Universities, Institutes and Similar.

CSIRO Australia, Adelaide, Eastern Queensland, Queensland, Australian Gov. Defence, Gov Western Australia, Rio de Janeiro Brazil, UNICAMP, Canadian Gov. Defence, Gov. Canada*, Simon Frazer, Alberta, Toronto, York Univ. Toronto, UCT Chile, UCV Chile, IMSC India, TIFR India, IMS Japan, Kobe, Univ. Tokyo, GIST Korea, ENS

Malaysia, UNAM Mexico, Massey NZ, UNAC Peru, KMITL Thailand, NCTU Taiwan, TNRC Taiwan, MOET Venezuela, FPT Venezuela.

1-27 August 2006

US Universities, Institutes and Similar.

Auburn, Berkeley*, Duke, Florida Atlantic, Indian River CC, Johns Hopkins Medical Institute, Johns Hopkins SPH, Johns Hopkins Applied Physics, Miami, MU Ohio, New York Univ., Mississippi, Princeton, Penn State, Purdue, Rice, Stanford, SUNY Stonybrook, Syracuse, Texas A and M, UAH, UALR, UC Irvine, UC Ok, UC Reno, U Mass, Maryland, Michigan, UMSL, North Carolina Charlotte, UNCG, Nevada Reno, UNT, South Carolina, Texas, Kentucky, Washington, Wisconsin, WKU, Argonne, US Gov BLM, Brookhaven, Lawrence Livermore, NASA (Arc), NASA (Larc), NIH, US Courts, US PTO, USAAF AFMC, US Army Heidelberg, US Army Monmouth, US Military DISA, US Navy NUWC.

Europe and Asia Minor: Universities, Institutes and Similar.

Belgian Royal Military Academy, Ghent, UPC Belgium, Zurich, EPF Lausanne, Geneva, Charles Univ. Prague, Palacky, German Synchrotron Facility (DESY), Max PLAnck Stuttgart, Max Planck Mainz, Max Planck MPQ, TUBS Germany, TU Chemnitz, Bonn, Dortmund, Duisburg, Goettingen, Heidelberg, Jena, Kiel, Leipzig, Munich, Oldenburg, Saarland, Stuttgart, SDU Denamrk, AENA Spain, Madrid 2, Zaragoza, UPCO Spain, CNRS Marseilles France, French national particle lab., INPL Nancy, IRIT France, OBSPM, SHOM, Strasbourg, AUTH Greece, UTH Greece, Iskon Hungary, Dublin Institute for Advanced Studies, Weizmann, INFN LNF Italy, NGI Italy, Milan Tech., PT Lithuania, Groningen, UPC Netherlands, Utrecht, UIO Norway, AON Poland, Krakow, Warsaw, Lisbon Tech., Joint Institutes for Nuclear Research Russia, Utrecht, Joseph Stefan Institute Slovenia, Agirc Univ. Slovakia, Cankaya Univ. Turkey, Metu Turkey, Kiev Ukraine, Odessa, BBSCR, Cambridge, Dundee, Durham, EPSRC, Imperial, Manchester, maths Oxford, phys Oxford, RCUK, Sheffield, York, British Gov. NICS.

Rest of World: Universities, Institutes and Similar.

UNT Argentina, MQ Australia, New South Wales, Queensland, Sydney, Australian Gov. Defence, Canadian Gov. Defence, Gov. Canada, Canadian NCF, US Navy Military, US Navy NUWC, Montreal Tech., Toronto, Waterloo, IIIT India, IMSC India, Hue Univ Japan IMS Japan, Kyoto, Titech, Tokyo, Postech Korea, Auckland, Massey NZ, UNESCO, Univ Philipines, NU Singapore, PSU Thailand, Academica Sinica Taiwan, NTNU Taiwan, UFG, UFPB, UNICAMP Brazil.

1 - 30 August 2006: Major US and European Corporations.

Amazon, Ask, ATK, Bank of America, Boeing, Botevgrad, Cisco, Credit Suisse, General Electric, General Electric Power, IBM Almaden, IBM UK, IBM US system, Intel, Layered Technologies*, Microsoft, Mobile Gas, Motorola, Philips, Picsearch, Raytheon, Shell US, Sun, Xerox, Kenko, Maxell, Nikon.

1-14 September 2006

US Universities, Institutes and Similar

BARA Arizona, Berkeley, Buffalo, Caltech, Carnegie Mellon, Colorado, Columbia, Duke, EHC, Florida Atlantic, Harvard, Iowa State, Kent State, Colorado School of Mines, MIT, Michigan State, Notre Dame, NEBO, New York Institute of Technology, Ohio State, Oklahoma, Princeton, Penn State*, Purdue, Radford, Rice, Rush, Rutgers, SWRI, Texas A and M, TMC, Texas Tech, Tufts, UAH, Chicago, South Carolina, Urbana Champaign, Kentucky, U Mass, North Carolina Charlotte, UNCOR, UNI, Texas Arlington, Texas Dallas, Kentucky, Vermont, Washington, WFU, US Gov. FAA, Los Alamos, US Gov WASA, Oak Ridge, St Lucie Co., USAAF ANG.

Europe and Asia Minor: Universities, Institutes and Similar.

ASN Graz, KG Graz, KLU Austria, FED Belgium, CERN, EPF Lausanne, ETH Zurich, Charles Univ. Prague, Geman Synchrotron Facility, HU Berlin, Max Planck MPE, Max Planck MPPMU, TU Berlin, TU Munich, Heidelberg, Leipzig, Muenster, Saarland, Schwabia, Stuttgart, BNAA Denmark, CSIC Spain, Rioja, UPC, UPV, HTV Finland, UTU Finland, INRIA France, Jussieu IPGP France, Jussieu Maths, Unilim France, Brest, MLV France, TEE Greece, Jerusalem, CNR IFAC Italy, INFN Padua, Rome 1, Milan Tech., Udine, Florence, Unimib Italy, Naples, Padua, Pavia, TU Delft, UIT Norway, AGH Poland, Krakow, Szczecin, Warsaw, UA Portugal, Porto, NSU Russia, Siberian Academy of Sciences Novosibirsk, HHO Turkey, KSU Turkey, Kiev Ukraine, Aberystwyth, Birmingham, Bristol, Queen's College Cambridge, Heriot Watt, Maths Oxford, OULS Oxford, Wolfson College Oxford, Swansea College, Central England, Univ. College London, Wolverhampton, BBC, Ceredigion, Belgrade, NS Yugoslavia.

Rest of World: Universities, Institutes and Similar.

East Queensland, Melbourne, UOW, Australian Gov. Defence, Gov. Brazil CNEN, Gov Brazil IRD, UNICAMP, Alberta, Moncton, Gov. Canada, UNAL Colombia, U Cuba, IUCAA India, KAIST Korea, Gov.

Mexico, Univsity system Malaysia, BUU Thailand, NCTU Taiwan, Academica Sinica Taiwan, VUT South Africa.

Major US Corporations
   Amazon, Amersham, Apple, Boeing, Browning, Ford, General Electric, Honeywell, IBM, Intel, Layered Technologies*, Lockheed Martin, Microsoft, Motorola, Raytheon.

15 - 29 September 2006

US Universities, Institutes and Similar.
   AFIT, Alaska, Berkeley, Buffalo, COD, Cooper, Cornell, Florida Atlantic, Georgia Tech, Harvard, Johns Hopkins Applied Physics, Kansas State, Louisiana Tech, MCW, Mercyhurst, MIT, MNSU, Montgomery College, North Carolina State, NEU, NIU, North Dakota, Northwestern, New York Univ., Oregon State, Pratt, Princeton, Penn State*, Purdue, Rutgers, San Jose State, Stonybrook, Texas A and M*, Tufts, UC Davis, Central Florida, Chicago*, UC Irvine, UCLA*, UC Santa Barbara, Florida, UMBC, UMCES, Michigan*, UNL, South Carolina, Utah, UWM, Vermont*, Washington*, WIU, Xavier, US Gov BOP, US Dept. of Energy, US FAA, Los Alamos, NASA, NIST, NOAA, NREL, Oak Ridge, US Gov PPPL, USAAF Elmsdorf, US Navy supship, Jefferson National Laboratory.

Europe and Asia Minor: Universities, Institutes and Similar.
   ASN Graz, UIBK Austria, TU Vienna, Belgian RMA, BOL Bulgaria, CERN, EB Zurich, HU Berlin, Freiburg*, Heidelberg*, Leipzig, Munich*, Saarland, Stuttgart, Tuebingen, Niels Bohr Institute, UT Estonia, UAM Spain, Zaragoza, Valencia*, SPT Finland, Jussieu, Synchrotron Soleil, Poitiers, Tours, Salonica Greece, NTUA Greece, Tellas, Abacus Hungary, Konkar Croatia, ELTE Hungary, Szeged, Univ College Dublin, Jerusalem, Technion, Aruba Italy, INFN Rome 1, INFN Sardinia, Bari, Florence, PT Lithuania, HVA Netherlands, Protagonist Netherlands, TNO, TU Delft, Eindhoven, Twente, MAC Poland, UJ Poland, Szczecin, UMA Portugal, Gazvyaz Russia, Joint Institutes for Nuclear Research Russia, Lipetsk, NSC Russia, Chalmers Sweden, KTH, UMU Sweden, UU, KCLJ Slovenia, Gymzh Slovakia, EPSRC, Huddersfield, Lancaster, Leverhulme Trust, Manchester, Oxford, Swansea, Tower, York SJ, Belgrade.

Rest of World: Universities, Institutes and Similar.
   Gov. Australia, Gov. Brazil*, UNICAMP, Alberta Research Council, Math Canada, McMaster, YRBE Ontario, Alberta, Toronto, UTA Chile, EAFIT Colombia, UO Cuba, IIIT India, Gov. India, RRI India, Chuo

Univ. Japan, Hokudai, Kyoto, Osaka, Univ. Tokyo, Chiba-C, NAIST Japan, KAIST Korea, PDN Sri Lanka, Gov. Mexico, IPN Mexico, Auckland* NZ, KMITL Thailand, KU Thailand, NUST, NTU Taiwan, UNET, HSPH Venezuela, VUT South Africa.

US, Europe, Japan: Major Corporations

Amazon, Browning, General Electric, Goodyear, IBM Almaden, IBM UK, IBM Germany, Intel, Layered Technologies*, Lucent, Microsoft, Motorola, Northrop Grummond, Siemens, Sony, Sun.

1 - 14 October 2006

US Universities, Institutes and Similar

Arizona, Berkeley, Buffalo, Cal Polytech, Cal Tech, CCSU, Colorado*, Drake, Duke*, Harvard, Hawaii, Princeton Institute for Advanced Studies, IC, Indiana, LETU, LTU, LUC, Maine, Miami, MIT*, NCAT, North Dakota*, New York Univ*, Ohio State, Penn State*, Purdue*, SJSU, Stanford, Stonybrook, Texas A and M*, TCU, California system, Chicago, UCLA, UC Santa Barbara*, UC Santa Cruz, UC San Diego, Florida, Iowa, Urbana Champaign*, Michigan, Minnesota* North Carolina Charlotte, U Penn., Utah, Virginia, Washington, Wisconsin*, Yale, Lawrence Berkeley, NASA (Jpl), St Lucie County, US Army ARO, US Military DLA Headquarters, US Naval Research Laboratory, US Military USCG.

Major US Corporations.

Adobe, Amazon, Boeing, IBM Almaden, Layered Technologies*, Lockheed Martin, Midland Steel, Motorola, Northrup Grummond, Philips, Raytheon.

Europe and Asia Minor: Universities, Institutes and Similar.

Belgian Royal Military Academy, Ghent, Intech Bulgaria, EB Zurich, Charels Univ. Prague*, Palacky, FU Berlin, Max Planck MPPMU, TU Darmstadt, Bielefeld, Frankfurt, Hamburg, Heidelberg*, Karlsruhe, Koeln, Leipzig, Munich, Saarland, Tuebingen*, KU Denmark, UDG Spain, Madrid 2, UV, JYU Finland, TUT Finland, Jussieu LISIF, Louis Pasteur, Polytechnique system, SHOM, Unilim France, Lemans, Poitiers, Tours, Tellas Greece, BMF Hungary, Haifa, Aruba italy, Rome 2, INFN Turin, INFN Trieste, Bologna, Pisa, Rome 1, Turin, KUN Netherlands, Groningen, Eindhoven, Twente, Amsterdam, NTNU Norway, UJ Poland, Poznan, Szczecin, Warsaw, Wroclaw, UMA Portugal, Lisbon Tech., Iasi Romania, SUSU Russia, Joint Institutes for Nuclear Research Russia, NSC Russia, ACS Slovakia, Agric. Univ., Slovakia, Bilkent, Gantep Turkey, Kkarkov Ukraine, Kiev, Odessa, Bucks. County Council, Bristol, Brunel,

City, Durham, Edinburgh, Imperial, NCL, Oxford, Sheffield, Southampton, Sunderland, Sussex, Univ. College London, Warwick, York, Belgrade.

Rest of World: Universities, Institutes and Similar.
Murdoch, QUT, Tasmania, UNICAMP Brazil, Dalhousie Canada, Manitoba, McMaster, Montreal, Saskatchewan, Toronto*, York Univ. Toronto, UNAL Colombia, IED Hong Kong, Indonesia, Hiroshima, Kyoto, Osaka, Tokyo, Chonnam, Chungbuk Korea, RUH Sri Lanka, KAIST Korea, Gov. Mexico, IPN Mexico, Massey NZ, MSUIIT Philipines, BUU Thailand, KU Thailand, CHU, NUST, NCTU*, NTNU Taiwan, MOET Venezuela.

15 - 30 October 2006

US Universities, Institutes and Similar.
Arizona State, Auburn, Boston, BYUH, Caltech, Central Michigan, Carnegie Mellon*, COFC, Columbia, City Univ New York, Duke*, GCU, Georgetown, Hutch CC, Kansas, Lehigh, Maine*, Mississippi State, Northwestern*, Oregon State, Penn State*, PUPR, Purdue*, Rowan, South Carolina, SP College, SPSU, Stanford, STSCI, SUNY Stonybrook, Texas Tech, Tufts, UAH, UC Davis, Central Florida, Chicago, UCLA*, UC San Diego, Urbana Champaign, U Mass Dartmouth, Michigan*, U Penn, USU, Utah*, Texas, Virginia*, Washington*, Wisconsin*, Wright, Xavier, Yale*, YSU, Gov. California, NASA (JPL), NIH, US Navy NCSC, US Navy NUWC, US Navy Spawar.

Europe and Asia Minor: Universities, Institutes and Similar.
Vienna, KU Leuven, RMA Belgium CERN, Basel, Freiburg, Geneva, Charles Univ. Prague, MUNI Czechia, Palacky, FU Berlin, HU Berlin, Albert Einstein Institute, Max Planck SB, RWTH Aachen, TUBS, Augsburg, Bielefeld, Bonn, Duisburg, Essen, Kaiserslautern, Leipzig, Marburg, Munich, Rostock, Saaralnd, Schwabia, Siegen, Wuerzburg, Aarhus, CVT Denmark, Danish Tech*, UB Spain, UDL, UV, UDL, UGR, HTV Finland, NOOS Frace, RAIN France, Strasbourg, Poitiers, Patras, Budapest University Campi, Szeged, Trinity College Dublin, Milan, Pavia, Rome 1, Turin, RU Netherlands, FFI Norway, UIB Norway, Lublin, ONI Portugal, IS Romania, FICP Russia, Aeroflot, Joint Institutes for Nuclear Research Russia, NSC Russia*, Gov Tatarstan, Physto Sweden, Gov. Turkey, VSTU Ukraine, Kharkov, bath, Cambridge, DL, Durham, Edinburgh, Imperial, Lancaster, Hertford College Oxford, Queen's Univ Belfast, Susses*, Ceredigion, British Ministry of Defence, National Health Service, Belgrade.

Rest of World: Universities, Institutes and Similar.

Gov. Argentina, Adelaide, Victoria, AEI Canada, Carleton, McGill, Alberta, Toronto, Waterloo, York Univ Toronto. PUC Chile, Biobio Chile, Gov Colombia, UO Cuba, KGV Kong Kong, UI Indonesia, NITK India, IACS India, Hue Univ. Japan, Osaka, Saga, Tohwa, Univ Tokyo, KAIST Korea, ENSMA Malaysia, Gov. Mexico, UNI Peru, NUS Singapore, CHU, DWU. KH, NTNU, TPC, YLC Taiwan, FHUCE Uruguay, ULA Venezuela, UFC, UFAL, UFG, UFMG, UNICAMP Brazil.

US and Europe: Major Corporations

Amazon, Bank of America, Daimler Chrysler, Honeywell, IBM Almaden, IBM Australia, IBM Britain, Jensen, Layered Technologies, Lockheed Martin, Microsoft, Motorola, Philips, Siemens (various countries), Raytheon.

1- 14 November 2006

US Universities, Institutes and Similar.

Arizona, Austin CC, Brown, Carnegie Mellon, Columbia, Cooper, Duke, Florida State*, Harvard, Indiana, JMU, Louisiana State, Colorado School of Mines, MIT, North Dakota, Northwestern, Oklahoma State, Princeton*, Penn State, South Carolina, South Florida State, Stanford, Univ California system, Central Florida, UC Irvine, UCLA, UC Reno, UC Santa Barbara, Urbana Champaign, U Mass., UMBC, Michigan*, Minnesota*, UNO, UPF, UPRA, South Carolina, Utah, UU, Virginia*, Williams, Yale, NASA Ames, California Dept. of Transport, Fermilab, Lawrence Berkeley, Oak Ridge, Virginia, USAAF Robins, US Army Ace, US Navy NMCI, US Navy NUWC, US Navy Spawar.

Europe and Asia Minor: Universities, Institutes and Similar.

Belgian RMA, Ghent, Belgian Dept of Energy, Charles Univ Prague, Palacky, KFA Juelich, Max Planck Garching, Max Planck MPIM, Augsburg, Bonn, Kaiserslautern, Leipzig, Mainz, Munich, Saarland, Aarhus Denmark, UC3M Spain, Valencia*, Helsinki, Turku Finland, CIUP France, ENS France, Polytechnique system, Bordeaux 1, Paris Sud, Brest, F Comte, Poitiers, Rennes 1, Budapest Campi Hungary, Trinity College Dublin*, Technion Israel, INFN Milan, Trieste, Pisa, Eindhoven, Utrecht, NTNU Norway, ICM Poland, PW Poland, UMCS Lublin, Lisbon Tech, PSTU Russia, Agric Univ. Slovakia, Siberian Academy of Sciences Novosibirsk, Metu Turkey, CAI Cambridge, physics Cambridge, particle Cambridge, particle Edinburgh, Essex, Hull, Liverpool, Liverpool JM, Nottingham, phys Oxford, Portsmouth, Queen Mary College, Queen's

Univ. Belfast, Sheffield, Swansea, Central England, National Health Service.

Rest of World: Universities, Institutes and Similar.

Latrobe, Gov. Australia ACT, UFG, UFMG, UFPB, UFRGS, UNICAMP Brazil, AEI Canada, TCC Ontario, Montreal, Quebec, Toronto, Victoria, ITP China, UP Cuba, IUCAA India, Tohoku Univ. Japan, Tokyo, KAIST Korea, Gov. Mexico, UDG Mexico, UITM Malaysia, TOPNZ New Zealand, UNAC Peru, HLC Taiwan, NTU Taiwan, UCT South Africa.

US and Europe: Small Selection of Major Corporations

Amazon, AOL, AT and T, Boeing, Botevgrad, Dell, Hewlett Packard, Kodak, Microsoft, Motorola, Northrop Grummond, Sun, Siemens.

15 - 29 November 2006

US: Universities, Institutes and Similar.

Arizona State*, Berkeley, Brandeis, Caltech, Cornell*, Dartmouth, Harvard, Hawaii, Indiana, LUC, Maine, Colorado School of Mines, North Carolina State, Northwestern, new York Institute of Technology, Princeton, Penn State, Purdue, Rochester, Stanford, St Edward's, Texas A and M, Central Florida, UC Santa Barbara, U Mass., Michigan*, Chapel Hill, UNLV, Nebraska Omaha, Virginia, Xavier, Yale, US Dept. of Energy, Los Alamos, NASA (JPL), Oak Ridge, US Treasury, US Military AFOSR, US Army Research Laboratory, US Army Stewart.

Europe and Asia Minor: Universities, Institutes and Similar.

RMA Belgium, Free Univ Brussels, Ghent, CERN, Palacky, EMBL Heidelberg, FH Brandenburg, TU Dresden, Heidelberg, Karlsruhe, Leipzig*, Munich, Saarland, CVT Denmark, EHU Spain, Helsinki, UTU Finland, French national particle lab., F Comte, Poitiers, UPS Toulouse, TEE Greece, Patras, Zsem Hungary, Trinity College Dublin, Univ. College Cork, BGU Israel, Pisa*, Turin, CNRC Netherlands, Groningen, Free Univ Amsterdam, UIB Norway, UWM Poland, Gov. Poland Ministry of Defence, Lublin*, Lisbon Tech., PSTU Russia, URC Russia, Siberian Academy of Sciences Novosibirsk, Chalmers Sweden*, IRFU Sweden, Agric. Univ. Slovakia, NSK Russia, Omu Univ Turkey, Kiev* Ukraine, particle Cambridge, Chester, Durham, physics Oxford, Queen's Univ. Belfast, Southampton, St Andrew's, British Gov GSI, Hillingdon Council, Belgrade.

Rest of World: Universities, Institutes and Similar.

CSIRO Australia, UFG, Unesp, UNICAMP Brazil, Manitoba, Montreal, New Brunswick, Victoria, IGIDR India, JNCASR India, BARC India, IMS Japan, Osaka, Saitama, Tsukuba, Univ. Tokyo, Hokkaido, Tokyo, KAIST Korea, UOS Korea, TOPNZ New Zealand, PSU Thailand, KH, MLC, NCTU, NTHU, NTTU, TP, TPC, YCRC, YLC Taiwan, Univ Zululand.

US and Europe: Small Selection of Major Corporations.

Amazon, Amgen, British Aerospace Engineering Systems, Boeing, Botevgrad, Chevron, Texaco, Honeywell, IBM Britain, Intel*, Layered Technologies*, Microsoft, Mobile Gas, Motorola, Turner, Siemens.

1- 15 December 2006

US Universities, Institutes and Similar.

Boston College, BJU, Brigham Young, Caltech, Albany, CHOP, Carnegie Mellon, Collins College, Columbia, Depaul, Florida International, IUP, Lamar, Maine, Maricopa, MIT, Michigan State, New York Univ., ODU, Mississippi, Oklahoma, Pepperdine, Pittsburgh, Penn State, South Dakota State, Texas A and M, UCAR, Central Florida, UC Santa Barbara, Urbana Champaign, Maryland, Michigan, UMR, Chapel Hill, USU, Kentucky, UTM, Vanderbilt, Vermont, Wisconsin, Washinton St. Louis, Brookhaven, US FAA, NASA (Jpl), USAAF PLH, USAAF Robins, USAAF Tinker, Carolinas Health Care.

Europe and Asia Minor: Universities, Institutes and Similar.

STH Pfeilheim Austria, KU Leuven Belgium, RMA Belgium, BAS Bulgaria, EPF Lausanne, Charles Univ. Prague, HU Berlin, RWTH Aachen, TU Berlin, TU Munich, Augsburg, Bonn, Freiburg, Cologne, Leipzig, Mainz, Munich, Paderborn, Saarland, Ulm, Leon Univ Spain, Ovi, URJC, HUT Finland, ENS Lyon France, Jussieu IHP, Marseilles, Poitiers*, ADSL Hungary, NUIM Ireland, Univ College Cork, Aruba Italy, INFN Milan, TurinTech., TIN Italy, Genoa, Messina, Milan, Pavia, Rome 1, Rome 3, Turin, Trento, School werkplek Netherlands, TNO, Utrecht, UIB Norway, UIO Norway, UJ Poland, Warsaw, MPEI Russia, Krasnoyarsk, Sakhalin, Chalmes Sweden, KTH, Umea, Gymzh Slovakia, TT net Turkey, Aberystwyth, Birmingham, New Hall Cambridge, Newton Institute Cambridge, particle Cambridge, Cardiff, chem Cardiff, Liverpool, Manchester, Porstmouth, Reading, Southampton, Swansea, UKC, Warwick, National Health Service.

Rest of World : Universities, Institutes and Similar.

Swinburne Australia, Sydney, Australian Gov. Scita, UNICAMP, AEI Canada, McMaster, Ottawa, Toronto, TJ China, Norte Colombia, UO Cuba, BARC India, Tenet India, Hokudai, IMS, Titech, Univ Tokyo*, KAIST Korea, UAM Mexico, Massey NZ, Otago, NCTU, NSYSU, NTCU, TPC Taiwan.

US and Europe: Small Selection of Major Corporations

Adobe, Amazon, Dell, IBM Almaden, Layered Technologies*, Lockheed Martin, Microsoft, Motorola, Northrop Grummond, Raytheon, Siemens, Sun, Wells Fargo.

13 - 29 December 2006

US Universities, Institutes and Similar.

Brigham Young, Carnegie Mellon, CNM, Collins College, CSUN, Duke, FDU, Johns Hopkins Applied Physics, Lousiana, Maine, Missouri, North Dakota, Ohio State, Oklahoma, Penn State, Rutgers, Syracuse, Akron, California System, UC Davis, Urbana Champaign, UNK, USU, Wayne State, Western Washington, NASA (JPL), USAAF Patrick, US Navy NMCI, UNESCO.

Europe and Asia Minor: Universities, Institutes and Similar.

AKH Vienna, AIC Austria, Belgian RMA, Ghent, ETH Zurich, Freiburg, Charles Univ. Prague, Palacky, HU Berlin, FFA Juelich, Augsburg, Bonn, Dortmund, Cologne, Leipzig, Munich, GLD Denamrk, UGR Spain, Juan Carlos, Valencia, HTV Finland, Metz, Poitiers, Marseilles, Patras Greece, Unified Campi Hungary, Weizmann Israel, CNR IGI Italy, INFN Calabria, Genoa, Twente, Karkow Poland, Kielce, Porto Portugal, TIM Romania, Joint Institutes for Nuclear Research Russia, Sakholin, Chalmers Sweden, Umea, Bilkent and Gazi Turkey, Lvic Ukraine, Poltava, Aberystwyth, Birmingham, Cardiff, chem Cardiff, Cranfield, Imperial, physics Imperial, Liverpool, Portsmouth, St Andrews, Swansea, UKC, Belgrade.

Rest of World: Universities, Institutes and Similar.

New South Wales, Sidney, UNICAMP Brazil, Sao Paolo, Alberta Research Council, Quebec, Toronto, City Univ. Hong Kong, Indian Statistical Institute, Kyoto, Univ Tokyo, KAIST Korea, UITM Malaysia, INP Mexico, Otago NZ, AIT and BUU Thailand, NCTU, NCU, NSYSU, TP2RC Taiwan.

US and Europe: Small Selection of Major Corporations.

Amazon, IBM Connecticut, IBM North Carolina, Intel, Layered Technologies*, Microsoft, Mobile Gas, Motorola, Siemens, Unilever.

END OF YEAR 2006

1 - 16 January 2007

US Universities, Institutes and Similar.

Austin College, Berkeley, Buffalo, Brigham Young, CNM, Drexel, EDP College, Harvard, Johns Hopkins Applied Physics, MIT, North Dakota, Northwestern, Princeton, Penn State*, Rice, UAH, UC Davis, UCLA, UC Santa Cruz, Kentucky, U Penn., Washington, Yale, Los Alamos, Jefferson Particle Laboratory, USAAF Wright Patterson, USAAF Monmouth, US Military MDA, US Military NGA.

Europe and Asia Minor: Universities, Institutes and Similar.

TU Vienna, HIB Liebenau, Vienna, Belgian RMA, Bulgarian Naval Academy, CERN, EPF Lausanne, Charles Univ. Prague, Palacky, KFA Juelich, TU Berlin, Heidelberg, Cologne, Leipzig, Copenhagen, Tartu Estonia, UPM Spain, US Spain, Helsinki, CEA France, Paris 13, Poitiers, ACN Greece, Szeged, Miskolc, Jerusalem, CNR ITB, INFN Pavia, Calabria, Rome 1, Palermo, Gov. Latvia, PRMG Lithuania, Leiden, TU Eindhoven, Twente, Utrecht, UIO Norway, ICM Poland, Krakow Univ., WSE Poland, KTH Stockholm, UMU Sweden, Joseph Stefan Institute Slovenia, Siberian Academy Novosibirsk, UTEL Ukraine, Aberystwyth, Cambridge, Cardiff, Durham, Exeter, Leeds, Sheffield, St Andrews, Swansea, Central England, Univ. Wales Institute Cardiff, BBC, Ceredigion, Swansea Educational Network, Welsh Office, National Health Service.

Rest of World: Universities, Institutes and Similar.

UBA Argentina, Western Power Australia, Flinders, Laurentian Canada, McMaster, TPL Toronto, British Columbia, Univ. Toronto, Waterloo, IITM India, Riken Japan, HanYang Korea, KAIST, Sejong, Sook Myung, Yonsei Korea, IPN Mexico, BUU Thailand, KU Thailand, YZU Taiwan, Unexpo Venezuela, USP, UNICAMP Brazil, Gov. Bhutan.

Small Selection of Major US and European Corporations.
   Amazon, General Electric, Layered Technologies, Mobile Gas, Motorola, Siemens, Fujitsu (US).

Most Read Papers: 72, 63, 75, ...
Most Read Articles: Lar Felker's book, Wikipedia rebuttal, ECE Article, MWE Scientific Career, Complete Works of MWE, Rotating Solenoid Exp., Space Energy, Munich Workshop slides.

Leading countries: USA, Russia, Italy, France, Belgium, Germany, Netherlands, ... 82 countries in two weeks.

15 - 30 January 2007

US Universities, Institutes and Similar.
   Auburn, Brigham Young, CCC, Drexel, Georgia Tech*, GCC, GCSU, HACC, Harvard, Iowa State, Johns Hopkins Applied Physics, Miami, North Dakota, Northwestern, Oklahoma State, Princeton, Penn State, Purdue, Rice, Syracuse, Texas A and M, South Florida, Texas*, Vanderbilt, Wednet, Wisconsin, Xavier, Yale, Fermilab, Los Alamos, Lawrence Livermore, US Gov WSF, New York City Gov., USAAF AFSPC, USAAF Wright Patterson, US Navy NMCI, US Military NGA.

Europe and Asia Minor: Universities, Institutes and Similar.
   HIB Liebenau, Charles Univ. Prague, Palacky, German Federal Government Bundestag, German Synchrotron Facility, FU Berlin, HU Berlin*, IU Bremen, Leitstern, MP Stuttgart, MP Dresden, TU Dresden, Augsburg, Bonn*, Cologne, Leipzig, Oldenburg, Stuttgart, Copenhagen, UT Estonia, Madrid Autonomous, Ovi, UPM*, Helsinki, CNRS LPN, Paris 13, Poitiers, TEE Greece, Weizmann, CNR IGI, CNR IM, INFN Bologna, Turin Tech., Calabria, Naples, Pisa, KTU Lithuania, Leiden, RU, Delft, Twente, Utrecht, UIO Norway, ICM Poland, PK Poland, WSE Poland, Krakow, Poznan, Warsaw, Porto Portugal*, RDS net Romania, NSC Russia, Gotland Sweden, Edus Slovenia, Agric Univ Slovakia, Ankora Univ. Turkey, Damtp Cambridge, Trinity Cambridge, Cardiff, City Univ. London, Durham, Imperial, Keele, Lancaster, Leicester, Leeds, Liverpool, Nottingham, Sheffield, Southampton*, Swansea, Swansea College, Univ Wales Institute Cardiff, BBC, Welsh Assembly, Gov Clackmanan Scotland, National Health Service, Parliament, Jersey Schools.

Rest of World: Universities, Institutes and Similar.

Australian National, Griffith, MQ, Melbourne, New South Wales, Gov. Rio Branco Brazil, UFPB, UNICAMP, Sao Paolo, National Research Council of Canada, NS Health Canada, Simon Fraser*, Montreal, Western Ontario, IITM India*, IUCAA India, PRL India, Univ. Tokyo, Waseda, Hanyang Korea, KAIST Korea, Sejong, SNU, UM Malaysia, Gov. Malaysia, Otago NZ.

Leading Interest for January 2007: Siemens Company, Welsh Office, Poitiers, Swansea, Univ Montreal, Brigham Young, Exeter Univ., Simon Fraser, Keele Univ.

1 - 14 February 2007

US Universities, Institutes and Similar.

Auburn, Buffalo, Brigham Young, Duke, Harvard, JSU, MTU, North Carolina State, Northwestern, New York Univ., Ohio State*, Philau, Penn State, Purdue, Renselaer, Stanford, UALR, Chicago, UCSC, Florida, Urbana Champaign, Minnesota*, North Texas,Utah, Texas Dallas, Wyoming, Wayne, Western Florida, Wisconsin, Xavier, NASA (JPL), USAAF Patrick, US Army Belvoir.

Europe and Asia Minor; Universities, Institutes and Similar.

Belgian RMA, Liege, Leuven, EPF Lausanne, Charles Univ. Prague, IU Bremen, Bochum, Augsburg, Bonn, Kassel, Kaiserslautern, Koeln, Konstanz, Leipzig, Munich, Wuppertal, CVT Denmark, Valencia, Spanish Parliament, CNES France, Nancy Metz, ENS Lyon, Strasbourg, Lemans, Lille 1, Montpellier 2, Poitiers, Patras, Technion, Weizmann, European Space Agency HQ, Milan Tech, SISSA, Calabria, Pomona, Rome 2, Eindhoven, Twente, Utrecht, UIO Norway, MIMUW Poland, Koszalin, UC Portugal, Joint Institutes for Nuclear Research Russia, BTH Sweden, KTH Sweden, Agric Univ. Slovakia, Dogus Univ. Turkey, Odessa, Bath, Sidney Sussex Cambridge, Cardiff, Durham, Edinburgh, Glamorgan, Imperial, Liverpool, Jesus College Oxford, LMH Oxford, Queen Mary London, Sheffield, Swansea, Univ. College London, Ceredigion, Welsh Office, Parliament.

Rest of World: Universities, Institutes and Similar.

INPE, UFF, UNESP, UNICAMP, Sao Paolo Brazil, McGill Canada, Ontario, British Columbia, Saskatchewan, Toronto, JMI India, PRI India, Tokyo Tech., Tokyo, YITJC, KRSU Korea, KAIST Korea, Postech, Sejong, UAM Mexico, UET Taxila Pakistan, Andes Colombia, Chula Thailand.

Leading Papers: 63, 76, 50, 15, 72.

Leading Articles: Lar Felker's book (virtual best seller, well over a million hits).

Leading Countries: USA, Germany, Britain, France, Canda, Belgium, ... 70 countries.

15 - 27 February 2007

US Universities, Institutes and Similar.

Arizona State, Berkeley*, Brown, Buffalo*, Brigham Young, Caltech, Colorado, CU, Denver, Denver, Drake, Florida International, HACC, Harvard, Maine, Missouri, MIT, Michigan Tech., New Mexico State, Northwestern, Ohio State, OSU, Princeton, Penn State, Rice, Rooseveldt, SJSU, SMU, Stanford, Swarthmore, TCCD, Temple, TENET, TJU, TMC, Tufts, California system, UC Davis, Central Florida, UCLA, UC Santa Barbara, UC San Diego, Delaware, Iowa*, Urbana Champaign, U Mass., Maryland, Michigan, UMSL, Utah, Vermont, Washington*, Whitman, Wisconsin, Wright, NASA (Ames), Argonne, US Gov. BOL, Lawrence Berkeley, NIST, NOAA, USAAF AFOSR, USAAF Headquarters, US Navy NMCI, US Navy NUWC.

Europe and Asia Minor: Universities, Institutes and Similar.

JKU Austria, Free Univ. Brussels, CERN, Basel, Palacky, German Synchrotron System, KFA Juelich, Max Planck SB, Max Planck MPQ, TU Berlin, TU Dresden, Hannover, Cologne, Leipzig, Munich, Paderborn, Rostock, Stuttgart, Barcelona, UPV Spain, Jussieu INSP, SHOM, F. Comte, Marseilles, Poitiers, Univ. College Dublin, Milan Tech., Trieste, Milan, Padua, LUB Latvia, Twente, NTNU Norway, AMU Poland, IFJ Poland, Krakow, Univ Warsaw, Poznan, Wroclaw, Lisbon Tech., Omsk, IRFU Sweden, UU Sweden, Agric Univ. Slovakia, Bilkent Turkey, Boun, Metu, Selcuck Turkey, Edinburgh, C251 EPSRC, Heriot Watt, Imperial, Loreto, New College Oxford, SERS Oxford, Southampton, UKC, Warwick, BBC, Ceredigion, Cumbria.

Rest of World: Universities, Institutes and Similar.

New South Wales, Gov. Brazil, UNICAMP, Gov. British Columbia, Canada, McGill, McMaster, Memorial Newfoundland, Alberta, Toronto, IITM India, Nagoya Univ. Japan, Oska, Ritsumei, Yamagata, Kaist Korea, Academica Sinica Taiwan.

Leading Interest for Feb. 2007: Astra Zlin, Univ. Paderborn, KAIST, Poitiers, EPSRC, FGAN, QSC, Jian Gxi, Poitiers, Comtes, Siemens, SB Redevelopment.

1- 15 March 2007

US Universities, Institutes and Similar.

Brigham Young, Clark, Embry Riddle, Florida International, Georgia Tech., GRCC, Princeton IAS, Maine, MIT, NAPS, Northern Illinois, Ohio State, Mississippi, Princeton, Penn State, Rensealer, Rutgers, Stanford, Stonybrook, SUNY Stonybrook, Texas A and M, UCLA, Dallas, Houston, Urbana Champaign, Michigan, Minnesota, Chapel Hill, South Florida, Washington*, Wednet, Wisconsin, Western Washington, US Dept of Energy, Fermilab, Lawrence Livermore, NASA (dfrc), NASA (Jpl), NASA (Jsc), St Lucie Co., US Navy NAVO, US Navy NUWC.

Europe and Asia Minor: Universities, Institutes and Similar.

TU Vienna, UIBK Austria, Univ. Vienna, CERN, Belfort, Charles Univ. Prague, Palacky, German Synchrotron Facility. HU Berlin, TUBS, TU Darmstadt, Bonn, Kaiserslautern, Cologne, Leipzig, Regensburg, Stuttgart, Univ Granada Spain, UJI, UPV, UV, VTT Finland, AC Nantes Brittany, French national particle lab., INPG France, Lemans, Poitiers, Dublin City Univ., BGU Israel, Jerusalem, Weizmann, CNR Iriti, CNR ITB, INFN Bologna, INFN Florence, SISSA, Trieste, Bologna, KUN Netherlands, RU, Utrecht, Free Univ. Amsterdam, UVT, UW Poland, Warsaw, UC Portugal, KTH Sweden, LTH Sweden, Agric Univ. Slovakia, Odessa Ukraine, Brookes Institute, Queen's Cambridge, particle Cambridge, Coventry, Imperial, Manchester, NCL, Nottingham, Hertford College Oxford, SERS Oxford, Warwick.

Rest of World: Universities, Institutes and Similar.

ANU Australia, Michelangelo Brazil, UNICAMP, Memorial University Newfoundland, YRBE Canada, Simon Fraser, York Univ. Toronto, Quebec, Puhep India, Kobe Japan, Chiba, Osaka, Tokyo, UNAC Peru, Massey NZ, SIT NZ, Kaist Korea, NCTU, NTU, TNIT Taiwan, FPT Venezuela, PTT Yugoslavia.

16 - 30 March 2007

US Universities, Institutes and Similar.

Boston College, Berkeley, Boston Univ., Brigham Young, CSULB, Duke, Florida State, Johns Hopkins Applied Physics, NOAO, New York Univ., Pittsburgh, Penn State, Purdue, Rutgers, Texas A and M, Tarleton,

UAF UAH, U Mass., South Carolina, Texas, Wisconsin, Yale, US Dept. of Energy, Fermilab, Los Alamos, NASA (Gsfc), USAAF Lajes, US Military USMC.

Europe and Asia Minor: Universities, Institutes and Similar.
Plovdiv, CERN, Charles Univ. Prague, Plzen, Palacky, Leitstern, Max Planck MIS, TGZ Ilmenau, Cologne, Leipzig, Munich*, Oldenburg, Stuttgart*, CVT Denmark, KU Denmark, Andalucia Govt., UCM, USC Spain, JYU Finland, ENIB France, French particle lab., SHOM France, Strasbourg, Montpellier 2, Poitiers*, Athens, GSRT Greece, Dublin City Univ., BGU Israel, Jerusalem, CNR IMM Italy, ENI Italy, INFN Florence, INFN Naples, Bologna, TY Eindhoven, Utrecht, INIC Norway, Royal Society of Chemistry, Krakow, Donn TU Ukraine, Koszalin, Lisbon Tech., Tuiasi, MTS-NN Russia, KIPT Kharkov Ukraine, BITP Kiev, East Anglia, Birmingham, particle Cambridge, Havering SFC, Imperial, Leeds, Liverpool, Sheffield, Swansea, Welsh Office, NHS*, PTT Yugoslavia.

Rest of World: Universities, Institutes and Similar.
Queensland Australia, McMaster Canada, Memorial Univ. Newfoundland, Queen's Univ., Royal Military College of Canada, TJ China, Escuelaing Colombia, Andes Colombia, IITM India, Hokudai, Kishou, Mexico, Malaysia, Massey NZ, UNAC Peru, UC Venezuela, MCRNS Korea, NCTU, PSG Korea, NCTU Taiwan.

During March 2007 visits from 85 countries led by US, France, Britain, Germany, Czechia, Belgium, Italy, ...

Leading Interest: Palacky, Motorola, Texas A and M, KAIST, Poitiers, Multivac, Theoretical Physics Munich, Poitiers, Pittsburgh, Varian, Clark Univ.

1 - 15 April 2007

US Universities, Institutes and Similar.
APU, Berkeley, Brown, Brigham Young, Caltech, Clark, Columbia, Duke, Harvard, Johns Hopkins, MIT*, North Dakota, Northwestern, New York Institute of Technology*, Philau, Pittsburgh, Penn State, Purdue, Rochester, Southern Illinois, Texas A and M, Urbana Champaign, U Mass., Maryland, Texas Dallas, UU, Vanderbilt, Vermont, Washington, Los Alamos, Lawrence Berkeley, St. Lucie County, US Gov PS, USAAF Wright Patterson, US Military DTRA, US Naval Research Laboratory.

Europe and Asia Minor: Universities, Institutes and Similar.

MUNI Czechia, Frankfurt, Cologne, Leipzig, Munich, Oldenburg, Paderborn, UTU Finland, French national particle lab., Jussieu CHUPS, Poitiers*, Univ College Cork, CNR IFAC Italy, Bologna, TNO Netherlands, Utrecht, NTNU Norway, Krakow, Dubna, KTH Sweden, Ljubljiana Slovenia, Uniba Slovakia, HUN Turkey, Metu, Manchester, Napier, SPC Oxford, Brighton and Hove, British Ministry of Defence.

Rest of World: Universities, Institutes and Similar.

CQU Australia, Monash, Gov. Brazil, UENF, UNEP, UNICAMP Brazil, EPL Canada, Toronto, Kyoto, Postech Korea, Massey NZ, Otago, NCU, NTU, SCU Taiwan, FPT Venezuela.

Leading Papers: 63, 81, 2, 77, 9, 76, 12, 70, 68, 6.
Leading Articles: Felker book, ECE, Hehl rebuttal, workshop overview.
Leading countries: USA, France, Czechia, Belgium, Britain, Japan, Australia, ... 58 countries

16 - 29 April 2007

US Universities, Institutes and Similar..

Berkeley*, Biola, Buffalo, Brigham Young, CHW, Cornell, Denison, Harvard, North Carolina State, Oregon State, ORU, Penn State, Purdue, Reed, Rice, South Florida State, Stanford, Houston, Urbana Champaign, Chapel Hill, UPS, South Florida, Worcester Polytechnic Institute, West Virginia, US Gov EEOC, Jefferson Particle Lab, Oak Ridge, US Gov.SRS.

Europe and Asia Minor: Universities, Institutes and Similar.

Charles Univ. Prague, MUNI Czechia, Studenten Wohnheim, Bonn, Frankfurt, Heidelberg, Cologne, Munich, Oldenburg, Paderborn, Stuttgart, CVT Denmark, Barcelona, UCM, UDL, UPV Spain, UTU Finland, Paris, Diderot, Poitiers, INFN Bologna, NGI Italy, Tuscany Gov., Padua, Sardinia, TNO Netherlands, Twente, Utrecht, NTNU Norway, Lisbon Tech., KBSU Russia, KTH Sweden, UMU Sweden, UU Sweden, Anadolu, Ege, Metu Turkey, Kiev Ukraine, Bristol, BSFC, Trinity Cambridge, particle Cambridge, Cardiff, Durham, Heriot Watt, Imperial*. Kent, Jesus College Oxford, York, BBC, Ceredigion, Parliament, Belgrade.

Rest of World: Universities, Institutes and Similar.

New South Wales Government, UFSJ Brazil, UFSC, Sao Paolo, UNICAMP, New Brunswick, Biobio Chile, AMT China, IITM India, Oita Univ. Japan, Waseda, Chiba, BUAP Mexico, UAM Mexico, Massey NZ, Otago.

April 2007 Summary

Leading Papers: 63, 15, 16, 17, 19, 11, 76, 81, 6, 9, 8, 2, 13, 12, 77, 10, 23, 7, 4, 83, 50.
Leading Articles: Felker book, ECE, Workshop overview, Space Energy, Hehl rebuttal, Workshop history, Pendergast, ...
Leading Countries: USA, Canada, France, Britain, Australia, Belgium, Russia, ... 68 countries.
Leading Interest: KAIST, Brigham Young, UQAC, Bologna, Swansea, Motorola, UPV, Stanford, Vanderbilt, Columbia, ...

THREE YEARS OF HIGH QUALITY FEEDBACK RECORDED.

---

1 - 16 May 2007

US Universities, Institutes and Similar.
   Boston College, Berkeley, Brigham Young, Colorado, Columbia, Duke, Harvard, Indian River CC, Johns Hopkins Applied Physics, Kansas State, Kettering, MIT, Princeton, PUPR, Stanford, Temple, UAF, Delaware, Iowa, North Carolina Grensborough, Vanderbilt, WWC, Argonne, DHS, OSIS, St Lucie County, US Courts, USAAF Loughlin, US Army YUMA, US Navy NMCI, Naval Research Laboratory.

Europe and Asia Minor: Universities, Institutes and Similar.
   KF Graz Austria, Charles Univ. Prague, German Synchrotron Facility (DESY), Max Planck SB, Max Planck MPIMF Heidelberg, TU Berlin, Bielefeld, UCM, UPC, UPV, UV Sapin, HUT Finland, CNRS CRHEA France, INSA Lyon, Perpignan, Jerusalem, INFN Florence, Bologna, Messina, Twente, Utrecht, FUW Poland, ICM Poland, UA Protugal, Lisbon Tech., Ljubljiana, Agric Univ. Slovakia, Bilkent, Metu, Uludeng Turkey, Kiev, ARTS Britain, BSFC, Particle Cambridge, Coleg Sir Ga^r, EPSRC, Imperial, Leeds, NCL, Swansea, Gov. Yugoslavia.

Rest of World: Universities, Institutes and Similar.
   ANU, MQ Australia, Defence Australia, UQAC Canada, IITM India, Ibaraki Japan, Osaka, Aizu, Univ. Tokyo, KAIST Korea, Auckland NZ, Massey, NCU Taiwan, FPT Venezuela, UGTO Mexico.

Lead Papers: 63, OO161, OO591, 83, 84, OO593, OO585, OO582, 8, ...

Lead Articles: Felker book, ECE, Munich Workshop Slides, Overview, Space Energy, Pendergast, Numerical Methods, Workshop History, ...
Most Interest: Luxembourg Ministry of Education, KAIST, Brigham Young, ... 66 countries.
Leading countries: USA, Britain, Belgium, France, Italy, Australia, ...

16 - 30 May 2007

US Universities, Institutes and Similar.

Brigham Young, Caltech, Georgia Tech*, Johns Hopkins Applied Physics, Kansas State, Maine, Rutgers, Stanford, Urbana Champaign*, Maryland, Toledo, NASA (JPL), St Lucie County, US Gov. PS, Gov. Virginia, US Army Amrdec, US Army NVL, US Navy NMCI.

Europe and Asia Minor: Universities, Institutes and Similar.

TU Vienna, Plovdiv Bulgaria, EPF Lausanne, Zurich, Palacky, Max Planck Heidelberg MPIMF, TU Berlin, Bayreuth, Frankfurt, Freiburg, Cologne, Leipzig, Munich, CVT Denmark, UDL Spain, Ovi, UPM, UPV, JYU Finland, Oulu, AC Amiens, CNRS Grenoble, French national particle lab., Poitiers, CSI Italy, INFN LNF, Turin Tech., ICTP Trieste, Bologna, Leiden, Utrecht, Oslo City Gov. Norway, ICM, PW, US Poland, UA Portugal, UMU Sweden, UU Sweden, Agric Univ., Slovakia, SDU Turkey, Kiev, Aberystwyth, BSFC, Exeter, Glamorgan, Balliol College Oxford, Worcester College Oxford, Porstmouth, Southampton, Sussex, Univ. College London, Ceredigion, British Gov. Energis, Welsh Office.

Rest of the World: Universities, Institutes and Similar

UNICAMP*, Gov. British Columbia, Quantum Foundations Canada, Laval, TJ China, UO Cuba, IACS India, TIFR India, Himeji Japan, Ibaraki, Aizu, KAIST, Pusan Korea, Auckland, Philipines, AIT Thailand.

May 2007 Summary of Interest.

Lead Papers: 63, OO591, OO161, 84, 83, 2, OO591, OO572, OO594, OO401e, ...
Lead Articles: Felker book, ECE, Workshop overview, Pendergast, Space Energy, ...
Lead Interest: Luxembourg Energy Ministry, Brigham Young, Fukuoka, Motorola, Towy Aberystwyth, ...
Lead Countries: USA, Britain, France, Belgium, Italy, germany, Australia, Japan, Canada, ... 73 countries in the month.

1 - 15 June 2007

US Universities, Institutes and Similar.
   BVB, Brigham Young, GAC, LLUMC, North Dakota, Northwestern,
Purdue, Rice, Roosevelt, Stanford, Texas A and M, UC Davis, Chicago,
UC Irvine, UC Santa Barbara, Idaho, Urbana Champaign*, UNL, South
Carolina, VCU, Wisconsin, US Gov FDA, Lawrence Berkeley, USAAF
Edwards, US Army AMRDEC, USAAF Monmouth.

Europe and Asia Minor: Universities, Institutes and Similar.
   TU Vienna, FUNDP Belgium, CERN, Charles Univ. Prague, RWTH
Aachen, Siemens, STW Bonn (students), Cologne, Leipzig, Schwabia,
UCM Spain, Helsinki, SHOM France, Paris 3, Poitiers, NTUA Greece,
Mathematics Institute Croatia, ELTE Hungary, INFN Bologna, SISSA
Italy, TIN Italy, ICTP Trieste, Pisa, Sardinia, Utrecht, IFJ Poland, Krakow,
Lisbon Tech., NSC Russia, KTH Sweden, LU Sweden, Joseph Stefan
Institute Slovenia, Ljubjliana, Aberystwyth, Bangor, Cardiff, Hull,
Imperial, Leicester, Sussex, Swansea, Univ College London, BBC,
Ceredigion, Welsh Office.

Rest of World: Universities, Institutes and Similar.
   Sydney, UNICAMP, Gov. Alberta, Toronto, TIFR India, Japan, KAIST
Korea, Sogang Korea, NU Singapore.

Lead Papers; 63, 25, 21, 22, 24, 50, OO421, 16, 23, 2, 1, ...
Lead Articles: (Felker book published by Abramis and removed from site),
ECE, Workshop Overview, Filtered feedback data, Hehl rebuttal,
Pendergast, Workshop History, Space Energy, Galaxies, Royal Decrees on
Civil List pension, ...
Lead Countries: USA, britain, Japan, France, Belgium, Germany, Italy, ...
60 countries.

16 - 29 June 2007

US Universities, Institutes and Similar.
   Brigham Young, Caltech, Cornell, Florida Tech., GPC, Iowa State*,
Johns Hopkins Applied Physics, MIT, Ohio State, Purdue*, Rowan,
Rutgers, Stanford, SUNY Stonybrook, SWMED, Texas A and M,
Chicago, UC San Diego, Illinois Chicago, UTUC, Minnesota, New
Mexico, Wayne State, US Gov. EDC, US Dept. of Energy, St Lucie
County, US Gov. PS, USPTO, USAAF Hill, US Army peoavn, US
Military BTA, US Anvy NMCI.

Europe and Asia Minor: Universities, Institutes and Similar.

BTC Net Bulgaria, ETH Zurich, Charles Univ. Prague, geman Synchrotron Facility, STW Bonn (students), TU Berlin, Cologne, Leipzig, Rostock, CSIC Spain, UPV, USC, French national particle lab., Poitiers, INFN Florence, Pisa, Uniud Italy, Groningen, Utrecht, UIO Norway, European Union, Warsaw, Uninho Portugal, SMTU Russia, Aberystwyth, Bradford*, BSFC, EPSRC, Lancs NGFL, Queen Mary, Southampton, Ceredigion, Welsh Office.

Rest of World: Universities, Institutes and Similar.

Monash, UFMG Brazil, McGill Canada, Usherbrooke, Toronto*, UDEA Colombia, City Hong Kong, Indian Statistical Institute, IUCAA, TIFR India, Tokyo, KAIST Korea, Sogang Korea, Massey NZ, UNAC Peru, NU Singapore.

June 2007 Summary

Lead countries: USA, Britain, France, Japan, Germany, Belgium, Australia, Italy, Hong Kong, Netherlands, ...
Lead papers: 63, 85, 25, 23, 15, 21, 22, 12, 1, 24, 9, 2, 8, 50, 77, 16, 76, 5, 3, 6, 82, 83, 43, 56, 11, 52, 10, 84, 61, 14, 81, 58, 75, 17, 26, 68, 79, 80, 28, 73, 13, 18, 51, ... all papers read.
Lead articles: ECE (Spanish), Hehl rebuttal, Munich overview, HSD6, UNCC Saga 1, Space Energy, Universe3, HSD2 Bangor, ... All articles read.

1 - 15 July 2007

US Universities, Institutes and Similar,

Carleton, Colorado, Cornell, CWRU, Florida Atlantic, Georgia Tech., Hawaii, MU Ohio, Rutgers, Texas A and M, UCAR, UC Davis, UCLA, UC San Diego, Urbana Champaign, New Mexico, U Penn., Washington, Fermilab, USAAF Keesler, US Navy NUMC, AOL, Cray, Lockheed Martin, Microsoft, Mobile Gas, Motorola.

Europe and Asia Minor: Universities, Institutes and Similar.

TU Vienna, ETH Zurich, Charles Univ. Prague, Palacky, Max Planck Halle, RWTH Aachen, Students Bonn, TU Dresden, Duisburg, Giessen, Goettingen, Heidelberg, Cologne, Leipzig, Mainz, Paderborn, CVT Denmark, ILL Grenoble, French national particle lab., OBSPM, Poitiers, ELTE Hungary, INFN Bologna, INFN Sardinia, ICTP Trieste, Calabria, Perugia, Pavia, Luxembourg Parliament, TU Delft, Eindhoven, Utrecht,

MTS-NN Russia, IRFU Sweden, Aberystwyth, Edinburgh, Imperial, SBS Oxford, VPN Oxford, Southampton, Warwick, Ceredigion.

Rest of World: Universities, Institutes and Similar.
    Adelaide, CDU Australia, Monash, Melbourne, UNICAMP, UIO Cuba, IIFR India, Kanazawa Japan, Auckland, Otago, UNI Peru, Mahidol Thailand, SU Thailand, Academica Sinica Taiwan.

Lead papers: 63, 76, 13, 15, OO108, OO563, 2, 23, 21, 68, 81, 83, 85, 77, OO32, OO462, 86, OO572, 20, 11, 12, 17, 25, 53, OO401e, 18, 19, 5, 8, 9, 3, 16, 26, 6, 4, 50, 56, 61, 78, 7, 29, 10, ... all papers read.
Lead countries: USA, Australia, France, Canada, Germany, Britain, Belgium, ... (61 countries).

16 - 30 July 2007

US Universities, Institutes and Smilar.
    PAS Arizona, Brigham Young, Colorado, Cornell*, OPAC Cornell, Johns Hopkins, Johns Hopkins Applied Physics, MIT, NPS, Ohio State, Rice, Roosevelt, Texas A and M, Urbana Champaign, Utah, Washington, Argonne, NSF, Santa Barbara, USPS, Jefferson National Laboratory.

Europe and Asia Minor: Universities, Institutes and Similar.
    ARCS Austria, Vienna, Max Planck MPIM Bonn, TU Darmstadt, Leipzig, JYU Finland, KFKI Hungary, Pavia, PT Portugal, Twente, Warsaw, Agric University Slovakia, Aberystwyth, Bristol, Oxford, Surrey, Warwick, BBC, British Gov. Energis.

Rest of World: Universities, Institutes and Similar.
    Gov. Brazil, Gov. Canada IC, Perimeter, Alberta, Kanazawa Japan, Kyoto, PKNU Korea, NU Singapore.

July 2007: Summary of Interest.

Lead Papers: 63, 89, 76, 15, 2, 85, 68, 83,13, 61, OO108, 77, 1, 12, 54, 23, 9, 11, 21, 53, 86, 6, 27, 81, OO595, 16, 8, 57, 5, 70, 3, 52, 4, 72, 18, 55, 7, OO401a, 17, 58, 78, .. All papers read.
Lead articles: Workshop overview, filtered feedback data, Galaxies (Sp), Space Energy, Pendergast, Numerical Solutions, UNCC, IBM Code, Workshop History, Universe3, Galaxies, Workshop introduction, HSD2, all articles read.
Lead countries: USA, Canada, Australia, France, Britain, Germany, Belgium, ... 72 countries.

1- 15 August 2007

US Universities, Institutes and Similar.
ACCD, Auburn, Berkeley, OPAC Cornell, Duke, Emory, Embry Riddle, Indiana, MIT, Urbana Champaign, Nevada Las Vegas, Washington, USAAF Robins, US Army Research Laboratory, USMA Army, NASA (LARC), NASA (SSC), Sandia, Jefferson, Marshall Medical, Biphan, Chiron, Epson, General Electic, IBM (Britain), Philips.

Europe and Asia Minor: Universities, Institutes and Similar
Vienna, BOL Bulgaria, CERN, KFA Juelich, RWTH Aachen, Cologne, Stuttgart, Tuebingen, HUT Finland, ILL Grenoble, European Space Agency ESTEC, Institute of Physics, UW Poland, UBI Portugal, Oilnet Russia, Physto Sweden, Aberystwyth, Churchill Cambridge, Particle Cambridge, Astronomy Cardiff, Imperial, Oxford, Swansea, Warwick, Ceredigion, Coventry, British Gov., DERA, NHS.

Rest of World: Universities, Institutes and Similar.
ANY, ACT Australian Gov., UERJ Brazil, Gov Canada, Biobio Chile, IUCAA India, BARC India, IPN Mexico, Massey NZ, Otago.

Lead Papers: 63, 90, OO582, 2, 61, OO108, 89, 85, OO595, 87, 76, 1, 50, 92, 83, 13, 82, 7, 9, 15, 88, 49, 77, OO589, OO594, 52, 6, O500, 23, 7, 88, 21, OO421, OO591, 19, 25, 53, 91, OO501, OO585, 11, 12, 68, 8, 72, OO483, OO563, OO593, 24, 74, 17, 58, 67, ... all papers read.
Lead Articles: ECE (Spanish), Workshop overview, Black Hole Catastrophe, ECE, Galaxies, Filtered feedback, Space Energy, Pendergast Crystal Spheres, Workshop history, UNCC Saga 3, IBM Code, UNCC Saga 1, UNCC saga 2, HSD2, HSD6, Royal Decree, ... all articles read.
Lead countries: USA, Canada, Netherlands, Britain, France, Germany, Switzerland, Italy, ...
61 countries.

15 - 30 August 2007

US Universities, Institutes and Similar.
ACCD, Arizona State, Berkeley, Brigham Young, Columbia, Drexel, Florida Atlantic, Johns Hopkins Applied Physics, Missouri State, MIT, Ohio State, Penn State, Stanford, Minnesota, New Mexico, California Gov., Santa Barbara, USPS, US Army Research Laboratory, US Military USACE, US Naval Research Laboratory, US Navy NUWC, Biophan, IBM Torolab, Microsoft.

Europe and Asia Minor: Universities, Institutes and Similar.

CERN, Charles Univ. Prague, Palacky, KFA Juelich, Siemens, Cologne, Tuebingen, HUT Finland, CNRS Grenoble, AUTH Greece, NTUA Greece, Dublin Institute for Advanced Studies, Aruba Italy, CNR Bologna, ENEA Italy, Utrecht, Omsk, Norrkoping Sweden, UMEA Sweden, Joseph Stefan Institute Slovenia, Gyte Turkey, EKO Ukraine, Aberystwyth, Engineering Cambridge, EBI, Imperial, Swansea.

Rest of World: Universities, Institutes and Similar.

ANU, New South Wales, Gov. Canada*, Chinese visits, IACS India, European Space Agency, Tohoku, Univ. Tokyo, KAIST Korea, SNU Korea, Otago, RP Singapore, Mahidol Thailand.

August 2007 Summary

Lead Papers: 63, 2, 89, 90, 76, 87, 61, 85, OO582, OO595, 15, 92, 17, 77, 1, 9, 50, 8, 6, OO108, 23, 52, 19, 13, 25, 91, 83, OO585, 82, 88, 55, 16, 22, 70, 11, 21, 53, 49, 81, OO500, 57, 58, 72, 3, 12, 54, 60, 24, OO583, 4, 5, 74, 75, ... all papers read.
Lead articles: ECE (Spanish), ECE, Munich overview, Galaxies, Filtered feedback data, Space Energy, HSD27, Munich History, UNCC 1, Crothers, Pendergast, ... all articles read.
Lead Countries: USA, France, Canada, Netherlands, Britain, Germany, Italy, ... 70 countries.

1 - 15 September 2007

US Universities, Institutes and Similar.

BSU, Drexel, Florida International, Georgia Tech., Missouri, North Dakota, Oregon State, Rutgers, Texas A and M, Truman, Texas Tech., Delaware, Illinois Chicago, UMB, UMBC, Michigan, South Florida, Washington, Wisconsin, Oak Ridge, OSHA, St Lucie County, US Army NGB, US Army ACE, Microsoft, Northrop Grummond.

Europe and Asia Minor: Universities, Institutes and Similar.

UPT Albania, Charles Univ. Prague, Max Planck MPE, Siemens, Erlangen, Cologne, UAM Spain, UIB, UPV, Spanish Parliament, CNRS Grenoble, French national particle lab., INPG, Poitiers, Tel Aviv, Technion, ICTP Trieste, Groningen*, Delft, Eindhoven, ACN Warsaw, NIPME Romania, TKSK Russia, Pentacom Slovakia, Aberystwyth,

Birmingham, BSFC, Cardiff, Durham, Lancaster, SPC Oxford, British Met. Office, UKIM Macedonia.

Rest of World: Universities, Institutes and Similar.
Flinders, Queensland, Sao Paolo, Americas Chile, China, U Cuba, Indian Statistical Institute, Kamazawa Japan, KAIST Korea, CCU Taiwan.

Lead Papers: 63, 93, 89, 15, 13, 9, 87, 76, 17, 2, 92, OO421, 21, 6, 23, 8,61, 85, 91, 1, 22, 4, 58, 3, 68, 74, 50, 53, 65, 12, 90, 66, 88, 25, 54, 81, 83, 86, 10, 57, 78, 7, 82, 20, 26, 31, 51, 70, 71, 77, 84, 27, 29, 32, 35, 43, 46, 52, 55, 5, 62, 64, ... all papers read.
Lead Articles: Munich Workshop Overview, ECE (Spanish), ECE, Pendergast Crystal Spheres, Galaxies, Workshop details, Workshop history, Space Energy, Galaxies, ... All articles read.
Lead Countries: USA, China, France, Germany, Britain, Australia, Italy, Mexico, Netherlands, Brazil, ...

16 - 29 September 2009

US Universities, Institutes and Similar.
Berkeley, Berry, Drexel, North Dakota, Philau, Pittsburgh, Rochester, Renselaer, Rutgers, South Florida State, Florida, Houston, Michigan, Weber, NASA (JPL), Oak Ridge, Santa Barbara Gov., USAAF Eglin, USAAF ACE, US Navy NMCI.

Europe and Asia Minor: Universities, Institutes and Similar.
Vienna, Pi Group Belgium, Albert Einstein Institute, Siemens, Giessen, Magdeburg, CVT Denmark, TLC Spain, ILL Grenoble, French national particle lab., Nice, Lemans, Poitiers, Mathematics BME Hungary, Jerusalem, CNR Rimini Italy, INFN Padua, AREA Trieste, Unimib Italy, Groningen, Kluwer, NTNU Norway, UVT Romania, Lipetsk, Hacetepe Turkey, Aberystwyth, University of Wales Registry, Hull, Swansea UKIM Macedonia.

Rest of World, Universities, Institutes and Similar.
Monash, Gov. Canada, Virtual Community Canada, China, UCLV Cuba, UO Cuba, City Univ. Hong Kong, Univ. Tokyo, Toshiba, KAIST Korea, Otago, RP Singapore, CCU Taiwan, Capetown South Africa.

September 2007 Summary
Lead papers: 93, 2, 63, 9, 89, 13, 15, 6, 23, 8, 76, 87, 17, 21, 92, 91, 12, 54, 22, 53, 3, 4, 50, 61, 85, 90, 16, 55, 52, 68, 10, 51, 58, 71, 26, 7, 82, 19, 25, 27, 5, 65, 74, 57, ... all papers read.

Omnia Opera (OO) Papers: 585, 591, 108, 593, 595, 592, 446, 6, 177, 461, 381, 472, 99, 95, 462, 582.
Lead articles: Workshop overview, ECE (Sp), Crystal Spheres (Kerry Pendergast), Galaxies (Sp), ECE, ... all articles read.
Lead Countries: US, China, Italy, France, Germany, Britain, Netherlands, Australia, ... (79 countries).

1 - 15 October 2007

US Universities, Institutes and Similar.
   Buffalo, California Polytechnic, Caltech, CSU Pomona, FHSU, Florida International, Princeton IAS, Illinois Tech., Indiana, Johns Hopkins, Ohio State, Oklahoma, Purdue, Renselaer, Syracuse, Texas Tech., UC San Diego, UFP, U Mass., New Mexico, U Penn*, UPR, Vermont, Santa Barbara Gov., St Lucie County, USAAF Wright Patterson, US Military AMSAA, US Military NGA.

Europe and Asia Minor: Universities, Institutes and Similar.
   Vienna, ETH, Charles Univ. Prague, KFA Juelich, Albert Einstein Institute, TU Darmstadt, TU Dresden, Stuttgart, KU Denmark, UCA, UP, UV, Spain. Oulu Fnland, CNSR Marseilles, French national particle lab., SHOM, Angers, Poitiers, Thessalonika, BGU Israel, ENI Italy, INFN LNF, Milan Tech., Unimo Italy, Leiden, Groningen, NTNU Norway, Institute of Physics, IFJ Poland, Krakow, Debryansk, Gov. Slovenia, Accross (Britain), Trinity Cambridge, Univ. Wales Registry, Imperial, King's College London, Maths Oxford, East Anglia, British Gov. Energis, Parliament.

Rest of World : Universities, Institutes and Similar.
   Deakin Australia, UFCG Brazil, Rio de Janeiro, Perimeter, British Columbia, HA China, UNAL Colombia, Meiji Univ. Japan, Osaka, Waseda, KAIST Korea, UNAM Mexico, PUK South Africa.

Lead papers; 93, 63, 2, 93 extra plots, 15, 1, 77, 13, 76, 95, ... all papers read.
Lead articles: Space Energy, Workshop overview, Galaxies, Workshop history, ... all articles read.
Lead Omnia Opera: 591, 381, 498, 501, 724, 16, 552, ...
Lead Countries: USA, Brazil, France, Britain, Germany, Russia, Netherlands, ... (59 countries).

15 - 30 October 2007

US Universities, Institutes and Similar.
   Buffalo, Caltech, Drexel*, Duke, Faytech CC, Harvard*, Missouri State, Michigan State, NKU, Penn State, SIU, Stanford, Syracuse, TMC, Nevada Las Vegas, Toledo, William and Mary, Western Washington, US Gov. BLM, St Lucie County, USAAF Edwards, Argonne, US Navy NMCI.

Europe and Asia Minor: Universities, Institutes and Similar.
   Vienna, NMV Belgium, Steorn Belgium, Charles Univ. Prague, Palacky, Max Planck Dresden, Bonn, CVT Denmark, UJI, UMH, Ovi Spain, ENST France, French national particle lab., Poitiers, Bio Academy Greece, INFN Turin, Gov. Tuscany, Groningen, Utrecht, HIO Norway, NTNU Norway*, IFJ Poland, Krakow, WSPIZ Poland, MTS-NN Russia, Bristol*, Imperial*, Leeds, Manchester, Sheffield, Sussex, NHS.

Rest of World : Universities, Institutes and Similar.
   Gov. Canada, YRBE Canada, Kyoto, Meiji, Osaka, KAIST Korea, NU Singapore, CU, NDMTSGH, NTU, YM Taiwan.

October 2007 Summary

Lead Papers: 93, 2, 63, 43, 23, 15, 76, 9, 13, 89, ... all papers read.
Documents read from site: 869.

1 - 15 November 2007

US Universities, Institutes and Similar.
   Central Michigan, Colby, CWRU, Drexel, Georgia Tech., IUP, Maine, MIT, New York Univ., Oklahoma, Princeton, Rice, Rochester, Texas A and M*, UC Davis, Florida, Illinois Chicago, Iowa, Urbana Champaign, UMR, UPR, UPRM, Utah*, Vanderbilt, Vermont, Oak Ridge, Santa Barbara Gov., St Lucie County, US Navy NMCI, Allegran, Mobile Gas, Motorola.

Europe and Asia Minor: Universities, Institutes and Similar.
   EPF Lausanne, Siemens, TU Cottbus, TU Munich, Freiburg, Cologne, Danish Tech., UCM, Ovi Spain, HTV Finland, ESIEE France, Jussieu LISIF, Poitiers, ELTE Hungary, Trinity College Dublin, Univ College Dublin, INFN Bari, VU Lithuania, MITS Latvia, Groningen, Utrecht, Free Univ. Amsterdam, Kluwer, NTNU Norway*, FUW Poland, Lisbon, Vladivostok, HJK Military Turkey, Aberystwyth, Bristol, particle

Cambridge, CWC, Dundee, Imperial, Lancaster, Liverpool, Oxford, Sheffield, Univ. College London, UMIST, Wimbledon, Gwynedd Conty Council.

Rest of World: Universities, Institutes and Similar.
ANU, UWS, UFG Brazil, British Columbia, Toronto, UTFSM Chile, DU India, IIIT India, Kyoto, Kyushu, Osaka, Tokyo, MRT Sri Lanka, CHULA Thailand, NTHU Taiwan, NTUST, Academical Sinica, TP1RC, IDV Taiwan.

Documents read: 856.
Lead Countries: USA, France, Britain, Germany, Czechia, Italy, Canada, ... 68 countries
Lead Papers: 26, 87, 21, 2, 89, 23, 43, 98, 8, 93, 47, 13, 14, 44, 17, 15, 1, 97, 67, 6, 45, 83, 9, 18, 85, 91,51, 99, 19, 25, 29,41, 42, 46, 92, 93 extra plots, 49, 63, 70, 90, 10, 22, 60, 24, 28, 53, 54, 88, 11, 37, 38, 5, 68, ... all papers read.
Lead OO Papers: 591, 421, 431, 461, 401c, 389, 563, 393, ...
Lead Articles: Space Energy, Workshop overview, Galaxies (Spanish), Crystal Spheres (Pendergast), UNCC Saga 1, Filtered feedback, UNCC international condemnation, ... al articles read.

15 - 30 November 2007

US Universities, Institutes and Similar.
Boise State, SUNY Buffalo, Colby, CSU Pomona, Johns Hopkins Medical Institute, Johsn Hopkins, New Mexico Tech., North Dakota, NRAO, New York Univ., Ohio State, Purdue, Rhodes, Rutgers, Syracuse, UCLA, UC Riverside, Florida, Houston, Urbana Champaign, Maryland, U Penn., Utah, Texas, Washington, Yale, NASA (Grc), NASA (JPL), Oak Ridge, Gov. Santa Barbara, St Lucie County, USAAF Brooks, US Army YUMA, US Navy NMCI.

Europe and Asia Minor: Universities, Institutes and Similar.
ARCS Austria, KU Leuven, Steorn Belgium, Charles Univ. Prague, Palacky, Jacobs Univ Germany, Siemens, TU Munich, Heidelberg, Cologne, UCM Spain, Ovi Spain, CNAM France, Poities, AUTH Greece, NTUA Greece, Trinity College Dublin, INFN Milan, Siemens Italy, RU Netherlands, Twente, Amsterdam, NTNU Norway, Gov. Poland, Rzeszow Poland, Lisbon, GJH Slovakia, Agric. Univ. Slovakia, Aberystwyth, Bristol*, Girton College Cambridge, particle Cambridge, Imperial*, Lancaster, Manchester, Univ. College London, York, Camden Council.

Rest of World : Universities, Institutes and Similar.

Monash Australia, Gov. Canada, British Columbia, ITP China, CEC European Community, Shiraz Iran, Tokyo*, KAIST Korea, NU Singapore, NSPO Taiwan.

November 2007 Summary.

Documents read from Site: 876.
Lead Papers: 2, 89, 87, 43, 21, 26, 63, 99, 76, 23, 6, 98, 8, 93, 9, 1, 67, 85, 45, 83, 10, 14, 15, 17, 13, 97, 46, 28, 72, 93 plots, 49, 51, 70, 91, 38, 54, 29, 11, 19, 6, 90, 40, 77, 92, 18, 18, 25, 41, 53, 96, ... all papers read.
Lead articles: Workshop overview, Space Energy Spanish, Galaxies (Sp), Crystal Spheres (Pendergast), UNCC Saga 1, Space Energy, Galaxies, ECE, UNCC international condemnation, ... all articles read,.
Lead OO papers: 421, 108, 461, 592, 431, 499, 372, 563, ...
Lead countries: USA, France, Italy, Czechia, Britain, Germany, Canada, ... 76 countries.

1 - 15 December 2007

US Universities, Institutes and Similar.

Arizona, Athens, Auburn, Benedictine, Boise State, Boston, Drexel, HMC, Kansas State, Louisville, Louisiana State, Metrostate, Miami, MIT, NEU, New York Univ., Oberlin, Ohio State*, Penn State, Purdue, Reinhardt, Semo, UC San Diego, Illinois Chicago, U Mass., Nevada Reno, Utah*, Texas, Washington, Teale County California, St Lucie County California, USAAF Monmouth, US Army Natick, Southwestern Laboratories.

Europe and Asia Minor: Universities, Institutes and Similar.

Siemens Austria, Liege, Charles Univ. Prague, Palacky, German Synchrotron Facility, Siemens, TU Berlin, Erlangen, Cologne, Munich, Ovi, BNF France, ENS Cachan, Polytechnique system, Strasbourg, Poitiers, Patras, Turin Tech., Amsterdam*, NTNU Norway, Lisbon, MTS-NN Russia, UU Sweden, Kiev, particle Cambridge, Hull, Imperial , Manchester, physics Oxford, Queen Mary London, Sheffield, Ceredigion.

Rest of World : Universities, Institutes and Similar.

Monash, South Australia, UNICAMP, McGill, Shiraz Iran, Gakushuin Japan, Hokudai, KAIST Korea, NU Singapore, NCU Taiwan, PUK South Africa.

Documents Read of Site: 864

Lead Papers: 6, 2, 76, 101, 54, 51, 63, 43, 87, 8, 21, 99, 45, 93, 1, 32, 44, 85, 10, 27, 55, 67, 28, 41, 40, 46, 90, 93 extra plots, 96, 3, 5, 4, 7, 95, 14, 33, 64, 65, 77, 15, 17, 23, 47, 70, 98, 9, ... all papers read.
Lead articles: Workshop overview, Space Energy Spanish, Crothers Criticisms, UNCC Saga 1, Workshop history, UNCC Saga 3, Galaxies (Spanish), Crystal Spheres (Penddergast), ... all articles read.
Lead Countries: USA, France, Britain, Germany, Italy, Russia, Belgium, ... (70 countries).

15 - 30 December 2007

US Universities, Institutes and Similar.
   Bucknell, ECE, New Mexico State, Oregon State, Princeton, U Mass., US Gov. Dept of Transport, NASA (CDSCC), NOAA, St Lucie County, US Army YUMA, Southwestern Labs., Mobile Gas, Motorola.

Europe and Asia Minor: Universities, Institutes and Similar.
   Leuven, EPF Lausanne, Geneva, Max Planck MPPMU, Cologne, Luebeck, Ulm, Elion Estonia, UHP Nancy, Nantes Brittany, Poitiers, ACN Greece, SISSA Italy, Padua, Amsterdam, NTNU Norway,Gdynia, ONI Portugal, Lisbon Tech., IYTE, SDU Turkey, Glasgow, Imperial, Lancaster, Manchester, British Gov Energis.

Rest of World: Universities, Institutes and Similar.
   UNICAMP, British Columbia, Montreal, UNB Canada, RE Korea, NCTU, NTU, NCU Taiwan.

Summary December 2007

 NEW RECORD of 11.268 gigabytes downloaded in a month
Documents Read off Site: 874.
Lead countries: USA, Italy, France, Britain, China, Russia, Gemany, Czechia, Belgium, Brazil, ... (74 countries).
Lead papers: 6, 2, 103, 43, 76, 63, 102, 54, 101, 51, 99, 87, 85, 9, 21, 45, 93, 10, 67, 55,44, 46, 8, 89, 32, 17, 27, 7, 96, 90, 98, 12, 61, 15, 28, 40, 50, 77, 93 plots, 37, 42, 95, 41, 4, 65, ... all papers read.
Lead Articles: Workshop overview, New Energy (Spanish), Crothers criticisms, Crystal Spheres (Pendergast), Workshop history, ... all articles read .
Lead OO Papers: 108, 594, 461, 421, 75b, 401e, 591.
Lead Historical Source Documents (HSD): 31, 6, 2b, 1, 5, ...

January 1 - 16 2008

US Universities, Institutes and Similar.
   DHCP, Boston Univ., CHW, Johns Hopkins Applied Physics, Maine, Renselaer, Urbana Champaign, U Mass., UMKC, UPF, Utah, Brookhaven, NASA (ARC), NASA (JPL), Santa Barbara Gov., St Lucie County, Gov. West Virginia, USAAF Robbins, US Army NVL, US Army YUMA, Dupont, Microsoft, Northrop Grummond, Raytheon.

Europe and Asia Minor: Universities, Institutes and Similar.
   CERN, ETH Zurich, HS Heilbronn, Juelich, Siemens, Cologne, DTU Denmark, UCM Spain, Helsinki, EBS Cachan, INSA Rennes, Poitiers, Rennes 1, OTE Greece, EFZG Hungary, KFKI Hungary, NUI Galway Ireland, Tel Aviv, INFN Bari Italy, Eindhoven, VU Netherlands, NTNU Norway, RFJ Poland, Lisbon, ISPCJ Romania, KNC Russia, Chalmers Sweden, Aberystwyth, Churchill College Cambridge, King's Cambridge, Biology Cambridge, Hull, Imperial, Sheffield, Swansea.

Rest of World: Universities, Institutes and Similar.
   ANU, Gov. Australia BOM, Gov. Canada, Perimeter, Univ Alberta, Univ. Toronto, Kuyshu, KAIST, Postech, National Univ.Mexico, USON Mexico.

Number of Documents read off the site: 860.
Lead papers: 103, 43, 102, 2, 54, 9, 63, 6, 46, 89, 56, 101, 1, 83, 99, 10, 21, 12, 45, 27, 61, 76, 87, 15, 62, 7, 93, 17, 47, 55, 90, 92, 11, 44, 50, 65, 40, 52, 5, 33, 66, 67, 70, ... all papers read.
Lead OO Papers: 461, 594, 108, 352c, 401c, 352b, 401e, 401b, 572, 421, 352d, 401a, 75a, ...
Lead Articles: Workshop Overview, Crystal Spheres (Pendergast), Space Energy, UNCC Saga 1, Workshop History, ECE Engineering Model, Galaxies Spanish, Galaxies, ... all articles read.
Lead Countries: USA, Russia, France, Britain, China, Germany, Italy, Canada, (67 countries).

15 - 30 January 2008

US Universities, Institutes and Similar.

Caltech, CHW, Central Michigan, CSUN, Harvard, Hawaii, Indiana, Kansas State, Kent State, Maine, MIT, New York Univ., Stevens Tech., SUBR, Chicago Illinois, UCOK, UC San Diego, Idaho, Minnesota, Oregon, Utah*, Wisconsin, Lawrence Berkeley, Santa Barbara, St Lucie County, USAAF Robins, US Army NVL, US Navy NMCI.

Europe and Asia Minor: Universities, Institutes and Similar.

UIBK Austria, Vienna, Siemens Austria, Leuven, Ghent, ETH Zurich, Charles Univ. Prague, German Synchrotron Facility, Jacobs Univ., RWTH Aachen, Siemens Germany, Hamburg, Karlsruhe, Cologne, Munich, DTU Denmark, UAB, Madrid 3, Unav, UV Spain, ENS Cachan, Poitiers, Rennes, EAP Greece, NU Galway, ISB Iceland, INFN Bari, NGI Italy, TU Delft, NTNU Norway, Gliwice Poland, Koszalin, Lisbon, Porto, RDS Romania, HC Russia, KTH Sweden, Hacettepe Turkey, ITU Turkey, Aberystwyth, East Anglia, Bolton, AST Cambridge, Imperial, Oxford-Man, physics Oxford, Queen's Univ. Belfast, Swansea, BBC.

Rest of World: Universities, Institutes and Similar.

ANU, Gov. of Canada, Toronto, China, Pune India, Kerala India, Hokudai, Kyoto, KYU Tech., Univ. Tokyo, KAIST Korea, Postech, TOPNZ New Zealand, RP Singapore, PUK South Africa.

Summary for January 2008.

Documents Read off site: 894.
Lead papers: 103, 2, 63, 43, 102, 46, 54, 9, 6, 89, 94, 99, 55, 1, 21, 93, 56, 12, 10, 17, 76, 101, 15, 45, 61, 67, 83, 11, 92, 37, 85, 50, 62, 65, 87, 44, 7, 27, 28, 33, 40, 47, 59, 57, 66, 8, 70, ... all papers read.
Lead OO Papers: 461, 108, 594, 352c, 401c, 352b, 401e, 401b, 75b, 572, 421, 352d, 401a, ...
Lead articles: Crystal Spheres (Pendergast), Workshop overview (Eckardt), Space Energy (Spanish), Workshop History, ECE Engineering Model, UNCC Saga 1, Galaxies (Spanish), Space Energy, Galaxies, ... all articles read.
Lead Countries: USA, Italy, Russia, France, Britain, Germany, Canada, China, ... (76 countries).

1 - 15 February 2008

US Universities, Institutes and Similar.

Boston College, Harvard, Missouri, North Carolina State, Northwestern, Ohio State, Purdue*, Stanford, Minnesota, South Florida,

Utah, Washington, NASA Ames, St Lucie County, USAAF Centaf, US Navy and Army, Jefferson Laboratory, Southwestern Laboratories, Honeywell, Lockheed Martin, Mobile Gas, Motorola, Philips.

Europe and Asia Minor: Universities, Institutes and Similar.
ULG Belgium, Zurich, Charles Univ. Prague, Bosch, Siemens, TU Berlin, Hannover, Cologne, Regensburg, Stuttgart, EHU Spain, INPG France, Jussieu LPTHE, Nice, Poitiers*, Rennes 1, Tellas Greece, NUI Galway, Univ College Dublin, Tel Aviv, CNR Bologna, INFN Calabria, NGI Italy, Florence, Utrecht, Royal Society of Chemistry, PK Poland, Torun, Warsaw, PU Russia, KTH, LIU Sweden, Agric. Univ. Slovakia, Aberdeen, Aberystwyth, Teifi Aberystwyth, engineering Cambridge, Lancaster, Liverpool College, Nottingham, Queen's Univ. Belfast, UMIST, Ceredigion.

Rest of World: Universities, Institutes and Similar.
Waterloo, Andes Colombia, Osaka, Yokohama, KAIST, CRI New Zealand.

Documents Read from Site: 539.
Lead Papers; 43, 94, 63, 103, 37, 100a, 89, 17, 23, 100d, 104, 14, 50, 1, 6, 99, 67, 85, 83, 21, 100c, 102, 100e, 10, 15, 28, 12, 31, 5, 96, 56, 45, 4, 90, 93, 98, 93 graphs, 2, 30, 40, 47, 61, 92, 97, 9, 38, 3, 42, 49, 55, 65, 88, 91, 11, 1, 20, 32, 33, 64, 66, all papers read.
Lead Articles: Crystal Spheres (Pendergast), Space Energy (Spanish), Galaxies, ECE Engineering Model, Workshop Overview, Workshop History, ... all articles read.
Lead OO Articles: 421, 585, 582, 572, ...
Lead Countries: USA, France, Netherlands, South Korea, Germany, Britain, Hong Kong, Canada, ... (58 countries).

15 - 27 February 2008

US Universities, Institutes and Similar.
Brown, Central Michigan, Eastern Michigan, Eastern New Mexico (ENMU), Harvard*, Johns Hopkins Applied Physics, NRAO, Scripps College, Temple, Illinois Chicago, Chapel Hill, U Penn., Utah, US Gov. BLM, Oak Ridge, US Gov PS, US Military NGA, Amazon, Boeing, General Electric, Honeywell, Lockheed Martin, Motorola.

Europe and Asia Minor: Universities, Institutes and Similar.
Charles Univ. Prague, Gov. Czech Republic, Jacobs, Pius Hospital, TU Berlin, Cologne, Munich, Gov. Andalucia, UPC Spain, NOOS France,

Lyon 1, Poitiers*, Tel Aviv, Rome 1, Unimib Italy, Utrecht*, NTNU Norway, Torun Poland, Kamisc Russia, Selcuk Turkey, Data Group Ukraine, Teifi and Towy Aberystwyth, Engineering Cambridge, Trinity Cambridge, particle Cambridge, Cardiff, HEFCW, Imperial, Kent, Physics Oxford, Queen Mary London, BBC, British Gov NICS, Bwrdd yr Iaith, Belgrade.

Rest of World : Universities, Institutes and Similar.
    ANU, Flinders, Western Ontario, Tsinghua China, U Cuba, PRL India, IIFR India, KAIST Korea, Postech Korea.

February 2008 Summary

Documents downloaded from site: 561.
Lead countries: USA, France, Britain, Netherlands, Germany, Canada, India, ... (71 countries).
Lead papers: 43, 94, 63, 103, 93, 37, 100a, 17, 89, 104, 100b, 14, 67, 1, 6, 99, 9, 100d, 23, 50, 12, 102, 15, 21, 28, 61, 100e, 47, 76, 100c, 93 extra plots,54, 85, 90, 105, 56, 83, 3, 40, 10, 31, 5, 65, 2, 13, 4, 7, 45, 64, 92, 96, 98, ... all papers read.
Lead OO Papers: 421, 585, 582, ...
Lead Articles: Crystal Spheres (Pendergast), Space Energy (Spanish), Galaxies (Spanish), Engineering Model, Workshop History, Galaxies, ECE Overview, ... All articles read.

1 - 17 March 2008

US Universities, Institutes and Similar.
    Caltech, Columbia, CSULB, Florida International, Hartford, Johns Hopkins, MIT*, Mississippi State, North Carolina State, Ohio State, Ohio Univ., Oklahoma State, Princeton, Stanford, Texas A and M, UCLA, UMBC, Michigan, Utah, Washington, US Gov BLM, St. Lucie County, USAAF Charleston, USAAF AERO, Honeywell, Lockheed Martin, Motorola, Northrop Grummond.

Europe and Asia Minor: Universities, Institutes and Similar.
    Free Univ. Brussels, Linz Austria, ETH Zurich, Charles Univ Prague, Film Akademie Germany, HU Berlin, Siemens, Bremen, Duisberg, Frankfurt, Jena, Leipzig, UNED Spain, US Spain, Helsinki, IAP France* , Poitiers*, Toulouse*, DUTH Greece, KFKI Hungary, Univ College Dublin, Jerusalem, ENI Italy, Unimo Italy, NTNU Norway*, PK Poland, Torun, Wroclaw, UCV Romania, Siberian Academy of Sciences, Physto Sweden, GTSI Slovakia, Hacettepe Turkey, Teifi and Towy Aberystwyth,

Cardiff, UCV Edinburgh, Hopwood, Lampeter, Lancaster, Leicester, Univ. College London, Western England, UWIC, Ceredigion.

Rest of World: Universities, Institutes and Similar.
ANU, Toronto, Western Ontario, UDEA Colombia, Kyoto*, Nihon, KAIST, Ctech South Africa, UL South Africa.

Documents Downloaded from Site: 606.
Lead Countries: USA, Britain, Italy, France, Yugoslavia, Germany, Czechia, ... (66 countries).
Lead Papers: 43, 106, 67, 94, 89, 63, 17, 6, 28, 76, 105, 45, 30, 65, 93, 54, 99, 100a, 85, 24, 47, 57, 23, 100a, 85, 24, 47, 57, 23, 56, 103, 44, 50, 61, 49, 59, 60, 31, 35, 40, 52, 53, 55, 102, 104, 42, 51, 10, 15, 36, 37, 46, 58, 32, 33, 62, 87, 25, 26, 29, 2, 34, 3, 48,64, 92, 9, 13, 16, 19, 21, 22, 7, 93 extra plots, 14, 1,20, ... all papers read.
Lead articles: Space Energy (Spanish), Workshop History, Crystal Spheres (Pendergast), Galaxies (Spanish), Space Energy, ECE Engineering Model, ... all articles read.

15 - 31 March 2008

US Universities, Institutes and Similar.
Berkeley, Buffalo, Harvard, Northwestern, New York Institute of Technology, SFUSD, UCHSC, UCLA, Florida, Urbana Champaign, Marietta Georgia Gov., St Lucie County, USAAF Charleston, USAAF Robins*, Southwestern Labs., Mobile Gas, Motorola, Northrop Grummond, Raytheon,

Europe and Asia Minor: Universities, Institutes and Similar.
UTS Austria, Eth Zurich, Charles Univ. Prague, Siemens, Hohenheim, Karlsruhe, CSIC Spain, UGR Spain, Lisif France, Poitiers*, KFKI Hungary, Jerusalem, Technion, Unimo Italy, Pavia, Groningen, UCV Romania, CN Russia, MTS - NN Russia, Bath, GRE, Napier, Sheffield, Swansea.

Rest of World: Universities, Institutes and Similar.
Tasmania, AEI Canada, Gov. Canada, HA China, IMSC India, TU Japan, Kaist Korea, Postech Korea, NU Singapore, UJ South Africa, Gov. South Africa.

Summary of March 2008

Documents Downloaded from Site: 709.

Lead Countries: USA, Britain, France, Italy, Yugoslavia, Germany, Czechia, ... (71 countries).
Lead Papers: 43, 94, 67, 106, 63, 6, 89, 93, 105, 17, 100a, 28, 1, 76, 45, 65, 30, 54, 103, 104, 7, 99, 37, 61, 15, 40, 56, 24, 44, 50, 85, 47, 57,102, 23, 10, 55, 9, 2, 32, 42, 52, 60, 26, 31, 49, 14, 21, 22, 35, 36, 53, 59, 90, 93 extra plots, 25, 29, 3, 46, 4, 51, 5, 92, 12, ... all papers read.
Lead articles: Space Energy (Spanish), Workshop History, Crystal Spheres (Pendergast), Galaxies (Spanish), Space Energy, ECE engineering Model. .... all articles read.

1 - 15 April 2008

US Universities, Institutes and Similar.

Boston Univ., Caltech, Clemson, Colorado*, Cornell, Hartford, Kansas State, Maine, New York Univ., Princeton, Purdue, Rutgers, Stanford, UC Santa Cruz, Florida, UGA, Nevada Las Vegas, West Virginia, US Gov. NOAA, Sandia, St. Lucie Co., USDA, USAAF Pope, USAAF Robins, US Army Belvoir, US Archives, Southwesten Labs., Boeing, CSE International, Dow Corning, Intel, Microsoft, Motorola.

Europe and Asia Minor: Universities, Institutes and Similar.

United Aran Emirates University, Vienna, ETH Zurich, Gov. Czech Republic, Siemens, Bonn, Hannover, Potsdam, UPV Spain, ILL Grenoble France, Orleans, Poitiers, AUTH Greece, Tellas Greece, Technion Israel, ENEA Italy, IOL Italy, Pisa, Twente, NVT, NTNU Norway, Krakow, KTH Sweden*, UMEA Sweden, Ljubjliana Slovenia, Aric Univ. Slovakia, Iyte Univ. Turkey, Towy Aberystwyth, Bath, Manchester, Nottingham, UWIC.

Rest of World: Universities, Institutes and Similar.

Melbourne, British Columbia*, York Univ. Canada, UMB Colombia, KAIST Korea.

15 - 30 April 2008

US Universities, Institutes and Similar.

Berkeley, Brandeis, Boston, Carnegie Mellon, Colorado*, Harvard, Indian River CC, Kansas State Univ.*, Ohio State, Ohio Univ., Pittsburgh, Princeton, Purdue, Rutgers*, SUNY Stonybrook, Texas A and M, UC San Diego, UDC, U Mass Medical, Hamilton County Ohio, US Gov WARA, USAAF Maxwell, USAAF Robins*, USAAF Wright Patterson*, US Navy NMCI, St Vincent, Southwestern Laboratories, Bank of America, Eastman, General Electric, Intel*, Microsoft*, Motorola*.

Europe and Asia Minor: Universities, Institutes and Similar.

Vienna, CERN*, ETH, Charles Univ. Prague, Palacky, German Synchrotron Facility, Siemens*, TU Berlin, TU Darmstadt, Erlangen, Freiburg, Hamburg, Heidelberg, Konstanz, Munich, Potsdam, Saarland, UPV Spain, Valencia, Helsinki, CNRS Grenoble, INSA Toulouse, Orleans, Poitiers*, Tellas Greece, ESS Croatia, European Union CEC, CSI Italy, ENEA Italy, INFN LNF, IOL Italy, Calabria, KLM Netherlands, Groningen, Twente, Utrecht, NTNU Norway*, Lisbon Tech., MTS-NN Russia, Chalmers Sweden, UU Sweden, Kiev, damtp Cambridge, Robinson Cambridge, Trinity Cambridge, particle Cambridge, Edinburgh, Glasgow*, Huddersfield College, Imperial*, King's College London, Lancaster, Leeds, Manchester, Napier, NISS, Queen Mary College, Queen's Belfast, Sheffield, Univ. College London, UHI, Univ London Computer Cente, Warwick, British Gov. Energis, Ministry of Defence, NHS.

Rest of World: Universities, Institutes and Similar.

Melbourne, Sydney, UFMG Brazil, Perimeter Canada, Saskatchewan, Toronto, Waterloo, Indian Statistical Institute, KAIST Korea, CENAM Mexico, UPM Malaysia, Otago MZ, Gov Peru, EDU Singapore.

Summary for April 2008

Documents downloaded from site: 897
Lead countries: USA, Britain, France, Switzerland, Mexico, Canada, Germany, Hungary, ... (79 countries).
Lead papers: 15, 94, 109, 54, 39, 106, 100a, 43, 102, 105, 37, 81, 40, 93, 6, 103, 61, 110, 63, 76, 67, 53, 83, 92, 104, 1, 30, 47, 49, 90, 100c, 12, 2, 100b, 55, 99, 9, 7, 88, 89, 91, 56, 25, 74, 98, 100d, 100e, 17, ... All papers read.
Lead articles: Space Energy (Spanish), Workshop details, Lindstrom 1, Workshop Overview, Workshop History, Lindstrom 2, Crystal Spheres (Pendergast), Crothers paper, ECE French, Space Energy, ... all articles read.

Complete Paper Survey Carried out on 18th April 2008, in order of most read papers.

109, 39, 54, 94, 106, 37, 102, 100a, 43, 40, 81, 105, 61, 53, 6, 93, 63, 30, 67, 103, 47, 83, 12, 25, 92, 15, 32, 17, 45, 90, 1, 49, 56, 44, 22, 36, 55, 74, 7, 42, 5, 26, 29, 34, 50, 88, 91, 104, 16, 2, 41, 64, 69, 73, 78, 14, 18, 23, 24, 28, 60, 65, 82, 10, 13, 19, 20, 21, 38, 48, 75, 8, 95, 97, 97, 98, 99,

100c, 101, 33, 62, 68, 70, 79, 80, 85, 87, 100b, 100d, 100e, 3, 4, 51,52, 77, 86, 89, 27, 31, 57, 59, 66, 35, 46, 58, 71, 84, 72.
Complete ECE Educational Article Survey, 18th April 2008, in order of most read. Space Energy (Spanish), Workshop Details, Overview and History, Crystal Spheres, Space Energy, Galaxies (Spanish), Workshop Introduction, ECE (French), 93 extra plots, ECE (English), ECE (Italian), Galaxies, ECE (Chinese), ECE (German), ECE (Spanish), Black Hole Catastrophe, Paper 93 Maxima Code, Effects of Homogeneous Current of ECE Theory, Galaxy Simulations, ECE (Russian).

END OF FOURTH YEAR OF RECORDING HIGH QUALITY FEEDBACK

---

1 - 15 May 2008

US Universities, Institutes and Similar.
   Alaska, Auburn, Colorado, Columbia, Cornell, Georgia Tech., Lehigh, MIT, Michigan State, New York Univ., Purdue, SCCNC, Akron*, UC Davis, U Dallas, Houston, Univ. Michigan School of Medicine, UWM, Washington*, US Gov. CDC, NOAA, Santa Barbara, St. Lucie County, USAAF Robins*, US Army Hood, US Army Rucher, US Navy NMCI, Duke Energy, Intel, Mobile Gas, Motorola.

Europe and Asia Minor: Universities, Institutes and Similar.
   CSC Austria, Charles Univ. Prague*, Palacky*, Max Planck Heidelberg, Siemens*, Hamburg, Hannover, Paderborn, Poitiers, Trinity College Dublin*, SISSA Italy, Rome 3, Twente, NTNU Norway*, Wroclaw, EB2 Portugal, RDS Net Romania, IFSAB Sweden, EDUS Slovenia, Agric. Univ., Slovakia, Iyte Turkey, Metu, SDU Turkey, Imperial, Manchester, UWIC, York*, British Gov. Energis.

Rest of World: Universities, Institutes and Similar.
   UOW Australia, UFMG Brazil*, McMaster Canada, HA China, EAFIY Colombia, BARC India, RRI India, Hiroshima, Nara Su, NU Singapore.

15 - 30 May 2008

US Universities, Institutes and Similar.
   Brigham Young, Caltech, CHOP, Colorado, Columbia, Iowa State, MIT, MTU, MU Ohio, North Dakota, Oneanta, PDX, Renselaer, UC Irvine, Texas Arlington, Texas, Washington*, Los Alamos, USAAF

Robins*, US Army USAR, US Navy NMCI, Southwestern Labs., Boeing, Botevgrad, Energetiq, Lockheed Martin, Microsoft, Mobile Gas, Motorola, Northrop Grumman.

Europe and Asia Minor: Universities, Institutes and Similar.
Medical Univ. of Vienna, Ghent, Charles Univ. Prague, Siemens*, Mainz, Mannheim, UGR Spain, Gov. Spain, Bordeaux, Poitiers, Trinity College Dublin*, ENEA Italy, AO Careggi, CWI Netherlands, NTNU Norway*, Poznan, Porto, ISSR Russia*, Chalmers Sweden, LUTH Sweden, Umea Sweden, UU Sweden, SAU Slovakia, Aberystwyth, Cardiff, Univ Wales registry, Exeter, Kent, Liverpool, Liverpool Derby-Rath, Nottingham, Belgrade.

Rest of World: Universities, Institutes and Similar.
Canadian engineering net (Kitchener, Malton, Montreal, Ottawa, Quebec, Sherbrooke, St Hubert, Sudbury, Toronto, Hamilton, Windsor, London), Canadian Defence Dept., UCDSB Ontario, Univ Calgary, UCF Cuba, UO Cuba, PUCE Equador, Univ Goa India, Shiraz Iran, UNAM Mexico, UITM Malysia, UCT South Africa, Witwatersrand South Africa.

Summary for May 2008

Documents downloaded from site: 942.
Lead Countries: USA, France, South Africa, Germany, Britain, Italy, Argentina, Mexico, Canada, ... (76 countries).
Lead papers: 15, 53, 49, 68, 94, 65, 64, 45, 43, 103, 110, 111, 93, 109, 105, 63, 21, 20, 106, 37, 67, 104, 100a, 9, 102, 76, 56, 50, 54, 6, 12, 47, 90, 40, 7, 99, 61, 55, 30, 2, 16, 3, 4, 66, 4, 60, 92, ... All papers read
Lead Articles: Space energy (Spanish), ECE Engineering Model, ECE, ECE (Spanish), ECE (French), Workshop history, Crystal Spheres,Galaxies (Spanish), Space Energy, Galaxies, Lindstrom 1, Crothers 1, Crothers 2, ... All articles read.

Paper Survey carried out on 27th May 2008, in order of most read.
15, 53, 49, 58, 94, 65, 64, 45, 43, 110, 103, 111, 109, 105, 63, 106, 21, 1, 93, 67, 37, 104, 100a, 102, 14, 76, 9, 50, 54, 56, 12, 90, 6, 40, 47, 99, 7, 61, 55, 2, 16, 17, 20, 30, 3, 42, 4, 92, 57, 60, 66, 25, 28, 41, 5, 19,23, 38, 44, 101, 11, 26, 33, 34, 36, 52, 59, 69, 70, 89, 8, 96, 10, 32, 39, 51, 91, 95, 98, 100d, 22, 31, 35, 58, 75, 87,100b, 100e, 24, 29, 46, 74, 82, 83, 84, 18, 27, 62, 71, 79, 97, 13, 48, 72, 73, 77, 78, 80, 85, 86, 88, 81.

1 - 15 June 2008

US Universities, Institutes and Similar.
Berkeley, Brigham Young, Colorado, Georgia Tech., Johns Hopkins Applied Physics, MIT, Princeton*, Rutgers, SJSU, UN Irvine, Vanderbilt, Vermont, Washington, Wednet, Fermilab, NASA (JSC), USAAF Robins*, US Army Tobyhanna, US Navy NAVO, Cisco Corporation, Microsoft, Mobile Gas, Motorola*.

Europe and Asia Minor: Universities, Institutes and Similar.
ASN Vienna, Schotten Gymnasium Austria, Gov. Geneva, Charles Univ. Prague*, Palacky, FH Aachen, RWTH Achen, Siemens, TU Berlin, Aragon, Avarra, Valencia, CNRS Grenoble, CUDL Lille, Jussieu IHP, Jussieu Sisyple, Renault, Univ-mlv France, Marseilles, Mulhouse, Poitiers, INFN Naples, SNS Italy, Groningen, Delft, Amsterdam, Utrecht, NTNU Norway, Gdynia, Innovative Group Romania, Svoyo Russia, Umea Sweden, Cambridge, engineering Cambridge, Imperial*, Nottingham, Swansea, Ceredigion County Council, British Ministry of Defence.

Rest of World: Universities, Institutes and Similar.
CNEA Argentina, RMIT Australia, Queensland*, UFG Brazil, UNICAMP Brazil, Carleton Univ. Canada, Canadian ECE engineering network, Canadian Dept. of Defence, British Columbia, Calgary, Saskatechewan, Toronto, Fukui Univ. Japan, Tohoku, KAIST Korea, UNAM Mexico, USON Mexico, UPM Malaysia.

15 - 29 June 2008

US Universities, Institutes and Similar.
Brigham Young, Cornell, Georgetown, MEC, Missouri, UC San Diego, Minnesota, Urbana Champaign, UNLV, UTMB, US Gov BLM, St. Lucie County, USAAF Robins*, US Navy NMCI*, South Western Laboratories, Intel, Macmillan, Microsoft, Motorola, Northrop Grumman, Raytheon.
Europe and Asia Minor: Universities, Institutes and Similar.
ARCS Austria, Liege, Charles Univ. Prague, Palacky, Helmholtz Foundation Munich, KFA Juelich, Siemens*, Cologne, Siegen, Wuerzburg, SDU Denmark, Ovi Spain, TUT Finland, JPP France, Poitiers, ACN Greece, ENEA Italy, INFN Bari, Perugia, TU Delft, Utrecht, NTNU* Norway, PEA Antonio Macedo Portugal, RDS Net Romania, Joint Institutes for Nuclear Research Russia, Omsk, Ege Turkey, Glamorgan, Imperial, Keele, Luton, Sheffield, Swansea College, UWIC, Birmingham City Council.

Rest of World: Universities, Institutes and Similar.

CNEA Argentina, Gov. Victoria Australia, Gov. Brazil, Canadian ECE Engineering Network, Gov. Canada Defence, McGill, ACI Ontario, Montreal, EMSSA Chile, Biobio Chile, Cauca Colombia, DIBRYU India, BARC India, Niha Univ. Japan, UNAM Mexico, Otago NZ.

Summary for June 2008

Articles downloaded from website: 1000
Lead Countries: USA, Germany, South Africa, Canada, France, Poland, Britain, Finland, ... (78 countries).

Lead Papers: 63, 93, 47, 45, 94, 43, 6, 113, 112, 114, 14, 1, 76, 2, 21, 103, 109, 100a, 111, 40, 58, 102, 12, 61, 90, 44, 50, 54, 56, 105, 37, 7, 99, 16, 23, 65, 25, 67, 70, 88, 3, 11, 104, 15, 57, 13, 51, 66, 68, 81, 19, 62, 92, 100b, 30, 5, 96, 69, 71, 10, 110, 18, 55, 74, 87, 41, 83, 53, 60, 89, 4, 72, 98, 17, 22, 33, 36, 42, 46, 75, 8, 95, 101, 106, 20, 24, 28, 31, 32, 35, 38, 64, 108, 26, 59, 77, 9, 29, 34, 39, 52, 91, 100e, 49, 82, 86, 78, 80, 84, 85, 97, 48, 100d, 27, 73, 79, 100c, ... all papers read.
Lead Articles: ECE Engineering Model, Space Energy (Spanish), ECE (Spanish), ECE and Spacetime, Workshop Histroy, Crystal Sspheres, Galaxies (Spanish), ECE, ECE (French), ECE (Italian), Space Energy, Crothers 1, ... all articles read.

1 - 15 July 2008 : Armorial Bearings awarded, 7th July, and TGA Gold Medals and Prize.

US Universities, Institutes and Similar.

Berkeley, Brigham Young*, Caltech, FISK, Georgia Tech, Johns Hopkins Applied Physics, MIT, Rice, Tufts, U Mass., Union, Nevada Las Vegas, Wright, US Gov. BLM, Fermilab, NASA (gsfc), US Gov PS, USAAF Robins*, US Navy NVL, US Military Central Command Iraq*, US Navy NMCI, San Diego, South West Laboratories, Philips, Daimler Chrysler, Geneal electric, MacMillan, Mobile Gas, Motorola*, Northrup Grumman, Scana, Siemens (US).

Europe and Asia Minor: Universities, Institutes and Similar.

Vienna, Gov. Belgium, Sofia, ETH Zurich, Charles Univ. Prague, Palacky, Fabiankeil, FU Berlin, Siemens, Karlsruhe, Siegen, Bonn, BCN Spain, RAN, UCLM, UDC, UPC, USC Spain, DOD Finland, FREE France, renault, Poitiers, Patras, Tel Aviv, Antibiotics Italy, INRIM Italy, Turin Tech., Rupa, Ukim Macedonia, Morganit Netherlands, NTNU Norway, Warsaw, Lisbon Tech., Tubitak Turkey, Kharkov, Damtp

Cambridge, Hull*, NCL*, SIHE, Univ. College London, UWIC, Warwick, BBC, Macmillan, Ceredigion.

Rest of World : Universities, Institutes and Similar.
Gov. Canada, Western Ontario, Canadian engineering network, IVIC Venezuela.

15 - 30 July 2008

US Universities, Institutes and Similar.
Brigham Young, Cornell OPAC*, Dixie, Embry Riddle, GPC, Georgia State, Harvard, Princeton*, Purdue, Stanford*, SUNY Stonybrook, UCOK, Minnesota, UPR, Oak Ridge, St Lucie County, USAAF Robins*, US Navy NUWC, Apple, Chevron, Conoco, Philips, Intel, Mobile Gas, Motorola*, Rockwell Collins, Siemens Transport, Southwestern Laboratories.

Europe and Asia Minor: Universities, Institutes and Similar.
Siemens Austria, BAS Bulgaria, ETH Zurich*, CVUT Czechia, ATB Potsdam, FU berlin*, HU Berlin*, Max Planck Garching, Siemens*, Augsburg, Frankfurt, Karlsuhe*, Cologne, Barcelona, UCM, UIB, US Spain, EBS Cachan France, Renault, Nice, Poitiers, OTE Greece*, Combined Budapest Campi, Tel Aviv, ICTP Trieste, Calabria, Unimib, NTNU Norway*, NIPNE Romania, TE Group Russia, Towy, Edinburgh, Hull*, Kingston, Swansea, Swansea College, BBC, Guardian.

Rest of World : Universities, Institutes and Similar.
ISA Argentina, Griffith Australia, GU Australia, Toronto, Western Ontario, Canadian ECE Engineering network, NRC Canada, ENAP Chile, Chiba Univ Japan, UAM Mexico, Umich Mexico, UPMX Mexico.

Summary of July 2008

Documents downloaded from site: 919.
Lead Countries: USA, Canada, Germany, France, Britain, Australia, South Africa, ... 82 countries.
Lead Papers: 15, 94, 115, 105, 63, 113, 89, 93, 56, 21, 93 extra plots, 76, 61, 12, 14, 1, 43, 103, 114, 6, 7,82, 90, 67, 37, 99, 100a, 112, 2, 108, 30, 53, 65, 92, 109, 10, 50, 111, 110, 13, 23, 40, 57, 64, ... All papers read.
Lead articles: Spacetime Devices, ECE Engineering Model, Space Energy (Spanish), ECE and Spacetime, ECE (Spanish), Crystal Spheres, ECE, ECE (Italian), Space Energy, Galaxies (Spanish), Workshop History, Lindstrom 1, ECE (French), Spacetime Devices (Spanish), Armorial Bearings and Badge, Royal Decree, Crothers 2, ... all articles read.

1 - 15 August 2008

US Universities, Institutes and Similar.

Arizona State, Brigham Young, Cornell OPAC, Florida Atlantic, Florida International, Princeton IAS*, MIT, SJSU, UC Irvine, UPR, US Airforce Academy, Xavier, New York City CCEA, US Gov. SSA, St Lucie County, USAAF Brooks, USAAF Wright Patterson*, USAAF Ribins, US Army Amedd, US Navy NMCI*, US Navy NUWC, APEX Telescope, Boeing*, IBM, Mobile Gas, Motorola, Rockwell Collins.

Europe and Asia Minor: Universities, Institutes and Similar.

Liege, ETH*, GBH Hannover, Siemens, RUC Denmark, UCM Spain, CSNRS LPN, Brest, Poitiers, Trinity College Dublin*, ICTP Trieste, ENT Latvia, Krakow, Gdynia, TRE Sweden, Metu Turkey, Towy Aberystwyth, particle Cambridge, Hull*, physics Oxford, Swansea, Univ. College London, UWIC, Ceredigion, Ministry of Defence, National Library of Wales.

Rest of World: Universities, Institutes and Similar.

UNR Argentina, Gov. of Queensland, Canadian ECE engineering network, McGill, Queen's, UTA Chile, MMD India, Titech Japan, UNAM Mexico, USON Mexico, UMP Malaysia.

15 - 29 August 2008

US Universities, Institutes and Similar.

Brigham Young, Colorado College, Texas A and M, Temple, UC Irvine, Illinois Chicago, Urbana Champaign, Nevada Reno, Washington, US Gov FDA, US Gov. SSA, US Dept. of State, St. Lucie County, USAAF Robins*, US Navy NMCI, AT and T, General Electric, Hewlett Packard, Motorola, Rockwell Collins.

Europe and Asia Minor: Universities, Institutes and Similar.

Basel, Charles Univ. Prague, GBH Hannover, Siemens, TU Clausthal, Bremen, RUC Denmark, Savoie, PGSM Hungary, Jerusalem, INFN Florence, ICTP Trieste, LTH Sweden, UU Sweden, SPP Slovakia, Cardiff, Liverpool, Nottingham*, Paisley.

Rest of World: Universities, Institutes and Similar.

AMC Australia, Melbourne, UNB Brazil, UNICAMP, Sao Paolo, Canadian Engineering Net, Canadian Gov., McGill, NCF Canada,

Queen's, UO Cuba, PRL India, ITESM Mexico, UMICH Mexico, UMP Malaysia.

Summary for August 2008

Number of documents downloaded from site: 521
Lead countries: USA, Britain, Germany, Canada, France, Taiwan,..... (73 countries).
Lead papers; 45, 94, 116, 107, 115, 117, 21, 2, 63, 93, 1, 89, 103, 113, 61, 76, 12, 90, 7, 109, 112, 114, 93 plots, 11, 43, 92, 15, 3, 105, 108, 5, 10, 111, 19, 56, 100a, 106, 110, 14, 23, 67, 8, 16, 37, 47, 68, 9, 100b, 104, 13, ... All papers read.
Lead articles: Space Energy (Spanish), Crystal Spheres, ECE and Spacetime, ECE Engineering Model, Galaxies (Spanish), ECE, Black Hole criticism by Crothers, Space Energy, Galaxies, Workshop overview, ECE (French), ECE (Italian), Armorial Bearings, Crothers 1, ECE (German), Crothers 2, Lindstrom 1, ... all articles read.
One Year Survey of Ten Most Read ECE Papers (in each month all papers were read)

Sept 08: 117, 94, 14, 47, 113, 1, 21, 107, 115, 93, ...
Aug. 08: 45, 94, 116, 107, 117, 115, 21, 2, 63, 1, ...
Jul 08:   15, 94, 115, 105, 63, 113, 21, 89, 93, 56, ...
Jun 08:   63, 93, 47, 45, 94, 43, 6, 113, 112, 114, ...
May 08: 15, 53, 49, 68, 94, 65, 64, 45, 43, 110, ...
Apr 08: 15, 94, 109, 54, 39, 106, 100a, 43, 102, 105, ...
Mar 08: 43, 94, 67, 106, 63, 6, 89, 100a, 93, 105, ...
Feb 08: 43, 94, 63, 103, 37, 93, 100a, 17, 89, 104, ...
Jan 08: 103, 21, 2, 43, 63, 102, 46, 54, 9, 89, ...
Dec 07: 6, 2, 103, 43, 76, 63, 102, 54, 101, 51, ...
Nov 07: 2, 89, 87, 43, 21, 26, 63, 99, 76, 67, ...
Oct 07: 93, 2, 63, 43, 23, 76, 93 extra plots, 15, 9, 13, ...

Six Month Survey of Most Read Articles (in each month all articles were read).

Sept 08: ECE (Dutch), Space Energy (Spanish), ECE (Spanish), ECE (Italian), Workshop History, ECE, ECE and Spacetime, ...
Aug 08: Space Energy (Spanish), ECE (Spanish), Crystal Spheres, ECE Engineering Model, ECE, Black Hole criticisms, Space Energy, Galaxies, ...
Jul 08: Spacetime Devices, ECE Engineering Model, Space Energy (Spanish), ECE and Spacetime, ECE (Spanish), Crystal Spheres, ECE, ECE (Italian), ...

Jun 08: ECE Engineering Model, Space Energy (Spanish), ECE (Spanish), ECE and Spacetime, Crystal Spheres, Workshop History, ECE, Galaxies (Spanish), ...

May 08: Space Energy (Spanish), ECE Engineering Model, ECE, ECE (Spanish), ECE (French), Workshop History, Crystal Spheres, Galaxies (Spanish), ...

Apr 08: Space Energy (Spanish), Workshop Details, ECE, Workshop History, Crystal Spheres, Crothers 1, ECE (French), Space Energy, ...

1 - 15 September 2008

US Universities, Institutes and Similar.

Espol, Bloom Univ., Brigham Young, Columbia, City Univ. New York, CWRU*, Haystack, IUPUI*, MEC, North Carolina State, North Dakota, OCIW, Radford, SJSU, SUNY Stonybrook, Syracuse, Illinois Chicago, Michigan, Washington, US Gov DHS, US Gov SSA, USAAF Robins*, US Navy Norfolk, US Navy NMCI*, American Institute of Physics, Hewlett Packard, Intel, Mobile Gas, Motorola, Raytheon, Rockwell Collins.

Europe and Asia Minor: Universities, Institutes and Similar.

Harrow School, Salzburg, CERN, Neufchatel, Siemens, Saarland, Ulm*, UPM, USC, UV Spain, HTV Finland, Renault, Poitiers, MTV Hungary, Weizmann*, Vilnius Lithuania, CNR IENI Italy, INFN Naples, UNFN Turin, NTNU Norway, KM Russia, SNT Slovenia, GTS Slovakia, particle Cambridge, Cardiff, Hull, Imperial*, Meirion Dwyfor, Swansea, Univ. College London, UMIST, UWIC, Ceredigion.

Rest of World: Universities, Institutes and Similar.

CSIRO Australia, Gov. Queensland, UFFS Brazil, UFMG Brazil, Alberta, Canadian Engineering Network, UTA Chile, HA China, EDU Equador, BHU India, UM Iran, Titech Japan, KAIST Korea, Yonsei Korea, USON Mexico, TUT South Africa, UKZN South Africa.

16 - 29 September 2008

US Universities, Institutes and Similar.

Augsburg, BEREA, Berkeley, Brigham Young, Clemson, Colorado, CSLB, City Univ New York, Embry Riddle*, Florida State, Georgia Tech, Marlboro, ODU, Princeton, Purdue, Renselaer, Rutgers, Texas A and M, TCE, Tufts, UC Santa Cruz, Vermont, Washington, US Gov. BNL, Gov GSA, NASA (larc), US Gov. SSA, St Lucie County, USAAF Kirtland, USAAF Robins*, US Army AMEDD, US Navy NMCI*, US Navy

NUWC, Southwestern labs., Adobe, IBM, Intel, Motorola*, Mobile Gas, Rockwell Collins.

Europe and Asia Minor: Universities, Institutes and Similar.
Gleidorf Gymnasium Austria, BOL Bulgaria, HS Niederrhein, Max Planck MPPMU, Siemens, Bonn, Hannover, Karlsruhe, Ulm, Madrid 2, USC Spain, HTV Finland, Siemens France, Lorraine, Orleans, Poitiers, OTE Greece, MTV Hungary, Jerusalem, ENI Italy, INAF Rome, Pavia, European Parliament Luxembourg, NTNU Norway, Warsaw, Paradigm Axis Portugal, Joint Institutes for Nuclear Research Russia, KTH Sweden, SNT Slovenia, BCU, Bristol, Cambridge*, Cardiff*, Gateshead, Hull, Lancaster, UMIST.

Rest of World: Universities, Institutes and Similar.
Siemens Australia, Swinburne, New South Wales, Schools Network Australia, Sherubtse Bhutan, Defence Canada, McGill, Quebec, British Columbia, Calgary, Canadian ECE Engineering Network, Institute of High Energy Physics China, KAIST, Victoria Univ Wellington NZ, TYRC Taiwan.

Summary of September 2008

Documents downloaded from site: 1642.
Lead countries: US, Canada, Britain, Netherlands, France, Italy, Germany, ... (85 countries).
Lead Papers: 117, 118,63, 55, 94, 43, 1, 37, 14, 93, 21, 113, 47, 107, 12, 8, 115, 100a, 100b, 11, 15, 76, 30,56, 2, 61, 67, 102, 13, 51, 65, 7, 12, 109, 23, 3, 82, 44, 6, 99, 103, 105, 111, 17, 38, 89, 90, ... all papers read.
Lead articles: Space Energy (Spanish), Workshop History, ECE (Spanish), Galaxies (Spanish), ECE, ECE and Spacetime, Crystal Spheres, Galaxies, ... all articles read.

1 - 15 October 2008

US Universities, Institutes and Similar.
Arizona, Berkeley, Binghampton, Brigham Young, COD, CSU Pomona, Louisiana Tech., MST, North Dakota, SCAD, SIUE, Stanford, Chicago, US Gov. DHS, NIH, US Gov SSA Gov. Virginia, USAAF Robins*, US Navy NUWC, American Institute of Physics, Saddlebrook Schools, Motorola, Northrop Grumman, Rockwell Collins.

Europe and Asia Minor: Universities, Institutes and Similar.

Vienna, Charles Univ. Prague, Siemens, Heidelberg, Regensburg, MVB Denmark, CNAM France, ENS Cachan, Bordeaux, Poitiers, NTUA Greece, Tellas Greece, RUPA Italy, Genoa, Erasmus MC Netherlands, VU Netherlands, UIO Norway, IPP Portugal, UCV Romania, Lobnya Russia, UU Sweden, King's College Cambridge, particle Cambridge, Imperial, Lancaster, Liverpool, St Hugh's Oxford, SIHE, St Andrews*, Staffordshire, Westminster, National Library of Wales.

Rest of World: Universities, Institutes and Similar.

Gov. Argentina, Melbourne, New South Wales, Queensland, UNICAMP, Canadian ECE engineering network, Perimeter Institute, British Columbia, Waterloo, York Univ. Toronto, UCN Chile, UDEA Colombia, KAIST Korea.

15 - 30 October 2008

US Universities, Institutes and Similar.

ACCD, Brigham Young, CHW, Columbia*, Harvard, Mercyhurst, Middlebury, Colorado School of Mines, MIT, Northwestern, Mississippi, Penn State, Richmond, Rochester Tech, STCHAS, Stonybrook, Texas Tech., Florida, Urbana Champaign, Kentucky, YC, US Gov BNL, US Navy NUWC, Microsoft, Motorola, Northrop Grumman, Rockwell Collins.

Europe and Asia Minor: Universities, Institutes and Similar.

Vienna, KU Leuven, BAS Bulgaria, German Synchrotron Facility, HS Niederrhein, Max Planck Dresden, RWTH Aachen, Siemens*, Heidelberg*, Leipzig, Regensburg, MVB Denmark, Rioja Spain, Bayonne France, Poitiers*, Iskon Croatia, Combined Campuses Budapest, Tel Aviv, INFN Sardinia, BTV Latvia, UIB Norway, UTCN Romania, ETA network Sweden, Josef Stefan Slovenia, Aradolu Turkey, ILT Kharkov, BSFC, Wolfson College Oxford, particle Cambridge*, Hull, Imperial, Liverpool*, Meirion Dwyfor, Paisley, St Andrews*, Univ. College London, Westminster.

Rest of World: Universities, Institutes and Similar.

Queensland, Gov. Brazil, FEE UNICAMP Brazil, Gov. British Columbia, Canadian Defence, MD Robotics Canada, Waterloo*, York Univ Toronto, Canadian ECE Engineering Network, Univ. Andes Colombia, UO Cuba, BHU India, KAIST Korea, RUH Sri Lanka, Auckland NZ.

Summary of October 2008

Number of documents downloaded from site: 1279.
Lead countries: USA, France, Canada, Russia, Britain, Germany, Italy, Brazil, ... (75 countries).
Lead papers; 63, 94, 43, 47, 92, 118, 93, 117, 113, 21, 120, 37, 61, 107, 111, 115, 100a, 112, 89, 76, 103, 104, 2, 99, 119, 15, 100d, 93 extra plots, 14, 82, 109, 75, 100b, 110, 8, 90, 100c, 114, 7, 5, 98, 11, 13, 45, 65, 74, ... all papers read.
Lead articles: Space Energy (Spanish), Galaxies (Spanish), Workshop history, ECE (Spanish), Crystal Spheres, ECE and Spacetime, The Life of Myron Evans, ECE Engineering Model, ... all articles read.

1 - 16 November 2008

US Universities, Institutes and Similar.
Austin CC, BGSU, Biola, Caltech, Columbia*, Cooper, Florida Atlantic, Florida International, Johns Hopkins Applied Physics, Louisville, Maricopa, MIT, MSOE, MTU, NOVA, Oklahoma State, Princeton*, Rochester, Renselaer, Rutgers, Syracuse, Texas A and M, Toronto* (on edu), UC Davis, Chicago*, Iowa, Urbana Champaign, UMKC, Minnesota, Montana, UMSL, Omaha, South Florida, Utah, UTEP, Washington, William and Mary, Yale, US Gov. NOAA, US Gov USPTO, USAAF Robins*, Southwestern Labs., Adobe, Duke Energy, Raytheon.

Europe and Asia Minor: Universities, Institutes and Similar.
TU Vienna, UIBK Austria, KU Leuven, ACS Bulgaria, Fraunhofer network, Siemens*, Students Bonn, TU Darmstadt, TU Hamburg, Bonn, Jena*, Mainz, Munich, Unity Media Group, Danish Tech., GVA Spain, UAB Spain, ENS Montpellier France, French national particle lab., INRA Nancy, Poitiers*, Thessalonika, Trinity College Dublin, TU Eindhoven, Twente, Utrecht, HIO Norway, Wroclaw, NSC Russia, LU Sweden, SNT Slovenia, GTS Slovakia, AU Slovakia, ITU Turkey, Metu, Uludag Turkey, Astronomy Cambridge, high energy Hull, Imperial, Lancaster, Liverpool, Manchester, Magdalen Oxford, math Oxford, phys Oxford, East Anglia.

Rest of World : Universities, Institutes and Similar.
Monash, UFPB Brazil, UFRJ, UNICAMP, Engineering College Colombia, Canadian ECE engineering network, UCLV Cuba, UO Cuba, UH Cuba, BARC India, AMD Gov. India, BARC Gov. India, UM Iran, Keio,Tohoku Japan, KAIST Korea, SK, KLN Sri Lanka, BOP Tech NZ, Otago NZ*, NU Singapore, HCMUT Venezuela,

US Universities, Institutes and Similar.

Arizona State, Bates, Berkeley, Boston Univ., Caltech, Columbia*, Cornell*, CSU Chico, CWRU, Duke, ECPI, Eastern Kentucky, Embry Riddle, Lane, Lehigh, MIT, MNSCU, Montgomery, MS State, MTU, North Dakota, NEU, Oklahoma State*, Princeton*, Rutgers, Santa Rosa, South Carolina, Swarthmore, Syracuse, Texas A and M*, UC Irvine, UCLA, Illinois Chicago, UMSL, UMS Medical, North Carolina Charlotte, Texas, VCU, Weber*, WIU, West Michigan, WVU, US Gov FDA, Oak Ridge, St Lucie County, USAAF Misawa, USAAF Robins*, US Army Ace, US Navy NMCI*, Berkeley, Botevgrad, Gulf Aero, Honeywell, Microsoft, Qinetiq, Raytheon.

Europe and Asia Minor: Universities, Institutes and Similar.

OENB Austria, Free Univ. Brussels, KU Leuven, Ghent, Fujitsu Siemens, MUNI Czechia, Palacky, German Synchrotron Facility, Heidelberg, Fraunhofer system*, FU Berlin, Albert Einstein Institute, Max Planck IPP, Max Planck Heidelberg, Siemens Inc., Bielefeld, Hamburg, Leipzig, Munich, Stuttgart, CSI Spain, GVA, UCM, Jaen, UPV Spain, Helsinki, CNAM France, Lille 1, Mulhouse, Paris 5, Poitiers*, NTUA Greece, Combined campuses Budapest, Technion Israel, Astro Italy*, ENEA Italy, INFN Bologna, ICTP Trieste, Bologna, Perugia, TEO Latvia, RU Netherlands, Groningen, Kluwer, UIO Norway, FUW Poland*, Krakow, Torun, Wroclaw, NSC Russia, LU Sweden*, Physto Sweden*, SU Sweden, SNT Slovakia, TUKE Slovakia, Novosibirsk Russia, Kiev Ukraine, Brunel, Damtp Cambridge, WLAN Durham, Imperial*, Leicester, Manchester, Meirion Dwyfor, Nottingham, Southampton, South Devon, St Andrews*, Strathclyde, European Space Agency.

Rest of World: Universities, Institutes and Similar.

UNC Argentina, CNEA Argentina, ANU, UNICAMP Brazil, ECE Canadian engineering network, Dalhousie, Gov. Quebec, Perimeter Institute, Queen's, Manitoba, Toronto, Waterloo, UCLV Cuba, UO Cuba, ZNU Iran, Hokudai, Tokyo, KAIST Korea, USON Mexico, PUCR Peru, Academica Sinica Taiwan, HCMINT Venezuela.

Summary of November 2008

Number of Documents downloaded from site: 1203
Lead countries: USA, Canada, Russia, France, Germany, Mexico, Britain, Italy, ... (84 countries).

Lead papers: 122, 117, 64, 116, 94, 121, 43, 41, 110, 113, 114, 25, 1, 76, 74, 99, 112, 102, 75, 109, 81, 120, 30, 52, 86, 93, 82, 90, 67, 85, 89, 91, 104, 95, 2, 88, 92, 37, 63, 103, 115, 61, 80, 21, 7, 98, 118, 93 extra plots, 105, 106, 42, 8, 56, 60,100a, ... all papers read.

Lead articles: Spin Connection Resonance in Spacetime, Space Energy (Spanish), ECE (Spanish), Galaxies (Spanish), SCR in Spacetime (Spanish), ECE Engineering Model, The Life of Myron Evans, Crystal Spheres, ... All documents read.

Complete Paper Survey Carried out on 19th Nov. 2008 (in order of most read).

117, 64, 116, 94, 43, 25, 121, 110, 76, 75, 113, 1, 114, 102, 74, 112, 99, 90, 2, 67, 93, 120, 21, 61, 63, 7, 89, 81, 82, 91, 92, 100a, 118, 80, 109, 115, 95, 93 extra plots, 120 tests, 37, 60, 78, 105, 3, 56, 88, 104, 14, 15, 20, 52, 86, 8, 106, 111, 119, 65, 73, 103, 11, 71, 97, 30, 44, 45, 40, 49, 57, 70, 9, 100c, 101, 12, 38, 41, 42, 72, 100b, 13, 17, 26, 32, 35, 53, 54, 58, 5, 66, 84, 100d, 108, 16,19,28, 34, 47, 48, 59, 62, 77, 83,107, 22,39, 68, 98, 10, 23, 29, 36, 4,50, 51,79, 96, 18,24, 55, 69, 87,27,33,46,6,85, 31.

Top Twenty ECE Article Survey 19th Nov. 2008 (in order of most read)
1) ADevices for Spacetime Resonance@ by Horst Eckardt.
2) ASpace Energy Devices@ (Eckardt), Spanish version by Alex Hill.
3) ECE Spanish (originally by Eckardt and Felker).
4) EEC English.
5) AGalaxies@ by Horst Eckardt, (Spanish).
6) ECE Munich Workshop organized by Horst Eckardt (History slides).
7) ECE Engineering Model Slides by Horst Eckardt.
8) AThe Life of Myron Evans@ by Kerry Pendergast.
9) ANumerical Methods for the Resonance Solution of the Coulomb Potential@, by Douglas Lindstrom.
10) ASpace Energy@ by Horst Eckardt.
11) ACrystal Spheres@ by Kerry Pendergast.
12) AECE and Spacetime@ by Horst Eckardt.
13) ECE Italian.
14) Spanish translation of (1) by Alex Hill.
15) ABlack Holes, Unicorns and All that Jazz@ by Stephen Crothers.
16) ECE French.
17) Eckardt Theory of Special Relativity, correcting errors by Einstein.
18) IJTP 1962 by Stephen Crothers.
19) IJTP 1960 by Stephen Crothers.
20) Criticims of Ricci flat concepts by Stephen Crothers.

US Universities, Institutes and Similar.

Arizona, Arizona State, Athens*, American Univ. Cairo, Bard, Berkeley, Boise State, Boston Univ., Brigham Young, Carnegie Mellon, Colorado College, Cornell, Florida International, Georgia State, Indiana, JMU, Juniata, Kettering, KSBE, Louisville, Louisiana State, Maine, MIT, Morgan, North Carolina State, North Dakota, New York Tech., New Mexico State, UCLA, U Connectictut, UC Santa Barbara, UC San Diego*, New Mexico, Oregon, Texas, Vanderbilt, Williams, WVNC, Yale, US Dept. of Education, US Army Tacom, US Army Tobyhanna, US Naval Research Laboratory*, Southwestern Laboratories, Goodrich, Hewlett Packard, IBM*, Microsoft*, Raytheon.

Europe and Asia Minor: Universities, Institutes and Similar.

Vienna, Linz, TU Vienna, Gov. Geneva, Palacky, Jacobs Univ., KFA Juelich, Bochum, RWTH Aachen, Siemens*, Giessen, Heidelberg, Jena, Cologne, Leipzig, Potsdam, Schwabia, Stuttgart, CSIC Spain, UPV, Vigo, HUT Finland, TUT Finland, ISFN France, LPTL Jussieu, Marseilles 3, Strasbourg, Poitiers*, ADSL hungary, ELTE Hungary, Trinity College Dublin, Univ. College Dublin, Tel Aviv, Weizmann, HI Iceland, INFN Bari*, Subria, Pisa, VGTU Lithuania, BAT Software Latvia, VU Netherlands, Delft, Lodz, Wroclaw, ASTRAL Romania, INFIM Romania, UAB Romania, Kirov, MTS-NN Russia, Chalmers Sweden, ORU, SU, UU Sweden, Metu Turkey, Aberystwyth, Cardiff, Durham, Glamorgan, Heriot Watt, Imperial, Manchester*, Meirion Dwyfor, MMU, NCL, Oxford*, Sheffield, St Andrews*, Univ College London*, British Gov. NICS, Belgrade.

Rest of World: Universities, Institutes and Similar.

Swinburne Australia, UERJ Brazil, Canadian ECE Network, Gov. Quebec, MD Robotics Canada, Alberta, British Columbia, Toronto, Waterloo, ESPL Ecuador, Veccal India, Chuo Univ. Japan, Metropolitan Univ. Japan, Shimare Japan, Titech, KEK , NAIST Japan, KAIST Korea, UASLP Mexico, USON Mexico, Knox STC NZ, PERN Pakistan, NTHU Taiwan.

15 - 30 December 2008

US Universities, Institutes and Similar.

AUC Egypt, Cornell, Indiana, Kansas State, Ohio State, Oklahoma State, UCSC, Houston, US Naval Academy, Yale, Jefferson Particle Laboratory, Oak Ridge, St Lucie County, USAAF Robins, US Army

PICA, US Army Tactical Command, US Army Tobyhanna, US Army ASACE, US Navy NMCI, US Naval Research Laboratory, US NAVY NUMC, Honeywell, Intel*, Mobile Gas, Motorola.

Europe and Asia Minor: Universities, Institutes and Similar.
Linz, Geneva, ETH Zurich*, Charles Univ. Prague, Palacky, Siemens, TU Berlin, Munich, Saarland, NBT Denmark, UPV Spain, Poitiers, Savoie, ADSL Hungary, Szeged, RETE Tuscany, European Parliament, RU Netherlands, Utrecht, KKI Krakw, Lodz, Poznan, NIPNE Romania, Kirov, MTU Network Russia, Chalmers Sweden, Gotland Sweden, SU students Sweden, UU students Sweden, Hacetteppe Turkey*, KU Turkey, Metu*, Odessa Ukraine, NTNU Norway, Bristol, DOW students Cambridge, ESRC, Manchester, Meirion Dwyfor, St Andrews*, Ceredigion, Staffordshire Council, NHS.

Rest of World: Universities, Institutes and Similar.
NLA Australia, Canadian Defence, Canadian ECE engineering network, Gov. Newfoundland, Biobio Chile, Tsinghua China, UGM Indonesia, JNU India, IUCAA India, Gov India, PRL, PRI India, Tokyo, Hitachi, KAIST Korea*, ITESM Mexico, USON Mexico, NUS, CCU, Academica Sinica Taiwan.

Summary of December 2008

Number of documents downloaded from site: 1595.
Lead countries: USA, France, Germany, Canada, Britain, Australia, South Korea, ... (82 countries).
Lead Papers: 120, 124, 118, 64, 123, 94, 122, 113, 110, 115, 43, 76, 54, 114, 121, 41, 103, 112, 99, 88, 81, 111, 95, 93, 109, 21, 74, 82, 14, 61, 98, 89, 102, 105, 1, 104, 91, 15, 90, 100a, 92, 75, 85, 52,56, 106, 119, 30, 107, 116, 37, 40, 34, 60, 6, 84, 53,7, 80, 13, 17, 20, 63, 97, 57, ... all papers read.
Lead articles: Spacetime Devices, 2DFE Part One (Lindstrom), Life of Myron Evans, Workshop History, Space Energy (Spanish), ECE Engineering Model, ECE (Spanish), Galaxies, Spacetime Devices (Spanish), ECE and Spacetime, Galaxies (Spanish), ECE, Felker3 (Spanish), Crystal Spheres, Crothers IJTP1962,........ all articles read.

END OF 2008

1 - 15 January 2009

US Universities, Institutes and Similar.

City Univ New York, Florida Atlantic, Harvard Students, Northwestern, Ohio State, Princeton, Purdue, Rice, Texas Medical Center, UC Davis, UC Irvine, Illinois Chicago, Michigan, U Penn, US Naval Academy, Washington, Wisconsin, Argonne*, Fermilab, NAS Jet Propulsion, St. Lucie County, USAAF Robins*, USAAF Wright Patterson*, US Navy NMCI, US Naval Research Laboratory, US Navy NUWC, Carnegie Tech Organization, Boeing, IBM Thomas J. Watson, Intel, Microsoft, Mobile Gas, Motorola, Raytheon.

Europe and Asia Minor: Universities, Institutes and Similar.

Vienna, OMA Belgium, Ghent, ITO Switzerland, Charles Univ. Prague, Palacky, Brandenburg, FU Berlin, Albert Einstein Institute, Max Planck MPE, Siemens, TU Berlin, TUNS, Bayreuth, Bremen, Hamburg, Jena, Leipzig*, UPM Spain, USC, HTV Finland, JYU Finland, CEA France, CNRS Grenoble, ENS Lyon, Strasbourg, Potiers, Rheims, AERA Greece, AUTH Greece, NTUA Greece, Tellas, Tel Aviv, Weizmann, ARVI Lithuania, ONS Student Net Netherlands, RU Netherlands*, UIO Norway, Lisbon, Minho Portugal, ATRAL Romania, FOI Sweden, LTH Sweden, Bilkent Turkey, Metu*, Kiev, MK Ukraine, Aberdeen, Aberystwyth*, Jesus College Cambridge, physics Cambridge, Trinity Cambridge, Trinity Hall Cambridge, EPSRC, Glasgow, Hull, Imperial, Liverpool, Meirion Dwyfor, Maths Oxford, VPN Oxford, Solent, TVU, West England, Wellcome Trust, BBC, Dudley Council, SNT Slovenia., AUB Lebanon.

Rest of World: Universities, Institutes and Similar.

Gov. Argentina, EQ Australia, Queensland, UFPR Brazil, UNICAMP, Canadian ECE engineering network, Gov. Quebec*, Alberta, UV Chile, Tsinghua China, RCIL India, Saitama Japan, Victoria Univ Wellington NZ, NU Singapore, CCU, NCU Taiwan.

15 - 30 January 2009

US Universities, Institutes and Similar.

SUNY Buffalo, Brigham Young, Colorado State, Case Western Reserve*, Embry Riddle, Evergreen, Georgia Tech., GMU, GPC*, Harvard*, Indiana, Lousiana State*, Colorado School of Mines, NAU, Mew Mexico State, ODU, Rowan, Sabanci, Stanford*, Toronto (on edu), Arkansas, Central Florida, Chicago, UC Irvine, UCLA, U Connecticut, UC Santa Barbara, U Mass Dartmouth, Chapel Hill, New Hampshire, Yale,

Argonne, Fermilab, USAAF Robins*, US Army Amrdec, US Army ARO, US Army Cecer, Southwestern Labs, Adobe, Boeing*, Bose, Botevgrad*, Honeywell, Microsoft*, Mobile Gas, UPS.

Europe and Asia Minor: Universities, Institutes and Similar.

Graz, JKU Austria, Liege, VKI Belgium, IMEC Belgium, EPF Lausanne, ETH Zurich, Belfort, Charles Univ Prague*, Palacky, Albert Einstein Institute, Max Planck Heidelberg, Siemens, TU Freiburg, Bremen, Frankfurt, Hamburg, Heidelberg, Karlsuhe*, Kassel*, Leipzig, Munich, Osnabrueck, Saarland, Unity Media Group, Danish Tech., Gov. Andalucia, MDE Spain, UB, UCM, UPM, UPV*, USC, UVA* Spain, HTV Finland*, HUT, UTU Finland, INSA Lyon, Univ-Ag France, Poitiers, AUTH, OTE Greece, Thessalonika, ISKON Croatia, NYME Sopron Hungary, HUJI Israel, INFN Florence, INFN Milan, INFN Rome 2, SISSA Italy, AREA Trieste, Unimo Italy, Sannio, Turin, European Parliament, IAE Netherlands, TU Delft, Free Univ. Amsterdam, UIO Norway, AP Krakow, IFMPAN Poznan, TU Szczecin, Lisbon Tech., ATRAL Romania, IGCTI Romania, Joint Institutes for Nuclear Research Russia, NRSL Russia, LTH Sweden, Physto, SU Sweden, Uni-Mb Slovenia, Ege, Gyte, Metu Turkey, MK Ukraine, Towy Aberystwyth, Jesus College Cambridge, physics Cambridge, Queens' College Cambridge, Prifysgol Cymru, Edinburgh, Glasgow, King's College London, Leeds*, Meirion Dwyfor, chemistry Oxford, Magdalen Oxford, offices Oxford, phys Oxford, University College Oxford, Queen Mary College, SEDC Sheffield, Southampton*, TVU, Univ. College London, ULL, Univ. Wales, Warwick, British Gov. GSI, AUB Lebanon.

Rest of World: Universities, Institutes and Similar.

Sherjah, United Arab Emirates, CNEA Argentina, Monash, Gov. Brazil, UFF Brazil, UNB, UNICAMP, Sao Paolo Brazil, Canadian ECE Net, Gov. Quebec, McGill, McMaster, NS Canada, Perimeter Institute, Davison College, British Columbia, Toronto, Victoria, UV Chile, ITP China, City Univ. Hong Kong, DNS India, Amrita, BHU, DU, NITRKL India, Gov India RCIL, Osaka, Tokyo, Sony Company, Itchihuahua Univ. Mexico, IPN, UASLP, UNAM, USON Mexico, UTHM Malaysia, ADDU Philipines, NUS Singapore, NCTU, NCU, NSYSU, Academica Sinica Taiwan.

Summary for January 2009

Number of Documents downloaded from site: 1187.
Lead countries: USA, Canada, Germany, France, Britain, Japan, Mexico, Australia, ... (80 countries).

Lead papers: 126, 125, 94, 110, 41,43, 116, 99, 103, 76, 112, 115, 88, 123, 124, 108, 91, 47, 74, 122, 98, 1, 77, 111, 81, 100a, 104, 105, 109, 37, 82, 95, 102, 10, 114, 90, 71, 93, 113, 92, 18, 84,117, 30, 56 107, 34, 86, 106, 11, 12, 21, 2, 38, ... all papers read

Lead articles: AThe Life of Myron Evans@ by Kerry Pendergast, Douglas Lindstrom paper 2, Space Energy (Spanish), ECE (Spanish), Workshop history, ACrystal Spheres@ by Kerry Pendergast, Galaxies (Spanish), ... all articles read.

Complete Paper Survey carried out on 23rd January 2009 (in order of most read).

94, 41, 110, 43, 125, 103, 99, 88, 115, 116, 124, 112, 76, 123, 47, 122, 77, 91, 1, 111, 98, 122 (Spanish), 114, 105, 37, 81, 82, 95, 104, 107, 109, 10, 18, 30, 102, 113, 90, 92, 93, 106, 56, 15, 34, 74, 75, 11, 21, 40, 60, 61, 84, 96, 120, 117, 38, 71, 7, 12, 45, 89, 108, 16, 54, 20, 27, 59, 79, 9, 14, 28, 32, 3, 63, 78, 121, 17, 2, 39, 65, 6, 86, 101, 13, 23, 49, 53, 80, 97, 119, 29, 33, 57, 62, 70, 73, 8, 100a, 118, 26, 46, 50, 51, 5, 67, 35, 55, 100d, 19, 22, 25, 31, 36, 4, 52, 58, 87, 100c, 100b, 44, 69, 85, 24, 42, 48, 66, 100e, 64, 72, 68, 126. All papers read.

1 - 15 February 2009

US Universities, Institutes and Similar.

Arizona*, Arizona State, Athens, Brigham Young, Caltech, Colorado State, Columbia, Cornell, CSU Fresno, Case Western Reserve, Duke*, Evangel, Evergreen, Florida Atlantic, Georgia Tech., Harvard, Illinois Tech., Louisiana State*, Maricopa, Colorado School of Mines, MIT*, MS State, Michigan State*, North Dakota, Old Dominion*, Ohio State*, Oregon State, Oklahoma, PDX, Penn State, Purdue, Radford, Trinity College, Californian system of universities, UCLA, UC Santa Barbara*, UC San Diego, Delaware*, Urbana Champaign, U Mass Dartmouth, Maryland*, Minnesota, US Naval Academy, Texas*, Washington, Fermilab, NASA (LARC), Santa Barbara, St. Lucie County, USAAF Robins, US Army APGEA, US Army ARC, Berkeley Library, Algonquin College, Botevgrad*, IBM (Britain), Microsoft*, Motorola*, Philips.

Europe and Asia Minor: Universities, Institutes and Similar.

Antwerp, KU Leuven, Sofia Bulgaria, ETH Zurich, CERN, VSB Czechia, KFA Juelich, Siemens, TU Berlin, Erlangen, Karlsruhe, Cologne*, Leipzig*, Unity Media Group, UGR, UV Spain, JYU Finland, CEA France, Strasbourg, Brest, Poitiers*, Rennes 1, Demokritos Greece, Patras, Sopron Hungary, ENEA Italy, INFN Florence*, IRES Piemonte, SISSA Italy, Pisa, Pavia, Twente Netherlands, Utrecht, Amsterdam, Kluwer, NTNU Norway, UIB Norway, UMK Poland, UTCCLUJ

Romania, Physto Sweden, SU Sweden, Joseph Stefan Slovenia, Ljubjliana, TUKE Slovakia, Metu, SDU Turkey, MK Ukraine, Brunel, BSFC, Astronomy Cambridge, Magdalen Cambridge, Queens' Cambridge, Glamorgan, Imperial*, Manchester, Meirion Dwyfor, Nottingham, Open, Merton Oxford, Robots Oxford, Plumpton, UMIST, Wolverhampton.

Rest of World: Universities, Institutes and Similar.

CSIRO Australia, Gov. British Columbia, Canadian ECE Network, Lakehead Univ., IAW Ontario, Queen's, British Columbia*, Toronto, Engineering Univ. Colombia, TUE Colombia, RCIL Gov. India, IUAC India, Kitami Univ. Japan, Osakafu Univ. Japan, Univ. Tokyo, KAIST Korea, IACH Mexico, USON Mexico, UTHM and USPI Malysia, NU Singapore, Chula Thailand, PSU Thailand, CCU, NCU Taiwan, LUZ Venezuela, SUN South Africa, Witwatersrand South Africa.

15 - 27 February 2009

US Universities, Institutes and Similar.

Alaska*, American Physical Society, Arizona, Berkeley*, Boston Univ., CHOP, Cornell, Case Western Reserve, Duke, Green River, Harvard* Princeton Institute for Advanced Studies, Iowa State, Kent State, Lehigh, Maricopa, MIT, MNSCU, MTU, NEBO, Oakton, ODU, Ohio, Penn State*, SDSU, SEMO. SIU, SJSU, St Cloud State, SUAGM, TCU, Toronto (on edu), UALR, Californian system*, UCLA*, U Conn., Florida, Houston, Urbana Champaign*, UMBC, Maryland*, UMSL, New Mexico, UPRM, Utah, Texas*, Kentucky*, Washington, Wisconsin, US Gov. DHS, IRS, St Lucie County, USAAF Robins, US Army Red River, AERO Organization, Southwestern Labs., Adobe, Boeing, Botevgrad, Honeywell, Hewlett Packard, Mobile Gas, Motorola.

Europe and Asia Minor: US Universities, Institutes and Similar.

UIBK Austria, Gov. Geneva, Charles Univ. Prague, BSH Germany, Jacobs Univ., KFA Juelich, Siemens*, SLUB Dresden, Cologne*, Leipzig*, Ulm, EHU, UPCT, UPM, UVA Spain, OBSPM France, Paris South*, Strasbourg, Poitiers*, Demokritos Greece* NTUA Greece, Trinity College Dublin, Tel Aviv, CNR IFAC Italy, ELSAG Italy, Modena, Pavia, MPE Latvia, Twente*, VU Netherlands, Kluwer, Lodz, Warsaw, Wroclaw, UC Portugal, Kaliningrad Russia, SU Sweden*, UU Sweden, KI Slovenia, Hacettepe Turkey, Aberystwyth, PWF Cambridge, Wolfson Cambridge, Trinity Cambridge, particle Cambridge, Durham*, Imperial*, Kent*, Lancaster, Student Halls London, Manchester, Meirion Dwyfor, Balliol Oxford, Mansfield Oxford, Merton Oxford, physics Oxford, QAA, Queen

Mary College London, Southampton, Strathclyde, Swansea, ULST, Ceredigion, Met. Office, National Health Service*, Belgrade.

Rest of the World: Universities, Institutes and Similar.

Reed Elsevier Australia, CSIRO, Defence Australia, Melbourne, Gov Brazil, UNICAMP, Sao Paolo, Canadian ECE Network, Gov. Quebec, McGill, IAW Ontario, Alberta, British Columbia*, Western Ontario, UTC Ecuador, CET, IITM, Gov. India, IMSC India, Daika Univ Japan, Hokudai, Ibaraki, JEC, Kyoto, Nihon, KAIST Korea, UNAM Mexico, UPM Malaysia, Chula Thailand, Capetown South Africa.

Summary of February 2008

Number of documents downloaded from site: 1184.
Lead Countries: USA, Canada, Britain, Germany, France, Australia, Japan, ... (77 countries).
Lead Papers: 94, 110, 103, 14, 41, 126, 99, 17, 74, 112, 2, 43, 114, 104, 109, 88, 15, 122, 102, 113, 115, 76, 32, 63, 111, 45, 47, 98, 107, 67, 56, 89, 125, 21, 86, 1, 7, 12,82, 28, 37, 61, 71, 105, 120, 81, 116, 124, 59, 122, 57, 108, 85, 10, 121, 23, 24, 25, 54, 55, 6, 8, 92, 93, 117, 100a, 123, 16, ... all papers read.
Lead articles: AThe Life of Myron Evans@ by Kerry Pendergast, ECE (Spanish), Space Energy (Spanish), ASpecial Relativity@ (Horst Eckardt), Galaxies, Workshop History, Galaxies (Spanish), Spacetime Devices (Spanish), ECE Second Current (Douglas Lindstrom), ECE and Spacetime, Douglas Lindstrom 2nd Numerical Paper, Felker4 (Spanish), Space Energy, ECE Engineering Model slides, Douglas Lindstrom 3, ...   All articles read.

1 - 15 March 2009

US Universities, Institutes and Similar.

Arizona, Bard, Brigham Young, CSU Pomona, Georgia Tech., Harding, Haverford, Iowa State, IPFW, Lafayette, Louisana State, Maine, MIT*, Northwestern, Princeton*, Stanford, St Lawrence, Texas A and M, Texas A and M K, Californian university system, UCLA, UGA, Kentucky, U Mass*, Minnesota, US Naval Academy, Utah, Virginia, Wisconsin, Western Kentucky, Xavier, Brookhaven, US Gov. Dept. of Health, FEMA, NASA, NOAA, St Lucie County California, Adobe, Boeing, Botevgrad, Hewlett Packard, Motorola, Varian, USAAF Edwards, USAAF Robins, US Military DISA, US Naval Research Laboratory.

Europe and Asia Minor: Universities, Institutes and Similar.

UTBK Austria*, KU Leuven, Palacky, RWTH Aachen*, Siemens*, TUBS, Cologne, SDU Denmark, IAC, UCM, UPCO, UPV Spain, OBSPM France, Strasbourg*, Poitiers, Savoie, TEE Greece, KBD Croatia, SYSQ Hungary, Tel Aviv, ICTP Trieste, Bologna, Unimib, Naples, Pavia, IT Lithuania, KUN Netherlands, Leiden, Twente, Amsterdam, Kluwer*, FUW Poland, US, TU Szczecin, UMK Poland, Argoncilhe Portugal, Nite Romania, Kursk Russia, MTU, Rostov, TSCU tech Russia, Gazi, ITU, Metu Turkey, MK Ukraine, Odessa Ukraine, Anglia, Bangor, Churchill Cambridge, Girton Cambridge, Cardiff*, Prifysgol Cymru, Glamorgan, Heriot Watt, Imperial*, King's College London, Kent*, Nottingham, CCC Oxford, Queens' College Oxford, SJC Oxford, Portsmouth, Rutherford Appleton, Sheffield, Swansea, Univ. College London*, East Anglia*, UMIC, University of Wales, Ceredigion, British Gov. GSI, Hampshire County Council, Trafford Council.

Rest of World: Universities, Institutes and Similar.

UNICAMP, Sao Paolo, Brazil, UNSL Argentina, National Research Council of Canada, Perimeter Institute Canada, Canadian ECE Network, Simon Fraser, British Columbia, Bishops, Regina, Toronto, Waterloo, UO Cuba, DHU, IITM, CAT Gov. India, NARL India, Tohoku, Tsukuba, Tokyo, Yamaguta Japan, KAIST Korea, Open University Sri Lanka, IPN Mexico, UPM Malaysia, Victoria Univ. Wellington NZ, MSUIT Philipines, PIE Pakistan, NU Singapore*, TKU Taiwan, LUZ Venezuela.

15 - 30 March 2009

US Universities, Institutes and Similar.

Brandeis*, Brigham Young, City Univ New York, Drexel*, Evergreen, ISI, Lehigh, MIT*, North Dakota, NIU, Northwestern, Oklahoma, Pittsburgh, Penn State, Rochester, SIU, SMU, Stonybrook, SUNY Stonybrook, Texas A and M K, Toronto (on edu), Tulane*, California system, Chicago, UCSC, Delaware, U Mass., U Mass. Dartmouth, Maryland, Michigan*, UNCW, Nevada Reno, Texas, UWM, Virginia*, Vermont, Wisconsin, Yale, NASA JPL, NOAA, St Lucie County, USAAF Robins*, US Central Command, US Military HPC, Boeing*, Botevgrad*, CIA, Ford, Microsoft*, Motorola, Panasonic, Varian*.

Europe and Asia Minor: Universities, Institutes and Similar.

STL Austria, Charles Univ. Prague, Palacky, German Synchrotron Facility, Jacobs Univ.*, Max Planck MIS, QSC, RWTH Aachen, Siemens*, TU Darmstadt, Freiburg, Jena*, Konstanz, Marburg, Potsdam, Wuerzburg*, KU Denmark, UPV Spain, Helsinki Finland, HUT, TUT

Finland, CEA France, Institut Laue Langevin, INSA Lyon, OSPM France, Polytechnique System, Marseilles 3, Poitiers*, Savoie, Sopron Hungary, Jerusalem, Technion, SISSA, ICTP Trieste, Turin, Free Univ Amsterdam, UIO Norway, TKE, UMK, Wroclaw Poland, Astral, NIPNE Romania, ETK Russia, UU Sweden, RZS-HM Slovenia, Ankara, Bilkent, Gyte, Hacettepe, KU Turkey, Gov Ukraine, Kiev, Aberystwyth, Birmingham, Bristol. BSFC, PWF Cambridge*, particle Cambridge, Edinburgh, Essex, Imperial, King's College London, Lancaster, Liverpool, Luton, Meirion Dwyfor, NERC Swindon, QMW, Roehampton, St Andrews, Swansea*, Univ. College London, East Anglia, York*, York City Council, NHS, Belgrade, SQU Oman*.

Rest of World: Universities, Institutes and Similar.

UAEU Argentina, ANU Australia, Sidney*, Presidency Univ. Bangla Desh, UFMG, UFPB, UFRJ, UNICAMP Brazil, Carleton Univ. Canada, McGill, Mount Royal, Gov. Nova Scotia, Canadian ECE Network, ACI Ontario, Perimeter Institute*, SLC Quebec, Simon Fraser, Alberta, British Columbia, Calgary, Laval, Institute of High Energy Physics China, UDEA Colombia, Univ. Andes Colombia, City Univ Hong Kong, IITM India, RCIL Gov. India, IMSC India, Hokudai, Osaka, Tohoku*, Tsukuba, Hitachi Company Japan, KIAST Korea, Jfn Sri Lanka, UCTO Mexico, UTM Mexico, UITM Malaysia, UPM Malaysia, Auckland NZ, Victoria Univ. Wellington New Zealand, PIE Pakistan, NU Singapore, NTHU, NTU Taiwan*, Academica Sinica Taiwan, Witwatersrand South Africa.

Summary of April 2009

Number of documents downloaded from site: 1506.
Lead Countries: USA, Canada, France, Britain, Germany, Australia, Italy, ... (85 countries).
Lead papers; 121, 94, 122, 116, 54, 88, 113, 43, 99, 67, 115, 55, 76,103, 114, 91, 112, 41, 32, 126, 127, 122 (Spanish), 104, 23, 56, 63,105, 125, 15, 45, 61, 20, 109, 17, 111, 102, 107, 21, 65, 22, 74, 106, 110, 39, 81, 13, 82, 119, 14, 120, 118, 124, 93, 47, 49, 11, 19, 8, 98, 123, 24, 37, 53, 40, 70, 95, 12, 34, ... (All papers read).
Lead Articles: ECE, AThe Life of Myron Evans@, ECE and Spacetime, Space Energy, Galaxies, Space Energy (Spanish), Proof One, Special Relativity, Galaxies (Spanish), ECE Second Current (Lindstrom), Felker4 (Spanish), ECE (Spanish), Crothers IJTP1960, Workshop Overview, Felker5, ... all articles read.

US Universities, Institutes and Similar.

Arizona State, Brown, Brigham Young, Caltech, Colorado, Columbia, Duke, Elmira, Florida International, Georgia Tech., Harvard, Indiana, Johns Hopkins Applied Physics, Karunya, Kent State*, Los Rios, Louisiana, MIT*, MS State, MTU, North Dakota, Northwestern*, New York Univ., Oakton, Missisippi, Penn State, Stanford, Tabor, Texas A and M, Toronto (on edu), California system, UC Santa Barbara, IC Santa Cruz*, UGA, Houston, Illinois Chicago, Urbana Champaign, U Mass Dartmouth, Maryland*, Minnesota, Nevada Reno, U Penn, UPRM, Washington, WSU, Gov Alaska, Argonne, Brookhaven, Dept Health, UPTO, USAAF Hill*, USAAF Robins, US Army AMRDEC, US Army ACE, US Military HPC, US Navy NAVO, Jefferson Particle Lab., Botevgrad, Microsoft, Motorola, Raytheon.

Europe and Asia Minor: Universities, Institutes and Similar.

Linz, Siemens Austria, Lausanne, SLU Czechia, Antenne Thueringen, FU Berlin, LRZ Munich, Siemens, TU Dresden, Jena*, Karlsruhe, Kaiserslautern, Leipzig, Muenster*, UAB, Jaen Spain, Helsinki, HUT Finland, CIUP, ENSTA, Schneider Electric France, Strasbourg, Poitiers*, Patras, Trinity College Dublin*, Tel Aviv, ISC Italy, ENEA, Trieste, CSS Latvia, AUB Lebanon, RU Netherlands, Utecht, NTNU Norway, SQU Oman, Wroclaw, Porto*, Lisbon Tech., NI Russia, Joint Institutes for Nuclear Research Russia, Siberian Academy of Sciences Novosibirsk, Chalmers Univ. Sweden, HIK Sweden, Institute of High Energy Physics Russia, Birkenhead*, BCU, Hull, King's College London, Liverpool, Meirion Dwyfor, Open, Oxford, Queen's Univ. Belfast, SHU, York, York City Council.

Rest of World: Universities, Institutes and Similar.

UNS Argentina, Sydney, Western Australia, Gov. South Australia, UFG Brazil, McGill Canada, Memorial Univ. Newfoundland, Alberta, British Columbia, IMSC India, Yamaguchi, KAIST Korea*, SNU Korea, UITM Malaysia, NU Singapore.

US Universities, Institutes and Similar.

Brandeis, California Polytech., Clemson, Colorado, Colorado State, ECU, Emory, Florida Atlantic, Hawaii, IUPUI, Johns Hopkins Applied Physics*, KCTCS, Louisiana, Louisiana State*, Miami, MIT*, Michigan State, New Jersey Institute of Technology, NMT, North Dakota, New York Univ., Ocean, Mississippi, Oregon State, PBA, Penn State, Purdue,

Rochester Tech., Rutgers, Stanford, Stonehill, Stonybrook*, SUNY Stonybrook, Texas A and M, UCAR, UC Irvine, UCLA*, U Conn*, UC Santa Barbara, UC Santa Cruz, Urbana Champaign, U Mass Dartmouth*, Maryland*, Michigan*, Minnesota, Chapel Hill, UPRM, Texas Arlington, Utah, Texas, Vermont, Wisconsin, Worcester Polytechnic Institute, Washington State*, Gov. Alaska, Gov. California, Dept. Health, Dept of Energy, Federal Aviation Authority, NASA, St Lucie Co., California, USAAF Hill*, US Army Medical Division, US Navy Fleet Command, US Navy NMCI, US Military NGA, Honda (US), Botevgrad, CIA, Honeywell, Lockheed Martin, Motorola*.

Europe and Asia Minor: Universities, Institutes and Similar.
   Vienna, Siemens Austria, Ghent, Digital Systems Bulgaria, Freiburg, FW Bonn, German Government, FU Berlin*, IPHT Jena, Siemens*, Bonn, Goettingen, Heidelberg, Jena, Konstanz, Muenster*, Stuttgart*, CSIC, Jaen, Avarra, UPV, Valencia, UVA* Spain, HTV, JSP, JYU, Finland, CIUP France, Grenoble, ECP, Bordeaux 1, Poitiers, Patras, AMIS Croatia, ISKON Hungary, INFN Bari, INFN Bologna, Calabria, Padua, Turin, INIT Lithuania, RU Netherlands*, Wartburg, AUB Lebanon, FFI Norway, Students PW Poland, Nasza Klasa Poland, Wroclaw, RACAI Romania, Lipetsk, SGS Network Sweden, Autocentre Ukraine, Kharkov Ukraine, East Anglia, BSFC, CAI Cambridge, Churchill College Cambridge, PWF Cambridge, Cardiff, Hull, Imperial, Meirion Dwyfor, Southampton*, St. Andrews, Univ College London, Hampshire County Council, NHS.

Rest of World: Universities, Institutes and Similar.
   UNSJ-CUIM Argentina, Gov. Argentina*, MQ Australia, QUT Australia, Brandon Univ. Canada, Canadian ECE Network, Humber, McGill, Alberta, British Columbia, Montreal, Waterloo, Polytechnic Colombia, IITM India, ITAP, IIAP India, Tohoku Japan, Yonsei Korea, UITM Malaysia, UPM Malaysia, Christchurch Polytech, NZ, PUCP Peru, UMAC Peru, PERN Pakistan, NU Singapore, NTU Taiwan, Capetown South Africa.

Summary of May 2009

Number of Documents Downloaded from Site: 1638
Lead Countries: USA, Canada, France, Germany, Britain, Bulgaria, Mexico, Netherlands, Australia, ... (83 countries).
   Lead Papers: 41, 116, 94, 17, 99, 104, 20, 103, 78, 43, 27, 113,102, 128, 59, 67, 107, 110, 112, 98, 95, 101, 122, 75, 114, 1, 109, 127, 129, 21, 76, 91, 81, 88, 2, 40, 47, 111, 123, 130, 63, 120, 10, 56, 61, 37, 45, 4, 54, 7, ... all papers read.

Lead Articles: AThe Life of Myron Evans@ (Pendergast), Felker4 (Spanish), Space Energy (Spanish), Galaxies (Spanish), Workshop History, ECE, Indices (Spanish), Felker5 (Spanish), Spacetime Devices (Eckardt), ECE and Spacetime, 122 (Spanish), Lindstrom 1, Felker2 (Spanish), Space Energy, Second Current (Lindstrom), Workshop Details, Notes 107 (1 and 2), Felker7 (Spanish), ECE (Italian), Proof 1, Cold Current Wave Equations (Lindstrom), Felker13 (Spanish), Black Hole refutation by Crothers, 120 Metrics, Basic Hypotheses (Spanish), ECE Engineering Model Linear Version, ... all articles read.

END OF FIFTH YEAR OF HIGH QUALITY FEEDBACK

---

1 - 15 May 2009

US Universities, Institutes and Similar.

Berkeley, Boston Univ*, Butte, Calvin, Columbia, FHSU, Indian River State College, Macomb, MIT*, MS State, Mount Holyoke, Notre Dame, New Mexico Tech., New York Univ., Ohio, PDX, Princeton, Rose Hulman, Rowan, Renselaer, Rutgers, SMCM, Stanford, Stonybrook, Temple*, UC Irvine, U Conn., UC Riverside, UC Santa Barbara, UC Santa Cruz, UC San Diego, Urbana Champaign, U Mass Dartmouth, UMBC, Maryland*, Michigan, North Texas, US Naval Academy, Texas Arlington*, Texas, Texas PB, Washington State*, NASA (gsfc), NASA (jpl)*, NIH, Oak Ridge, Gov. Santa Barbara, St Lucie County, USAAF Robins, US Navy NAVO, US Navy NMCI*, AERO, Hopkintown Library, Boeing, Botevgrad, CIA, Lockheed Martin, Mobile Gas, Motorola, Wells Fargo.

Europe and Asia Minor: Universities, Institutes and Similar.

Graz, Vienna, Liege, Leuven, Lausanne, Charles Univ. Prague, Palacky, Bremen, Niederrhein, Jacobs Univ., LRZ Munich, MH Hannover, Albert Einstein Institute, Max Planck Stuttgart, Max Planck MPL*, Siemens, Students Bonn, TUM, Bremen, Duisberg, Heidelberg*, Karlsruhe, Munich, Wuerzburg, Birke Gymnasium Denmark, TTU Estonia, UT Estonia, Aragon, TSAI, UCM, UPV Spain, Helsinki, HTV, JYU* Finland, ECP France, ENSTA France, OBSPM France, Marseilles, Poitiers, Sopron Hungary, T UMTS Hungary, NUIM Ireland, INFN Naples*, ICTP Trieste, Pisa, YU Jordan, INIT Lithuania, Laisvas Latvia, AUB Lebanon, SQU Oman, Institute of Mathematics Macedonia, Leiden,

Twente, Wroclaw, RACAI Romania, Sun NSK Russia, Umea, ITU Turkey, Aberystwyth, Bath, Newton Institute Cambridge, particle Cambridge, Durham, Heriot Watt, King's College London, Liverpool, Luton, NCL, Nottingham, Churchill College Oxford, Portsmouth, Rutherford Appleton, Southampton, St Andrews, Sunderland, Sussex, Swansea, Union, BBC*, NHS.

Rest of World: Universities, Institutes and Similar.

UNS Argentina, Reed Elsevier Australia, Monash*, MQ, Western Australia, UFPA Brazil, UNICAMP, Shell Canada, British Columbia, Regina, Toronto, Waterloo, Central Chile, UNAL Colombia, Andes, IITM India, MNIT, IMSC India, Hokkaido, Osaka, Tokyo, KAIST Korea, NUL Lisotho, Otago NZ, UNAC Peru, PERN Pakistan, UJ South Africa.

15 - 30 May 2009

US Universities, Institutes and Similar.

Buffalo, Brigham Young, California Polytechnic, Caltech, Carnegie Mellon, Cornell, Depaul, Georgia Tech., Harvard, Johns Hopkins Applied Physics, Lake View College, MIT, Notre Dame, New Mexico Tech., North Western*, Nova, Stanford, Syracuse*, California System, Chicago*, UC Santa Barbara, UC San Diego*, Illinois Chicago, Michigan, Minnesota, U Penn, US Naval Academy, Texas, Vanderbilt, Washington, US Dept of Health, NASA (JPL), Princeton Plasma Physics Lab., Santa Barbara, Virginia, USAAF Robins, US Navy NMCI, Adobe, AOL, Botevgrad, Duke Energy, Harris, Honeywell, Hewlett Packard, Microsoft*, Motorola, Scana.

Europe and Asia Minor: Universities, Institutes and Similar.

Vienna, Liege, EKK Bulgaria, CERN, Charles Univ. Prague, FU Berlin, KFA Juelich, Albert Einstein Institute, Siemens*, Karlsruhe, Kiel, Magdeburg, Regensburg, Unity Media Group, KOU Estonia, IFAE, Ovi Spain, Helsinki, ENS Lyon, French national particle physics lab., INIST, Poitiers, Rennes 1, AMIS Croatia, Magnet Ireland, Weizmann, ENEA, INFN Rome 1, INFN Turin, Milan Polytech*, ICTP Trieste, Bari, Calabria*, Catania, Unimb, Pisa*, Perugia, Rome 1, Turin, YU Jordan, VU Lithuania, NGN Latvia, NTNU Norway, Torun* Poland, Warsaw, Lisbon, Lisbon Tech, Moscow, SU Sweden, ITU, Iyte, Metu Turkey, BPC, Newhall Cambridge, Durham, Essex, Leeds, Manchester, physics Oxford, SAC, Univ, College London, BBC, British Aerospace Engineering, Unilever.

Rest of World: Universities, Institutes and Similar.

UNS Argentina, EQ Australia, Monash, Newcastle, Gov. South Australia, Gov. Brazil*, PUC-RIO Brazil, UFMG, UNICAMP Brazil, AEI Canada, British Columbia, EICAT Canada, HDSB Canada, Canadian ECE Network, McMaster, Queen's, Alberta, ENAP chile, UNAL Colombia, Shiraz Iran, Kumamoto Univ. Japan, Tohoku, UNAM Mexico*, UPM Malaysia, Victoria Univ. Wellington NZ, PUCP, UNAC Peru, PERN Pakistan, CHULA Thailand, NTU Taiwan.

Summary of June 2009

Number of documents downloaded from site: 1621
Lead Countries: USA, Germany, Canada, Britain, France, Italy, Australia, Mexico, Netherlands, Argentina, ... (90 countries).
Lead Papers: 94, 107, 131, 99, 110, 103, 37, 67, 114, 130, 76, 104, 41, 88, 112, 116, 122, 28, 105, 127, 128, 113, 123, 63, 111, 129, 95, 5, 43, 4,7, 102, 11, 1, 30, 75,80, 109, 54, 6, 14, 18, 20, 98, 82, ... all papers read.
Lead Articles: AThe Life of Myron Evans@, ECE (Spanish), Space Energy Devices (Spanish), ECE and Spacetime, ECE (Dutch), ECE Engineering Model, Felker13 (Spanish), 122 (Spanish), ECE Second Cold Current, Black Hole Criticisms, Felker3 (Sp), Special Relativity (Eckardt), Space Energy, Proof One (Spanish), Spacetime Devices, Crystal Spheres, Proof One, ECE (French), Galaxies, Third Lindstrom Numerical Paper, Felker12 (Sp), Index Summary (Spanish), Proof Five (Spanish), Proof Three (Spanish), Proof Two, Cold Current Wave Equations (Lindstrom), ... all articles read.

1 - 15 June 2009

US Universities, Institutes and Similar.

Brigham Young, Caltech, Georgia Tech*, Harvard, Haverford, Princeton Institute for Advanced Studies, Kansas, New York Univ., Purdue, Texas A and M, UAB, UC Davis, Florida, Houston, North Carolina Wilmington, Nevada Reno, US Naval Academy, Washington, West Michigan, US Dept of Health, NIH*, NIST, NOAA, Sandia*, St Lucie County, US Army NVL, US Military NGA and OSD, AERO, Boeing*, Botevgrad, Duke Energy, IBM Thomas J. Watson, Intel, Lockheed Martin, Microsoft, Motorola, Northrop Grumman, Philips, Rockwell Collins.

Europe and Asia Minor: Universities, Institutes and Similar.

TU Vienna, Univ. Vienna, Liege, KU Leuven*, BTC Bulgaria, CERN, Charles Univ. Prague, Bosch, German Synchrotron Facility, Bonn students, Siemens, TU Dresden, Umwelt, Heidelberg*, RUC Denmark, IFAE Spain, UGR Spain, CNRS Grenoble, IAP France, French national particle lab., INRA Avignon, Paris Sud, Poitiers, Patras, ADSL Hunagry, INFN Mib Italy, Turin, SISSA, Pisa, ECN Netherlands, RUG, Twente, Utrecht, UW Poland, Lodz, Torun, Warsaw, Zgora, Lisbon Tech., TVR Romania, SBB Russia, Ege, Gantep Turkey, Kharkov, Kiev Ukraine, CAI Cambridge, Cranfield, Heriot Watt, Imperial, Llys Fasi*, Keble College Oxford, University College Oxford, Southampton*, Swansea, Warwick, York, BBC, Ceredigion.

Rest of World: Universities, Institutes and Similar,

ANU Australia, Gov. Brazil, Concordia Univ. Canada, Canadian ECE Network, Gov Canada IC, McGill, McMaster, Guelph, Simon Fraser, York Univ. Toronto, UTFSM Chile, Tsinghua China, Gov. Colombia, Polytechnic Hong Kong, MRI, IMSC, IIFR India, Tsukuba Japan, Sony, KAIST Korea, ICTO, USON Mexico, PUCP Peru, Chua, RMUTI Thailand, NTHU*, NTU* Taiwan.

15 - 28 June 2009

US Universities, Institutes and Similar.

Austin CC, Brigham Young, EPCC, Harvard*, Illinois Tech., IRSC, Northwestern, Princeton, Purdue, Rice, SUNY Stonybrook, Texas A and M, California system*, UC San Francisco, Houston, UTEP, YSU, US Gov. Dept. of Health, NASA (GSFC),NIH, NIST, NOAA, Botevgrad, Duke Energy, Ford, Fujitsu (US), Siemens (US), IBM*, Motorola, St John's Preparatory School New York City.

Europe and Asia Minor: Universities, Institutes and Similar.

Vienna, KU Leuven, CERN, ETH Zurich, Charles Univ. Prague, Palacky, Bosch, FGAN Germany, FH Cologne, Fraunhofer system, FU Berlin, Max Planck MPE, Max Planck Stuttgart, RWTH Aachen, Siemens*, Theysohn, Giessen, Hannover, Jena, Wuerzburg, Aarhus Denmark, UCM, UPV, UVA Spain, Polyechnique system, Marseilles 3, UHP Nancy*, OTE Greece, Mathos Croatia, NLG Hungary, SNR, SNS Italy, Florence, Pisa, Toscania, FI Latvia, RUG Netherlands, Wroclaw, Porto Portugal, NIPNE Romania, PSN Russia, UMU Sweden, Agric. Univ. Slovakia, Gyte, Tubitak Turkey, BBSRC, Glasgow*, Hull, Imperial*, Lancaster, Napier, Nottingham, Oxford*, Queen Mary London, St. Andrews, Sussex, Swansea*, Warwick, York, Hampshire County Council, Peterborough Schools.

Rest of World: Universities, Institutes and Similar.

East Queensland, Woolongong, Sao Paolo Brazil, Gov British Columbia, Canadain ECE network, Niagra Falls Hydro, Perimeter Institute, Saskatchewan, Victoria, JNCASR, RCIL India, Chonnan Sri Lanka, KAIST Korea, Itchihuahua Mexico, ITAM, UMICH, USON Mexico, Auckland NZ.

Summary for July 2009.

NEW MONTHLY RECORD OF 25,007 INDIVIDUAL VISITS combined with www.atomicprecision.com

Number of documents downloaded from site: 1627.
Lead Countries: USA, Thailand, France, Germany, Britain, Australia, Mexico, ... (81 countries).
Lead Papers: 107, 94, 102, 103, 133, 110, 131 (Spanish), 41, 132, 54, 112, 114, 116, 99, 43, 63, 45, 131, 61, 1, 15, 17, 21, 11, 6, 18, 56, 2, 47, 53, 76, 120, 105, 118, 130, 20, 67, 104, 128, 129, 58, 57, 75, 13, 52, 122 (Spanish), 111, 113, 115, 124, 7, 37, 8, 108, 122, 123, 12, 26, 59, 4, 65, ... all papers read.
Lead articles: Space Energy (Spanish), Space Energy, The Life of Myron Evans, Spacetime Devices, Space Program, Felker13 (Sp), Felker14 (Sp), ECE Engineering Model, Galaxies (Sp), Workshop History, ECE and Spacetime, Criticsm of Black Holes, Lindstrom 2D Cold Current, Spacetime Devices (Spanish), Dvcpra (Lindstrom), Galaxies (Eckardt), Special Relativity (Eckardt), Indices (Spansish), ... all articles read.

1 - 15 July 2009

US Universities, Institutes and Similar.

Bowdoin, Brigham Young, Buffalo, Carnegie Mellon, Columbia, CSUSB, Drexel*, Florida Tech, Indian River State College, Northwestern, New York Univ., Purdue, California system, Delaware, Florida, Illinois Chicago, Urbana Champaign, Michigan, North Carolina Charlotte, UNLV, US Naval Academy, Texas, Worcester Polytechnic Institute, Lwrence Livermore, PNL, Princeton Plasma Lab., St Lucie County, USGS, USPS*, Gov. Virginia, USAAF Edwards, US Navy NMCI*, Boeing*, Botevgard*, Honeywell, IBM*, Lockheed Martin, Microsoft, Motorola*, Northrop Grumman.

Europe and Asia Minor: Universities, Institutes and Similar.
Medical Univ. Vienna, TU Graz, CERN, Charles Univ. Prague, Siemens*, RWTH Aachen, Bielefeld, Dortmund, Jena, Ulm, Wuerzburg*,

KOU Estonia, UCM, UMA Spain, HTV Finland, Poitiers, KBD Croatia, Haifa, INRIM Italy, SISSA, Kluwer, UIO Norway, IFPAN Poland, GDA Poland, Porto, Siberian Academy of Sciences Novosibirsk, Agric Univ. Slovakia, UPJS Slovakia, Metu Turkey, BBSRC, Durham, Galsgow*, Hull, Lancaster, Leicester, Manchester, Meirion Dwyfor, Swansea, Ceredigion, British Gov. Energis.

Rest of World: Universities, Institutes and Similar.
   GBAGN Argentina, UNUCAMP, UFPB Brazil, Canadian ECE Network, British Columbia, Manitoba, UFRO Chile, Kyoto Tech., Massey NZ, NTHU Taiwan.

15 - 31 July 2009

US Universities, Institutes and Similar.
   Carnegie Mellon, Cornell*, CSUSB, Dartmouth, Drexel*, Georgia Tech*, GSU, Harvard, Indian River, Johns Hopkins Medical Institute, Mercy, Providence, Penn State*, Rochester Tech, Stonybrook, TCE, UAB, UCLA, U Conn., Maryland, US Naval Academy, Vermont, Washington, Wisconsin, Los Alamos, NIH, NOAA*, NSF, PNL, Princeton plasma physics lab., Sandia, USAAF Edwards, Eglin, Robins, US Army WVA, US Navy NMCI*, Beverly Hills, Cumberland*, Jefferson, Boeing, Honeywell, IBM Thomas J Watson, Lockheed Martin*, Microsoft*, Raytheon, Siemens Transport.

Europe and Asia Minor: Universities, Institutes and Similar.
   TU Vienna, Ghent, CERN*, Palacky, Bosch, FH Cologne, FU Berlin, Max Planck IA Heidelberg, Max Planck IPKS Dresden, Siemens, Siemens Transport, TU Dresden, Bielefeld, Bremen, Frankfurt, Heidelberg, Karlsruhe, Cologne, Stuttgart, Aarhus Denmark, Niels Bohr Institute, Aragon, CSIC, UMA, UV Spain, OBS Besancon, Avignon, Poitiers, Combined Campuses Budapest, Jeusalem, Weizmann, US Iceland, ENI Italy, INRIM Italy, ICTB Trieste, Calabria, Florence 1, Rome 1, Rome 3, Utrecht, Kluwer, AMU Poland, UMK Poland, Gov. Qatar, MT S-NN Rusia, Siberian Academy Novosibirsk, Samarlan, KTH Sweden, Agric Univ. Slovakia, UPJJ Slovakia, Kiev, Bath, BBSRC, Cardiff, Imperial, Leeds, Nottingham, Southampton, Strathclyde, Swansea, Univ. College London, National Library of Wales.

Rest of World: Universities, Institutes and Similar.
   Gov. Argentina, IIT Adelaide, UFPB, UNICAMP Brazil, Sao Paolo, McGill Canada, Canadian ECE network, McMaster, Perimeter, Gov. Quebec, British Columbia, Toronto, Waterloo, NITRKL India, MUI Iran,

KAIST Korea, JFN Sri Lanka, AUT NZ, UDEA Colombia, Hokudai, Kyoto, Kyoto Tech., PERN Pakistan, SUN South Africa, Capetown, Witwatersrand, LUZ Venezuela.

Summary of August 2009

Number of Documents downloaded from site: 1650
Lead Countries: USA, France, Canada, Germany, Britain, Saudi Arabia, Australia, Italy, ... (92 countries).
Lead Papers: 107, 94, 43, 57, 102, 103, 110, 112, 41, 99, 133, 61, 131 (Spanish), 114, 63, 113, 1, 76, 104, 30, 134, 120, 131, 67, 123, 59, 95, 101, 119, 124, 15, 32, 40, 12, 130, 49, 8, 118, 126, 127, 132, 56, 129, 2, 47, 75, 122 (Spanish), 111, 115, 116, 117, 34, 46, 28, 96, 108, 109, 122, 13, 98, 87, 105, 37, 6,106, ... All papers read.
Lead articles: Spacetime Devices, ECE Popular Lecture by Horst Eckardt, Workshop Slides, Spacetime Devices (Spanish), ECE Engineering Model (updated version), Space Energy, Life of Myron Evans, Spacetime Devices 2, Space Program, Crystal Spheres, ECE and Spacetime, Felker4 (Sp), ECE (Spanish), Crothers Black Hole criticism, F13 (Sp), Galaxies (Sp), ECE, Space Indices , Special Relativity (Eckardt), F11 (Sp), ... all articles read.

Combined Site Total of 22309 individual visits.

August 1 - 15 2009

US Universities, Institutes and Similar.
   Berkeley, Drexel, Indiana, Indian River, Louisville, MIT*, NAU, Penn State, Rutgers, SIUE, Stanford, Texas A and M, Chicago*, U Conn., UC Santa Cruz, U Michigan, Univ. North Carolina Asheville*, Univ. North Carolina Charlotte*, U Penn, Texas Arlington, Utah, Washburn, Washington, Brookhaven, Fermilab, US Gov SSA, St Lucie County, US Gov. PS, USAAF PLH, USAAF Robins, US Military HPC, US Navy NMCI*, Botevgrad, Central Intelligence Agency, Duke Energy, Rockwell Collins.

Europe and Asia Minor: Universities, Institutes and Similar.
   Wienkaw Austria, CERN, ETH Zurich, Basel, Charles Univ. Prague, Palacky, UNBR Czechia, Max Planck MPIA-HD, Ruhr Univ. Bochum, Siemens, Freiburg, Karslruhe, Cologne, CSIC Spain, Gov. of Andalucia, Bayonne, CNAM France, French national particle lab., Nice, OLEANE, Reunion, Univ. College Dublin, ICTP Trieste, Turin, RU Netherlands,

UIO Norway, UMK Poland, ISKRA-MIS Slovenia, Gyte*, Kharkov, Kiev, Glasgow,particle Glasgow, Nottingham, physics Oxford*, Swansea, NHS.

Rest of World: Universities, Institutes and Similar.
    UNC Argentina, CSIRO Australia, Gov. Brazil CNEN, UNB Brazil, Gov. Canada Defence, Toronto, Western Ontario, UPB Colombia, IIITA, IITM, RCIL Gov, India, IMSC India, MUI, Shiraz* Iran, USON Mexico, CCU Taiwan, UWC South Africa.

 Summary of August 2009 : NEW MONTHLY RECORD of 27527 INDIVIDUAL VISITORS.

For www.aias.us there were 75980 hits from 11145 visitors, downloading 5.02 gigabytes and generating 43287 previews from 98 countries. The leading countries were: USA, Britain, France, Germany, Australia, Canada, Brazil, ... There were 16382 visits for www.atomicprecision.com, (also known as www.unifiedfieldtheory.info) making a new record of 27527 individual visits. The total number of hits for the ECE sites was 137,610

The lead ECE papers were: 43, 23, 131 (Spanish), 135, 8, 2, 133, 134, 76, 41, 4, 60, 63, 14, 88, 45, 51, 7, 21, 54, 59, 18, 30, 123, 1, 13, 11, 20, 67, 94, 134 (Spanish), 107, 47, 6, 12, 132, 131, 90, ... all papers read.
    The lead articles were: Life of ME, SE Devices (Spanish), Crothers' criticisms of Ricci flat theory, ECE (Spanish), Spacetime Devices, ME Biography 3, Felker 4, Felker 10, Space Program, ECE, Space Energy, ECE Engineering Model, Crystal Spheres, Galaxies (Sp), ECE and Spacetime, Felker 6, DvcprA, ECE Public Lecture, UNCC Saga 1, Exp. Refutation of Heisenberg, Felker 13, Spacetime Devices 2, Spacetime Devices (Spanish), Felker 2, Felker 3, Cosmo1, Felker 5, Etherington Report, Crothers 1, paper 135 (Spanish), Crothers presentation ot German Physical Society, Yarman criticisms, Workshop History, Edyn3, Crothers 2, Galaxies, ECE Second Current (Lindstrom), Armorial Bearings, Numerical Note 2, Definitive Proof 3 (Spanish), ... All books and articles read.

August 15 - 30, 2009.

US Universities and Similar.
    Berkeley, Binghampton, Boston Univ., Caltech, Clemson*, Columbia, Drexel, Duke, Florida State, Georgia Tech.*, Georgetown, Harvard, Indian River, Missouri, MIT, North Dakota, Oklahoma State, Princeton*, Penn State, Purdue*, Rice, Rutgers, Southern Illinois Medical, Texas Tech., UA,

Conencticut, UC Santa Barbara, UC Santa Cruz, UC San Diego, Minnesota, UNCC*, U Penn*, UPF, Vermont, US Gov. USPTO, US Navy DT, US Navy NMCI, Jefferson National Laboratory, Beverley Hills, Boeing, Botevgrad, Duke Energy, Hewlett Packard, Intel, Nokia.

Europe and Asia Minor: Universities and Similar

KU Leuwen Belgium, EKK Bulgaria, EPF Lausanne, Charles Univ Prague, UNBR Czechia, Palacky, Max Planck MPI-SB, Max Planck Dresden, Siemens, Frankfurt City Council, Bremen, WH Stuttgart, SDU Denmark, XAV-DC Europe, ENTPE France, INRA Nancy, UHP Nancy*, Tel Aviv, Weizmann, CSI Italy, ENI Italy, INFN Rome 1, INFN Turin, RU Netherlands, Eindhoven Netherlands, UIO Norway, PW Students Poland, US Poland, Lodz, UMK Poland, MSU Russia, Jozef Stefan Institute Slovenia, ITU Turkey, Metu Turkey, Birmingham, Bristol, particle Cambridge*, EBI, Edinburgh, Imperial, Liverpool JM*, Lancaster*, NERC Swindon, NWKC, Open University, RHUL, Southampton, NHS*.

Rest of World: Universities and Similar.

UNSJ Argentina, CSIRO Australia, Monash, Queensland, Sydney, ESM Brazil, UFMG, UFPB, UNICAMP Brazil, MD Robotics Canada, Alberta, British Columbia*, Waterloo, ATCI Ivory Coast, ENAP Chile, SJTU China, Andes Colombia, UPB Colombia, EIMA Cuba, BHU India, IUCAA India, Gov. India, RRI India, TIFR India, Shiraz Iran, Tohoku Japan, Tokyo, UFC Japan, UASLP Mexico, UGTO, UNAM Mexico, AUT New Zealand, Chuba, KKU Thailand, CCU*, NCU, NTU Taiwan, Gov. Venezuela, Capetown South Africa.

Summary of September 2009: NEW MONTHLY RECORD OF 28,265 VISITORS.

Up to Sept. 29th 2009 for www.aias.us there were 73,067 hits from 12,725 visitors, reading 1,668 documents and downloading 6.336 gigabytes and 39,828 pages. There were visitors from 93 countries, led by USA, France, Canada, Germany, Britain, Portugal, Australia, ... For www.atomicprecision.com there were 63,859 hits from 15,540 visitors, downloading 2.815 gigabytes and previewing 40,315 pages. This makes a total of 136,926 hits from 28,265 visitors, downloading 9.151 gigabytes and previewing 80,143 pages.

All 136 ECE source papers were read, the most read were:
11, 94, 107, 100, 4, 8, 2, 1, 41, 74, 76, 43, 54, 124, 67, 38, 44, 81, 7, 18, 13, 15, 17, 136, 26, 82, 63, 80, 110, 118, 123, 119, 47, 133, 27, 49, 60, 23, 111, 120, 21, 28, 30, 45, 46, 116, 130, 111, ...

All Articles were read, the most read were:

Myron Evans Biography 3 by Kerry Pendergast, ECE Engineering Model, Spacetime Devices, Space Energy Devices (Spanish), Space Energy, The Life of Myron Evans, Overview of ECE Theory, 135 (Spanish), Space Programme, Galaxies (Spanish), ECE and Spacetime, Felker4 (Spanish), Felker5, Paper 131 Notes 8 (Spanish), Felker6, 135, Felker10, Curvature Based Propulsion, Horst Eckardt ASpecial Relativity@, Workshop History, Space Devices (Spanish), ECE Popular Lecture, 122 (Spanish), Proof 5 (Spanish), Hehl Rebuttal, Lindstrom 2-D Current, ECE (Spanish), ACrystal Spheres@, ASolutions of the ECE Vacuum Equations@ by Lindstrom and Eckardt, Lindstrom 2, AThe Antisymmetry of Cartan Geometry@, ...

September 1 - 15 2009

US Universities, Institutes and Similar.
   Arizona, Berkeley*, Chapman, Colorado State, Columbia, Cornell, City Univ. New York, Earlham, Florida Atlantic, Florida State*, Georgia Tech*, Harvard students, Johns Hopkins Medical Institute, Kansas State, Colorado School of Mines, Missouri, MIT, Missouri Univ Science and Technology, Northern Arizona, North Carolina State, Notre Dame, Northern Illinois, Northwestern, Naval Postgraduate School, New York Univ., Penn State, Purdue, Stanford, St Cloud State, Texas A and M, Central Florida, UC Irvine, Georgia, Houston, Illinois Chicago, Illinois Urbana Champaign*, Maryland*, Minnesota Twin Cities*, New Mexico, Northern Texas, Utah, Texas, Virginia*, Vermont*, Washington, Weber, Western, Wisconsin, Fermilab, Lawrence Berkeley, NASA, St Lucie County, US Treasury, US Army Korea, US Navy Navair, Radical Designs Organization, Boeing*, Hewlett Packard, Microsoft*, Northrop Grumman, Perkin Elmer.

Europe and Asia Minor: Universities, Institutes and Similar.
   UKLIBK Austria, Ghent, EPF Lausanne, Charles Univ. Prague, MUNI Czechia, Palacky, BSB Munich, Humbolt Berlin, INP Greifswald, Albert Einstein Institute, Max Planck MPE, Max Planck Muelheim, Bielefeld, Karlsruhe*, Konstanz, Munich, SDU Denmark, UBU Spain, Madrid Complutensian, UPV, UVA Spain, CEA France, EC Lyon, Ensee IHT France, French particle laboratory, INRA, IRAM, Poitiers*, NTUA Greece, IRB Croatia, BME Hungary, Trinity Colleeg Dublin*, Astro Italy, CSI Italy, ENEA, ENI*, Fiera Milan, INFN Naples, Turin, Gov Kyrgyz Republic, FI Lithuania, LMT Latvia, TU Eindhoven, Univ. Lisbon, Brasov Romania, Univ Belgrade, NSC Russia, Univ Stockholm physics, Uppsala students, Aberystwyth, Brunel, Cambridge engineering, chemistry Cardiff,

Glasgow, Imperial physics, Kent, Lancaster, Hertford College Oxford, physics Oxford, UMIST, Medway Council, Parliament, Vale of Glamorgan School.

Rest of World: Universities, Institutes and Similar.

CNEA Gov. Argentina, Central Queensland, Latrobe, Monash, QUT Australia, Melbourne, APH Gov. Australia, Univ Western Australia, UFPB Brazil, Unicamp*, AEI Canada, NRC Canada IIT, NSERC Canada, Montreal Tech. Physics, British Columbia, Quebec Trois Rivieres, Victoria, York Univ. Canada*, PUC Chile, Biobio Chile, UFRO Chile, Institute of High Energy Physics China, Gov. Colombia ICFES, Amrita India, IITM, ISICAL, IMSC, PRL Ondia, Kogaguin Univ. Japan, Kyoto, Yamagata, KEK Japan, RUH Sri Lanka, UASLP Mexico, UITM Malaysia, Auckland NZ, NU Singapore, LUZ Venezuela.

September 15 - 29 2009.

US Universities and Institutes

Arizona, Boston University, Brigham Young, Chapman, Colorado, Duke, Florida Atlantic, Florida International, Florida State, Haverford, Indian River State College, Johns Hopkins University, Lousiana State, Mayo Clinic, Mesa State, Middlebury, Colorado School of Mines, Missouri, MIT*, Montana, Mississippi State, North Dakota, Northern Illinois, New Mexico State, Northwestern, Palomar, Renselaer, Stanford*, Stonybrook, Texas A and M, Texas Tech., University of California System, UC Santa Cruz, Delaware, Houston, Illinois Urbana Champaign*, U Mass., Chapel Hill*, North Texas, Vanderbilt, Virginia, Vermont,Wayne, Wisconsin, Washington Univeristy St. Louis, Yale; City of El Paso, Kentucky Gov., NIST, NOAA, Robins Airforce Base, US Army AMRDECC, US Navy UAR, US Military Command, Jefferson National Laboratory, COMRH Organziation, MECDC, W3; Boeing, Honeywell, IBM Thomas J. Watson*, Microsoft, Perkin Elmer, Philips, Raytheon.

Europe and Asia Minor: Universities, Institutes and Similar.

University of Sofia, ETH Zurich*, Charles Univ. Prague, Palacky, HU Berlin, LRZ Munich, Ruhr Univ. Bochum, RWTH Aachen, TU Darmstadt, Bonn, Freiburg, Hannover, SDU Denmark, Salamanca, Vigo*, ECP France, ENS Lyon, INRA Clermont, Strasbourg*, Poitiers*, HOL Greece, OTE Greece*, UOC Greece, Patras, Gibraltar, BME Hungary, HEP Hungary, IRB Hungary, INFN Florence, Trieste, ICTP Trieste, Unicas, Pavia, Rotterdam, TU Eindhoven, Utrecht, NTNU Norway, PW Students Poland, URC Russia, NSC Russia, KTH Sweden, LU Sweden,

Cambridge*, Conel, Imperial, Lancaster, Liverpool, Sheffield, UMIST, HSHA Organization, British Aerospace.

Rest of World: Universities, Institutes and Similar.
UNLP Argentina, Gov. Argentina, Australian National University, Queensland*, Sao Paolo, Unicamp Brazil, NRC Canada, Montreal Tech., Gov. Quebec, British Columbia, Quebec Trois Rivieres, Toronto, UDEA Colombia, UCLV Cuba, City Univ. Hong Kong, Amrita, IUCAA India, Gov. India BARC, ZNU Iran, Tsukuba Univ. Japan, KEK Japan, UITM Malaysia, NU Singapore, NCKU Taiwan, NTU Taiwan.

Summary of October 2009. NEW MONTHLY RECORD OF 29,135 VISITORS

For www.aias.us there were 85,475 hits from 13,786 visitors, reading 1,629 documents, downloading 9.043 gigabytes, generating 50,571 previews from 87 countries, led by USA, France, Germany, Australia, Britain, Canada .... For www.atomicprecision.com there were 60,515 hits from 15,349 visitors downloading 2.630 gigabytes, giving a total of 145, 990 hits from a record number of 29,135 visitors, downloading 11.673 gigabytes.

The most read source papers were: 100 (introduction), 54, 11, 41, 43, 4, 8, 2, 131, 25, 67, 76, 136, 94, 52, 56, 81, 74, 88, 20, 47, 7, 80, 26, 123, 12, 1, 82, 122, 42, 6, 2, 61, 44, 18, 3, 93, 116,121, 28, 32, 15, 45, ... all source papers read.
The most read articles were: Myron Evans Biography by Kerry Pendergast, Felker3, Felker5, Space Energy, Numerical Solutions, Workshop overview, Spacetime Devices, Lindstrom2, 135 (Spanish), Galaxies (Spanish), Introduction (Spanish), ECE and Spacetime, Felker4, Engineering Model, Workshop History, Ricci Concepts (Crothers), Johnson Magnets, Felker10, Felker13, Overview of ECE Theory, Felker15, Crystal Spheres, Refutation of Heisenberg, Eckardt Special Relativity, Felker6, Balck Holes Univorns and All that Jazz (Crothers), Felker2, ...

October 1 - 15 2009

US Universities and Institutes
Arizona, Arizona State, Boston University*, Buffalo*, Central Michigan*, Colorado, Columbia, Cornell, Case Western Reserve, Evergreen, Florida International, Fitchburg State College, Georgia Tech., Harvard, Haystack, Iowa State, Illinois, Indiana Purdue Indianapolis, Los

Rios, Missouri, MIT*, Michigan Tech., New York University, Oregon Institute of Technology, Texas Pan American, Princeton*, Penn State, Purdue, Rowan, Renselaer*, St Mary's Maryland*, St. Thomas, Texas Southwestern Medical Center, Texas A and M, Chicago, UC Irvine, UC San Diego, Indonesia (edu system), Maryland, Michigan, North Carolina, UNCC*, New Mexico, Oregon, Pennsylvania, Pittsburgh Medical Center, South Florida, Utah, Vermont, Washington, Wisconsin, WIU*, William and Mary, Lawrence Berkeley, NIH, NSF, SRS, SSA, St Lucie County, Robins Airforce Base, West Virginia Army, SUMY, Algonquin College, IBM Thomas J. Watson*, Raytheon.

Europe and Asia Minor: Universities, Institutes and Similar.

Univ. Vienna, CERN, Charles Univ. Prague, Palacky, FH Bochum, IPH Jena, Albert Einstein Institute, Bayreuth, Frankfurt, Hamburg, WIAS Berlin, AAU Denmark, SBU Denmark, UCM Spain, Canaries, Leon, US Spain, Spanish Parliament, JYU Finland, EMA France, Poitiers*, UPS Toulouse, OTE Net Greece, ELTE Hungary, NUIM Ireland, INFN Turin, Genoa, KLM Netherlands, Leiden, RU Groningen, UIB Norway, AGH Poland, Warsaw, UC Portugal, UP Portugal, Brasov Romania*, NI Russia, Uppsala Sweden, Josef Stefan Institute, ITU Turkey, Metu Turkey, Birmingham City, Buckinghamshire New, Cambridge*, Cardiff, Glasgow, Imperial, King's College London, Manchester, Newcastle, Nottingham, Trinity College Oxford, Rutherford Appleton, Scottish Funding Council, Sussex, Swansea, Ceredigion, NHS*.

Rest of World: Universities, Institutes and Similar

UNJU Argentina, ADFA Australia, Queensland, Sydney, Gov. British Columbia, Gov. Canada, Queen's, MacMaster, NCF, Alberta, British Columbia, Guelph, Waterloo, PUC Chile, Gov. Colombia, Dendai Univ. Japan, Univ Tokyo, KAIST*, UAEM Mexico, UNAM*, NU Singapore, NTNU, NTU, TPC Taiwan, Uniwest South Africa.

October 15 - 30

US Universities, Institutes and Similar

American International College, Andrews, Arizona, Arizona State*, Caltech*, Carson Newman, Colorado State, Cornell, Florida International, Hawaii, Johns Hopkins, McDaniel, Middlesex CC, MIT, Michigan State, New York University, Mississippi, Penn State*, Purdue, Rutgers, Stanford, Stonybrook, Toronto (edu), Arkansas Monticello, U Chicago, Central Oklahoma, UC Santa Barbara, Florida*, Houston, Iowa, Urbana Champaign, Maryland, North Carolina Charlotte and Wilmington, Nebraska Lincoln, New Mexico, North Texas, Pittsburgh Medical, Texas

El Paso, Texas*, Vermont, Washington, Western Illinois*, Yale; Brookhaven, Federal Oversight of Energy (US Gov.), Lawrence Berkeley, Pension Guaranty, Social Security, St Lucie County, Gov. Virginia, Army Elmendorf, EPG, PICA< Airforce HPC, MDA, Boeing, Honeywell, Hewlett Packard, IBM Thomas J. Watson*, Intel, Northrop Grumman*.

Europe and Asia Minor: Universities, Institutes and Similar

Ghent, CUNY, Palacky, Bosch, GWDG Germany, KFA Juelich, LMU, Siemens, Clausthal, Kiel, Muenster, Stuttgart, Tuebingen, Weimar, WIAS Berlin, Aarhus Denmark, IAC Spain, UB, Complutensian Madrid, Spanish Parliament, JYU Finland, Cirad France, Polytechnique, UHP Nancy*, Poitiers*, ONDSL Greece, ZABA Croatia, Atomki Hungary, UCS Ireland, Tel Aviv, Technion, Pisa, Telfort Netherlands, NTNU Norway, US Poland, UWM Poland, Gov. Poland, HBF Poland, UC Portugal, JINR Russia, DCN Romania, UU Sweden, Jozef Stefan Slovenia, Bilkent, Ege, Marmara, Metu Turkey, Aberystwyth, Birmingham City, Bristol, Cambridge*, Exeter, Imperial, Leicester, Napier, Nottingham, Oxford, Queen Mary, Southampton, St Andrews, UMIST, West Sussex CC.

Rest of World: Universities, Institutes and Similar.

UNJU Argentina, ADFA Australia, East Queensland, Queensland, UWS, Gov. New South Wales, AEI Canada, Defence Canada, Lakehead, Perimeter, Alberta, British Columbia, Montreal, Waterloo, UTFSM Chile, UDEA Colombia, UCF, UCLV, UO Cuba, Hong Kong, ITS Indonesia, Amrita, NITRKL, IIT Karagpoor, MSC, ESA India, Kyoto, Tohoku, Tokyo Japan, UGTO, UNAM, USON Mexico, Auckland NZ, Chula Thailand, LUZ Venezuela, Univ West South Africa.

Summary of November 2009

For www.aias.us there were 89849 hits from 13510 visitors, reading 1651 documents, and downloading 8.792 gigabytes of material with 54837 previews. There were visitors from 94 countries, led by USA, India, France, Germany, Czechia, Britain, Japan, Canada, ...

For www.atomicprecision.com there were 59613 hits from 13725 visitors, downloading 2.92 gigabytes.

The total is therefore 149462 hits from 27235 visitors, downloading 11.71 gigabytes.

The most read source papers were:

8, 41, 25, 88, 126, 21, 125, 127, 138, 4, 123, 82, 128, 130, 11, 63, 20, 43, 81, 119, 1, 85, 107, 129, 61, 18, 33, 60, 80, 3, 67, 124, 47, 76, 110, 12,

39, 74, 98, 117, 56, 7, 2, 13, 14, 94, 23, 28, 115, 133, 116, 22, 75, 55, 120, 59, 15, 17, 32, 54, 30, 42, 68, ... All source papers read.

The most read articles were: Biography by Kerry Pendergast, Spacetime Devices, Space Energy, Spacetime Devices (Spanish), Felker 5, Crystal Spheres, ECE Levitron, Felker 3, Felker 4, ...

November 1 - 15

US Universities, Institutes and Similar

Boston State, Boston, Brigham Young University*, Caltech, Chapman, Central Michigan*, Coe College, Colorado, Colorado State, Columbia, Cornell, Catholic University of America, Delaware Tech., Duke, Georgia State, Manhattan, MIT*, Montana, North Carolina State, Nebraska Wesleyan, North Dakota system, Penn State, Purdue, Rutgers, Stonybrook, Syracuse*, Taft College, Tulsa CC, Akron, Chicago, UC Santa Barbara, UC Santa Cruz, Florida, Illinois Urbana Champaign, Maryland, North Carolina Wilmington, North Florida, Nebraska Lincoln, Oregon, Texas, Tennessee Martin, Vermont*, Washington, Wisconsin, Washington State, Yale; Los Alamos, NASA*, US Dept. of Justice, Gov. Virginia, Wright Patterson Airforce Base, US Navy NMCI*, Boeing, IBM, Lockheed Martin, Northrop Grumman, Raytheon.

Europe: Universities, Institutes and Similar.

UIBK Austria, Free Univ. Brussels, Ghent, ETH Zurich, Charles Univ Prague, Palacky, Albert Einstein Institute, PII Berlin, STW Bonn, Dortmund, Freiburg, Hannover, Karlsruhe, Cologne, Wuerzburg, DTU Denmark, SDU Denmark, ETV Estonia, Barcelona, Canaries, UPC, Vigo, Valencia, EFFI Finland, PUV Finland, CNRS Grenoble France, CROUS Limoges, Polytechnique, Poitiers*, Tel Aviv, Technion, SISSA Italy, Pavia, VPU Lithuania, Skopje, Groningen, Eindhoven, NTNU Norway, PK Poland, UTL Portugal, LIU Sweden, Astro Slovakia, Metu* Turkey, Data Group Ukraine, Birmingham, Cambridge (Corpus Christi, St John's, Queen's, St. Edmunds, particle), Durham*, Liverpool, Medical Research Council, Oxford (Bodleian, Chemistry, Keble, Mathematics, Merton, St Anne's), Queen's Univ. Belfast, Southampton, Sussex, East Anglia, National Academy of Sciences of Armenia.

Rest of World: Universities, Institutes and Similar.

MDP Argentina, Griffith Univ. Australia, RMIT Australia, UFMG, UNICAMP Brazil, McMaster Canada, Queen's, British Columbia, PUC Chile, UCN, UTA Chile, UNAL Colombia, Salesianas Equador, Gov. Ecuador, HU, BARC, IUCAA India, Kyoto, Meiji, Toyota Corp., Chungbuk Korea, KAIST Korea, JFN Sri Lanka, UGTO, UNAM Mexico, MMU Malaysia, Chula Thailand, NCTU, Academica Sinica Taiwan.

November 15 - 29

US Universities and Similar.

Arizona, Berkeley*, Bryn Mawr, Boston Univ., Colorado, Columbia*, Cornell*, CSU Pomona, Catholic Univ America, Case Western Reserve, Dodea, Drexel, Duke, Geneseo, Iowa State, Johns Hopkins, MIT, Minnesota State Mankato, Nebraska Wesleyan, Ohio State, Princeton*, Penn State, Purdue, SUNY Stonybrook, Saginaw Valley State*, Texas A and M, Trinity College, Central Florida, Univ Chicago*, Univ. Connecticut, UC Riverside, Florida, UMB Maryland, Maryland, Michigan, Minnesota*, North Carolina Wilmington, South Florida, Utah, Texas El Paso, Texas, William and Mary, Washington, Wisconsin*, West Virginia, US Dept. of Energy, NASA*, St. Lucie County, Gov. Virginia, US Army Bliss, US Army USACE, US Navy NAVO, NMCI*, IBM (Thomas J. Watson), Raytheon.

Europe and Asia Minor: Universities, Institutes and Similar.

KU Leuven, ETH Zurich, CERN, CUNI*, Max Planck MPQ, RWTH Aachen, Jena, Konstanz, Muenster, Regensburg, Unity Media, Helsinki, Oulu, CEA France, OBSPM, Polytechnique, UJF Grenoble, Poitiers*, UCD Ireland, HUJI Israel, Tel Aviv*, CNR Isc., Italy, Siemens Italy, Trieste, Genoa, Unile Italy, Modena, Turin, KTU Lithuania, UKM Macedonia, Leiden, Eindhoven, RU Netehrlands, Delft, Amsterdam, IFJ Poland, UTCLUJ Romania, JINR Russia, KTH Sweden, DFU Turkey, Kiev, Cambridge*, Durham*, Imperial, Kent, Lancaster, Napier, Oxford*, Sheffield, Strathclyde, Swansea, UMIST*.

Rest of World: Universities, Institutes and Similar.

MDP Argentina, UNC, UNCU, CNEA Gov. Argentina, Griffith Australia, Gov. Queensland, Montreal, British Columbia, Saskatchewan, Toronto, Waterloo, Univ Chile, UCN Chile, HA China, UDEA Colombia, IMSC India, Tohoku, Tokyo* Japan, ITESO Mexico, UGTO, Chula Thailand, THU Taiwan, Free Univ. Transvaal South Africa.

Summary of December 2009

For www.aias.us there were 76372 hits from 10910 visitors, downloading 8.248 gigabytes from 88 countries, led by USA, Russia, France, Britain, Germany, Canada, ... During 2009 for www.aias.us there were 906,048 hits from visitors downloading 84.864 gigabytes.

For www.atomicprecision.com there were 59116 hits from 11434 visitors downloading 2.776 gigabytes.

So the total was 135,488 hits from 22,344 visitors, downloading 11.024 gigabytes.

The most read source papers were: 58, 94, 41, 11, 54, 4, 139, 8, 25, 82, 43, 88, 1, 30, 12, 44, 138, 28, 55, 98, 32, 47, 18, 15, 26, 56, 14, 61, 67, 93, 21, 123, 2, 76, 7,116, 13, 94, 137, 16, 42, 81, 19, 40, 60, 9, 126, 3, 49, 63, 197, 53, 110, 88, 20, 45, 57, 31, ...

The most read articles were: Crothers IJTP 1960, ECE and Spacetime, Space Energy, Felker3, Sapcetime Devices, Overview of ECE Theory, MWE Biography, ...

December 1 - 15

US Universities and Similar.

Austin Community College, Brown, Ball State, Boston, Caltech, Clemson, Colorado, Columbia, CSU Fresno, CU Denver, Duke*, Florida International, Georgia Tech., Gettysburg, Harvard*, Johns Hopkins, Lousiana State, Maine*, Missouri, MIT*, MS State, Michigan State, Northern Illinois, New Mexico State, Northwestern, Naval Postgraduate School, Ohio State, Purdue, Seattle University, Texas A and M, New Hampshire, Vanderbilt, Virginia, Worcester Polytechnic, Washington State*, Yale, US Dept. Education, Federal Aviation Admin., Hanford, Idaho National Laboratory, National Institutes of Health, Treasury, Gov. Virginia, Arnold AF Base, Hill AF Base, US Military Amrdec, US Army DRUM, US Navy NAVO, US Military USMC*, IBM Bluebird and IBM Thomas J. Watson, Intel, Lockheed Martin, Microsoft.

Europe: Universities, Institutes and Similar.

Ghent, Charles Univ. Prague, HU Berlin, Max Planck IKG, Max Planck MP1, Siemens, TU Berlin*, Freiburg, Kaiserslautern, Mainz, Saarland, Aragon, UPC Spain, UV Spain, JYU, TUT Finland, CNRS Grenoble, Paris 1, Perpignan, Poitiers*, HUJI Israel, ICRA Italy, INFN Bologna, Naples, Perugia, Tvente Netherlands, UIB Norway, Univ Nis Serbia, AOF Russia, Ege, Metu Turkey, Cambridge, Imperial*, Kent, Lancaster, Oxford, Univ College London*, UIST, Warwick, BBC*, NHS, Jozef Stefan Institute Slovenia.

Rest of World: Universities, Institutes and Similar

UNR Argentina, CNEA Gov. Argentina, ME Gov Argentina, UFF Brazil, UFBR Brazil, Unicamp Brazil, Gov British Columbia Canada, McMaster, Univ British Columbia, Univ Saskatchewan, UDEA Colombia, IMSC India, Kyoto, Tokyo, Waseda, USON Mexico, SP Singapore, CHULA Thailand, NCU, NCTW, NSYSU Taiwan, FING Uruguay, HCMIST Venezuela.

December 15 - 30

US Universities, Institutes and Similar

Colby, Colorado, CSU Fresno, Duke, Harvard*, Hawaii, MIT, SMC, Texas A and M*, UC San Diego, Univ Michigan*, Los Alamos, NASA, Santa Barbara, USBR, Virginia, Navy NAVO, US Military USMC, Boeing, General Electric, IBM Bluebird, IBM Thomas J. Watson, Microsoft.

Europe and Asia Minor: Universities, Institutes and Similar.

AIT Austria, EPF Lausanne, Reinhardt Microtech, Charles Univ. Prague, Jacobs, RWTH Aachen, Bonn, Karlsruhe, Cologne, Siegen, HTV Finland, HUT Finland, CNRS Grenoble, Poitiers, HUJI Israel, Technion, Weizmann, ICTP Trieste, Pavia, Perugia, Utrecht, ITU Turkey, Metu Turkey, Cardiff, Manchester, Oxford, Univ. College London, UEL, West Dunbarton.

Res of World: Universities, Institutes and Similar.

CSU Australia, MacMaster, Alberta, British Columbia, Montreal, Western Ontario, Engineering Univ. Colombia, CMI India, IUCAA India, KEK Japan, CNU South Korea, UNA Mexico, NCU Taiwan.

END OF 2009

Summary of January 2010: TWO NEW RECORDS, HITS and GIGABYTES DOWNLOADED

For www.aias.us there were 113506 hits from 14465 individual visitors, downloading 10.06 gigabytes of material and generating 71331 previews, reading 1687 documents from 86 countries, led by US, Russia, France, Italy, Germany, Canada, Australia, Austria, Japan, Britain, ...

For www.atomicprecisison.com there were 58960 hits from 11765 visitors, downloading 3.17 gigabytes of material.

The total is therefore 172466 hits from 26230 visitors downloading 13.23 gigabytes of material.

The most read papers were: 94, 11, 107, 43, 26, 4, 41, 8, 110, 118, 117, 119, 98, 25, 1, 116, 87, 100, 89, 37, 28, 120, 85, 93, 104, 122, 91, 92, 33, 44, 113, 96, 108, 40, 54, 99, 46, 61, 88, 76, 111, 136, 142, 102, 104, 106,

97, 123, 82, 34, 6, 10, 67, 7, 88,115, 70, 101, 138, 12, 131, 64, 18, 36, 50, 63, 74, 32, 39, ... all 143 source papers read.

The most read articles were: Kerry Pendergast biography, Spacetime Devices, Space Energy, Edyn3, Overview of ECE Theory, ECE Engineering Model, Special Relativity, Felker3, Device Dev., Space Energy (Spanish), ECE and Spacetime, Space Programme, Crystal Spheres, Film Script, Ages Without Knowing, Workshop History, ECE Levitron, Lindstrom 2, ...

1 - 16 January

US Universities, Institutes and Similar.
    Berkeley, Caltech, Central Michigan, Dartmouth, Dixie, Florida International, Georgia Tech. Iowa State, Maine, MIT, North Dakota, Ohio State, Ohio University, Oregon State, Oklahoma, Princeton*, Stanford, Stonybrook*, SUNY Stonybrook, UC Davies, U. Chicago, Central Oklahoma, UC Santa Cruz, UC San Diego, Missouri Kansas City, North Texas, Tennessee Martin, Virginia, Wright; Fermilab, House of Representatives, Los Alamos, Oak Ridge, Santa Barbara, USPTO, Amredec, Army Research Laboratory, Navy*, OSD; Boeing, IBM Thomas J. Watson, Lockheed Martin, Microsoft*.

Europe and Asia Minor: Universities, Institutes and Similar.
    UIBK Austria, Vienna, Brussels, EPF Lausanne, Geneva, CUNI* Czechia, Palacky*, GWDG Germany, RWTH Aachen, Siemens, Clausthal, Freiburg, Bremen, Dortmund, Frankfurt, Hannover, Karlsruhe, Cologne, Leipzig, Siegen, SDU Denmark, EHU Spain, Leon, UPC, UV, CEA France, CNRS Grenoble, OBSPM France, Paris Psud, Poitiers, HOL Greece, BME Hungary, ENI Italy, INFN Florence, NGI Italy, Milan, Utrecht, Wroclaw, OSU Russia, FEJK Sweden, Metu Turkey, SDU Turkey, OD Ukraine, Birmingham City, College of North West London, Durham, Liverpool, Sheffield Hallam, Swansea, UWIC, Warwick, Ceredigion.

Rest of World: Universities, Institutes and Similar.
    CSIRO Australia, QUT, NB Canada, Perimeter, Alberta, UBC*, UQTR, Toronto, IIIT India, BARC, IMSC, RRI, TIFR India, Kyoto Japan, Osaka, Tsukuba, SSU South Korea, Yonsei, UNAM Mexico, Tecsup Peru, IOBM Pakistan, QAU, NCTU, NSYSU Taiwan.

15 - 30 January

US Universities, Institutes and Similar.

Berkeley*, Bowdoin, Caltech, Central Connecticut State, Central Michigan*, Colorado*, Columbia, Creighton, Dixie, Drexel, Denver University, George Washington University, Illinois, Indian River, Johns Hopkins, Louisville, Louisiana State*, Maine, MIT, Michigan State*, Ohio State*, Princeton, Penn State*, Purdue*, Rice, Rochester, Stanford, SUNY Stonybrook, UC Davis, Chicago*, UC Irvine, U Connecticut, U Michigan, North Carolina Greensboro, U Penn., Utah State, Texas Arlington, Utah, Kentucky, Yale; Center for Disease Control US Government, St Lucie, Virginia, Hill AF Base, Army Research Laboratory, Army Eustis, BAE Systems, Boeing, General Electric, IBM, Lockheed Martin, Microsoft, Volvo.

Europe and Asia Minor: Universities, Institutes and Similar

UIBK Austria, Linz, Free Univ Brussels, EPF Lausanne, Charles Univ Prague, SLU Czechia, Palacky, FU Berlin, Siemens, TU Berlin, Darmstadt, Augsburg, Dortmund, Freiburg, Heidelberg, Magdeburg, Oldenburg, Regensburg, SDU Denmark, TTU Estonia, EHU Spain, UAM Spain, UPC, UV* Spain, CEA France, CEA Saclay, CNRS Grenoble, ENS Lyon, INSERM Toulouse, Nancy, Limoges, Poitiers, DWTH Greece, Dublin City Ireland, ICTP Italy, INFN Bologna, INFN Milan, Naples, NFRAN Netherlands, RU Netherlands, Groningen, NTNU, UIO Norway, Ksiezyc Poland, UMA Portugal, Astral Romania, Electroshield Russia, MKS Tombov, UR Russia, LIU, LTH Sweden, ITU Turkey, Cambridge, Royal Society of Chemistry, Glasgow*, Newcastle upon Tyne, Nottingham, Oxford, Southampton, Swansea, University College London*, Warwick, Conwy, Gwynedd, Ceredigion.

Rest of World: Universities, Institutes and Similar

Queensland Australia, IMPA Brazil, Gov British Columbia, Brock, NCF, YRBE Ontario, Queens, British Columbia*, Quebec Trois Rivieres, Toronto, IITM India, Gunma CT Japan, KYU Tech, Osaka*, Tokyo, Honda, KAIST, SSU, Yonsei S. Korea, UTHM Malaysia, Otago NZ, Panama, IOBM Pakistan, CYUT Taiwan.

Summary of February 2010

For www.aias.us there were 75864 hits from 11381 visitors from 94 countries, downloading 5.59 gigabytes and 1686 documents and generating 46309 previews.

For www.atomicprecision.com there were 52659 hits from 10276 visitors, downloading 3.35 gigabytes.

This gives a total of 128523 hits from 21657 visitors, downloading 8.94 gigabytes.

The most read papers were: 94, 11, 43, 131(Spanish), 107, 41, 140(Sp), 1, 46, 4, 74, 44, 140, 26, 8, 137(Sp), 63, 119, 39, 82, 141(Sp), 136(Sp), 142, 67,61, 141, 52, 122(Sp), 116, 118, 42, 104, 31, 47, 100, 120, 25, 56, 7, 102, 142(Sp), 10, 123, 103, 98, 12, 40, 57, 133(Sp), 49, 80, 110, 96, 143(Sp), 131, 113, 136, ... all papers read.

The most ready articles were: Felker3, Kerry Pendergast biography, Felker5, Space Energy, Spacetime Devices (Spanish), Lindstrom2, OO563, Crystal Spheres, ...

1 - 15 February

US Universities, Institutes and Similar
   Arizona*, Bellevue, Berkeley*, Buffalo*, Caltech, Colorado, Columbia*, Cooper, Case Western Reserve, East Michigan, Florida State, Gustavus Adolphus College, Georgia Tech., Harvard, Hudson Valley Community College, Indiana, Johns Hopkins, Kansas State, Lowell*, Missouri, MIT*, Morgan, New York University, Ohio State, Penn State, Renselaer Polytechnic Institute, Rutgers, Texas A and M, Central Florida, Chicago, UCLA*, UC Riverside*, Houston, Illinois Urbana Champaign, U Mass., Minnesota, Univ Nebraska Lincoln, U Penn., South Carolina, New Mexico, Vermont, Washington University St. Louis, Fermilab., NASA, NOAA, Eglin AF Base, Patrick AF Base, Army Tacom, Navy Navair, Navy NAVO, IBM Thomas J. Watson, Dow.

Europe and Asia Minor: Universities, Institutes and Similar.
   Vienna*, EPF Lausanne*, Charles Univ.Prague, Bosch, Max Planck Dynamics, Siemens, TU Chemnitz, TU Hamburg, Bonn, Konstanz, Luebeck, Stuttgart*, WH Stuttgart, GVA Sapin, Complutensian Madrid, UPV, UV, UVA Spain, OBS Nice, Perpignan, Poitiers*, NTUA Greece, TAU Israel, ICTP Italy, INFN Mib., INFN Pisa, Turin Tech., SISSA, Turin, Telfort Netherlands, TU Eindhoven, UJ Poland*, Torun, UTL Portugal, MKS Tambov Russia, MTS-NN Dynamic Russia, Troitsk, HV Sweden, KTH Stockholm, UMU Sweden, Students UU Sweden, Slovak Radio, Bristol*, Cambridge*, Carmel College, Cranfield, Durham, Imperial, Leeds, Leeds Met., Student Halls London, Oxford*, Southampton, West England, York, Ministry of Defence, Rendcomb College.

Rest of World: Universities, Institutes and Similar.

Univ. Western Australia, Gov. Queensland, Gov. British Columbia Canada, Gov. Canada IC, Perimeter, Alberta, British Columbia, Laval, Quebec Trois Rivieres, UDEA, UNAL Colombia, UAI Indonesia, RCIL India, Oita Univ Japan, KAIST, Yonsei, Samsung S. Korea, KKU Thailand, NTU Taiwan, Univwest South Africa.

15 - 27 February

US Universities, Institutes and Similar

Arizona, Berkeley, Buffalo, Caltech*, Cornell, Florida Atlantic, Georgia Tech., Hawaii, Iowa State, James Madison, Kansas State, Lousiana Tech., Lowell*, Maricopa, Marshall, Mercer, MIT*, Michigan State, North Carolina State, North Dakota, Ohio State*, Oklahoma State, Oregon State, Pomona, Penn State, Renselaer, Rutgers, St Petersburg College, Susqeuhanna, Texas A and M, Tufts*, Alabama Birmingham, Central Florida, Chicago, Connecticut, UC Santa Barbara, UC Santa Cruz, Illinois Urbana Champaign, Michigan, Chapel Hill*, New Mexico, U Penn., Texas*, William and Mary, Robins AF, US Airforce Kuwait, US Navy NAVO, St Lucie County, Honda (US), IBM*, Microsoft, Yonkers Public Schools.

Europe and Asia Minor: Universities, Institutes and Similar.

Medical Univ. Vienna, Free Univ. Brussels, Charles Univ. Prague, Palacky*, Bosch, Jacobs Univ., Max Planck Heidelberg, TU Chemnitz, Ilmenau, Freiburg, Potsdam, Stuttgart, GVA Spain, UCM, Zaragoza, UPV, Academica Finland, French Particle Laboratory, Bourgogne, Poitiers, AUTH Greece, Patras, FSB Croatia, Tel Aviv, ICTP Trieste, INAF, INFN Bari, INFN Pisa, INFN Rome 1, RU Groningen, TUEindhoven, KIS Russia, VSI Russia, FEJK Sweden, SU Sweden, Kharkov, Ege Turkey, Birmingham, Cambridge*, Lancaster, Nottingham*, Oxford*, Reading, Swansea, Warwick, British Gov. Energis*,

Rest of World: Universities, Institutes and Similar.

UNLP, UNT Argentina, Flinders, Melbourne, Western Australia, Alberta, British Columbia, Calgary*, Trois Rivieres, Toronto, UNAL Colombia, UAI Indonesia, Kigali Institute Rwanda, II Sermali India, Gov. India, Hiroshima, Tohoku, KAIST S. Korea, Postech, Yonsei, Samsung Korea, OUSri Lanka, IPICYT Mexico, UASLP, UGTO Mexico, UTHM Malaysia, KMITL Thailand, NTCU Taiwan.

Summary of March 2010

For www.aias.us there were 92305 hits from 12631 visitors, downloading 7.37 gigabytes, generating 59165 previews, reading 1691 documents from 94 countries, led by US, Russia, Britain, France, Austria, Australia, Japan, Germany, ...

For www.atomicprecision.com there were 45791 hits from 10848 visitors, downloading 4.39 gigabytes.

For www.upitec.org there were 6597 hits from 394 visitors, downloading 0.08 gigabytes.

This gives a total of 144693 hits from 23873 visitors, downloading 12.52 gigabytes.

The most read source papers were: 145, 94, 67, 43, 41, 111, 11, 25, 1, 107, 131, 140 (Spanish), 4, 8, 116, 118, 42, 26, 46, 19, 12, 81, 142, 119, 61, 85, 131, 38, 63, 7, 101, 110, 35, 140, 44, 65, 99, 21, 115, 122 (Sp), OO653, 18, 54, 56, 74, 113, 114, 82, 141, 31, 40, 104, 39, 52, 102, 103, 91, 97, 144 (Sp), 122, 17, 49, 117, 83, 144, 3, 50, 57, 137 (Sp), 141, 10, 136, 92, 98, ... all papers read.

The most read articles were: Biography by Kerry Pendergast, Felker 3 (Sp), Space Devices (Sp.), Space Energy, Sapcetiem Devices, Overview of ECE Theory, all articles read. All notes and Omnia Opera papers read.

March 1 - 15

US Universities, Institutes and Similar.

Arizona State College, Brockport, Boston University, Buffalo, Caltech, City Colleges of Chicago, Colorado, Cornell, CUNY, Denver University*, Duke, Emry Riddle, Florida International, Florida State, Harvard, Haverford, Hawaii, Indian River, Colorado School of Mines, Missouri, MIT*, North Dakota, Oregon State, Rochester, Renselaer, Rutgers, SP College, Stockton, Stonybrook, SUNY Stonybrook, Texas A and M, UC Riverside, UC Santa Barbara, Delaware, Girona (edu system), U Mass Dartmouth, Minnesota, Northern Colorado, Southern California, Toledo, Virginia, Wisconisn, West Virginia University, NASA JSC, NIST, NOAA, St Lucie County, US Airforce AFOSR, Eglin AF, Kirtland AF, Lackland AF, US Army Stewart, NMCI Navy*, Northwestern Memorial Hospital, IBM Bluebird, IBM Thomas J. Watson, Lockheed Martin, Northrop Grumman.

Europe and Asia Minor: Universities, Institutes and Similar.

TU Graz, TU Vienna, UIBK Austria, KU Leuven, Belfort Switzerland, CUNI Prague, Palacky, Max Planck SB, STW Bonn, TU Berlin, Heidelberg, AU Denmark, EHU Spain, UAB, UAM, Uned, Univ Canaries, Ovi, UPV, USC, UVA Spain, Paris 13, Poitiers*, ATT, TEE Greece, BME Hungary, GN Gov. Ireland, INFN Bologna, INFN LNF, Milan, Pavia, Rome 1, Utrecht, UVT Netherlands, Kluwer, NTNU Norway, Warsaw, UTL Portugal, JINR, Ankara, Gantep Turkey, ITL Ukraine, Teifi and Towy Aberystwyth, Birmingham Central, EPSRC, Glamorgan, Imperial, Lancaster, Liverpool, Oxford*, Queen's Univ., Belfast, Sheffield, Southampton, Swansea College, UCL*, Univ London Computer Centre, Warwick, York, British Gov. Energis.

Rest of World: Universities, Institutes and Similar.

MDP Argentina, Gov. Argentina, East Queensland Australia, Tasmania, UFPA Brazil, McGill Canada, Perimeter, British Columbia, Laval, Quebec Trois Rivieres, Waterloo, UTP Colombia, Engineering University Colombia, Pacific Univ Ecuador, JNU India, BARC, CAT India, Chonnam S. Korea, KAIST*, Yonsei S. Korea, RUH Sri Lanka, UAEM Mexico, UTHM Malaysia, SQU Oman, Dakkar Univ., NCKU, NTHU Taiwan, SUN South Africa.

March 15 - 30

US Universities, Institutes and Similar

Arizona, Ball State, Boston, Brigham Young, Central Michigan, Duke, Harvard, Haverford, Hawaii, Missouri, MIT*, North Carolina State, Northern Illinois, Penn State, Purdue, Rice, Renselaer*, Rochester, Rutges, South Carolina, Scripps, Sterling, Sisternia, SUNY Stonybrook, Minnesota*, New Hampshire, Nevada, Pennsylvania, Utah, Wisconsin River Falls, Vanderbilt, Vermont, Washington St Louis; Fermilab, Hamilton County Ohio, Laurence Berkeley, US Gov. WOAA, National Research Council, South Carolina Dept of Health, St. Lucie, US Mil AFOSR, Eglin AF Base, US Army ACE, US Navy NAVO, US Navy NMCI*, Boeing, Dow, Honeywell, IBM Bluebird and Thomas J. Watson, Intel.

Europe and Asia Minor, Universities, Institutes and Similar

TU Vienna, UIBK Austria, Vienna, Free University Brussels, CERN, Belfort, Palacky*, Max Planck Informatik, Siemens*, Bonn, Karlsruhe, Kaiserslautern, Rostock, AAU Denmark, UMA Spain, Canaries, UPV, UVA*, Paris South, Poitiers*, AUTH Greece, Bio Academy Greece, ELTE Hungary, INGV Italy, Pisa, Turin, SU Lithuania, Leiden, RUG

Netherlands, Eindhoven, Amsterdam, NTNU Norway, WSZP Poland, Torun, ECRO Romania, TBC Romania, NI Russia, UU Sweden, Jozef Stevan Slovenia, Ege Turkey, Poltava Ukraine, Cambridge, Durham, Imperial, Lancaster, Univ. College London, York*, Ministry of Defence, National Library of Wales.

Rest of World: Universities, Institutes and Similar
    UNJU Argentina, Australian National University, East Queensland, UFPR Brazil, UFPRE Brazil, McGill* Canada, Perimeter, British Columbia, Calgary, Quebec Trois Rivieres, Waterloo, UNAL Colombia, UTP Colombia, Riken Japan, KAIST S. Korea, NU Singapore, KKU Thailand, CYUT Taiwan, Capetown South Africa.

Summary of April 2010

For www.aias.us there were 85,128 hits from 11,619 visitors, downloading 8.56 gigabytes, 53,274 previews, and 1692 documents from 85 countries, led by USA, Germany, Russia, Austria, France, Canada, Britain, ...

The most read papers from www.aias.us were: 64, 11, 94, 41, 43, 25, 4, 8, 107, 81, 140(Sp.), 118, 131(Sp.), 142, 42, 104, 119, 110, 144(Sp), 142, 100, 101, 80, 146, 1, 74, 20, 37, 40, 54, 76, 26, 116, 63, 82, 88, 46, 108, 15, 2, 33, 61, 34, 39, 47, 67, 102, 85, 147, 27, 7, 117, 91, 48, 6, 96, 140, 141, 12, 3, 44, 50, 103, 120, 143, 75, 87, 88, 10, ... all source papers read.

The most read articles and books were: MWE Biography, Felker5, Felker3, Space Energy Devices (Spanish), Special Relativity, Spacetime Devices, Overview of ECE Theory, ... all articles and books read.

For www.atomicprecision.com there were 41,275 hits from 10,729 visitors, downloading 2.97 gigabytes.

For www.upitec.org there were 5117 hits from 352 visitors.

This makes a combined total of 131,520 hits from 22,700 visitors, downloading 11.63 gigabytes.

April 1 - 15

US Universities, Institutes and Similar
    Arizona State, Berkeley, Boston University*, Buffalo, Caltech*, Central Michigan, Colby, Columbia, Dept. of Defense Education Activity,

Drexel, Fordham, Florida State, Georgia Tech., Harvard, Louisiana Tech., Maine*, Mississippi State, Michigan State*, Northeastern, New Mexico Tech., Ohio State, Oregon State, Princeton*, Penn State, Rutgers, Tarleton, Univ. California system, UCLA*, Central Oklahoma, Connecticut, Univ. California Office of the President, Houston, Massachusetts, Michigan, North Carolina Charlotte, U Penn*, US Naval Academy, Vanderbilt, Virginia, Vermont*, Washington, Williams*, Western Illinois, New Mexico, Fermilab., NOAA, St Lucie County, Kessler AF Base, Wright Patterson, Army TACOM, Navy NAVO, Corning, Honeywell, Intel*, Northrop Grumman*, National Institute of Oceanography.

Europe and Asia Minor: Universities, Institutes and Similar.
TU Vienna, ETH Zurich, Belfort, Charles Univ. Prague, Palacky, Jacobs, RWTH Aachen, Siemens Company, TU Berlin*, TU Dresden, TU Muenster, AAU Denmark, Tarragona, UB Spain, UPM*, US, UV Spain, HTV, UTU Finland, Poitiers, Rennes 1, University College Dublin Ireland*, INFN LNF Italy, INFN Milan, Palermo, Telfort Netherlands, UIO Norway, UWM Poland, UTL Portugal, SBB Russia, Murmansk, LU Sweden, Ege Turkey, Aberdeen, Cambridge*, Manchester, Nottingham Trent, Oxford, Scottish Grid Service, Southampton.

Rest of World: Universities, Institutes and Similar.
UNSL Argentina, Australian National University, Curtin, Gov British Columbia Canada, Quebec Trois Rivieres, Toronto, Waterloo, CIGB Cuba, ITB Indonesia, Osakafu Japan, TUS, TUT, Chiba Japan, UNAM Mexico, UNP Peru, QAU Pakistan, NTHU Taiwan, AIMS South Africa, Capetown, Witwatersrand.

April 15 - 29

US Universities, Institutes and Similar.
Arizona, Caltech, Colby, Colorado, Colorado State*, California State Fresno, Fordham, Georgia Southern, Maine*, North Dakota, New Jersey Institute of Technology, New Mexico State, Northwestern, Naval Postgraduate School, Ohio, Ohio State, Portland State, Pittsburgh, Princeton*, Penn State, South Carolina, Stanford, The College of New Jersey, Texas Tech., Ubaguio, UCLA, UC Santa Cruz, Charleston, Univ. Iowa, U. Maryland, U. Michigan*, U. Minnesota, Chapel Hill, North Texas, U. Southern Mississippi, U Texas El Paso, U Texas, U Toledo, Vermont*, Wisconsin, Wittenberg, Washington Univ. St Louis, West Virginia, Argonne, Los Alamos, US Dept of Labor, NASA, NOAA, St Lucie County, Air National Guard, Edwards AF, Spangdahlem AF, Wright Patterson AF, US Army Tactical Command, Lake Agassiz Regional

Library, Spokane County Library District, Worthington Libraries, BMW Group, Chevron, IBM Thomas J. Watson, Intel.

Europe and Asia Minor: Universities, Institutes and Similar
KU Leuven, Tatra Czechia, Palacky, VUTBR Czechia, Jacobs Germany, QSC, Siemens, TU Berlin, Heidelberg, Mainz, Muenster, Tuebingen, EMT, UT Estonia, CSIC Spain, MCU, UGR, UNEX, UPC, US, UV* Spain, Helsinki, HTV Finland, CEA, INSA Lyon, Poitiers, Rennes 1 France, Forthnet Greece, Trinity College Dublin Ireland, Jerusalem College of Technology Israel, CNR IIT Italy, ICTP Trieste, NGI, SNS, Perugia Italy, TU Eindhoven, Twente Netherlands, NTNU Norway, ACN Warsaw, Kazan State, Omsk, Moscow State Univ. Lomonosov, UU Sweden, Teifi Aberystwyth, Rutherford Appleton, Imperial, Liverpool, Manchester, Napier, Nottingham Trent. Oxford*, Scotgrid, Sheffield, Sheffield Hallam, Sussex, Swansea College, Univ. College London, CEC Europe.

Rest of World: Universities, Institutes and Similar
UNT Argentina, Australian National University, Western Australia, UFPB Brazil, UNESP Brazil, Ericsson Canada, Gov. Newfoundland, U British Columbia Canada*, UQAC, Quebec Trois Riviers, Waterloo, UFRO Chile, Gov. Colombia, CIGB Cuba, CMI India, Tezu, Gov. India, RCIL, Bose, TIFRBNG India, ZNU Iran, Kyoto Japan, Osakafu, Tohoku, Tsukuba, MOD Japan, KAIST South Korea, RUH Sri Lanka, UNAM Mexico, Pieas Pakistan, KMITL Thailand, NCTU, NTU, Academica Sinica Taiwan.

Summary of May 2010

For www.aias.us there were 74963 hits from 12678 visitors, reading 1712 documents, downloading 7.432 gigabytes with 43126 previews. There were visitors from 100 countries led by USA, Russia, France, Austria, Germany, Britain, Australia, ...

For www.atomicprecision.com there were 39527 hits from 10229 visitors.

For www.upitec.org there were 4992 hits from 332 visitors

This makes a total of 119482 hits from 23239 visitors.

All source papers were read, the most read papers were: 150, 4, 111, 94, 148, 11, 41, 25, 43, 8, 142(Sp), 140(Sp), 131(Sp), 118, 123, 107, 110, 116, 88, 33, 137, 51, 54, 1, 44, 81, 47, 102, 141, 142, 104, 149, 143(Sp),

122(Sp), 17, 18, 2, 12, 101, 144, 42, 61, 74, 20, 3, Summary of Metric results, 136(Sp), 37, 7, 10, 19, 135, 26, 80, 119, 120, 139, 67, 6, 122, 147, 30, ...

All Documents were read, the most read being MWE Bio, Space Energy (Sp), OO563, Galaxies (Sp), Special Relativity, Spacetime Devices (Sp), Edyn3, Galaxies, Spacetime Devices, Felker5, Overview of ECE Theory, ...

May 1 - 15

US Universities, Institutes and Similar

Arizona, Biola, Brown, Cameron, Cerritos, Chapman, Claremont, Central Michigan*, Columbia, Cooper*, Cornell, Fairleigh Dickinson, Frank Phillips College, Galileo, Georgia College and State University, Harvard, Indian River State College, Indiana Purdue Indianapolis, Kent State, Liceo, MIT*, New York, Old Dominion, Oregon State, Portland State, Princeton*, Penn State*, Smithsonian, Southern Illinois, Southwest Research Institute, Central Florida, UC Riverside, UC Santa Barbara, Houston, Univ. Massachusetts, U Mass. Lowell, Minnesota, Nebraska Lincoln, U Pennsylvania, South Carolina, Texas, Virginia, Wisconsin, Washington Univ., St. Louis, JSC NASA. St Lucie County, NCSC Navy, Hewlett Packard, IBM Thomas J. Watson, Philips.

Europe and Asia Minor: Universities, Institutes and Similar.

Ghent, Charles Univ. Prague, MUNI Czechia, IFW Dresden*, Max Plank MPPU, RWTH Aachen*, TU Berlin, Cologne, Leipzig*, AU Denmark, Gov. Andalucia, Autonomous Madrid, Barcelona, Complutensian Madrid*, Granada, UPV, UV Spain, European Union Portal, HTV Finland, CEA France, UHP Nancy*, Poitiers*, UOC Greece, UCD Ireland, Tel Aviv Israel, Astro Italy, ITN CNR Italy, ICTP Italy, Calabria, Padua, Sannio, Turin, Trento, FI Lithuania, Language Institute Luxembourg, Joseph Bech Luxembourg, CSS Latvia, Telfort Netherlands, Free Univ. Amsterdam, UMK Poland, Belgrade Serbia, HC, HITV, NSMA Russia, UNU, UU Sweden, Jozef Stefan Slovenia, DEU Turkey, Metu Turkey, Data Group Ukraine, Birkbeck, Bristol, Cambridge, Derby, Edinburgh, Exeter, Leicester, Liverpool, Oxford*, Queen University Belfast, Rutherford Appleton, Sussex, Warwick, Cyngor Cefn Gwlad Countryside Council for Wales, Ceredigion, National Health Service, Llyfrgell Cenedlaethol Cymru (National Library of Wales).

Rest of World: Universities, Institutes and Similar.

UNJU Argentina, New South Wales, Newcastle*, Melbourne*, Queensland, South Queensland, UFP, UNB, USP Brazil, Perimeter

Canada, Hydro Quebec, British Columbia, Waterloo*, Western Ontario, UCN Chile, City Hong Kong, Polytechnic Hong Kong, UNS Indonesia, JNCASR, JMI, CMERI, IIFR India, Osaka, Ritsumei, Tokyo* Japan, OU Sri Lanka, CICY Mexico, UNAM Mexico, Swinburne Malaysia, PUCP Peru, Apiquitos Peru, QAU Pakistan.

May 15 - 30

US Universities, Institutes and Similar.
 Bard, Carleton, Carthage, Central Michigan, Georgia State, Harvard, Indiana, Kansas, Missouri*, Milwaukee School of Engineering, Northwestern, New York University*, Ohio State, Princeton*, State Univ. New York Stonybrook, Southwest Research Institute, UC Santa Barbara*, UC Santa Cruz, Girona, Tucson Medical Center, Utah, Texas, US DHS, NASA (ARC), NASA (HST), NOAA, St. Lucie County, IOR Navy, NMCI Navy, BAE Systems, IBM Thomas J Watson, Lockheed Martin, Microsoft*, Northrop Grumman.

Europe and Asia Minor: Universities, Institutes and Similar.
 Ghent, ETH Zurich*, Charles Univ. Prague, German Synchrotron Facility (DESY), FH Potsdam, Bremen, Erlangen, Hannover, Heidelberg*, Kassel, Potsdam, Regensburg, SDU Denmark, UAB* Spain, UCA, Madrid Complutensian, Avarra, UVA*, HTV, JYU, Oulu Finland, CEA France, ENS, French Nuclear Physics National Laboratory*, LPTHE Jussieu, Marseilles, Poitiers, Rennes 1, UPS Toulouse, Hebrew Univ Jerusalem, Tel Aviv Israel, INFN Catania, Florence, Turin, SISSA, Florence, AUB Lebanon, PT Luxembourg, KLM Netherlands, RU Nethelands, Groningen, Telfort Netherlands, Minho Portugal, UT Cluj Romania, Debryansk Russia, Physto Sweden, UMU, UU Sweden, NLB Slovenia, Kiev, Aberystwyth, Cambridge*, Leicester de Montfort, Exeter, Imperial*, King's College London, Liverpool, Manchester*, Nottingham*, Oxford*, Portsmouth, Queen Mary College London, Swansea, Univ East England, Fife Council.

Rest of World: Universities, Institutes and Similar.
 Australian National University, Griffith, Monash*, Queensland, Sydney, Western Australia, UFG Brazil, UFRGS, Sao Paolo, Gov British Columbia Canada, Calgary, Waterloo, UCN Chile, CHLA Colombia, UNAL Colombia, UO, INF, Sime Cuba, ITS Indonesia, VSSC India, Kyoto, Osaka Oskafu*, Hyogo, Tokyo* Japan, Postech, SNU Korea, USON Mexico*, NTHU, NTU Taiwan, Capetown South Africa.

Summary of June 2010

For www.aias.us there were 65,194 page views and 101,030 hits from 15,896 unique visits, downloading 8.664 gigabytes and reading 1,758 documents. There were visitors from 87 countries, led by USA, Russian Federation, Britain, France, Germany, Austria, Australia, Mexico, …….

For www.atomicprecision.com there were 341,218 hits, 13,015 pages, and 8,273 unique visits.

For www.upitec.org there were 4,121 hits, 558 pages and 267 visits.
This makes a total of 446,339 hits, 78,767 pages and 24,436 unique visits.

All 152 source papers were read. The most read papers were: 150(notes), 63, 150, 107, 4, 41, 139, 25, 149, 56, 140, 18, 140(notes), 149(notes), 88, 150(Spanish), 94, 8, 146, 102, 43, 26, 144(notes), 49, 54, 107, 145, 147, 60, 53, 124, 23, 50, 133, 148,57, 140(Spanish), 14, 19, 20, 52, 58, 122(notes), 21, 61, 11, 67, 136, 144, 55, 127, 152, 17, 64, 122 (notes2), 134, 137, 15, 130, 132, 16, 22, 24, 47, …….
All articles were read, the most read ones were: Space Energy, Felker3, Space Program, Ricci Concepts, ECE Engineering Model, ECE Levitron, Galaxies (Spanish), Spacetime Devices, Ages without Knowing, Biography, Spacetime Devices (Spanish).
The Three Talks were listened to, the most popular ones being: On the Deflection of Light by Gravity, Essay of UFT 139 and Film Script.

June 1 - 15

US Universities, Institutes and Similar
Caltech*, Central New Mexico, Columbia, Kansas University, Northwestern, Naval Postgraduate School, Ohio State, Pomona, Southern Oregon, Stanford, Texas A and M, Tufts, UC Irvine, UC Santa Barbara*, UC San Diego, Girona (edu), Lousiana Monroe, UMass Dartmouth*, Minnesota Twin Cities, Univ Pennsylvania, Utah, Wisconsin River Falls, Washington, Wisconsin, Washington State, Yale, Argonne National, Edwards AF, Army Amrdec, Monmouth, BAE Systems, General Electric, IBM Thomas J. Watson, Lockheed Martin, Raytheon.

Europe and Asia Minor : Universities, Institutes and Similar.
Linz, Vienna, Art of Technology Switzerland, Charles Univ. Prague*, Palacky, UPCE Czechia, FH Potsdam, LRZ Munich, Max Planck Garching, RZ Berlin, Ruhr Univ, Bochum, RWTH Aachen, Siemens, Frankfurt, Hannover, Cologne, Leipzig, Mannheim, Marburg, Siegen,

Wuerzburg, KU Denmark, UAM Spain, UCO, UGR, UJI, Unav*, Canaries, UPCO, US, UV* Spain, HTV Finland, HUT Finland, ENSCP, IRAM France, Mathematics Jussieu, OBS Nice, Polytechnique, Poitiers, UPMC France, AMIS Croatia, Tel Aviv, Weizmann Israel, INFN Catania, INFN Florence, Calabria, CSS Latvia, UNAM Mexico, GGZ Nijmegen, TU Eindhoven, UVA Netherlands, AGH Poland, US Poland, UMA Portugal, MTS-NN Russia, Saransk, Bradford, Cambridge, Rutherford Appleton, Durham, Liverpool, Napier, Newcastle upon Tyne, Oxford*, QMU London, Swansea, Univ. London Computer Centre, Gov. Coventry, Gov. Energis, Ministry of Defence*, Llyfrgell Cenedlaethol Cymru (National Library of Wales).

Rest of World: Universities, Institutes and Similar.

UNNE Argentina, HCDN Gov. Argentina, Flinders Australia, Monash*, Queensland, Walter and Eliza Hall Institute of Medical Research Australia, UFPB, UFPE, UFSC, USP Brazil, British Columbia, NRC Canada, Perimeter Institute, Queens Univ. Canada, Univ. British Columbia, New Brunswick, PUC Chile, Tokyo, Waseda Japan, SNY Korea, NUS Singapore*, SP Singapore, NTHU Taiwan, NTU Taiwan, UDSM Tanzania, Cape Town.

June 15 – 29

US Universities, Institutes and Similar.

Caltech, Columbia*, Cornell, Florida State University, Kansas*, Mayo Clinic, Missouri, Michigan State Unviersity*, North Dakota, New York University, Old Dominion, Rice, Stevens Tech., Tufts, U Mass Dartmouth, New Mexico, Utah, Tennessee Knocksville, Wisconsin River Falls, Valdosta, Argonne, Department of Energy, Fermilab, Lawrence Livermore, NASA*, NOAA, Charleston Airforce Base, Monmouth Army Base, Defense Threat Reduction Agency, Chevron, CIA*, Lockheed Martin, Microsoft, National Medical Clinic, Northrop Grumman, Jefferson National Laboratory.

Europe and Asia Minor: Universities, Institutes and Similar.

TU Vienna*, CERN, UPC Czechia, Fraunhofer, RWTH Aachen, TH Harburg, Heidelberg*, Kiel, Leipzig, Munich, Siegen, UAM, UB, UMA, UPV, UV* Spain, Crous Limoges France, INRIA, OBS Nice, Marseilles, Poitiers, Weizmann, ICTP Italy, INFN Florence*, Milan Polytechnic, Cassino, Modena, Turin, CSS Group Latvia, RTU Latvia, Telfort Netherlands, Utrecht, Free Univ. Amsterdam, GDA Poland, UMA Portugal, SBB Russia, KTS Russia, Kharkov, Kiev, Odessa Ukraine, Birmingham City, Cambridge, King's College London, Leeds, Luton SFC,

Oxford, Queen's Belfast, Southampton, Swansea, Univ. College London, UMIST, Warwick, Gov. Energis, NHS.

Rest of World: Univesities, Institutes and Similar.

Australian National University, Queensland, UFAL, UFPB, UPRJ, USP Brazil, Memorial University Newfoundland Canada, Queen's, British Columbia*, Victoria, Waterloo, York Univ. Toronto, UTFSM Chile, UNL Ecuador, Bose Institute India, Tokushima, Tokyo Japan, MRT Sri Lanka, CAP, UNAM Mexico, CCU, NTU, Government Taiwan, PUK South Africa, RUS South Africa.

Summary of July 2010

For www.aias.us there were 11,115 unique visits, 88,670 hits, 54,469 page views, 6.402 gigabytes downloaded, 1,892 documents read from 87 countries, led by US, Germany, Russia, Canada, France, Britain, Mexico, ...

For www.atomicprecision.com there were 41,985 hits from 8,263 visits, and 18,921 page views.

For www.upitec.org there were 3,830 hits, 2676 page views and 363 visits.

This gives a total of 134,485 hits, and 76,066 page views from 19741 distinct visits.

All source papers read, the most read were 153, 68, 64, 25, 94, 152, 4, 108, 131 (Sp.), 8, 41,140, 111, 54, 43, 107, 47, 142(Sp), 117, 67, 116, 11, 86, 82, 142, 75, 18, 30, 7, 144(Sp), 37, 119,150, 1, 33, 46, 110, 139, 40, 76, 118, 2, 49, 88, 102, 143(Sp), 12, 63, 17, 20, 21, 19, 26, 80, ...

All articles etc. read or listened to, the most popular being: Spacetime Devices, Felker 3 (Spanish), Pendergast Bio, Space Energy Devices (Sp), Johnson Magnets, Talk on the Deflection of Light by Gravitation, Galaxies, Special Relativity (Eckardt), Talk on the Complete Film Script of "The Universe of Myron Evans", Workshop History, Space Energy, Talk on the Introductory Essay of UFT 139, Overview of ECE Theory, Devices Section, ECE Engineering Model, Galaxies (Spanish), ECE and Spacetime, Space Devices (Spanish), Latest Family History, ...

July 1 – 15

US Universities, Institutes and Similar

Caltech, Carleton, Central Michigan, Harvard, Johns Hopkins Applied Physics, Lansing Community College, MIT, Princeton, Penn State, Space Telescope Science Institute, SUNY Stonybrook, Univ. California System, U Mass., Michigan, Chapel Hill, Nebraska Lincoln, Valdosta, Washington State, Federal Aviation Authority*, National Research Council, Gov. New York City, Oak Ridge, City of Cincinnati, USPTO, Gov. Virginia, Andrews AF Base, Wright Patterson AF Base, Naval Research Laboratory, Boeing, Hewlett Packard, Intel, IBM, Motorola, Panasonic.

Europe and Asia Minor: Universities, Institutes and Similar.

Liege, Ghent, CERN, Centro Projekt Czechia, MUNI Czechia, UPC* Czechia, Bosch Corporation Germany, FH Lausitz, Fraunhofer, Oberhausen, Siemens, TU Freiburg, Bonn*, Frankfurt, Marburg, Rostock, BWV Hamburg, Aragon Spain, UPV Spain, HTV, JYU Finland, Crous Limoges France, INPG France, Univ Paris Diderot, Poitiers, rennes 1, AUEB Greece, NTUA Greece, Patras, Trinity College Dublin Ireland, Hebrew Univ. Jerusalem Israel, DAUB Italy, INFN Florence, INFN Padua, ISPR Ambiente, Milan Polytechnic, SIF Bologna, Turin, Royal Society of Chemistry, Cronosit Latvia, CSS Group Latvia, KUN Netherlands, TU Eindhoven, Utrecht, AMU Poland, Lodz, Wroclaw, Minho Portugal, Badi Russia, Joint Institutes for Nuclear Research Russia, UFA Network Russia, KTH Sweden, NLB Slovenia, Antik Slovakia, PSG Slovakia, Uludag Turkey, Aberystwyth, City University London, Edinburgh, EPSRC, Imperial*, Liverpool, Manchester, Oxford, Queen Mary London, Swansea, Energis, National Health Service.

Rest of World: Universities, Institutes and Similar.

UNLP Argentina, Australian National University, RJ Gov. Brazil, UFC, UFF, Sao Paolo Brazil, Future Shop Canada, British Columbia*, Laval, Regina, Saskatchewan, York Univ. Toronto, Univ. Chile, UFRO Chile, UNL Ecuador, City Univ. Hong Kong, IIITA India, RRCAT Gov. India, Nihon Univ. Japan, Titech, Hyogo, Toshiba Company Japan, KAIST South Korea, OU Sri Lanka, IPICYT Mexico, RP Singapore, Chula Thailand, KMITL Thailand, LRU Thailand, Ccu Taiwan, NTU Taiwan, Capetown South Africa.

July 15 – 30

US Universities, Institutes and Similar.

Amherst, Buffalo, Central Michigan, Central New Mexico, Columbia, Florida International, Georgia Tech., Harvard, Hawaii, Kansas State, Lamission, MIT*, Naval Postgraduate School, Princeton, Radford,

Rutgers, Stanford, SUNY Stonybrook*, UC Irvine, UCLA, Iowa, Michigan, Valle, Nebraska Lincoln, South Carolina, Wisconsin Parkside, Washington, Wese Virginia, US Govt, Federal Aviation Authority (FAA), NASA, NIST, NOAA, NRC, SSA, Gov. Virginia, Andrews Airforce Base, Army PICA, Navy EHIS, Bavy Navair-Rdte, Navy Navo, Dow Corp., Honeywell.

Europe and Asia Minor: Universities, Institutes and Similar.

Vienna*, Basel, ETH Zurich, Charles Univ. Prague*, Bosch, HU Berlin, Max Planck MPI-HD, PDI Berlin, Bochum, Siemens, TU Darmstadt Electronics, TU Munich, Bonn, Cologne, Leipzig, Marburg, EMT Madrid, UB, Unicam, UV Spain, RTGI Europe, Helsinki, BNF France, RAIN France, Poitiers*, Rennes 1, Dublin City University Ireland, Tel Aviv University Israel, INFN Pisa, Univ. Pisa, Bari Tech., TIN Italy, Pavia, EPF Luxembourg, CSS Group Latvia, Dordrecht Netherlands, Erasmus MC, Telfort, NTNU Norway, Troll Fjord Norway, Wroclaw, MTS Russia, PU Russia, Anadolu, SUMY Ukraine, Classics Cambridge, Engineering Cambridge, Prifysgol Cymru*, Durham, Manchester, Newcastle upon Tyne, Bodleian Oxford, OULS Oxford, Physics Oxford, SERS Oxford, Southampton, Swansea College, Univesity of Wales*, British Gov. Energis.

Rest of World: Universities, Institutes and Similar.

GBA Gov. Argentina, ANU Australia, Eastern Queensland, Griffith, Western Asutralia, INPE Brazil, UFF, UFRJ Brazil, Memorial Newfoundland Canada*, Perimeter Institute, Montreal Polytechnic, British Columbia (DHSCP)*, Calgary, New Brunswick, Saskatchewan, Waterloo, UFRO Chile, UNL Ecuador, ITS Indonesia, CAT Gov. India, UM Iran, Kaiyodai Japan, Osaka, Tokyo, Massey New Zealand, AIT Thailand*, CCU Taiwan*, NCTU, Academica Sinica Taiwan, LUZ Venezuela, CPUT South Africa, TUT, UCT.

Diary Feedback Activity (blog stats) for www.aias.us

In 2010 to date there have been 14,289 diary viewings, the peak to date was on Thursday July 15th., 2010, there is interest in all aspects of the blog: science, literature, genealogy, history and social affairs such as local opposition to wind turbine developments.

Summary of August 2010

For www.aias.us there were 85,902 hits and 51,281 page views from 12,632 unique visits, downloading 10.069 gigabytes of material and reading 1,973 documents from 95 countries, led by USA, Russia, France, Germany, Britain, Italy, Mexico, Australia, ……

All UFT source papers were read, the most read papers were: 11, 25, 41, 43, 107, 153(Sp), 8, 140, 143, 154, 116, 11, 140, 87, 4, 54, 20, 150(Sp), 18, 82, 6, 150, 128, 94, 33, 110, 155, 26, 76, 149, 67, 63, 114, 153, 17, 111, 152, 123, 142, 38, 142, 148, 146, 29, 118, 127, 47, 120, 131, 1, 7, 119, 139, 2, 40, 58, 61, 105, …..

For www.atomicprecision.com there were 34,688 hits and 11,707 page views from 7,458 distinct visits.

For www.upitec.org there were 4,176 hits and 689 page views from 430 visits
This gives a total of 124,766 hits and 63, 677 page views from 20,520 distinct visits.

All other books and documents were read, the most read being: Johnson Magnets, Space Energy Devices (Spanish), Felker3 (F3, Spanish), Spacetime Devices, ECE (Spanish), Criticisms of the Einstein Field Equation, MWE Biography, ECE Engineering Model, Special Relativity, F5, Talk on Light Deflection, Overview of ECE, Talk on Introduction to UFT 139, Ricci Concepts, F1, Galaxies (Spanish), Universe of MWE (Script), SU(2) Theory (Spanish), H Bonding Plenary Paper, …….

August 1 – 15

US Universities, Institutes and Similar
Albany, Amrita, Austin CC, Berkeley, New Mexico CC, Georgia Tech., Hawaii, Iowa State, Kansas, MIT, Morgan, Michigan State, North Carolina State*, Olin, Rutgers, Stonybrook, Ubaguio, Iowa, New Mexico, Texas, Yale, NASA ARC, NSA GRC, Arnold AF, Hurlburt AF, Navy NAVO, Boeing, IBM Torolab, IBM Thomas J. Watson, Northrop Grumman, Raytheon.

Europe and Asia Minor: Universities, Institutes and Similar
Leobon Univ. Austria, WU Vienna, KU Leuven, BAS Bulgaria, CERN. EPF Lausanne, Zurich, Belfort, Max Plack Plasma Physics, Max Planck Garching, Max Planck Bremen, Siemens, TU Clausthal, Augsburg*, Goettingen, Heidelberg, Cologne, Konstanz, Wuppertal*, KIKU Denmark, Aragon Spain, GRN, IAC, UAB Spain, Suurmetra Finland, OTE Greece, UCD Ireland, ENI Italy, INFN Lecce Italy, CSS Latvia, Groningen, TU Eindhoven, Utrecht, Amsterdam Netherlands, HIVE Norway, UC, UT Lisbon Portugal, HITV, NCP Russia, NLB Slovenia, GTS Slovakia,

Kharkov Ukraine, Aberystwyth, King's College London, Lancaster, Nottingham, University College London, Churchill Cambridge, NHS.

Rest of World: Universities, Institutes and Similar.
    UNC Argentina, Australian National University*, Adelaide, Defence Govt. Australian Govt. Brazil, Concordia Univ Canada, Perimeter Institute, British Columbia*, Saskatchewan, Waterloo, IDS Indonesia, IITM India, MNIT India, Kokudai, Oskaka, Tokyo, Tohoku Japan, UNAM Mexico, PUCP Peru, Chula Thailand, NCTU, NTHU Taiwan, VDC Venezuela, Univ. Johannesburg, Witwatersrand, Free State South Africa.

August 15 – 30

US Universities, Institutes and Similar.
    Adventist International Philipines on edu, Bard, Boston, Columbia, Cornell*, CSU Fresno, Georgia College and State, Georgia State, Harvard*, Hawaii, Princeton Institute for Advanced Studies, Kent State, MIT*, Mohave, North Carolina State, North Dakota System, Oklahoma, South Carolina, Sam Houston State Carnegie Research Doctoral University, Southern Methodist, Texas A and M, Temple, Univ Georgia, Univ Iowa, Illinois Urbana Champaign*, Maryland, Michigan, North Carolina Charlotte, Utah, US Gov. FAA, Fermilab, NASA ARC, NIST, Sandia*, US Airforce AFSPC, US Arny NGB, US Army Tacom; Beoing, Corning, Dupont, Honeywell, IBM Rochester, IBM Thomas J. Watson, Northrop Grumman*, Jefferson Laboratory, Prince George's County Public Schools Maryland.

Europe and Asia Minor: Universities, Institutes and Similar.
    Vienna, WU Vienna, BTC Bulgaria, CERN, RIST Switzerland, Geneva, University of Zurich, CVUT Czechia, German National Synchrotron Facility (DESY), FH Munich, HS Regensburg, Max Planck Institute for Plasma Physics, Max Planck Institute Heidelberg, Ruhr Univ. Bochum, Siemens Company*, Goettingen, Kaiserslautern, Novo Denmark, Eurociber Spain, HGGM Spain, HTV, HUT Finland, Institut Laue Langevin Grenoble, IPSL France, Observatory Nice, Polytechnique, Poitiers, CYTA Greece, BME Hungary, ELTE Hungary, Milan Tech., CSS Latvia, CWI Netherlands, Leiden, Groningen, Telfort, Free Univ. Amsterdam, UIT Norway, CFT Poland, US Poland, UMA Portugal, NIPNE Romania, Univ Belgrade Serbia, HITV Russia, Lipetsk, NCP Russia, RZN Russia, NLB Slovenia, Iyte Turkey, ITL Ukraine, Poltava Ukraine, Aberystwyth*, Towy Aberystwyth, Anglia, Boston, Bristol, Cambridge, Edinburgh, Imperial, Leeds, London School of Hygiene and Tropical Medicine, Manchester, Oxford, Swansea, UMIST, York, NHS.

Rest of World: Universities, Institutes and Similar.

UNC Argentina, UNLP, Gov. Argentina, East Queensland Australia, Queensland, Sydney, Gov. New South Wales, Gov. Brazil, UFC, UFRGS Brazil, AEI Canada, Fraser Health, McMaster, Perimeter Institute, Saskatchewan, British Columbia*, UDEC Chile, UNAL Colombia, Andes Colombia, Gov. Ecuador, UGM Indonesia, IITM India, BARC Gov India, JAIST Japan, Kanazawa, UEC* Japan, KLN, OU Sri Lanka, Gov. Mexico, UASLP, UGTO, UNAM Mexico, UNI Nicaragua, Massey New Zealand, KKU Thailand. NCTU, NTU Taiwan, VTC Venezuela, Univ. Kwa Zulu Natal South Africa, Univ. Witwatersrand South Africa.

Diary Viewings for 2010 to August 30th: 15,434.
Summary of September 2010

For www.aias.us there were 74,658 hits, 42,000 page views, 12,704 distinct visits, 1,911 documents read, 9.243 gigabytes downloaded from 97 countries, led by US, Britain, Russsia, France, Germany, Australia, Austria, Mexico, ...

All UFT papers read, most read papers were: 155, 25, 41, 53, 156, 94, 140, 144, 131(Sp), 151(Sp), 43, 4, 116, 54, 20, 107, 37, 1, 88, 6, 29, 137(Sp), 64, 118, 67, 40, 145, 76, 152(Sp), 11, 32, 86, 18, 87, 143(Sp), 33, 47, 63, 81, 132(Sp), 150B, 24, 44, 80, 82, 148(Sp), 42, ...

All Articles and books etc. read: most read were: Spacetime Devices, Felker3 (Sp), Space Energy, Biography, Spacetime Devices (Sp), ECE (Spanish), Criticisms of the Einstein Field Equation, Overview of ECE Theory, Essay 7 on Summary of Advances, Latest Family History, Talk on the Defelction of Light by Gravity, ...

The Diary of www.aias.us set a record high on 12th September, and to date there have been18,226 diary viewings this year.

For www.atomicprecision.com there were 33,923 hits, 18,482 files read, 10,440 pages read, and 6,408 visits. For www.upitec.org there were 3,265 hits, 2,130 pages, 843, pages and 455 visits.

This gives a total of 111,846 hits, 53,283 pages read and 19,567 distinct visits.

1 – 15 September

US Universities, Institutes and Similar.

Alamo, Berkeley*, Boston University*, Caltech, Cleveland State, Columbia*, Cornell, Dartmouth, Fort Hays State, Georgia Tech*, George

Mason, Harvard, Illinois Institute Technology, Illinois, Illinois State, Karlsruhe IT (on edu), Kansas, Mayo Clinic, MIT*, Michigan State, Western Connecticut State, North Dakota, New Jersey Institute of Technology, Oklahoma State, Purdue, Richland, Rollins, South Florida State Univ.*, Southern Polytechnic State, Stanford, St. Cloud State, Texas A and M*, Texas State, U Connecticut, U Delaware, Univ Georgia Atlanta, Illinois Chicago, Illinois Urbana Champaign, Maryland, Massachusetts Lowell, Chapel Hill, Oregon, Utah, Western Carolina, Wisconsin*, Lawrence Livermore, NASA*, Oak Ridge, Sandia*, Kirtland AF, Monmouth Army, Tacom Army, Raytheon.

Europe and Asia Minor: Universities, Institutes and Similar.

Vienna*, Berger Logistik, Geneva, TUL Czechia, Audi, FH Giessen, Max Planck X Ray Astronomy, Siemens, TY Freiburg, Bonn, Duisburg, Frankfurt, Freiburg, Hannover, Heidelberg, Karlsruhe*, Konstanz, Mainz, Regensburg, BNAA Denmark, UAM Spain, Avarra, Unex, UPV*, UV Spain, Oulu Finland, EMN France, French Particle Laboratory, Lisif Jussieu, Observatory Nice, Avignon, Poitiers*, UOI Greece, Patras, UCD Ireland, Weizmann, INFN EMFCSC Italy, CSS Group Latvia, Utrecht, Telfort, NTNU Norway, Krakow, ESRI Portugal, PUB Romania, Belgrade, JINR, MSU Sweden, GU Sweden, UU Sweden, NLB Slovenia, Met* Turkey, LDS Ukraine, Aberystwyth, Cambridge, Imperial, Lincoln College Oxford, Pharm Oxford, Scotgrid, Sheffield, Swansea, UMIST, UWIC, Wolverhampton, British Library, National Library of Wales.

Rest of World: Universities, Institutes and Similar.

UNC, UNGS Argentina, Australian National University, Eastern Queensland, Monash, TCHE, UEL, UFMG Brazil, McGill Canada, Saskatchewan, Alberta, British Columbia, Quebec Trois Rivieres*, Waterloo, GE Chile, UTA, UNAL Chile, MNIT, PRL, PUHEP India; Fujitsu Japan, IIE Mexico, EPIE Peru, PUCP Peru, KKU Thailand, Capetown, UKZN, Government South Africa.

15 – 29 September

US Universities, Institutes and Similar.

Arizona State, Berkeley*, Buffalo, City Colleges of Chicago, Clarion, Clemson, Cleveland State CC, Cornell*, De Paul, Duke, Georgia Tech*, Illinois Institute of Technology, Illinois, Kansas*, Lousiana State, MIT*, Montana, Morgan, Missouri Univ. Technology, North Carolina State, New Mexico Tech., New York Univ.*, Old Dominion*, Princeton, Purdue, Reed, Richland, Rutgers, South Florida State, Simmons, Texas A and M, Thiagarajar (on edu), Delaware*, Kentucky, Maryland, Michigan, U Mass

Lowell, Minnesota Twin Cities, Texas Arlington, Utah, Texas*, William and Mary*, Wyoming, Villanova, Western Carolina, Washington State Pullman, Yale*, Fermilab, NIST, Oak Ridge, Jefferson Lab., Army Carlisle, Army Tactical Command, NAVY COT, Navy Med., Navy NAVO, Navy NMCI*, Boeing*, Honeywell, Lockheed Martin, Northrop Grumman, Qinetic, Raytheon, Varian.

Europe and Asia Minor: Universities, Institutes and Similar.

Univ Graz, TU Graz, Ghent, EPF Lausanne, Belfort, Geneva, Fraunhoffer, HMI* Germany, Jacobs Univ., Max Planck Augsburg, Max Planck Duesseldorf, Siemens, TU Dresden, Augsburg, Bonn, Erlangen, Frankfurt, Hamburg, Jena, Paderborn, Stuttgart, UPC Spain*, USC, UV*, HUT Finland, CNRS Grenoble, Georgia Tech Metz, Particle Physics National Lab., Azur Observatory, Paris Psud, UHP Nancy, Avignon, Lyon 1, Poitiers*, Toulouse 1, NTUA Greece, ONDSL Patras Greece, Ahive Hungary, Malev Hungary, Univ College Dublin (Excalibur and Pwyll), ENEA Bologna, ENEA Frascati, ICRA Italy, OSPFE Italy, Turin Tech., Florence, Milan, Modena, Pavia, Rome 1, Inturbo Lithuania, CSS Latvia, Utrecht, NTNY Norway, AGH, Wroclaw Poland, ESRI Portugal, BH Romania, NIPNE, PUB Romania, Belgrade Serbia, MSU Russia, GU, TDC Sweden, NLB Slovenia, Iyte Turkey, SDU Ukraine, Bristol, Cambridge, Imperial*, Sheffield, St. Andrews, Swansea*, West Thames, Bwrdd yr Iaith, Cumbria, Manchester School.

Rest of World: Universities, Institutes and Similar.

MDP Argentina, RIU, UNT Argentina, CSIRO Australia, Adelaide, Australian National University, Melbourne*, Sydney, BOM Gov Australia, Defence Gov. Australia, FURB, UEL, UFC, UFPE, USP Brazil, City of Winnipeg, McGill, Simon Frazer, Alberta, British Columbia*, Calgary, Quebec Trois Rivieres, Victoria, Waterloo, PUC Chile, UNAB Chile, UTFSM Chile, Valle Colombia, Hacienda Gov. Colombia, City Hong Kong, UST Hong Kong, BHU India, IITH India, Goa, BARC Gov India, TCIL Gov. India*, Kyushu, Mikon, Ricoh, SNU South Korea, Gov Mexico, UNA Nicaragua, Auckland NZ, EPIE – UNSA Peru, PUCP Peru, Meridian Philippines, UP South Africa, Witwatersrand South Africa.

Summary of October 2010

For www.aias.us there were 15,706 distinct visits, 79,592 page views, 11.23 gigabytes downloaded, 115,202 hits, 1,868 documents read from 102 countries, led by USA, Germany, France, Britain, Russian Federation, Mexico, Australia, Austria, …... All UFT source papers and documents were read, the talks being very popular. The most read UFT papers were:

160, 158, 25, 157, 41, 140, 43, 116, 4, 140, 94, 8, 88, 54, 156, 47, 11, 63, 144(Sp), 81, 10, 26, 1, 76, 107, 118, 93 (extra plots), 142, 59, 6, 143, 110, 131(Sp), 18, 33, 142(Sp), 65, 126, 149, 37, 128, 20, 66, 27, 51, 56, …………

The most read documents were:

Felker3 (Sp), Talk "Nobody is Perfect", Levitron, Space Energy (Spanish), Biography, Talk Essay 7, Space Energy, Spacetime Devices, Criticisms of the Einstein Field Equation, Special Relativity, Talk "On the Deflection of Light by Gravity", Talk "Universe of Myron Evans", ECE (Sp), Latest Family History, Galaxies (Sp), SU2 (Sp), Overview of ECE Theory by MWE, Talk Essay 4, ECE Engineering Model, Talk Essay 7, Talk Essay 8, Talk Introduction to Paper 139, Johnson Magnets, ECE Theory of H Bonding, Felker13(Sp), Felker14(Sp), ……….

For www.atomicprecision.com there were 6,078 distinct visits, 33,714 hits, 10,706 pages.

For www.upitec.org there 458 distinct visits, 797 pages and 3.02 hits.

This makes a total of 22,242 visits, 91,095 pages and 152,008 hits.

1 – 15 October

US Universities, Institutes and Similar.

California State Los Angeles, Dartmouth, Embry Riddle, Georgia Tech*, Georgia College and State, Harvard*, Howard Community College, Illinois Institute of Technology, Kansas, Lehigh, MIT, Missisippi State, Missouri University of Science and Technology, North Dakota, Oklahoma, Princeton*, Penn State*, Purdue, Richland, Rochester*, Southeast Missouri State, Stanford*, Stevens Tech, Alabama Huntsville, Akron, Arkansas Little Rock, Arkansas, California Irvine, California San Francisco, Florida, Maryland, Chapel Hill, Nebraska Lincoln, Pennsylvania*, US Naval Academy, Utah, Texas Dallas, Texas*, Tennessee Knoxville, Vanderbilt, Vermont*, Washington, William and Mary, Washington St Louis, Yale, District of Columbia Government, Federal Emergency Management Agency, NASA*, NOAA*, Department of Agriculture, Virginia, US Army PICA, Defense Information Systems Agency, US Marine Corps, IBM Almaden, IBM North Carolina, Microsoft, Northrop Grumman, Aerospace Corporation, Holland Public Schools.

Europe and Asia Minor: Universities, Institutes and Similar.

AIT Austria, Vienna, Free Univ. Brussels, EPF Lausanne, Belfort, Charles Univ. Prague, JCU Czechia, GSI, Max Planck Augsburg*, Max Planck Informatik, Ruhr Bochum, Siemens, Students Bonn, TU Berlin, TU Harburg, Dortmund, Giessen, Hannover, Cologne Geo, Cologne OC, Leipzig*, Regensburg, Saarland, Wuppertal, Unity Media Group, Cirque Denmark, Aragon, Almeria, Madrid Autonomous, Barcelona, Spanish National Education System (UNED), Catalonia Tech Barcelona, Seville, Valencia*, Spanish Xunta (Parliament), ENS Cachan, INRIA, Poitiers*, Rouen, Savoie, Gibraltar, NTUA Greece, UOA Greece, Patras, GN Government of Ireland, University College Dublin, Tel Aviv, INFN Florence, INFN Pisa, INFN Rome 1, Trieste, Florence, Genoa, Milan, Turin*, CSS Latvia, ONS Eindhoven, ONS Students, ROAC Netherlands, Telfort, TU Delft, NTNU Norway, UIO Norway, UIT Students Norway, Szcsecin Poland, Wroclaw*, Brasov Romania, SBB Serbia, DSl Avangard Russia, GTN Russia, HITV, Moscow State Russia, LFV Sweden, NLB Slovenia, GTS Slovakia, Novosibirsk (SU net), Bilkent Turkey, Metu*, Kiev Ukraine, Cambridge*, Coventry, Greenwich, Hull, Liverpool John Moores, Students Leeds, Students Luton, Nottingham, Sheffield, Swansea, Government of York, Essex Schools*, Sherfield School Essex.

Rest of World: Universities, Institutes and Similar.

MDP, UNLP Argentina, CNEA Gov. Argentina, Australian National, East Queensland, Melbourne*, Defence Gov. Australia, UNESP*, UNICAMP, Sao Paolo Brazil, AEI Canada, British Columbia, Manitoba, Quebec Trois Rivieres, Regina, Saskatchewan, Victoria, Waterloo; UCN, UNAB, UTFSM Chile; UDEA, UNAL, Valle Colombia; Jijel Algeria, IIS Ermohali, IITM, BARC Gov. India, RCIL Gov. India; IMSC* India, PRL India; Shiraz Iran, Kitosato, Osaka, Fujitsu*, Honda Japan; Chonnam South Korea, OU Sri Lanka, Unich, Unach, UNAM Mexico; UITM* Malaysia, Victoria Univ Wellington New Zealand, NTHU, NTU, Academica Sinica, YM Taiwan; Capetown*, NWU, UWC South Africa.

15 – 30 October

US Universities, Institutes and Similar.

Arizona*, Berkeley, Bowdoin, Boston University, Buffalo*, Columbia, Cornell*, Duke*, Florida International, Hawaii, Hudson Valley Community College Troy, Iowa State, Johns Hopkins*, James Madison, Kansas State, Lesley, Maine, Missouri, MIT*, Michigan State*, North Carolina State, New Haven, North Dakota, Ohio State, Ohio, Oregon State, Occidental College Los Angeles, Pittsburgh, Pomona, Princeton*, Penn State, Rochester*, Stanford, Texas Tech, Univ California System,

UC Fresno, Chicago*, UC Irvine, UC Riverside, UC Santa Cruz*, UC San Diego, Florida, Iowa, Illinois Urbana Champaign, Kentucky, Maryland, Nebraska Omaha, UT Dallas, UT El Paso, Texas*, Tennessee Knoxville*, Virginia, Washington, William and Mary, Yale*; US Bureau of Land Management, US Emergency Management Agency, Library of Congress, NASA, US Army, Army DCMA, US Navy Med., Ford, Intel, Kodak, Lockheed Martin, Microsoft*, Motorola*, Northrup Grumman, Gov. Aurora Colorado, Catholic Relief Services, Queen's Library, NYC.

Europe and Asia Minor: Universities, Institutes and Similar.

Austrian Institute of Technology, UIBK Austria, Klagenfurt, Vienna, KU Leuven Belgium, Ghent*, UU Net Belgium, ETH Zurich*, UZH Switzerland, CAC Grup Czechia, Charles Univ Prague, Brno Univ. Technology, BCA Germany, Jacobs, MHN Germany, Max Planck Institute for Extraterrestrial Physics, Ruhr Bochum, TU Chemnitz, TU Freiberg, Jena, Mannheim, Oldenburg, Wuerzburg, Aarhus Denmark, Kopenhagen, Tartu Estonia, Aragon Spain, AZTI Spain, Spanish Aerospace Technology Institute (INTA), Autonomous Barcelona, Autonomous Madrid, Granada, Polytechnic Catalonia, Polytechnic Valencia, Seville, Valencia, Bruno Kesller Foundation Trento, HUT Finland, ENS France, French National Particle Lab., Caen, Paris Diderot, Poitiers*, CYTA Greece, NTUA Greece (Hellenic National Research Foundation), Tel Aviv, EMFCSG of INFN Italy, Milan Polytech, Venice, Vilnius Lithuania, CSS Latvia, RU Netherlands*, Utrecht, UVT Netherlands*, PW Poland, Szczecin Poland, Wroclaw, INFIM Romania, UNS Serbia, Russian Academy of Sciences Institute of Chemical Physics, Dubna, HITV Russia, Moscow State University Lomonosov, Novgorod, Chalmers Sweden, NLB Slovenia, Bratislava Slovakia, NSK.SU, Tarassul Syria, Anadolu, Gantep, Hacettepe Turkey, Poltava Ukraine, Aberdeen, Students Aberystwyth, Cambridge*, Astronomy Cardiff*, Durham , Imperial*, Lancaster, Liverpool, Luton, Middlesex, Manchester Metropolitan, Oxford*, Sheffield, Southampton, Strathclyde*, Sussex, Swansea*, University College London, East Anglia, York, Ceredigion, East Dunbarton, York County Council, Llyfrgell Genedlaethol Cymru (National Library of Wales).

Rest of World: Universities, Institutes and Similar.

UNLP Argentina, Villa Maria Gov. Argentina, Eastern Queensland, Griffith, Monash*, Melbourne, Wollongong, Queensland, Victoria, Australian Defence Ministry, Physics UNICAMP, Sao Paolo*, AEI Canada, Gov. British Columbia, INO Canada, IREQ, McMaster, Perimeter Institute, Alberta*, British Columbia, Manitoba, Quebec Trois Rivieres, Saskatchewan, Toronto, Victoria, Waterloo*, UFRO Chile, UTFSM Chile, IHEP China, Xian Jiaotong China, UDEA, UNAL*, Andes*, UTP

Colombia, BHU India, BARC Gov. India, RCIL Gov. India, IIAP, IMSC, PRL India, Tehran Iran, Keio, Kitasako, Teikyo, Tohoku*, Tokyo* Japan, OU Sri Lanka, SLT Sri Lanka, IPN, UNAM*, USON Mexico, UITM Malaysia, USIL Peru, PERN Pakistan, NU Singapore, NDHU Taiwan, NWU South Africa, Capetown, Witwatersrand.

Estimated almost 700 university visits in October 2010. 20, 649 views of diary by 31st October; 9,306 postings.

Summary of November 2010

For www.aias.us there were 113,402 hits, 6.511 gigabytes downloaded, 16,357 distinct visits, 75,334 page views, 1,919 documents read from 103 countries, led by USA, Russia, Germany, France, Mexico, Britain, Colombia, Canada, ...

The most read UFT papers were 68, 4, 25, 41, 43, 140(Sp), 46, 8, 156, 67, 161, 154, 142(Sp), 107, 18, 76, 80, 110, 29, 63, 118, 131(Sp), 143(Sp), 21, 59, 144(Sp), 116, 149, 47, 94, 65, 88, 154, 20, 54, 81, 146, 145(Sp), 148(sp), 45, 152(Sp), 159, 2, 58, 93, 156(Sp), 142, 11, 32, ... All 164 papers read and in translation.

The most read papers and books by ECE colleagues were: Spacetime Devices, Felker 3 Spanish, Complete filmscript broadcast, Criticism of the Einstein Field Equation, Biography, Levitron, Johnson Magnets, Nobody is Perfect Broadcast, Spacetime Devices (Spanish), Galaxies (Spanish), ECE (Spanish), Latest Family History, Deflection of Light by Gravity Broadcast, Space Energy, ... all papers and books by colleagues read.

The most read essays and broadcasts were: Complete filmscript broadcast, Nobody is perfect broadast, Deflection of light by gravity broadast, Essay 7 (E7), pdf, UFT 139 Introduction broadast, E8 (broadcast), E7 (broadcast), E10 (pdf), E9 (pdf), E12 (pdf), E4 (pdf), E15 (pdf), E14 (Spanish, pdf), E15 (Sp, pdf), E9 (Sp, pdf), Poetry4 (P4, broadcast), E10 (Spanish, pdf), E13 (Spanish, pdf), P1 (broadcast), P3 (broadcast), P2 (broadcast), E10, E11, E13, E14, E15, E9 (broadcasts), P5, P6 (broadcasts).

For www.atomicprecision.com there were 6,646 visits, 12,681 page views and 35,246 hits.

For www.upitec.org there were 514 visits, 838 page views and 3,108 hits.

This gives a total of 23,516 visits, 88, 853 page views and 151,756 hits.

There was a total of 23,420 visits to the diary on 30th Nov. 2010 since monitoring of the blog started a few months ago.

US Universities, Institutes and Similar.

Arizona State, Austin Community College, Berkeley*, Boston University, Buffalo*, Brigham Young, California Polytechnic, Caltech, Colgate, Columbia, Cornell, Drexel, Duke*, Einstein Clinic, Embry Riddle, Florida State, Georgia Tech., Georgia Perimeter College, Iowa State, Indiana*, Indiana Purdue Indianapolis, Illinois Wesleyan, Johns Hopkins Applied Physics, Louisville, Maine, MIT, Mississippi State, Michigan State, U.S. National Academies, North Carolina State, North Dakota, New York University, Ohio State*, Ohio University, Penn State*, Quinnipiac, Renselaer*, Rochester, Rutgers, Seattle, Stanford, Stonybrook, Susquehanna, Texas A and M, Chicago, UC Irvine, UCLA, UC Riverside, UC Santa Barbara, Florida, Illinois Urbana Champaign, UMBC Maryland, Maryland, Michigan, Minnesota Twin Cities, Chapel Hill, Oregon, Pennsylvania, Utah, Texas, Wyoming, Virginia, Vermont, Washington, Wisconsin*, William and Mary, Yale; US Government Bureau of Land Management, Dept. of Energy, Federal Emergency Management Agency, Library of Congress, New York City, Sandia, US Airforce Centaf, US Airforce Wright Patterson, US Army PICA, US Navy NRLSSC, US Navy SPAWAR, Jefferson National Laboratory, Peak Organization, Austin ISD, General Electric, Honeywell, IBM, Microsoft*, Motorola, Northrop Grumman.

Europe and Asia Minor: Universities, Institutes and Similar.

UIBK Austria, Leoben Austria, Liege, KHK Belgium, CERN, EPF Lausanna, ETH Zurich, Neuchatel, UPOL Czechia, Bremen, GWDG Germany, HS Esslingen, Max Planck MIS, RWTH Aachen, Students BW, TU Berlin, TU Chemnitz, Erlangen, Frankfurt*, Freiburg, Hannover, Heidelberg, Cologne, Leizpig, Mainz, Marburg*, Padebron, Potsdam, Regensburg, Tuebingen, Aaurhus Denmark, Aragon, Valencia General, High Energy Institute of Spain (IFAE), UNED Spain, Leon, Ovi, Rioja, Catalan Polytech, Polytechnic Madrid, Vigo, Joensuu Finland, JYU Finland, ENS France, IHES, French Particle Lab., INSA Lyon, Strasbourg, Nancy*, Grenoble, Poitiers*, Rouen, Patras Greece, UCD Ireland, Hebrew Univ. Jerusalem, Technion Israel, HI Iceland, ECT Italy, ICTP, INFN Naples, Florence, Modena, Perugia, Pisa, Rome 3, MEN Luxembourg, CSS Latvia, Elekta Lithuania, Amsterdam, AMU Poland, GDA Poland, Akadem Russia, IKZ Russia, Chalmers Sweden, KTH Stokholm, LIU, LU Sweden, NLN Slovenia, Antik Slovenia, Tarassul Syria, Bilkent Turkey, Poltava Ukraine, East Anglia, Catherine's, Churchill, CSX, Newnham, Computing Service Cambridge, Durham, Glasgow, Glamorgan,

Huddersfield, Imperial, Liverpool, Nottingham, Churchill, New, Physics, St. Edmund Hall, St. Anne's, St Cats., Worcester Oxford, Royal College of Art, Southampton*, St Andrew's, Swansea, Univ. College London*, East Anglia, UMIST, York, British Library, National Health Service, Libraries NI.

Rest of World: Universities, Institutes and Similar.

UNGS, UNSJ Argentina, Eastern Queensland, Melbourne, Queensland, UFRN, UNESP*, Sao Paolo* Brazil, Concordia, McGill*, Alberta*, British Columbia, Quebec Trois Rivieres, Saskatchewan, Toronto, Waterloo*, Western Ontario, MOP Gov Chile, UNINE, URA, UTFSM Chile, ESPOL Ecuador, City Hong Kong*, UST, MNIT, IIT KGP, RCIL Gov., Bose India, Shibaura, Titech, Tsukuba, Tokyo*, UNAM Mexico, Auckland NZ, NTHU, Academica Sinica, TKU* Taiwan, Capetown, UJ South Africa.

15 – 29 November

US Universities, Institutes and Similar.

Amrita (on edu), Arizona, Arizona State, Baylor, Berkeley, Brigham Young, Brigham Young Hawaii, California Polytechnic, Caltech*, Central New Mexico, Colorado College, Columbia, Cornell, Sacramento State, Florida State, Georgia Tech., Georgia State, Harvard*, Johns Hopkins Applied Physics, Kansas, Marquette, North Carolina A and T, New York, Oregon State, Oswego, Oklahoma, Puget Sound, Reed, Rutgers*, Seton Hall, Southern Methodist, Univ Texas Southwestern Medical Center, Texas Tech., Chicago, UC Los Angeles, UC Santa Barbara, UC Santa Cruz, UC San Diego*, Illinois Chicago, UMBC Maryland, Michigan, Minnesota Twin Cities, Chapel Hill, Nevada Lincoln, Oregon, US Naval Academy, Texas*, Western Florida, Wyoming, Washington State, Yehiva; Lawrence Berkeley, Lawrence Livermore, Goverment of Montana, US Postal Service; Arnold Airforce Base, Army Research Laboratory, Microsoft, Motorola.

Europe and Asia Minor: Universities, Institutes and Similar.

UIBK Austria, Leuven, Ghent, CERN, ETH Zurich*, Geneva, Zurich, ECIC Czechia, Bremen (school), FBH Berlin, Ruhr Univ. Bochum, Siemens Company, TU Berlin, TU Darmstadt, Bremen (university), Duisburg, Heidelberg, Karlsruhe, Leipzig*, Mainz, Muenster, Paderborn*, Potsdam, Rostock, Stuttgart, Tuebingen; CSIC Spain, Madrid Autonomous*, Cordoba, Granada, Murcia, UNED Spain, Seville, Valencia, Sapnish Parliament (Xunta), HYV Finland, YOK Finland, AC Amiens,EC-M, OBSPM France, Oleane France, Evry, Poitiers*; NTUA

Greece, Tellas, Patras, FER Hungary, FSB Hungary, KFKI Hungary, Dublin City, DKIT, University College Dublin Ireland, Hebrew Univ., Jerusalem, SISSA Italy, Trieste, Modena*, Padua, Rome 1, UNO Norway, ITP Norway, Al Balqa Jordan, EX Forum Poland, Belgrade*, Kirov, Dubna, Joint Institutes Nuclear Research Dubna, Russian Space Science Internet, Chalmers Sweden, Lund* Sweden, Tarassul Syria, Itu Turkey, Metu*, SDA, ILYA Ukraine, Brunel, Corpus Christi College Cambridge, Eduram Cambridge, NAT Cambridge, Jesus College Cambridge, Cardiff, Durham*, Edinburgh, Glasgow, Kent, London Metropolitan, Manchester, Churchill College Oxford, Physics Oxford*, Queen Mary London, Rutherford Appleton, Southampton*, St. Andrews*, Government Shetland Islands, Bedford School, George Watson's College Edinburgh, Westminster School, Westminster Parliament.

Rest of World: Universities, Institutes and Similar

CNEA, Mecon Gov. Argentina, East Queensland, Monash, Wollongong, Queensland, UEFS, UEG, UFPA, UFRN* Brazil, Brock Canada, Lakehead, National Research Council, Siemens Canada, Saskatchewan, Alberta, British Columbia*, Toronto, Western Ontario; UCN UTFSM Chile, UDEA, UTP Colombia, Jijel Algeria, ESPOL Equador, ITS, UGM Indonesia, Kyoto, Shibaura, Totech, Tokyo, Chiba Japan, Fuji Xerox, KAIST* South Korea, UNAM Mexico, Tatiuc Malaysia, Auckland NZ, SQU Oman, Quitos Peru, LCWU, NUST, QAU Pakistan, NU Singapore, NDHU Taiwan, AIMS, Capetown, Pretoria South Africa.

Summary of December 2010

For www.aias.us there were 105,092 hits and 68,648 page views from 12,524 distinct visits. There were 6.766 gigabytes downloaded and 2,042 documents read from 98 countries, led by USA, Russia, Britain, Sweden, France, Ukraine, Austria, Germany, ... The twelve month running total of hits was 1,128,670 and there were 27,006 blog viewings during the year.

The most read UFT papers were: 24, 41, 43, 51, 4, 94, 168, 165, 107, 166, 157, 152, 164, 18, 155, 53, 161, 156, 20, 63, 142, 49, 7, 163, 54, 121, 21, 110, 149, 159, 61, 93, 131 (Spanish), 162, 46, 114, 140, 11, 2, 38, 100, 140 (Spanish), 30, 81, 116, ... All papers read.

The most popular talks were: Film script, On the Deflection of Light by Gravity, Nobody is Perfect, Introduction to 139, Attempts to measure photon mass, self inconsistencies of twentieth century physics, general relativity and particle scattering, geometry and conservation of linear momentum, covariant mass ratio, meaning of covariant mass ratio, Essays 4, 9, 7, and 8. All talks received an audience.

The most popular articles and books were: F3 (Spanish), Levitron, Spacetime Devices, Sapce Energy (Spanish), Mercury as Crystal Spheres, Space Energy, MWE biography, Galaxies, Latest Family History, F1, F2, CEFE, …….. All articles and books were read.

For www.atomicprecision.com there were 35,097 hits and 12,079 page views from 5,917 distinct visits.
For www.upitec.org there were 3,235 hits and 824 page views from 505 distinct visits.
This gives a total of 143,424 hits and 81,551 page views from 18,946 distinct visits.

1 – 15 December

US Universities, Institutes and Similar
Albany, Berkeley, Boise State, Buffalo, Caltech, Colorado, Cornell*, Central Washington, DBCC, Duke, Emory, Furman, Georgia Tech., Harvard*, Haverford, Illinois, Indiana, Johns Hopkins, Kansas State, Louisiana State, Mineral Area College Missouri, Montana, Northwestern, Redwoods, Richland, Renselaer, San Diego State, Southern Michigan, Stanford, Stonybrook, Mendez, SUNY Stonybrook, Tuskegee, Central Florida, Chicago*, UCLA, Florida*, Georgia, Maryland, Michigan, Minnesota Twin Cities, Pennsylvania*, Texas Arlington, Utah*, Texas, Washington State; Bureau of Land Management, US Gov., DHS, Library of Congress, WOAA, Social Security, US Patent Office, US Military DISA, US Navy NMCI*, Naval Research Laboratory, Ford, Honeywell.

Europe and Asia Minor: Universities, Institutes and Similar.
TU Vienna, UCL Belgium, Ghent, EPF Lausanne, CAS Czechia, Brandenburg, German Synchrotron Facility, HS Bremen, HS Karlsruhe, TU Darmstadt Electrical Engineering, Bayreuth, Erlangen, Heidelberg, Jena, Konstanz*, Leipzig*, Oldenburg, Potsdam, Regensburg, Wuerzburg, Sonderborg Denmark, Valencia Polytechnic, Univ. Valencia, Valladolid, Tontut Finland, Nancy Metz France, CIUP, ENSCP, OBS-MIP, OLEANE France, Polytechnique, Avignon, Lyon 2, Poitiers*, UCD Ireland, Ben Gurion Negev, Jerusalem College of Technology, INFN Bologna, Univ Calabria, Modena*, Pavia, Rome 1, VGTU Lithuania, Etat Luxembourg, CSS Latvia, FR – FANS Netherlands, Groningen, FFI Norway, AGH Poland, US Poland, Lodz Poland, UC Portugal, UBBCLUJ Romania, Nis University Serbia, Irkusk*, MTU Net Russia, Norrkoping Sweden, Lviv Ukraine, Aberystwyth*, Bristol, Wolfson College Cambridge, Cardiff, Edinburgh*, Imperial, Manchester, Maths Oxford, Oxford OULS, Oxford St Edmund Hall, Plymouth, Queen Mary, Royal Holloway London,

Shefield Hallam, Strathclyde, Swansea, East Anglia, British Library, Met Office, Llyfrgell Genedlaethol, Ynys Mo^n, Ankara, Novosibirsk.

Rest of World: Universities, Institutes and Similar.
Shariah Emirates, UNLP, UNSJ Argentina, Cordoba Gov. Argentina, Queensland, Victoria Australia, UFMG, UNICAMP Brazil, Gov. Canada, McGill, McMaster, memorial Newfoundland, Ontario, Queens, British Columbia*, Manitoba, Montreal, Quebec Trois Rivieres, Saskatchewan, Victoria, York Univ. Toronto, UNAL UTP Colombia, IDSn Gov. Colombia, IIS Ermoahli India, IITM India, Univ Nitw India, Kindai, Osakafu, Titech, Tokyo*, Chiba Japan, KAIST S. Kora, Ipicyt, UNAM, USON Mexico, NUST, QAU Pakistan, NCTU, NDHU, NTHU Taiwan.

15 – 30 December

US Universities, Institutes and Similar.
Ball State, CSU Ohio, Georgia Tech., Kansas, North Dakota System, Salk Institute, Ana Mendez, UCLA, Central Oklahoma, Florida*, Maine, Maryland, Michigan*, Texas Arlington, Utah, Vermont; Brookhaven, US Dept. of Energy, El Paso, IRS, Library of Congress, Miami Dade, USPTO, Andrews AF, Eglin AF, Robins AF*, Gordon Army, NGB Army, US Army Special Operations Command, Motorola*.

Europe and Asia Minor: Universities, Institutes and Similar.
TU Vienna, Ghent, Charles Univ Prague, HU Berlin, Siemens, TU Darmstadt, Heidelberg, Leipzig, Stuttgart, CNIC Spain, Madrid Complutensian, TSAI Spain, Laguna, Murcia, Leon, Barcelona Tech., Cesiege France, CIUP, Inserm, Oleane, Paris Psud, Lyon 2, Poitiers* France, Hebrew Univ. Jerusalem, Bologna, Modena, Naples, Pavia Italy, Groninegn TO Eindhoven Netherlands, AGH Poland, UJ Poland, UNS Serbia, Irkutsk, Jozef Stefan Institute, Ahievran Turkey, Hacettepe Turkey, Brunel, Heriot Watt, Southampton, European Space Agency.

Rest of World: Universities, Institutes and Similar.
ECU Australia, Wollongong, UFPE Brazil, British Columbia Canada*, Univ. Chile, National Univ Colombia, Gov. Equador, BARC Gov. India, RCIL Gov India, JEC Japan, Kyoto, Shimane, Tohoku, Chiba Japan, KMITL Thailand.

Summary of January 2011

For www.aias.us there was a new eight year record of 122,521 hits, an average of about 4,084 hits a day for thirty days. There was a technical outage on 1st Jan. so the total is estimated to be 126,605 hits. There were 80,350 page views, average of 2,678 a day for 30 days, so estimated total of 83,028, and 15,350 distinct visits, average of 512 a day for 30 days, so estimated total of 15,862, 7.738 gigabytes downloaded, 1,947 documents read from 93 countries led by US, Germany, France, Brazil, Russian Federation, Britain, Ukraine, Austria, ……

All UFT papers read, the leading papers were 166, 165, 25, 169, 41, 43, 54, 149, 140(Sp), 81, 107, 18, 4, 168, 150B, 28, 7, 1, 42, 8, 93, 27, 2, 65, 130, 131(Sp), 142, 161, 76, 11, 23, 170, 63, 35, 67, 140, 20, 37, 45, ….

Leading Talks were: Universe of Myron Evans, Nobody is Perfect, Deflection of Light by Gravity, Introduction to UFT 139, Essays 15, 10, 11, 4, 8, 12, 13, 7, 14, 9, …..

All articles read, leading articles were Sapcetime Devices, Device Dev, Levitron, Family History, Space Energy, …..

For www.atomicprecision.com there were 32, 235 hits, 10,228 pages downloaded , 4,919 distinct visits and 0.768 gigabytes downloaded.

For www.upitec.org there were 3,964 hits, 1.025 page views, 552 distinct visits and 0.368 gigabytes downloaded.

This gives a total of 162,804 hits, 94,281 page views, 21,333 distinct visits and 8.874 gigaybtes downloaded.

Dairy now records a total of 30,497 views from Jan. 2010. Average per day for 2011 is 116, 42340 a year.

1 – 15 January

US Universities, Institutes and Similar

Auburn, Caltech, Catholic Healthcare West, Colorado, Columbia, City University New York, Case Western Reserve, Florida State, Husson, Indiana, Johns Hopkins Applied Physics, Mayo Clinic, MIT, North Dakota, Ohio State, Plymouth, Penn State*, Purdue, Stanford, UC Davis, Univ. Chicago, Univ. California Irvine, Univ. California Santa Barbara, Univ. California San Diego*, Unity, Univ. Texas, Virginia, Argonne National Laboratory, Gov. California*, U. S. Dept. of Energy, US General Services Administration, US Bureau of Reclamation, US Geological Survey, US Army ARO, US Army Wiesbaden, US Navy NMCI, Lockheed Martin, Northrop Grumman, Raytheon.

Europe and Asia Minor: Universities, Institutes and Similar

TU Graz, Univ Insbruck, Linz, Vienna, Louvain, Free Univ Brussels, KU Leuven, OMA Belgium, CMI Czechia, Siemens, TU Darmstadt Physics, TU Munich, Dortmund*, Frankfurt, Heidelberg, Karlsruhe, Siegen, Wuerzburg, Herlufsholm Denmark, Parliament of Andalucia, Valencia, HTV Finland, JYU Finland, French National Particle Laboratory, LPTL Jussieu, Paris Psud*, F. Comte, Paris 13, Poitiers*, Savoie, KLTE Campi Hungary, Bar Ilan, Yel Aviv, INFN Naples, NGI Italy, SISSA Italy, Bologna, Medina, Milan, Turin, Luxembourg Sites, CSS Latvia, Univ Malta, RU Netherlands, TU Delft, NTNU Norway, UJ Poland, Astral Romania, Volgograd Russia, MGN Russia, MTS-NN Russia, Ankara Turkey, Kharkov Ukraine, Kiev Ukraine, Bath, Biotechnology and Biological Research Councils, Luton, Oxford Physics, British Government Energis.

Rest of World: Universities, Institutes and Similar

CNEA Gov Argentina, CSIRO Australia, Gov. Brazil, UEFS, UFF, UFPR Brazil, Concordia Canada, McGill*, Univ British Columbia, Univ of Northern British Columbia Prince George, Univ Quebec Trois Rivieres, LN China, PRI India, Shiraz Iran, Nagoya, Tokyo Japan, Sonora Mexico, Univ Engineering and Technology Lahore Pakistan, IT Ladkrabang Thailand, CHNA Taiwan, Academica Sinica Taiwan, VDC Vietnam.

15 – 30 January

US Universities, Institutes and Similar.

Austin CC, Berkeley, Bethany Lutheran, Brigham Young, Caltech, California College of the Arts, Allegheny County CC, Cooper*, CPE Moody Bible, California State Univ. Long Beach, City Univ. New York, Dartmouth, Georgia State, Harvard*, Hawaii, Johns Hopkins Applied Physics, MIT*, Michigan State, North Carolina A and T, North Dakota, Northwestern*, Naval Postgraduate School, Old Dominion, Oklahoma State, Pacific, Stanford*, Arkansas, UC Irvine, UC Santa Barbara, UC Sanat Cruz, UC San Diego, Delaware, Girona (on edu), Florida*, Maryland, Michigan, Pennsylvania, Texas Arlington, Univ Texas*, Tennessee Martin, Toledo, Vermont, Washington, Wednet, William and Mary, Washington Univ., St. Louis, Gov. El Paso Texas, US Food and Drug Administration, US Gov. NOAA, US Social Security, US Dept. Agriculture, Kirtland Airforce Base, Wright Patterson Airforce Base, US Navy Mediterranean, IS Navy NMCI, Honeywell, Lockheed Martin, Raytheon.

Europe and Asia Minor: Universities, Institutes and Similar.

Vienna, KU Leuven, EPF Lausanne, Univ. Zurich, Josefka Gymnasium Czechia, Max Planck Institute for Physics Munich, RWTH Aachen, Siemens, Augsburg, Bielefeld, Heidelberg, Jena*, Karlsuhe, Konstanz, Leipzig, Mainz, Ulm, Kollegienet Denmark, Aragon, Barcelona, Granada, Seville, HTV Finland, JYU Finland. IRCAM France, OLEANE France, Nice, Lille 1, Poitiers*, NTUA Greece, RU Netherlands, RU Groningen, TU Eindhoven*, NTNU Norway, Krakow Poland, UB Romania, UBBCLUJ Romania, SBB Serbia, Corbina Russia, HC, OSU Russia, Troitsk, LIU Sweden, UU Sweden, WLB Slovenia, Sik Slovakia, Metu Turkey, Anadolu Turkey, Kharkov Ukraine, Kiev Ukraine, Birmingham City, Bristol, Cambridge*, Durham, Edinburgh, Essex, Luton, Nottingham, Oxford (Keble, Merton, Trinity, physics), Southampton, Swasnea*, University College London, York, Ceredigion.

Rest of World: Universities, Institutes and Similar.

Gov. Argentina, Gov. Brazil, Concordia* Canada, McGill*, Memorial Univeristy Newfoundland, Perimeter Institute, Univeristy British Columbia, Calgary, Manitoba, Quebec Trois Rivieres, Saskatchewan, Toronto, Victoria, UTFSM Chile, SJTU China, UTP Colombia, IITM India, Shiraz Iran, NIIT Japan, UTM Malaysia, UET Pakistan.

Summary of February 2011

For www.aias.us there were 105,707, 74,230 page views, 12,596 distinct visits, 8.948 gigabytes downloaded, 1,963 documents read from 88 countries, led by USA, Russia, France, Germany, Australia, Canada, Austria, Sweden, Mexico, Hungary, Britain, Slovak Republic, …..

All UFT papers and articles by colleagues were read, all talks received an audience.

The most read UFT papers were: 92, 25, 159(Sp), 43, 41, 172, 4, 165, 94, 54, 107, 2, 18, 8, 50, 21, 140, 81, 19, 143, 149, 114, 140(Sp), 74, 96, 53, 88, 110, 104, 131, 142, 29, 76, 131(Sp), 45, 27, 61, 93, 99, 144(Sp), …….

The most read articles were Devices, Felker3 (Sp), Progress in ECE Theory, Levitron, Maxwel III, Maxwell, Devices (Sp), Family History, Special Relativity, Universe of ME.

The leading lectures were: "Nobody is Perfect", Universe of Myron Evans, Essay 7, Advances in ECE Theory, Deflection of Light by Gravity, Intro 139, 7, 4, 12, 13, 15, 11, 14, 19, 8, 9, 10, 20, 22, 21, 23, ……

From 1st January 2010 to 28th Februrary 2011 there have been 33,430 diary viewings, for February 2011 averaging 100 a day.

For www.atomicprecision.com there were 26,724 hits, 7,967 page views, 3,733 distinct visita, and 0.563 gigabytes downloaded.

For www.upitec.org there were 4,225 hits, 852 page views, 509 distinct visits and 0.680 gigaybtes downloaded.

This gives a total of 136, 656 hits, 83,049 page views, 16, 838 distinct visits and 10.191 gigabytes downloaded.

1 – 15 February

US Universities, Institutes and Similar

Berkeley*, Boise State, Caltech*, Colgate, California State Dominguez Hills, California State Long Beach, Duke, Harvard, Johns Hopkins*, Maine, Colorado School of Mines*, MIT*, New Mexico State, Naval Postgraduate School, Ohio State*, Oregon State, Owens, Penn State, Rice, Rockefeller, Santafe, Stanford, SUNY Stonybrook, Texas A and M*, UC Irvine, UC San Diego, Florida, Illinois Urbana Champaign, New Mexico*, Vermont, Washington, William and Mary, Washington St. Louis, US Bureau of Land Management, NASA, Sandia National Laboratory, US Social Security Administration, US Department of Agriculture, US Department of Justice, US Army NGB, Sill, Tactical Command, Amersham, Lockheed Martin*, Motorola, Northrop Grumman*, Aerospace Corporation, World Wide Web Consortium.

Europe and Asia Minor: Universities, Institutes and Similar

KU Leuven, CERN, EPF Lausanne, FU Berlin, Ruhr University Bochum, RWTH Aachen, Siemens, STW Bonn, TU Berlin, Bremen, Heidelberg, Karlsruhe, Stuttgart, Dronninglund Gymnasium Denmark, Vendia College Denmark, Ministry of Science Spain, Gov. Valencia Spain, TSAI Sapin, Barcelona Autonomous, Granada*, Seville, Salamanca, Vigo, Valencia, Europa, JYU Finland, AC Nancy Metz, French National Particle Laboratory, OBSPM France, Schneider Automation France, Cergy, Brest, Poitiers*, UTC France, NTUA Greece, Szeged Hungary, NUI Galway, Technion Israel, Weizmann, INFN Bologna, INFN Padua, Univ. Padua, Turin, Trieste, Duoman Centras Lithuania, CSSGroup Latvia, RU Netherlands, Wroclaw Poland, Telfort, Joint Institutes for Nuclear Research Russia, Chalmers Sweden, Umea, Uppsala, Bilkent Turkey, Kiev, Bishop Grosseteste University College Lincoln, Bristol, Huddersfield, Imperial, King's College London, Lancaster, British National Museum Consortium, Brasenose College Oxford, University College Oxford, Worcester College Oxford, Scotgrid system, St. Andrews, Sussex, University of West Scotland, Warwick, York, National Heath Service.

Rest of World: Universities, Institutes and Similar.

CNEA and CBA Gov. Argentina, Defence Australia*, Transport Queensland, Camarqa Govt. Brazil, UFPB, USP Brazil, McGill Canada*, Perimeter Institute, Alberta, British Columbia*, Laval, Quebec Montreal, Quebec Trois Riviers, Saskatchewan, Waterloo, HA China, UO Cuba, INF Cuba, UH Cuba, ULEAM Equador, Unita Equador, IIT Karanpur India, RCIL Gov. India, NARL India, Hokudai Japan, IMS, Osakafu, Tokyo*, Waseda, SQU Oman, Pieas Pakistan, NU Singapore, Chula Thailand, VC Venezuela, UTNG Mexico, UNAM Mexico, UITM Malaysia, Munin Venezuela.

15 – 27 February

US Universities, Institutes and Similar.

Bowdoin, Caltech, Claremont, Colorado, Columbia, Cornell, Florida State, Georgia College, Iowa State, Illinois, Johns Hopkins, Kansas State, Miami, Notre Dame, New Mexico State, Ohio State, Oregon State, Princeton, Penn State, Purdue, Smithsonian, Stanford, Texas Tech, Ubaguio, Central Florida, Chicago, Connecticut*, UC Santa Cruz, UC San Diego*, Delaware, Florida, Illinois Chicago, Illinois Urbana Champaign, Minnesota, U Penn, Puget Sound, Toledo, Yeshiva; US BNL, NIH, Oak Ridge, OSIS, USPTO, Wright Patterson AF Base, US Army Belvoir, US Army SWA, Lockheed Martin.

Europe and Asia Minor: Universities, Institutes and Similar.

TU Vienna, Ghent, ETH Zurich, MUNI Czechia, German National Synchrotron Facility, Jacobs, KFA Juelich, Siemens, Students Bonn, TU Berlin*, TU Chemnitz, TU Darmstadt, Heidelberg, Karlsruhe, UCO Pain, UPC*, UPM, UV*, ECP France, Mairie Quimper, OBSPM France, Strasbourg, nice, Poitiers*, UCSQ Hungary, BME Hungary, KFKI Hungary, Szeged, Trinity College Dublin Ireland, INFN Florence, INFN Naples, INFN Padua, Univ. Padua, Units, LTF Luxembourg, IU Netherlands, Leiden, RU Netherlands, Poznan Poland, Warsaw, UA Portugal, SBB Serbia, Corbina Russia, Yaroslav, BTH Sweden, Students UU Sweden*, HCK Slovakia, Odessa Ukraine, Students Aberystwyth, Bristol, Bolton, Churchill Cambridge, CSX Cambridge, Cranfield, Imperial*, Nottingham, Churchill Oxford*, Exeter Oxford, St Annes Oxford, St Hilda Oxford, Southampton, Sussex, York SJ.

Rest of World: Univeristies, Institutes and Similar.

CSIRO Australia, Melbourne, UWS, Defence, UFPR Brazil, UFSAM, USP Brazil, Concordia Canada, McGill*, British Columbia*, Laval,

Quebec Montreal, Quebec Trois Rivieres, Waterloo, UCLV Cuba, IITM India, Goa, IIT Karangpur, CAT Gov. India, IRDE India, NARL India, Kyoto*, Titech, Tohoku, Tsukuba, Tokyo, UTNG Mexico, SQU Oman, UOK Pakistan, KMITL Thailand, Betla Venezuela, CPUT South Africa, Witwatersrand.

Summary of March 2011

For www.aias.us there were 105,932 hits, 69,795 page views, 14,579 distinct visits and 7.917 gigabytes downloaded. 2009 documents were read from 96 countries, led by USA, Russia, Britain, Germany, France, Australia, Mexico, Austria, …..

The most read papers were: 143(Sp), 19, 25, 92, 43, 69, 18, 94, 41, 54, 107, 39, 66, 4, 7, 140(Sp), 45, 146, 1, 51, 67, 8, 144, 21, 110, 44, 63, 68, 116, 159, 174, 49, 99, 161, 11, 17, 140, 47, 159(Sp), 132, 158, 131(Sp), 59, 104, 170, 26, 142, 102, ….

The most popular talks were: Universe of Myron Evans, Nobody is Perfect, Deflection of Light by Gravity, Intro to UFT 139, 20, 7, 10, 4, 14, 7, 8, 9, 11, 13, 23, 12, 22, 15, 19, 21, 17, …

The leading articles by colleagues were: Felker3 (Sp), Latest Family History, Sapcetime Devices (Spanish), Levitron, Spacetime Devices, Space Energy, Galaxies (Sp), Overview of ECE Theory.

For www.atomicprecision.com there were 32,358 hits, 11,820 page views and 4,309 distinct visits, 5.99 gigabytes downloaded.

For www.upitec.org there were 5,414 hits, 864 page views and 545 distinct visits.

This makes a total of 143,404 hits, 82,479 page views and 20,433 individual visits, 13.91 gigabytes downloaded.

1 - 15 March

US Universities, Institutes and Similar.

Alfred, Bucknell, Caltech*, Colorado, Cornell, Dartmouth, Drexel, Duke, Elgin, Georgia Tech., Harvard, Princeton Institute for Advanced Studies, Illinois State Univ., Lehigh*, Louisville*, MIT*, Minnesota State Colleges, North Carolina State*, Ohio State*, Princeton, Penn State, Purdue, Stanford*, Trident Tech, Chicago, UC Santa Barbara, UC Santa Cruz, Delaware, Florida, Illinois Chicago, Toledo, Washington, Wisconsin, West Virginia, Federal Bueau of Prisons, Lawrence Livermore, NASA, NIH, NOAA, Sandia, US Navy Med., US Archives, Yonkers Public Schools NYC, SD Medical Center, Dupont, Honeywell, Lockheed Martin, Microsoft, Northrop Grumman, Raytheon.

Europe and Asia Minor: Universities, Institutes and Similar.

ESI Austria, Namur, KU Leuven, CERN*, ETH*, Charles Univ. Prague, Bosch, FU Berlin, Juelich*, Siemens, TU Darmstadt, Heidelberg, Mainz, Rostock, CBS Denmark, Aragon, UA Barcelona, Cordoba, Murcia, Zaragoza, Valencia Tech., Helsinki*, JYU Finland, CAM France, ISAE France, Polytechnique System, Paris Psud, Angers, Avignon, F. Comte, Poitiers*, Savoie, Forthnet Greece, Iskon Croatia, GN Gov. Ireland, NUI Galway, UCD Ireland, INFN Bari, INFN Turin, Bologna, Perugia, ONS Student Net Netherlands, TU Delft, FU Amsterdam, NTNU Norway, UI Oslo, AMU Poland, IFJ Poland, MIMUW Poland, Krakow, SBB Serbia, Corbina Russia, MTS-NN Russia, KTH Sweden, LU Sweden, UU Sweden, NLB Slovenia, SPSMT Slovakia, Gazi Turkey, Gyte Turkey, Kiev Ukraine, Aston, CSX Cambridge*, Dampt Cambridge, Coventry, Durham, Edgehill, Exeter, Hull, Manchester, Newcastle*, Bodleian Oxford, Christchurch Oxford, Physics Oxford*, Sheffield Hallam, Southampton, St. Andrews, British Library, BBC (thls), British Gov Energis and Ministry of Defence, NHS, National Library of Wales, Ysgolion Sir Ga^r (Carmarthen Schools).

Rest of World: Universities, Institutes and Similar

Melbourne, Queensland, UNESP Brazil*m Sao Paolo, McMaster Canada, Trinity Western, British Columbia*, Quebec Trois Rivieres, Toronto, Victoria, Waterloo*, UDEA Colombia, Andes*, RI Ministry of Education Cuba, Goa India, SKR India, Kyoto*, IPICYT Mexico, UNAM Mexico, LUMS Pakistan, NU Singapore, NCKU Taiwan, NCU*, NSYSU, NTHU*, Academica Sinica Taiwan, Univ Zulu South Africa, Witwatersrand.

15 - 30 March

US Universities, Institutes and Similar

Brandeis, Caltech, Colorado*, Colorado State, Cornell, Duke*, Eastern Kentucky, Elgin, IUPUT, Lehigh, MIT, Mount St. Mary's, Northeastern, New York Univ., Pittsburgh, Rutgers, Stanford*, SUNY Stonybrook, Syracuse, UC Santa Barbara, Univ. Iowa, UNC Asheville, New Hampshire, South Carolina, South Florida, Utah, Wisconsin, Yale, US Gov. Mine Safety and Health, National Institutes of Health, Oak Ridge, US Social Security; AFNOC Military, NMCI Navy, Dow, Honeywell.

Europe and Asia Minor: Universities Institutes and Similar

Medical Faculty Univ. Vienna, Brussels, ETH Zurich, Belfort, Geneva, HS offenburg, Max Planck Stuttgart, TU Berlin, TU Chemnitz, TU Darmstadt, Heidelberg, Jena*, Karlsruhe, Leipzig*, Aragon, EUSS Spain,

Madrid Autonomus, Granada, Canaries, Valencia Tech., Seville, Valladolid, Helsinki*, Kyla, Tontut Finland, AC Nancy Metz France, grenoble, INP France, INPG, Univ Psud France, Strasbourg*, UHP Nancy, F. Comte, Lille 3, Poitiers*, Rennes 1, Bar Ilan Israel, INFN Turin, Modena*, Perugia, CSS Group Latvia, RU Netherlands, Utrecht Netherlands, HIST Norway, UIO Norway*, European Trade Union Organization, AGH Poland, Wroclaw, PUB Romania, SBB Serbia, Chalmers Sweden, BPKS Russia, Corbina, Debryansk, Dubna, MTS-NN Russia, Novosibirsk, HCKSlovakia, Anadolu, Ankara, Gantep*, Uludag Turkey, SUMDU Ukraine, Brooke's, Brooklands, Cambridge, Cranfield, Durham*, King's College London, Kent, Lancaster, Leeds, Nottingham, Bodleian Library Oxford, University College Oxford, Palmers, Queen Mary Univ. London, Sussex, Univ. College London, British Library*, Ministry of Defence, National Health Service, Libraries in Northern Ireland, Llyfregell Gendlaethol Cymru, National Library of Wales.

Rest of World: Universities, Institutes and Similar.

Gov. Argentina, INTI Argentina, Australian National Univ., Univ. South Asutralia, Melbourne, Univ New South Wales, Queensland, PUC Rio Brazil, UFPA, UFRJ, UNESP Brazil, Campinas Brazil, McMaster Canada*, MTA*, YRBE Ontario, British Columbia*, Quebec Trois Rivieres, Waterloo*, York Univ. Toronto, TIE Chile, UFRO, U Major Chile, UTFSM, UDEA, UNAL*, Andes* Colombia, City Univ Hong Kong, BARC Gov. India, Osakafu Japan, SKKU S. Korea, Gov. Mexico, UMICH, UNAM Mexico, UNP Peru, UPAOTeens Peru, GCUF, IIU, UET, PERN Pakistan, KMITL Thailand, NTHU*, NTU* Taiwan, SUN, Capetown, Witwatersrand South Africa.

Summary of April 2011

For www.aias.us there were 93,561 hits, 13,688 distinct visits, 58,354 pages, 7.826 gigabytes downloaded, 2066 documents read from 94 countries, led by USA, Russian Federation, Gemany, France, Mexico, Britain, Australia, Canada, …..

The most read UFT papers were: 178, 179, 171, 25, 158, 41, 43, 175, 94, 18, 4, 149, 32, 107, 44, 1, 135, 8, 176, 35, 104, 57, 7, 140(Spanish), 124, 123, 159(Spanish), 47, 59, 131(Spanish), 118, 54, 134, 166, 33, 116, …..

The most read articles were: Space Program, Felker3 (Spanish), Spacetime Devices, Levitron, Latest Family History, Space Energy, Engineering Model, Devices, F1, F4, ECE (Russian), Special Relativity, Galaxies (Spanish), ……..

The most listened to or read talks / essays were: Nobody is Perfect (narrated by MWE), Deflection of Light by Gravity (MWE), Introduction to UFT 139 (MWE), Nobody is Perfect (narrated by Robert Cheshire), 7, 24(pdf), 4, 7(pdf), 19, 10, 8, 11, 20, 23, 17, 13, 14, 15, 9, 12, 22, 21, Universe of Myron Evans (narrated by Robert Cheshire), Universe of Myron Evans (narrated by MWE), …….

For www.atomicprecision.com there were 30,379 hits, 6.29 gigabytes downloaded, 4,176 visits and 9,224 pages downloaded.

For www.upitec.org there were 5,251 hits, 0.75 gigabytes downloaded, 632 pages and 929 pages.

There was a total of 129,191 hits, 14.87 gigabytes downloaded, 18,496 visits and 68,506 pageviews.

1 – 15 April

US Universities, Institutes and Similar.

Arizona State, Berkeley, Ball State, Boston University, Caltech, Calpoly CP, Drexel, Harvard, Howard, Illinois, Indiana, Lane College*, Lousiana Tech, Maine, MIT, Mississippi State, Midwestern State Wichita Falls Texas, Northern Arizona, Northwestern*, New York Univ., Oakland, Ohio, Oklahoma City, Pittsburgh*, Princeton, Penn State*, Redwoods, Rice, Richmond, Rhode Island School of Design, Rochester, Stanford, Stockton, Texas A and M, Arkansas, UC Santa Barbara*, UC San Diego*, Illinois Urbana Champaign, U Mass., Minnesota Twin Cities, Chapel Hill*, South Central Los Angeles, Vermont*, Washington, Wisconsin*, Yale, US Gov., DHS, Los Alamos, Lawrence Livermore, US Airfoce Afnoc and AFSP, US Army SWA, US Navy NMCI*, Naval Research Laboratory, Honeywell, Intel, Raytheon, Jefferson Laboratory, City of Des Moines, International Union of Crystallography, Municap Research and Services Center of Washington.

Europe and Asia Minor: Universities, Institutes and Similar

UIBK Austria, KU Leuven, ITC Bulgaria, EPF Lausanne, MUNI Czechia, German Synchrotron Facility, FU Berlin, Jacobs, KFA Juelich, Max Planck Plasma Physics, Siemens, Bonn*, Giessen, Karlsruhe*, Cologne, Leipzig*, Mainz, Muenster, Stuttgart, Aragon, Autonomous Barcelona, Autonomous Madrid, Barcelona, Granada, Canaries, Catalonia Polytech., Madrid Tech., Valenica Polytech., Seville, Santiago de Compostela, Tegascas, Valencia, Euroo, helsinki, Kyla, ASSO France, ENS2M France, French National Particle Laboratory, INPL Nancy, INSA Rouen, Poitiers*, Rouen, Patras Greece, UTH Greece, Amis Croatia, Iskon

Croatia, Medri Hungary, MPHZRT Hungary, Trinity College Dublin*, Univeristy College Dublin, Hebrew Univ Jerusalem, INFN Bari, INFN Rome 1, Calabria, Catania, Perugia, Pavia, Turin, Trento, MII Lithuania, CSS group Latvia, OBA Netherlands, NTNU*, UIO* Norway, Wroclaw* Poland, UTL Portugal, UPT Romania, Belgrade, Corbina* Russia, MTS Russia, Bionet NSC Russia, KTH Sweden, Josef Stephan Institute, NLB Slovenia, LSG Slovakia, Tarassul Syria, Ankara, DEU Turkey, Brooklands, Cambridge, Durham, Edinburgh*, Kent, Lancaster, Manchester, Nottingham, Gradacc Oxford, Queen's Univ. Belfast, Seffield, Sheffield Hallam, Stirling, East Anglia.

Rest of World: Universities, Institutes and Similar.

Citefa Gov. Argentina, CSIRO Australia, Monash, QLD, QUT, Melbourne, Queensland Australia, UFRJ, Sao Paolo* Brazil, McGill*, British Columbia*, Calgary, New Brunswick, Quebec Trois Rivieres, Toronto*, York Univ. Toronto Canada, Univ Chile*, UCN, UCT Chile, HA China, UNAL Colombia, BHU India, Indian Statistical Institute, Shahed Iran, Ibaraki, Junshin, Osaka, Osakafu, Yamagata Japan, UGTO, ITAM, USON, UIA Mexico, UTHM Malaysia, PUCP Peru, Kohat, LHR-NU, NUST, QAU Pakistan, NU Singapore*, NCTU, NSYSU, NTHU Taiwan, Hartrao, NWU, Capetown South Africa.

15 – 29 April

US Universities, Institutes and Similar.

Alabama Supercomputer Authority, Arizona State, Aurora, Baylor, Boise State, Caltech., Colorado, Columbia*, Cornell, California State Monterey Bay, Dartmouth, Dickinson, Dixie, Elon, Emory, Harvard*, Hawaii, Illinois*, James Madison, Kansas State, Louisiana Tech., Maricopa, Midlessex Community College, MIT, Missouri Science and Technology, Michigan Tech.*, Miami, Northern Arizona*, North Carolina State, NIT India (edu), Northwestern, Old Dominion, Ohio*, Pittsburgh, Penn State*, Purdue, Reed, Rensealer, St Petersburgh College Florida, UC Los Angeles*, UC San Diego, Florida, Houston, South Alabama, Illinois Chicago*, Illinois Urbana Champaign*, U Mass., Minnesota Twin Cities, New Hampshire, Dhaka (edu), Pennsylvania, South Carolina, Utah, Texas, Virginia, Vermont, Wallawalla (edu), Washington, Washington St Louis*, Yale*; NASA, NOAA, NMCI Navy*, General Electric, BMW Group, Hewlett Packard, Mount Michael Benedictine Abbey Nebraska, Wilcox Schools.

Europe and Asia Minor: Universities, Institutes and Similar.

Free Univ Brussels, EPF Lausanne*, ZCU Czechia, German Synchrotron Facility, Max Planck Halle, Max Planck Heidelberg, Siemens, TU Freiburg, TU Harburg, Heidelberg*, Jena, Karlsruhe*, Cologne, Magdeburg, Muenchen, Wuerzburg, The University of the Basque Country, Barcelona, Zaragosa, HTV Finland, ANPE France, ASSO France, CNRS Grenoble, INRA France, LISIF France, Jussieu, Strasbourg, Poitiers*, Reims, Trinity College Dublin, Technion Israel, INFN Florence, INFN Legnaro National Laboratory, Florence, KUN Netherlands, RU Netherlands, Utrecht*, UIO Norway, CFT Poland, UJ Poland, UTL Portugal*, Serbian Academy of Sciences, Belgrade, Corbina Russia, Kubang SM, Lipetsk, MGn Russia, MTS russia, MTU Net Russia, KTH Sweden, Antik Slovakia, Ankara Turkey, Gazi*, Kocaeli, Metu Turkey, Edinburgh, Imperial, Maths Oxford, VPN Oxford, Portsmouth, Swansea.

Rest of World: Universities, Institutes and Similar.

Gov. Queensland, UFPE Brazil, Sao Paolo, Gov. British Columbia, Perimeter Institute, British Columbia, Calgary, Quebec Trois Rivieres, Toronto, Waterloo, UFRO Chile, UTFSM Chile, INF Cuba, IPM Indonesia, ITS Indonesia, UNPAR Indonesia, IITM India, Minpaku Japan, Osaka, Tokyo Tech, Tokyo, SNU South Korea, UITM Malaysia, Auckland*, Kohat Pakistan, New Pakistan University, QAU Pakistan, NU Singapore, Chula Thailand, NTHU, NTU* Taiwan, RU South Africa, Capetown, Witwatersrand.

END OF SEVEN YEARS OF RECORDING

Summary of May 2011

For www.aias.us there were 110,015 hits, 73,150 page views, 16,301 distinct visits, 13.53 gigabytes downloaded, 2,117 files read from 98 countries, led by US, Germany, Russian Federation, Czechia, France, Italy, Britain, Australia, …. All paper, articles and essays read, all broadcasts heard.

The leading UFT papers were: 68, 25, 46, 33, 41, 18, 43, 58, 52, 110, 177, 49, 107, 140, 104, 94, 127, 54, 81, 29, 88, 118, 61,102, 88, 142, 1, 2, 35,175, 63, 149, 48, 91, 99, 7, 64, 116, 65, 159(Sp), 170, 171, 181, 20, 37, 11, 47, 53, 66, 131, 140(Sp), 168, 38, 8, ….

The leading articles were: Felker3 (Sp), Family History, levitron, Galaxies (Sp), Space Energy, Spacetime Devices, Spacetime Device (Sp), Johnson Magnets, EEC Engineering Model, ECE, Crothers Sharples, ...

The leading talks and essays were: Nobody is Perfect, Deflection of Light due to Gravity, Universe of ME (pdf), Nobody's Perfect (Robert Cheshire), 22, 23, 21, 17, Essay of UFT139, 15, 24, 7,13, ...

For www.atomicprecision.com there were 28,935 hits, 2.28 gigabytes downlaoded, 4,292 visits and 8,986 page views.

For www.upitec.org there were 4,738 hits, 0.76 gigabytes downloaded, 560 visits and 846 page views.

This gives a total of 143,688 hits, 82,982 page views, 16.57 gigabytes downloaded and 21,153 visits.

1 – 15 May

US Universities, Institutes and Similar

Arizona, Austin CC*, Binghampton, Brandeis, Ball State, Canyons, Caribbean, Cedarville, Cornell*, Gustavus Adolphus, Geneseo*, Hawaii, Illinois Institute Technology*, Indiana, Johns Hopkins Applied Physics, James Madison, Kent State, Lousiana State*, MIT, Montana, Michigan Tech., Northern Arizona. North Carolina State*, New Mexico Inst. Mining and Tech., Oklahoma Christian, Ohio State, Princeton, Purdue*, Rose Hulman, Southern Methodist, Stanford, Stevens Tech., UC Davis, UC Fresno, UC Irvine, Connectictut, UC Riverside*, UC Santa Barbara, UC San Diego, Delaware, Houston, Illinois Urbana Champaigne, Michigan, Minnesota Twin Cities, North Carolina Charlotte, South Florida, Washington, Winona Wisconsin; Kirtland AF, NMCI Navy*, Brookhaven, Los Alamos, National Institutes of Health, IBM*, Northrop Grumman, Rocky Vista University, Seton Organziation.

Europe and Asia Minor: Universities, Institutes and Similar.

TU Vienna, Liege, Neuchatel, Zurich*, CERN, MUNI Czechia, Max Planck Halle, Max Planck Stuttgart, Bochum, Aachen, Siemens*, Students Bonn, TU Berlin, TU Darmsadt students, Bonn, Hamburg, Hannover, Heidelberg, Karlsruhe, Cologne*, Leipzig, Muenster, Paderborn, SDU Denmark, Aragon, ICMAB, TSAI, UDC, Zaragosa, UPC, UPV, UVA, Parliament of Spain, Helsinki, HTV, JYU Finland, ENS France, ENST Brittany, INRIA, LPTHE Jussieu, SNECMA, Nice, Poitiers*, UPS Toulouse France, CYTA Greece, BHN Hungary, INFN Florence, genoa, Rome 1, CSS Latvia*, Leiden, RU Netherlands, Utrecht students, UVT Netherlands, NTNU Norway, Oslo, INFIN Romania, Akadem Russia, Corbina, MTS Russia, KTH Sweden, NLB Slovenia, GTS Slovakia,

Anadolu, Itu, Iyte Turkey, City of Kiev Ukraine, Teifi Aberystwyth, Bristol, Brooklands, Cambridge*, Cardiff, Durham*, Imperial*, King's College London, Lancaster, Leicester*, Leeds, Luton, Newcastle upon Tyne, Oxford*, Southampton, Strathclyde, Univ.College London, Ministry of Defence, Epilepsy Society, Llyfrgell Cenedlaethol Cymru (National Library of Wales), The Grid.

Rest of World: Universities, Institutes and Similar.

Adelaide, Australian National, East Queensland, Monash*, Marquette, Queensland, Ministyr of Defence Australia, UFC, UFRGS, UFSC Brazil, McGill, Perimeter, Simon Fraser, British Columbia*, Waterloo Canada, UNAL Colombia, RCIL Gov. India, Kanazawa, Kansai, Kyoto,Osaka*, Osakafu, Titech*,Tohoku Japan, UNAM Mexico*, Uniten Malaysia, National Unv Nicaragua, Victoria New Zealand, National Univ. Peru, CIIT, NUST Pakistan, National univ Rwanda, NCUT, National Univ Taiwan, NWU, SUN, Cape Town South Africa.

15 – 30 May

US Universities, Institutes and Similar.

Buffalo*, Caltech*, Colorado, Drexel, Harvard, LeHigh, Michigan State, North Carolina State, Ohio State*, Ohlone College, Oregon State, Pasadena, Penn State, Purdue, Rice, Texas Tech., UC Davis, UC Fresno, Chicago, UC Irvine, UCLA, UC Riverside, UC Sane Diego, Florida, Illinois Urbana Champaign, Michigan, Minnesota, North Carolina Wilmington, Phoenix, Puerto Rico, US Naval Academy, Utah State, Washington, Weber, Yakima Valley; Gov Georgia, Lawrence Livermore, Sandia, US Dept of Agriculture, Army SC, Army Tacom, Navy NMCI*, Naval Research Laboratory, East Texas Medical Center, Rocky Vista University, General Electric, Honeyweel, Intel, Northrop Grumman, Raytheon.

Europe and Asia Minor: Universities, Institutes and Similar.

TU Vienna, Vienna, Louvain, EPF Lausanne, German National Synchrotron Facility (DESY), IFW Dresden, Max Planck Heidelberg, Bochum, Siemens*, Students Bonn, TU Berlin, TU Freiburg, Bonn*, Hannover, Heidelberg, Karlsruhe, Kaiserslautern, Koblenz, Leipzig, Saarland, Tuebingen, Wuerzburg, Hamburg, College Net Denmark, Aragon, EHU Spain, GMV, MDE, TSAI* Spain, UCM, Granada*, UPV*, USAL, UV Spain, Spanish Parliament, Nancy Metz France, ESPCI, French National Particle Laboratory, ONERA, Polytechnique, F. Comte, Poitiers*, NTUA Greece, TUC, UOC Greece, JCT Israel, Emilia Romagna Italy, Ghislieri, INFN Bari, Rome 2, Genoa, Padua, Pavia, RU

Netherlands, UPC, Utrecht*, UVT Netherlands, UIB, UIO Norway, Poznan Poland, Lisbon, SBB net Russia, Corbina net Russia, Kirov, Saratov, MTS-NN Russia, Chalmers Sweden, Umea Sweden, NLB Slovenia, Birmingham City, CSX Cambridge, Girton, Newhall, physics, pwf* Cambridge, Imperial*, Keble, Lincoln, Mansfield, Wolfson, physics, vpn Oxford, Runshaw, UCL, BBC*, Energis.

Rest of World: Universities, Institutes and Similar.

UNLP Argentina, Australian National, Monash, Queensland, Melbourne, Western Australia, Queensland Government, Queensland, UFG Brazil, UFPA, Unicamp Brazil, McMaster*, Perimeter, British Columbia*, Waterloo*, Western Ontario Canada, UTFSM Chile, UDEA, UNAL, Andes Colombia, IMSC India, Shiraz Iran, Osakagu*, Shisuoka, Waseda Japan, KNU, Postech S. Korea, UAM, UOA, UNAM* Mexico, Auckland, Victoria New Zealand, GCUF, NUST, UAF, UET Pakistan, NU Rwanda, NU Singapore, NTHU, NTU* Taiwan, British Univ Uruguay, NWU, UP South Africa.

Summary of June 2011

For www.aias.us there were 96,141 hits, 8.752 gigabytes downloaded, 13,670 distinct visits, 58,361 page views, 2,154 documents read from 92 countries, led by US, China, Britain, Russia, Germany, Argentina, France, ……

The most read UFT papers were: 49, 43, 25, 41, 4, 60, 159(Sp), 52, 94, 140, 65, 110, 175, 18, 38, 157(Sp), 46, 107, 177, 20, 54, 37, 61, 169, 140(Sp), 11, 7, 81, 104, 116, 99, 147, 149, 184, 1,2, 102, 181, 118, ……. All papers read.

The most popular talks essays etc. were: Nobody is Perfect (HE and MWE), On the Deflection of Light by Gravity (MWE), Nobody's Perfect (RC), Universe of Myron Evans (pdf), Spacetime Devices, 15, 20, Universe of Myron Evans (RC), 17, 13, 23, 22, 14, 21, Intro to UFT 139, 19, 7, 9, 37 (pdf), 7, 8, …..

For www.atomicprecision.com there were 27,717 hits, 19,006 page views, 2,28 gigabytes downloaded, 4,603 distinct visits.

For www.upitec.org there were 3,393 hits, 2,142 page views, 0.40 gigabytes, 573 distinct visits.

This gives a total of 127,251 hits, 79,509 page views, 18,846 distinct visits and 11.43 gigabytes downloaded.

1 – 15 June

US Universities, Institutes and Similar

Austin CC, Buffalo, Columbia, Cornell, Drexel, Edison, Geneseo, Kansas State, Monash, Ohio, Oregon State, Pittsburgh*, Southern Methodist, Syracuse, Texas State, UCLA, UC Santa Barbara*, Illinois Urbana Champaign, Maryland, South Carolina, Texas, Lawrence Livermore, NSF, Sandia, Army Belvoir, Army South Carolina, Army Corps of Engineers, Army WES, National Geospacial Intelligence Agency, Chief of Army Reserve Pentagon, Raytheon, The Aerospace Corporation, Argonne National Laboratory Guest House, CBF YMCA, Los Angeles Public Library, Sacramento Public Library.

Europe and Asia Minor, Universities, Institutes and Similar

UC Liege, Charles Univ Prague, German Synchrotron Facility*, Bosch, FH Muenster, Fraunhofer*, Max Planck Heidelberg, Bocum, Siemens*, TU Darmstadt*, Erlangen, Frankfurt, Hannover, Karlsruhe, Cologne, Leipzig*, Muenster, Aarhus Denmark, UBU, UCA, UN*, UNEX, UPC*, UPV, UV, UCA Spain, Helsinki, Kyla Finland, CIUP, Particle Physics Laboratory, Polytechnique, Poitiers*, Savoie France, CRES Greece, Central Campi Budapest, Hebrew Univ. Jerusalem, HI Iceland, IPCF (CNR) Italy, INFN Bari, Pisa, Pavia, Turin, MII Lithuania, TU Eindhoven, Utrecht, UVT Netherlands, Mittal Steel Romania, SBB Serbia, Corbina Russia, MTU Net Russia, KTH Sweden*, NLB Slovenia, Gyte, Metu Turkey, Kiev, Birmingham City University, Broklands, Cambridge*, Cardiff*, Exeter*, Imperial*, Leeds, Luton, Keble Oxford, Maths. Oxford, St Hugh's Oxford, Strathclyde, Sussex, Swansea, National Botanical Garden of Wales.

Rest of World: Universities, Institutes and Similar

UNLP Argentina*, Government Argentina, Australian National, Queensland, RMIT Melbourne, Swinburne, UFAM, UFPEL, INF, UF Rio*, UFSC, UNICAMP, Sao Paolo* Brazil, Owen Sound Canada, British Columbia*, Waterloo, Western Ontario, Swiss Com Switzerland, UTA Chile, HA China, UDEA, Norte, Valle Colombia, MRI, Government, Bose, IMSC India, Shiraz Iran, Kokudai, NIFS, Shinsu Japan, KAIST, KHNP South Korea, UASLP, UNAM Mexico, Government Malaysia, Otago New Zealand, GCU, GCUF, New Univesity, TUF, UET Pakistan, NCTU, NTHU* Taiwan, NWU South Africa.

15 – 30 June

US Universities, Institutes and Similar.

Albany Medical Center, Augsburg, Berkeley, Bostin University, Buffalo, Camden CC, Colorado, Cornell*, Georgetown, Indian River, Lock Haven, Michigan Tech., NIT India (using edu), Oklahoma Christian, Potsdam, Purdue, Rutgers, Texas A and M, IC Irvine, UC Merced, Iowa, Maryland, Michigan, Chapel Hill, Texas, BNL, NIST, Airforce AFNOC, Army SMDC, NGA Military, Intel, Lockheed Martin, Northrop Grumman, Berkeley Public.

Europe and Asia Minor: Universities, Institutes and Similar.

SMC Austria, UC Liege, KU Leuven, ETH Zurich, Charles University Prague, Brandenburg,Bochum, Siemens, Cottbus, TU Darmstadt, Bonn*, Duisburg, Frankfurt, Jena, Kaiserslautern, Mainz, Regensburg, Tuebingen, Wuerzburg, Aragon, GMV, GVA, UAM-CISC, UGR, USAL, UV, UVA Spain, HUT Finland, SIMA, Avignon, Lyon 1*, Marseilles, Poitiers*, Savoie France, Aegean, NTUA Greece, NUI Galway, Trinity College Dublin Ireland, Tel Aviv, RM.CNR Italy, ICTP Trieste, INAF, Trieste, Perugia, Rome 1, Eindhoven, UVT Netherlands, HIST Norway, Porto Portugal, Russian Academy of Sciences Institute for Nuclear Research Troitsk, Ural, Corbina, Pskov, Saratov, HC,Vologda Russia, MAH Sweden, Anadolu, Ege Turkey, Cosmonova Ukraine, Astronomy Cambridge, Trinity College Cambridge, Cardiff, Coventry, Imperial*, Maths Oxford, Queen Mary London, Sheffield*, Swansea College, Univ College London, University of East Anglia, Wellcome Trust, Tormead School.

Rest of World: Universities, Institutes and Similar

Western Power Australia, ANU, Ballarat, Queensland Tech., Melbourne, Queensland, UFAM Brazil, Sao Paolo, Perimeter Canada, UBC*, TIE Chile*, UCT Chile, UDEA Colombia, UO Cuba, TIFT India, Kobe, TUS Japan, Tokyo, NIT Kenya, UAEH Mexico, UAEM, Government Malaysia, LUMS Pakistan, NCP, QAU, CMU Thailand, NCU Taiwan.

Summary of July 2011

For www.aias.us there were 78,144 hits, 6.799 gigabytes downloaded, 13,313 distinct visits, 46,198 page views, 2,036 documents read from 89 countries, led by USA, Russian Federation, Germany, Britain, France, Argentina, Australia, Mexico, …..

All UFT papers read, led by 189, 180, 187, 188, 25, 43, 186, 152, 167, 41, 140, 189-4, 166, 107, 177, 26, 94, 110, 157 (Sp), 11, 99, 142, 140(Sp), 52, 47, 159(Sp), 108, 7, 119, 44, 150, 1, 57, 64, 8, 42, 65, 118, ……

Leading Essay Broadcasts were: 3(RC), Nobody is Perfect (MWE), 3 (MWE), 28(RC), Universe Part 1(RC), 30(RC), 37(RC), 29(RC), 34(RC), 12(RC), 33(RC), 24(RC), 10(RC), 32(RC), 35(RC), 26(RC), 5(RC), 11(RC), 14(RC), 6(RC), 15(RC), 25(MWE), 9(RC), 19(RC), 17(MWE), 16(MWE), 22(MWE), 4(MWE), Introduction UFT 139 (MWE), ... Estimated 3,500 hearings of broadcasts.

Leading Articles were Spacetime Devices, Felker 3 (Sp.), Infinite Solenoid, Universe of Myron Evans, Latest Family History, GGltr1, ECE Popular, Space Energy, Workshop Overview, Workshop Slides, ECE Engineering Model, ...

For www.atomicprecision.com there were 27,333 hits,18,427 page views, 2.12 gigabytes and 4,948 distinct visits.

For www.upitec .org there were 4,152 hits, 1,770 page views, 0.47 gigabytes and 738 distinct visits.

This gives a total of 109,629 hits, 66,395 page views, 9.37 gigabytes and 18,999 distinct visits.

1 – 15 July

US Universities, Institutes and Similar

Andover, Buffalo*, Cape Fear, Columbia, Cornell*, Maine, MIT, Naval Postgraduate School, Oklahoma Christian, Princeton, SUNY Stonybrook, UCLA, University of California Office of President, UC San Diego, Illinois Urbana Champaign*, Massachussets Boston, Michigan*, Nebraska Lincoln, U Penn., Texas Arlington, Texas, Wisconsin, Western Illinois, William and Mary, Washington St Louis, NSF, Government of Virginia, US Airforce NOC, General Electric, Honeywell*, IBM (General), IBM Thomas Watson, Northrop Grumman, Siemens (US), East Texas Medical Center Tyler.

Europe and Asia Minor: Universities, Institutes and Similar

TU Graz, Vienna, BUS Bulgaria, CERN, Neuchatel Switzerland, Fraunhofer, Max Planck Heidelberg, RWTH Aachen, Siemens, TU Harburg, Hamburg, Heidelberg, Karlsruhe, Leipzig*, Tuebingen, IFAE Spain, UCM, UMA, UNED*, UPV Spain, Spanish Parliament*, Helsinki, LISIF Jussieu France, Angers, Poitiers*, UTG Greece, ICTP Trieste, INFN Naples, INFN Turin, Milan Tech., Calabria, Pisa, Rome 2, Turin, CSS Latvia, RU Netherlands, UVT Netherlands, Warsaw, Corbina Russia*, Siberian Academy Novosibirsk, LIY Sweden, NLB Slovenia, Mariboru Slovenia, SCC Slovakia, Anadolu Turkey, Brooklands, Cambridge, Cardiff, Coventry, Leeds, NSMS Oxford, Runshaw, Paragon School, Ceredigion, Libraries NI, Llyfrgell Genedlaethol Cymru.

Rest of World: Universities, Institutes and Similar.

Murdoch Univ Australia, SCU Australia, Gov. Australia, PUC Rio Brazil, UFG, UFPR, UFRJ Brazil*, GC Canada, McGill, UBC*, UNINE Chile, Univ Central Chile, UFRO, UP Cuba, RRI India, Kyoto, UGTO Mexico.

15 – 30 July

US Universities, Institutes and Similar.

Arizona, Arizona State, Boston University, Buffalo, Caltech, Central Michigan, Columbia, Cornell, Johns Hopkins, Kentucky Technical and Community College System, Mississippi State, North Carolina State, Oklahoma Christian, Texas A and M, UC Davis, Chicago, Illinois Urbana Champaign, Michigan, North Carolina Charlotte, New Mexico, Texas, Kentucky, Washington, NASA GRC, NSF*, Army peoavn, Army, Army Special Operations Command, General Electric, Microsoft, Raytheon, Fallon Clinic.

Europe and Asia Minor: Universities, Institutes and Similar

Vienna, BAS Belgium, CERN*, German Synchrotron Facility, DFKI, Siemens, TU Berlin, Hannover*, Heidelberg, Kiel, Leipzig*, Tuebingen, Tartu Estonia, Aragon, EHU, IFAE, UCM, UPC, UPCT, UPM, USC, UV Spain, Gov Canary Islands, FMI Finland, Joensuu Finland, OBSPM France, Oleane France, Polytechnique, RAIN France, Rennes 1, URG Greece, ADSL Hungary, Radio Hungary, HUJI Israel, INFN Padua, Modena, Rome 1, CSS Latvia, Tvente Netherlands, Utrecht, UVT, CAMK Poland, CM FARO Portugal, AMRES Serbia, Corbina Russia, JINR Russia, MTU Net Russia, SCC Slovakia, KIPT Kharkov, BBSRC Britain, Brooklands, Cambridge, Cardiff, Coventry, Imperial, Leeds, Nottingham, York, NNC (THLS), NSH, Essex Schools.

Rest of World: Universities, Institutes and Similar.

Australian National, Melbourne, BOM Gov Australia, Defence Gov Australia, McGill Canada, UBC, Montreal, UFRO Chile, HA China, UNAL Colombia, Gov. Colombia, UNL Ecuador, Gov. India, TIFRBNG India, Hirsohima, Hokudai, Kyoto*, Tokyo, UTHM Malaysia, Auckland, Apiquitos Peru, GU Pakistan, New Pakistan University, MCUT Taiwan, Tlabs South Africa, UFS South Africa.

Summary of August 2011

New diary or blog record of 582 visits on 30th August, 53,558 total on 31st August.

For <u>www.aias.us</u> there were 85,554 hits, 50,677 page views, 13,942 distinct visits and 13.96 gigabytes downloaded. A total of 2,218 documents were read from 99 countries, led by USA, Russian Federation, Britain, Germany, Mexico, Australia, Canada, Japan, ...

http://www.atomicprecision.com

The most read source papers were: 25, 43, 41, 26, 18, 4, 8, 159(Sp), 191, 19, 190, 52, 116, 140, 107, 140(Sp), 177, 170, 64, 54, 42, 7, 47, 49, 94, 11, 50, 44, 45, 118, 1, 20, 46, 65, 76, 108, 12, 51, 31, 3, 59, 119, 85, 81, 161, 28, 60, 161(Sp), 128, 189, 75, 115, 149, 38, 88, 142, 160, 33, 34, 37, 57, 15, 2, 67, ... all 193 source papers read.

The most read articles were: Felker 3 (Spanish), Overview of ECE Theory, New Autobiography, Spacetime Devices (Spanish), Latest Family History, Universe of Myron Evans, Johnson Magnets, MWE Curriculum Vitae. .... all articles read

The leading broadcasts were: On Light Deflection by Gravitation (RC), Nobody's Perfect (MWE), Nobody's Perfect (RC), 42 (RC), 41 (RC), 34 (RC), 9(RC), 7(MWE), 23 (MWE), 22 (MWE), 21 (MWE), Nikola Who? (RC), 25 (MWE), 26 (RC), 5(RC), 28(RC), 29(RC), Universe 1 (RC), 10 (RC), 19 (RC), 27 (MWE), 36 (RC), 39 (RC), 14 (RC), 24 (RC), 32 (RC), 35 (RC), 40 (RC), 8 (RC), ... All essays and broadcasts read and listened to in English and Spanish.

For <u>www.atomicprecision.com</u> there were 29,074 hits, 19,420 files, 5,022 distinct visits and 0.629 gigabytes downloaded.

For <u>www.upitec.org</u> there were 4,008 hits, 2,477 files downloaded, 638 individual visits and 0.051 gigabytes downloaded.

This is a total of 118,636 hits, 72,574 files, 19,602 distinct visits and 14.64 gigabytes downloaded.

1 – 15 August

US Universities, Institutes and Similar

Ave Maria, Colorado, Dartmouth, Georgia Tech., Georgia State, Harvard, James Madison, Maine, Mississippi State, North Carolina State, National Radio Astronomy Observatory, New York*, Old Dominion, Purdue, St Joseph's College, Stanford, Texas A and M, Tarrant County College, UC Riverside, UC Santa Cruz, UC San Diego, Michigan, Minnesota Twin Cities, South Carolina, Utah*, Texas*, Vermont, Wisconsin, Princeton Plasma Physics Laboratory, Gov. West Virginia, Iraq Central Command*, US Army DARPA, General Electric, IBM*, Microsoft, Northeast Illinois Council Boy Scouts.

Europe and Asia Minor: Universities, Institutes and Similar

Linz, Vienna, ETH Zurich, Belfort, Geneva, Bosch, German Synchrotron Facility, FH Duesseldorf, TU Darmstadt*, Erlangen, Hannover*, Jena, Karlsruhe, Kiel, Kaiserslautern, Cologne*, Leipzig*, Tuebingen, Munich, Aaraus Denmark*, Aragon, MDE Spain, UPC, UPV, UV, Strasbourg France, UJF Grenoble, OIV Croatia, Hebrew Univ Jerusalem, Turin, CSS Latvia, RU Netherlands, TU Delft, UPC Netherlands, Utrecht, NTNU Norway, Jagiellonian Poland, Warsaw, Minho Portugal, Corbina Russia, Sura Russia, NLB Slovenia, SCC Slovakia, Anadolu Turkey, Bath, British Geological Survey, Bristol, Cambridge, Edinburgh, Glasgow*, Lancaster, Manchester, Open University, Ministry of Defence, National Health Service.

Rest of World: Universities, Institutes and Similar.

UNT Argentina, Adelaide, Swinburne, Melbourne, UEMA Brazil, UFPE Brazil, Saskatchewan Schools, UBC*, York University Canada, Univ. Chile, UTFSM Chile, Andes Colombia, CAT Gov. India, Nagaohout Japan, UGTO Mexico, KNU South Korea, Auckland, ADZI Phillipines, NUR Rwanda, KMITL Thailand, SUN, TLABS, UL, Witwatersrand South Africa.

15 – 30 August

US Universities, Institutes and Similar

Colorado, Cornell*, Georgia Tech., Macon State, Maine, MIT, Mississippi State, North Carolina State, Ohio State, Princeton, Penn State, Purdue*, Texas A and M, Maryland, Michigan, Chapel Hill, New Hampshire, New Mexico, Urah, Texas, Toledo, West Michigan, US Gov. Brookhaven National Laboratory, NASA, Nuclear Regulatory Commission, Dept. of Agriculture, US Military DFAS, Navy NAVO, Marine Corps, BMW Group, General Electric, Honeywell, IBM, Intel, Lockheed Martin, Northrop Grumman, Rockwell Collins, Xerox, CHS Omaha, LA Alliance, UCSF Medical Center, W3 Organization.

Europe and Asia Minor: Universities, Institutes and Similar.

Vienna, Free Univ Brussels, BAS Bulgaria, Lausanne, Geneva, Fraunhofer, Max Planck Halle, Max Planck Heidelberg, Bochum, Siemens, TU Darmstadt, Bonn, Duisburg and Essen, Hannover, Heidelberg, Leipzig, Muenster, Niels Bohr Institute, Gov. Andalucia, MDE Spain, TSAI Spain, Navarra, French Particle Laboratory, UPS Toulouse, UTG Greece, EAP Greece, UOC Greece, HI Iceland, Pisa, Rome 3, RU Netherlands, UVT Netherlands, AMU Poland, Lodz,

Warsaw, UNL Portugal, Aiasi Romania, Corbina Russia, IRK, Sura, USTU Russia, Anadolu Turkey, Iyete, Cambridge*, Imperial*, Manchester, Sussex, Swansea, West of Scotland, Wales, The Glasgow Academy Scotland, York Schools, NTNU Norway, UU Sweden.

Rest of World: Universities, Institutes and Similar.

ANU* Australia, Latrobe, Melbourne*, Western Australia, Defence Australia, Gov British Columbia Canada, Fraser Health, Valleyfield Quebec, UBC*, Manitoba, Montreal, BSJ, Waterloo, UTFSM Chile, HA China, UNAL Colombia, TSS Gov. Colombia, PUCE Ecuador, BRCM India, Tezu, BARC, IIT Karanpur, RCIL Gov. India, UGTO Mexico, UMICH Mexico, UNAM Mexico, GOB Peru, GIKT Pakistan, NUST Pakistan, NCU Taiwan, NWU South Africa, Capetown, UKZN, NL South Africa.

Summary of September 2011

New diary or blog record of 682 visits on September 19th., total of 58,327 viewings since January 1st, 2010.

For www.aias.us there were 11,858 distinct visits, 78,226 hits, 40,721 page views, 10.026 gigabytes downloaded, 2,234 documents read from 97 countries led by USA, Britain, Russia, Germany, Mexico, Australia, France, Japan, ...

All UFT papers were read, in order: 193, 194, 150-B, 41, 25, 159(Sp), 140, 54, 4, 116, 188, 140(Sp), 49, 107, 166, 47, 142(Sp), 170(Sp), 175, 7, 114, 81, 157, 192, 110, 11, 169, 33, 2, 68, 12, 37, 3, 61, 144(Sp), 157(Sp), 24, 38, 59, 161(Sp), 166(Sp), 1, 20, 42, 102, 88, ...

All broadcasts were heard, all essays read, in order: On the Deflection of Light by Gravity (RC), Nobody is Perfect, Nobody's Perfect(RC), 30(RC), (RC), 7, Intro 139 (RC), 27, 41(RC), 42(RC), 20, 50(pdf), 7, 19, 35, 38(RC), 11(RC), Univ. ME1 (RC), 14(RC), 15, 16, 18, 25, 33, 36, 4, Nikola Who? (RC), Deflection of Light by Gravity(RC), 17, 10, 13, 21, 22, 23, 24(RC), 8(RC), 26(RC), 12(RC), 6(RC), 8, ...

Articles were read in order: Felker3 (Sp), Autobiography, Space Energy, Overview of ECE Theory, Johnson Magnets, Latest Family History, Felker 4, Space Devices (Sp), MWE CV,....

For www.atomicprecision.com there were 33,085 hits, 21,663 files, 5,977 distinct visits and 0.429 gigabytes downloaded.

For www.upitec.org there were 3,802 hits, 2,353 files, 589 distinct visits and 0.057 gigabytes downloaded.

This gives a total of 115,113 hits, 64, 737 files, 18,424 distinct visits and 10.51 gigabytes

1-15 September

US Universities, Institutes and Similar.

Arizona, Augusta State University, Boston College*, Berkeley*, Boise State, Colorado, Cornell*, CSU Dominguez Hills, Dartmouth, Fayete Community College, Florida International, Georgia Tech., Harvard, Iowa State, Illinois, Kent State, Lane Community College, Missouri, New Mexico State, Rensselaer, Stevens Tech., Texas A and M*, UC Davis, UC Irvine, UC Riverside, Maryland*, Chapel Hill, Mayaguez Dominican Rep., Univ Texas El Paso, Toledo, Wisconsin Milwaukee, Wisconsin, Worcester Polytechnic Institute, West Virginia, Lawrence Livermore, Library of Congress, Gov. Santa Barbara, Duke Energy, IBM.

Europe and Asia Minor: Universities, Institutes and Similar.

Linz, Salzburg, KU Leuven, BAS Bulgaria, CERN, Geneva, Charles Univ. Prague, Max Planck Augsburg, Max Planck Heidelberg, Bocum, Siemens, TU Darmstadt*, TU Dresden, TU Harburg, Bonn, Hannover*, Konstanz*, Leipzig, Potsdam, Wuppertal, CSIC Spain, TSAI, UAM*, UCM, UPV, UV Spain, Spanish Parliament, JUY Finland, Tontut Finland, ENST Bretagne (Llydaw), HIVA, French Particle Lab., INRETS, Montpellier, Poitiers* France, AUTH Greece, RM CNR Italy, INFN Naples, Milan Tech., SISSA, Modena Italy, RUG Netherlands, TU Eindhoven, Utrecht, European Southern Observatory, Poznan, Corbina, Yandex Russia, BTH Sweden, Jozef Stefan Institute Slovenia, HCK Slovakia, OD Ukraine, Aberystwyth, Birmingham Central, Cardiff, Durham, Imperial, Oxford*, Swansea, Univ. College London, British Library*, Neath Port Talbot, Reading County Council, Hertfordshire GFL, Llyfrgell Cenedlaethol Cymru.

Rest of World: Universities, Institutes and Similar.

UNLP Argentina, Australian National University, C.O. C. Australia, MacQuarie, Armidale, Defence, UFC Brazil, Colval Quebec Canada, Triumf Canada, UBC*, Quebec Trois Rivieres*, PUC Chile, Univ Chile, UNAB Chile, UDEA, UNAL, Valle Colombia, Baidu China, IIT Karangpur India, Government India, AFRI, PRL India, Osaka, TCU, TUAT Japan, MRT Ceylon, IIE, UGTO, UNAM, USON Mexico, IBA, LHR-NU, New PU Pakistan, NTHU Taiwan*, VDC Venezuela, SUN SA, UFS, Witwatersrand South Africa*.

US Universities, Institutes and Similar.

Arizona State, Berkeley, Boise State, Boston University*, Butler, Claremont, Columbia, Cornell*, Dartmouth, Duke, East Carolina, Fay Tech CC, Florida International, Florida State, Geneseo, Gettysburg, Harvard*, Illinois, Kamehameha, Missouri, MIT*, Mount St. Mary's Maryland, Missouri Univ Science and Technology, North Dakota, New Mexico State, New York University, Pomona, Princeton, Penn State*, Purdue, Ruritan Valley, Rice*, Rose Hulman, San Jose State, Stanford, Stonybrook, SUNY Stonybrook, Texas A and M, Central Florida, UC Riverside, UC Santa Barbara*, Delaware, Florida, Iowa, Illinois Urbana Champaign, Maryland, Michigan*, North Carolina Charlotte, South Florida, Utah*, Texas*, Vanderbilt, Washington, Wisconsin*, West Virginia Wesleyan; Federal Aviation Authority, NASA (JPL and JSC), Honeywell, Northrop Grumman*.

Europe and Asia Minor: Universities, Institutes and Similar.

UMH Belgium, CERN*, Belfort, Geneva, DESY Germany, HU Berlin, KFA Juelich, Max Planck Mathematics Leipzig, Max Planck Heidelberg, Bochum, Siemens, TU Berlin, TU Darmstadt, Dortmund, Konstanz*, Paderborn, Wuppertal, TSAI Spain, Barcelona, UPC, Madrid, Vigo, Helsinki Finland, HTV, LUT Finland, HIVA France, French Particle Laboratory, OBSPM, Bordeaux 1, Monpellier, Marseilles, Poitiers, UPS Toulouse, NTUA Greece, INFN Naples, Milan Polytetchnic, TU Delft, Utrecht (students), UVT, National Univ Norway, Cleveland Clinic, FUW Poland, Poznan, Lisbon, TM Romania, UPT Romania, Belgrade, Corbina System Russia, Yandex system Russia, Pskov, KTH Sweden, LTH Sweden, Kharkov, Kiev, Poltava, Aberdeen, Aberystwyth, Birkbeck. Cambridge*, Cardiff, Colchester FC, Cymru, Edinburgh, Higher Education Council for Wales, Imperial, Oxford (chemistry and mathematics), Portsmouth, Swansea, Univ College London, Wales, Warwick, British Library*, Neath Port Talbot, National Health Service, Llyfrgell Genedlaethol Cymru.

Rest of World: Universities, Institutes and Similar .

MDP Argentina, UNLP, Government, Adelaide Australia, ANU, Monash, Melbourne*, Queensland, UFF, UNICAMP Brazil, Brandon Canada, Gov. Canada, McGill, McMaster, Colval Quebec, UBC*, Sakatechewan, Victoria, HA China, Eafit Colombia, UDEA, UNAL*, Andes*, IITM India, MNIT, BARC, IIT Karangpur, Gov. India, Ritsumeia, TUA tech, IIE Mexico, UNAM, UTHM, Waikato New

Zealand, SQU Oman, New Pakistan University, NUST, QAU, Peshawar, NCTU Taiwan*, NITU, Academica Sinica, Gove Taiwan, Fmed Uruguay, NWU, Unisam UP Witwatersrand South Africa.

Summary of October 2011

For www.aias.us there were 90,011 hits, 7.98 gigabytes downloaded, 14,326 distinct visits, 52,805 pages downloaded, 2,246 documents read from 102 countries, led by USA, Australia, Russia, Canada, Germany, Spain, France, Britain, ...
The most read UFT papers were: 194, 177, 25, 64, 41, 43, 150-B, 118, 107, 18, 123, 140(Sp), 8, 159(Sp), 149, 175, 4, 142 (Sp), 57, 94, 116, 81, ...
The most heard essay broadcasts were (by Robert Cheshire unless otherwise indicated): 3 (Light Deflection by Gravitation), Nobody is Perfect (MWE), 39, 6, 9, 38, 5, Nikola Who?, 43, 47(MWE), 41, 44(MWE), 42, 7(MWE), 35, 49(MWE), 18(MWE), 32,19, 24, 27(MWE), 50(MWE), Universe of Myron Evans Part 1, 37, 4(MWE), 48(MWE), 8, 36, 46(MWE), Nobody's Perfect, 45(MWE), 17(MWE), 12, 20(MWE), 34, 49, 28, 14, 16(MWE), 25(MWE), 26, 29, 10, 15(MWE), 22(MWE), ...
The most read articles and books were: Felker3 (Spanish), Prose Autobiography, Spacetime Devices (Sp), Overview of ECE Theory, CV of MWE, ...
For www.atomicprecision.com there were 32,216 hits, 0.49 gigabytes downloaded, 11,642 distinct visits and 5,471 distinct visits.
For www.upitrec.org there were 5,353, hits, 981 pages, 634 visits and 0.07 gigabytes downloaded.
This gives a total of 127,580 hits, 65,428 pages, 20,431 visits, 8.54 gigabytes downloaded.

1 - 15 October

US Universities, Institutes and Similar
Arizona*, Boston College, Caltech, Central Michigan*, Colorado, Columbia*, Florida International, Georgia Tech., George Mason, Harvard*, Iowa State, ICCS Karachi (on edu), Illinois, India University Purdue Indianapolis, McMurry, MIT, North Carolina State, North Dakota System, Old Dominion, Ohio State*, Princeton*, Rochester Tech., Renselaer, Roanoke, Spelman, Stonybrook, Anna Mendez, Texas A and M, Arkansas, Colorado Springs, Chicago, UC Irvine*, President of the University of California, UC Santa Barbara, Delaware, Florida, Houston, Iowa, Illinois Urbana Champaign*, Massachusetts, Maryland, Michigan,

Kansas City, New Mexico, Southern California, South Dakota, Texas, Vanderbilt, Vermont, Wisconsin, Western Washington, Yale, Yeshiva; Argonne, Brookhaven, NASA*, NIST, Virginia, DLA Military, NMCI Navy, General Electric, IBM, Northrop Grumman, Raytheon, Rockwell Collins, Aerospace Corporation, Ohio History Central.

Europe and Asia Minor: Universities, Institutes and Similar

Royal Society, TU Vienna, ULB Belgium, EPF Lausanne, ETH, Geneva, Zurich, CU Prague, Max Planck Heidelberg, Siemens*, TU Cottbus, Leipzig, Magdeburg, Stuttgart, Tuebingen, SDU Denmark, Aragon, TSAI Spain, Madrid Autonomous, UCLM, UCM, UCO, Canaries, Jaen, US, USAL, UV* Spain, CNRS Grenoble France, ENS Cachan, French National Particle Lab., ISAE, OBSPM, RAIN, SNECMA, Lemans, Poitiers, CYTA Croatia, FESB Croatia, BME Hungary, NUIM Ireland, JCT Israel, Tel Aviv, Pisa Italy, BAU Jordan, Australian College of Kuwait, Utrecht Netherlands, UV Amsterdam, UIO Norway, FUW Poland, Jagiellonian Krakow, UC Portugal, Tuaisi Romania, Bucharest, UBBCLUJ Romania, SBB Serbia, KINR Kiev Ukraine, Odessa, Aberystwyth, Birmingham Central. Bristol, Cambridge*, Univ. Wales*, Cymru, Keele, Lancaster, Leicester, Manchester, Newcastle upon Tyne, Nottingham, Southampton*, University College London, UMIST, British Library*, East of England Broadband Network..

Rest of World: Universities, Institutes and Similar

MDP Argentina, Gov. Argentina*, Monash Australia, RMIT, Melbourne*, Sydney, Western Australia, PUC Rio Brazil, UFMG, UNESP, Unicamp, Sao Paolo*, Concordia*, Mount Royal, Colvac Quebec, UBC*, Calgary, Quebec Trois Rivieres, Toronto, Waterloo, UZH Chile, UDEA Colombia, ULEAM Equador, CMI India, IITM, JNCASR, VIT, Government, IMSC, IRDE, Iwate Japan, Ritsumei, IPICYT Mexico, UAEM Mexico, UITM Malaysia, NU Singapore, NCTU Taiwan*, NTU, VDC Venezuela, NWU Zambia.

15 - 30 October

US Universities, Institutes and Similar

Al Quds Jerusalem (on edu), Arizona*, Auburn, Augsburg, Austin CC, California Lutheran, California Tech., Carleton, Claremont, McKenna, Central Michigan, Colorado State, Cornell, Covenant, Florida State, Galileo, Georgia Tech., Georgetown, Hamilton, Iowa State, Indiana, James Madison, Kansas, Lousiana, Maine, MIT*, Michigan State, NIIT, Old Dominion, Oklahoma, Portland State, Pittsburgh, Princeton*, Purdue, San Jose State, Stanford*, Stevens Tech., Texas A and M, Arkansas, Chicago,

UC Irvine*, UCLA, Connecticut, North Texas, Puerto Rico, Texas Brownsville and Southmost, Texas*, William and Mary, Vermont*, Washington*, Washington St. Louis*, Yale, Brookhaven, District of Columbia, Federal Energy Regulatory Commission, NIST*, Airforce AFNOC, Navy NAVO, General Electric, Honeywell, Hewlett Packard.

Europe and Asia Minor: Universities, Institutes and Similar.

Linz, Vienna, Siemens Austria, ETH Zurich*, Zurich, Bonn, TU-BS Germany, Freiburg, Heidelberg, Karlsruhe*, Leipzig*, Regensburg, Tuebingen, AU Denmark, SDU Denmark, BDE Spain, Madrid Autonomous, UCA, Ovi, UPM, USAL*, Helsinki, JYU Finland, ENSM France, French Particle Laboratory, Mobius, OBSPM, SRR, Paris 10*, UHP Nancy*, Paris Diderot, Paris 1, Poitiers, FESB Croatia, Atomki Hungary, BME Hungary, University College Dublin Ireland, ENI Italy, ICTP Trieste*, INFN LNF, INFN Milan, INFN Padua, INFN Rome 2, Bologna*, BAU Jordan, RU Netherlands, TU Eindhoven, UCT*, Free Univ. Amsterdam*, WAU, SQU Oman, Jagiellonian Krakow, Krakow, Wroclaw, Corbina Russia, Yandex, ER Russia, Troitsk, Jozef Stefan Slovenia, Boun Turkey, SDU, SIIRT Turkey, Kiev Ukraine, Poltava, Aberystwyth*, Aston, BBSRC*, Bristol, Cambridge*, Glasgow, Imperial*, Liverpool, NCL, Oxford, RDG, South, Swansea, NHS.

Rest of World: Universities, Institutes and Similar.

ITBA Argentina, Adelaide, ANU, Queensland*, Melbourne, Sydney, UFC Brazil, UFPR, UFS, UNESP, UNICAMP, Sao Paolo, British Columbia, Colval*, Queens, UBC, Calgary, Toronto*, Waterloo, UGM Indonesia, Karazawa Japan, Kyushu, Reitaku, Korea, SNU Korea, MRT Sri Lanka, UMICH Mexico, UITM Malaysia, FUUAST Pakisatn, LHR-NU, NUST, NU Singapore, NTU Taiwan, UPEL Venezuela, UCT South Africa.

PAPERS READ BY LEADING UNIVERSITY STAFF AND STUDENTS: Nov 5th to Oct 2011

Nov 5th: Bonn UFT88

Nov 4th: Jena UFT116, DNLB UFT41, Northwestern UFT143, Maryland Levitron, Canaries Essay 49, Simres Canada UFT150-B, Sheffield UFT2.

Nov 3rd.: Gov. Brazil UFT127, Perimeter Institute UFT64, EPF Lausanne UFT88, UCT Chile UFT41(Sp), Max Planck AEI UFT 103, Baylor UFT25, Caltech Essay 44, Montana UFT82, Northwestern UFT143, UIC UFT111, Washington UFT110, Canaries Felker2 (Sp), ENS

France UFT110, Rennes Proof 2, UFT76, Argonne proof 4, CAMK Poland Ed Note 2, UTL Porugal UFT176, Imperial UFT103, Lancaster UFT118.

Nov. 2nd: TU Berlin UFT 162, Kassel Numerical Solutions, DTU Denmark Infinite Solenoid, Georgia Tech UFT140, NOAO UFT18, UC Riverside UFT41, Washington UFT133, BGU Israel UFT152, Pune India Essay 24, Calabria Essay 4, TU Eindhoven UFT81, Heriot Watt UFT148, Nottingham UFT41, RDG UFT80, Warwick UFT41, Capetown UFT114.

Nov. 1st: Gov Brazil UFT42, trinity UFT157, Iowa Essay46, Texas UFT25, Navy NMCI UFT153, Free Univ Amsterdam UFT88, Bath UFT175,Bristol UFT30, CSX Cambridge UFT41, Quen Maey UFT104.

Oct. 31st: Colorado UFT99, JSU UFTT18, North Dakota UFT40, SIUE Refutation Heisenberg, UC Irvine UFT59, Kentucky UFT156, ENS France UFT2, Delhi UFT25, Delft UFT7, Chem Delft Numerical Note 2, Panama UFT159 (Sp), Gov. Peru UFT44, Lampeter poetry, new Colleeg Oxford UFT151, UCL UFT41.

Oct 30th: Gov. Brazil Essay 43, Arkansas UFT178.

Oct. 29th: Students Bonn UFT18,Kralsruhe UFT73, UFT190, UC Riverside UFT39, Iowa UFT149, NTU Taiwan Essay 27.

Oct 28th: Queensland DS, TDC Denmark UFT107, Galileo UFT169(Sp), NIIT UFT158, UC Riverside UFT42, Minnesota Twin Cities UFT18, French Aprticle Lab ECE (French), SIIRT Turkey UFT76, BBSCR Britain UFT166, UPELV Venezuela Essay 24.

Oct. 27th: Linz UFT175, ETH UFT88, Auburn UFT158, California Lutheran UFT41, Vermont UFT116, Washington St Louis UFT99, ENSMP France UFT28, Brookhaven UFT110, INFN Padua Overview, INFN Rome 2 UFT114, Bologna UFT18, MRI Sri Lanka UFT166, LHR-NU Pakistan UFT151, Palestine Authority UFT25, Swansea Blog, Capetown Refutation of Heisenberg.

Oct. 26th: HMI Germany UFT174, TU-BS Germany Essay 15, Heidelberg UFT165, Al Quds Jerusalem Table; Caltech UFT81, Covenant UFT18, Georgia Tech Overview, Maine UFT146, Stevens Proof 1, U Mass UFT143, UNCC UNCC saga4, FERC US Gov. UFT25, Hyderabad UFT150-B, Free Univ Amsterdam UFT136, Krakow UFT32, Boun Turkey UFT101, Abeystwyth CV and Album 12 photographs. Queen's Cambridge UFT10.

Oct. 25th: UFC Brazil UFT104, Zurich UFT41,Tuebingen UFT175, Ecuador UFT140(Sp.), Chemistry Cornell Essay 11, Georgetown 't Hooft rebuttal; Chicago UFT41, Connecticut extensive UFT191; Florida UFT81, Minnesota UFT44, Iovi Spain UFT122, JUY Finland UFT175, NIST UFT41, Univ College Dublin Educational Note 2, ENI Italy UFT94, Marwan Morroco UFT19, UMICH Mexico Felker4 Spanish.

Oct. 24th: Queensland UFT158 and Essay 15, Gov Ecuador UFT170, Carleton Proof 1, Central Michigan UFT81, Old Dominion UFT175, UHP

Nancy UFT157 and UFT158, Wroclaw Poland HSD27, Aberystwyth CV and Essay20, CSX Cambridge UFT10, King's College Cambridge UFT10, Imperial UFT158, Oxford UFT175.

Oct. 23rd: Vienna UFT101, Sydney UFT25, Old Dominion UFT175, Minnesota UFT178, new Mexico Essay 11.

Oct. 22nd: Vienna UFT50, 54, 55, 63, 12, Hamilton UFT148, Princeton UFT41, NUST Pakistan UFT141 and UFT143, Capetown UFT166.

Summary of November 2011

For www.aias.us there were 73,439 hits, 5.794 gigabytes downloaded, 13,679 visits, 40,552 page views from 102 countries, led by US, Australia, Russian Federation, Germany, Mexico, France, Britain, ...

All UFT papers were read, led by 25, 54, 140, 41, 150B, 4, 140(Sp), 43, 166, 175, 159(Sp), 18, 107, 116, 11, 166(Sp), 88, 169, 13, 149, 110, 8, 142(Sp), ...

All broadcasts heard, led by 3, Nobody is Perfect, 49, 50, Universe of Myron Evans, 8, 47, 24, 30, 41, 4, 42, 9, 26, 17, 25, 45, 40, 48,37, 7, 44, 6, 35, Intro UFT139, Nicola Who?, 27, Nobody is Perfect, 39, 12, 20, 10, 11, 14, 16, 18, 21, 22, 33, 31, 38, 13, 15, 19, 23, 32, 43, 5, 29, 34, 46, 28, ...

All articles and books read, led by Felker5 (Sp), Autobiography, CV, ...

For www.atomicprecision.com there were 34,844 hits, 6,393 visits, 0.41 gigabytes and 13,037 pages.

For www.upitec.org there were 5,369 hits, 532 visits, 877 pages, 0.07 gigabytes.

The total was 113,652 hits, 20,604 distinct visits, 54,466 pages and 6.27 gigabytes

1 - 15 November

US Universities, Institutes and Similar

Arizona, Baylor, Berkeley, California Lutheran, Caltech*, Colorado State, Columbia, CSU Long Beach, Georgia Tech., Hampshire, Maine, Millersville, MIT, Montana, North Carolina A and T, North Carolina State, New Mexico State, National Optical Astronomy Observatory Kitt Peak, Northwestern, New York University, Ohio State, Oregon State, Occidental, Princeton, Penn State*, Purdue, Rutgers, Trinity, Texas Tech*, Colorado Denver, California Irvine, Connecticut, California Riverside, California San Diego, Indonesia (edu system), Illinois Chicago, Idaho,

Iowa, Illinois Urbana Champaign, Maryland, Michigan, Minnesota, North Carolina Charlotte, US Naval Academy, Texas Arlington, Texas, Utica, Kentucky, William and Mary, Washington*, Wisconsin, Washington St Louis Missouri, Yale, Youngstown State Ohio; House of Representatives, NASA, National Research Council, Argonne*, Army SWA, Department of Defense High Performance Computing Modernization, Navy NMCI, General Electric, Lockheed Martin, Raytheon, Argonne Guest House, American Welding Society, Mariemont Schools, Mid Hudson Regional Information Center, School District of Philadelphia, Trinity Lutheran School, Bend, Oregon.

Europe and Asia Minor: Universities, Institutes and Similar.

EPF Lausanne, Bern, Freiburg, Charles University Prague, Bosch Company, DRK Freiburg, Albert Einstein Institute, RWTH Aachen, Siemens Company*, TU Berlin*, Bonn, Erfurt, Hannover*, Jena, Karlsruhe, Kassel, Cologne, Leipzig*, Osnabrueck, Regensburg, Tuebingen, Danish Technical University, Copenhagen, Basque University, Spanish Ministry of the Environment (MAPA), Spanish Ministry of Health, Univ Barcelona Autonomous, Barcelona*, Juan Carlos Madrid, Spanish National Univ of Internet Education, Canaries*, Valencia Polytech., Seville, Valladolid*, Vigo, Helsinki, Jyvaskila, Tampere Tech Finland, Ecole Nationale Superieure* (ENS), ENS Cachan, ENS de Technique (ENSTA), French National Particle Laboratory (In2p3), Paris Observatory (OBSPM), Polytechnique de France*, Bordeaux 1, Poitiers, Rennes 1, Savoie, National Technical Univ. of Athens (NTUA), Ioannina University Greece*, Zagreb, Trinity College Dublin, Ben Gurion, Weizmann, INFN Turin, Milan Tech., Calabria, Radboud Netherlands, TU Eindhoven*, Free Univ. Amsterdam, BKBB Norway, CAMK Poland, Wroclaw, Lisbon Tech., Belgrade, Institute of Nuclear Research Russian Academy, Atomlink Russia, RSU Russia, yandex system Russia, Corbina system Russia, Lund Sweden, Gazi Turkey, Iyte Turkey, Marmara Turkey, Aberystwyth, Bath, Birmingham*, Bristol, Brooklands, Christ's Cambridge, CSX system Cambridge, Jesus College Cambridge, PWF system Cambridge, Selwyn College Cambridge, Univ Wales Registry, Durham, Edinburgh, Glasgow*, Hull, Heriot Watt, Imperial*, King's College London, Lancaster, Leicester, Luton, Nottingham*, Materials Oxford, Physics Oxford, Queens Oxford, Wadham Oxford, Portsmouth, Queen Mary, Reading, Hull, Sheffield, St Mary's University College Belfast, Strathclyde, Sussex, University College London, East Anglia, University of London Senate House Library, Warwick.

Rest of World: Universities, Institutes and Similar.

UNSA Argentina, Adelaide, Australian National University*, Flinders, Sydney*, UFMG Brazil, AEI Canada, Dalhousie, McGill, Netcampus Canada, Government of Ontario, Perimeter Institute, Alberta, British Columbia, Calgary, Quebec Trois Rivieres, Toronto, Waterloo, York University Toronto, PUC Chile, UNAL Colombia, Government of Colombia, UO Cuba, INF, OHC Cuba, Rmed, MICSF Cuba, Government of Ecuador, Bhabha Atomic Research Centre India, IOPB India, Tokyo, Waseda, SNU South Korea, UGTO Mexico, CIIT-WAH Pakistan, National University of Singapore*, Educational Institute Hong Kong, National Taiwan University, TCCPM Taiwan, African Institute for Mathematical Sciences Capetown, Central University of Technology Orange Free State, Nelson Mandela Metropolitan University Port Elizabeth, Capetown University, Witwatersrand, UC Venezuela.

15 - 29 November

US Universities, Institutes and Similar

Alaska, Arizona*, Arizona State, Austin, Biola, Brockport, Buffalo, Brigham Young, Central Michigan, Colorado, Colorado State, Columbia, Cornell*, Drexel, Harvard*, Illinois Institute of Technology, Millsaps, MIT, Monmouth College, New York University, Oberlin, Old Dominion, Ohio State, Oklahoma*, Pittsburgh, Princeton*, Purdue*, Regis, Rice, Renselaer, Rochester, Rutgers*, Stillman, UC Irvine, UC Los Angeles, Delaware, Girona (edu), Florida, Georgia, Michigan, New Mexico, South Carolina, Texas Arlington, Kentucky, Washington St. Louis, Yale*, Gov. California, Mecklenburg County, US Dept Agriculture, US Give High Performance Computing Modernization, Argonne Guest House, US Archives San Francisco, American Welding, Houston Independent School District, International Labour Organziation (UN Conference), Mid Hudson Information Center, BAE Systems (US), General Electric, Northrop Grumman.

Europe and Asia Minor: Universities, Institutes and Similar.

Vienna, ETH Zurich, Freiburg, FZU Czechia, TSE Czechia, VSPJ Czechia, VUTBR Czechia, Bosch, German Synchrotron, German Red Cross Freiburg, Max Planck MIS, Siemens*, Goettingen, TU Berlin, TU Darmsatdt, Bayreuth, Bonn, Giessen, Heidelberg, Karlsruhe, Leipzig*, Regensburg, Tuebingen, Ulm, Aarhus Denmark, Kollegienet Denmark, CSIC Spain, Andalucia Government, Barcelona, Juan Carlos III Madrid, Granada, Murcia*, Univ Miguel Hernandez, Santiago de Compostela, Valencia, Valladolid, Vigo, JYU Finland, Academie Lyon, ENST Bretagne, Paris Observatory, Poitiers*, Rennes1, Savoie, AUTH Greece,

NTUA Greece, UOI Greece, AMIS Croatia, Jerusalem College of Technology, International Centre for Theoretical Physics Trieste, SISSA Italy, Modena*, Pisa, CSS Group Latvia*, AMOLF Netherlands,CWI Netehralnds, NIKHEF Netherlands, TU Delft, Tilburg Univ., NTNU Norway, UIB Norway, Polish Academy Wroclaw, Coimbra Portugal, Technical Univ Lisbon, UAIC Romania, Academic Network of Serbia, INR Russia, yandex system Russia, Lund Sweden, Uppsala Sweden, Ljubljana Slovenia, Ege Turkey, Lviv, Odessa Ukraine, Aberystwyth, Brunel, St Catherine's Cambridge, Churchill Cambridge, Darwin Cambridge, CSX Cambridge, Newnham Cambridge, Sidney Sussex Cambridge, Selwyn Cambridge, Durham, Edinburgh, King's College London, Lancaster, Leeds, Manchester, Newcastle upon Tyne, Nottingham*, Keble Oxford, Lincoln Oxford, Magdalen Oxford, St Edmund Hall Oxford, Queen's Univ. Belfast, Southampton Solent, Southampton*, Sussex, University College London, York*.

Rest of World: Universities, Institutes and Similar.

UNLP Argentina, CNEA Gov. Argentina, Adelaide, Marquette, Sydney, Western Sydney, Gov. Brunei, LACTEC Brazil, UFF, UFMG, UFPA Brazil, Laurentian Canada, McGill, Perimeter Institute, Queens, Simon Frazer, Alberta*, British Columbia*, North British Columbia, Quebec Trois Rivieres, Toronto, Waterloo, NJU China, UNAL Colombia, Gov. Colombia, OHC Cuba, IED Hong Kong, ITB Indonesia, IITKGP India, Gov. India*, PRL India, Keio Japan, Titech, Tokyo Univ. Science, IPN Mexico, ITESM, UNAM* Mexico, UITM Malaysia, Otago New Zealand, New Univ Pakistan, KKU Thailand, KMITL Thailand, National Taiwan Univ., UDSM Tanzania, UC Venezuela.

Summary of December 2011

For www.aias.us there were 95,927 hits, 14.18 gigabytes downloaded, 15,334 visits, 60,938 page views, 2.372 documents read from 106 countries, led by US, Britain, Ukraine, Australia, Germany, Mexico, Netherlands, France, ...

All UFT papers read repeatedly, led by: 25, 43, 177, 200, 140, 41, 54, 150B,88, 175, 166,116, 118, 142(Sp), 57, 159(Sp), 107, 18, 8, 94, 149, 140(Sp), 104, 100, 20, 66, ...

All Broadcasts listened to repeatedly, led by: Deflection of Light by Gravity, Nobody is Perfect, 45, 43, 7, 42, 5, 38, 4, 41, 46, 19, 32, 49, 40, 27, 48, 33, Nikola Who?, 44, 30, 24, 30, 9, 36, 6, 20, 8, 50, 14, 23, 34, 35, 37, IntroUFT139, 28, 12, 15, 18, 21, 22, 25, 17, 11, 16, 29, 10, 13, 24, Nobody's Perfect, ...

Books and Articles read repeatedly, led by: UNCC Saga 4, Autobiography volume 1, My CV, ...

For www.atomicprecision.com there were 36,956 hits, 5,997 visits, 12,486 pages downloaded, 0.54 gigabytes.

For www.upitec.org there were 4,883 hits, 815 pages, 553 visits, 0.07 gigabytes.

The total was 137,755 hits, 21,884 visits, 74,239 pages, 14.79 gigabytes, 106 countries.

1 - 15 December

US Universities, Institutes and Similar

Arizona*, Biola, Caltech*, Calvin, Central Michigan, Colorado State, Columbia*, Cornell*, California State Univ Long Beach, Santa Barbara City College, Duke, Emory, Fort Hays State Univ., Florida State*, Georgia Tech., Georgia Southern, Hawaii, Harvey Mudd, Kent State, MIT*, Michigan State*, North Carolina State, New Mexico Institute of Mining and Technology, Northwestern, Old Dominion, Princeton*, Penn State, Purdue*, Reed, Rice, Southern Illinois Edwardsville, San Jose State, Stanford, Syracuse, Texas A and M, Trinity College, Reed, Rice, Arkansas, UC Irvine, UCLA, UC Santa Barbara, UC Santa Cruz, Illinois Chicago, Michigan*, Montana, Chapel Hill, North Texas, Oregon, Utah, Texas*, Toledo, Washington, Wheaton, Wisconsin*, Brookhaven, Fermilab, National Oceanographic and Atmospheric Administration, US Airforce AFNOC, US Army MARDEC, PAC, PICA, US Navy NMCI; IBM EMEA, Lockheed Amrtin, US Archives San Francisco, Austin and Houston Independent School Districts.

Europe and Asia Minor: Universities, Institutes and Similar

TU Graz, Ghent, CERN, EPF Lausanne*, ETH*, Geneva, Charles Univ Prague, DC Gymnaisum Czechia, Masaryk Univ., German Synchrotron Facility, Red Cross Freiburg, FH Furtwangen, FH Trier, FH Stralsund, Fraunhofer, Max Planck Heidelberg, Bochum, Siemens*, TGZ Ilmenau, TU Darmstadt, TU Munich, Bielefeld, Bonn, Hannover, Karlsruhe, Leipzig*, Mainz, Munich, Oldenburg, Potsdam, Rostock, Tartu Estonia, Aragon, Junta Andalucia, Murcia, Canaries, Vigo, Valencia, CBS CNRS France, ECP France, INP Grenoble, AMN Jussieu, Poitiers*, Rennes 1, Savoie, Tours, Ioannine Greece, Hebrew University Jerusalem*, Weizmann, Arcetri Observatory Italy, CNR Italy, Trento, SISSA Trieste, Rome 1, Barlaues Gymnasium Amsterdam, INS Eindhoven, TU Eindhoven, UJK Poland, Szczecin Poland, Coimbra Portugal, Minho Portugal, UBBCLUJ Romania, UPJ Romania, Corbina system Russia, JINR Russia, Gov el Salvador, Ankara Turkey, Kiev Uhraine, Bath,

Bristol, Oxford Brookes, Cambridge, Cranfield, Durham, Glamorgan, Heriot Watt, Imperial, Lancaster, Luton, Jesus College Oxford, MRI Centre Oxford Univ., Royal College of Paediatrics, Royal Holloway, Southampton, Strathclyde, Sussex, Wyggeston and Queen Elizabeth School Leicester, York University *, British Library*, Llyfrgell Cenedlaethol Cymru (National Library of Wales).

Rest of World: Universities, Institutes and Similar
   Jijel Algeria, UNLP Argentina, Toshiba Company Australia, Autsralian National, Chief Sceintific Advisor Australia, South Queensland, Victoria, Unicamp Brazil, Sao Paolo, Acadia Univ Canada, Gov. British Columbia, Brock, Dalhousie, McGill, Robarts Institute, Univ Alberta*, Univ British Columbia, Calgary, Manitoba, Moncton, Quebec Trois Rivieres, Victoria, High Energy Physics China, UNAL Colombia, IGCAR Gov. India, IUAC India, Kanazawa Univ Japan, Kyoto*, Tokyo, UEC Japan, Chiba, Japan Atomic Energy Agency, National Incident Management System Japan, Chonnam South Korea, UGTO Mexico, Otago New Zealand, UNP Peru, AU Pakistan, FUAAST Pakistan, NCP, UET, UMT Pakistan, HEC Gov. Pakistan, NU Rwanda, Chula Thailand, NCTU, NTU Taiwan, Tlabs South Africa, Capetown.

15 - 30 December

US Universities, Institutes and Similar
   Boston University *, Columbia, Florida State, Georgia Tech., Harvard Smithsonian, Northwestern, Ohio State, Princeton, Purdue, Stanford, Stevens Tech., Towson, UC Irvine, Florida,
   U Mass Lowell, Nebraska Lincoln, South Carolina, Texas, Wisconsin, Brookhaven, Fermilab, Los Alamos, US Airforce AFNOC, US Army AMRDEC, US Navy NMCI, General Electric, Intel, Lockheed Martin, Michigan Aerospace, Northrop Grumman, US Archive Organization, Institute for Defense Analyses, Northwestern University Libraries.

Europe and Asia Minor: Universities, Institutes and Similar
   Karl Francis University Graz, Liege, Leuven, CERN, ETH Zurich, Czech technical University Prague, Mazaryk University, German Synchrotron Facility, Red Cross Freiburg, Max Planck Heidelberg, RWTH Aachen, TU Darmstadt, Erlangen, Leipzig*, Mainz, Oldenburg, Tuebingen*, Spanish Ministry of Defence, Barcelona, Murcia, Polytechnic University of Catalonia, Polytechnic University of Madrid, Seville, Vigo, Valencia, Astrophysics Institute of Paris, Mairie Rueil Malmaison, Noopsis Mining Solutions France, Poitiers*, University College Dublin, Hebrew University Jerusalem, Tel Aviv University, Weizmann Institute*.

Helsinki, INAF Lambrate Italy, CSS Group Latvia, Daugavpils University Latvia, ONS Student net Netherlands,

Tilburg University Netherlands, Oslo, Royal Society of Chemistry, Pilicka, Szczecin, Coimbra Portugal, Aveiro Portugal, Corbina system Russia, Debryansk, National Nuclear Research University Moscow, Slovenian Technical University Bratislava, HUN Turkey*, Iyte*, OMU Turkey, Bristol, Brookes, Unix central Univeristy of Cambridge, Trinity College Cambridge, King's College London, Manchester, Southampton, Strathclyde, British Library*, Defence Science and Technology Laboratory.

Rest of World: Universities, Institutes and Similar

INTA Government of Argentina, Australian National University, UFMG Brazil, UFPA Brazil, Queen's University Canada, Univ British Columbia, Laval, Montreal, Quebec Trois Rivieres, Waterloo, Providencia Chile, Sernageomin Chile, Talcahuano Chile, Kanazawa Univ. Japan, Kyoto University, Tsukuba University, Japan Atomic Energy Authority, Korea University, Postech, National Univ of Peru, CIIT Pakistan, NCP, NUST, UET Pakistan, Tzing Hwa University Taiwan, National Taiwan University*, Academica Sinica Taiwan, Capetown University.

Summary of January 2012

For www.aias.us there were 93,176 hits, 15,501 visits, 51,581 pages downloaded, 7.715 gigabytes, from 96 counties led by US, Australia, Germany, Mexico, Ukraine, France, Britain, ...

For www.atomicprecision.com there were 41,125 hits from 6,533 visits, 13,120 pages downloaded and 5.84 gigabytes.

For www.upitec.org there were 5,955 hits from 558 visits, 910 pages downloaded and 1.21 gigabytes.

This gives a total of 140,256 hits from 22,592 distinct visits, 65,611 pages views and 14.77 gigabytes downloaded.

All UFT papers studied, led by 25, 19, 43, 17, 41, 203, 4, 205, 54, 140, 88, 76, 142, 140 (Spanish), 166, 107, 204, 75, 8, 1, 18, 37, 116, 110, 118, ...

All essay broadcasts heard, led by 3 (Light Deflection by Gravity), Nobody is Perfect, 5 (Criticisms of the University of Wales), 44, 25, 48, 41, 46, 43, 40, 42, 19, 34, 45, 37, 30, 35, 6, Universe of MWE Part 1, 38, 18, 8, 24, 49, 4, 12, 14, 29, 32, 9, 17, 16, 22, 23, 27, 33, 28, 10, 13, 21, 11, 15, 20, 26, 36, ...

All Article by colleagues read led by Felker 3 (Spanish), My Autobiography, Johnson Magnets, Levitron, Filtered Statistics, My CV, Felker 2 (Spanish), Spacetime Devices, ...

US Universities, Institutes and Similar.

Berkeley, Illinois, Indiana, Marshall, Colorado School of Mines, Missouri, MIT, Pittsburgh, Scripps, Southern Illinois, Stonybrook, SUNY Stonybrook, Texas A and M, Universities Corporation for Atmospheric Research, UC Davis, UC Riverside, UC Santa Cruz, Delaware, Florida, Ibadan (on edu), University of North Carolina Charlotte, New Mexico, Wesleyan, William and Mary, Western Washington, Brookhaven, NASA NDC, US Airforce AFNOC, US Army AFO, US Nave Marines Inernet, US Navy SPAWAR, IBM EMEA, Intel*, Microsoft, US Archives, City of Palo Alto, Fountain Fort Carson School District, Progress Place Organization, St Pauls Lutheran School Orange California, Worthington Libraries, State of Georgia, National Safety Technologies.

Europe and Asia Minor: Universities, Institutes and Similar.

Vienna*, Belgian Institute for Space Aeronomy, CERN, EPF Lausanne, Zurich University of the Arts, German Synchrotron Facility*, Red Cross Freiburg, FU Berlin, Technical University Schorndorf, Siemens, Student and Church Halls of Residence Bonn, TU Berlin, TU Darmstadt, Bielefeld, Hamburg, Hannover, Karlsruhe, Cologne*, Leipzig, Siegen, Niels Bohr Institute Copenhagen, IFAE Spain, Barcelona Autonomous, Madrid Autonomous, Madrid Complutensian, Zaragosa, Salamanca*, Valencia, TUT Finland, VTT Finland, CNRS Grenoble, Institute of Physics Grenoble, Paris Psud, Poitiers, Rennes 1. Tours, Aristotle University Greece, ATT School System Greece, Commissioner for Irish Lights, Hebrew University Jerusalem, Tela Aviv University, Weizmann Institute, INAF Lambrate, INFN Bari, SISSA Trieste, Milan, Padua, Stockholm University in Riga, Free University of Amsterdam, Tilburg, European Archive Paris and Amsterdam, Faculty of Physics Univ Warsaw, University of Mining and Metallurgy Krakow, Technical Univ Lisbon Portugal, Babes Bolya and Cluj Napoca Romania, Univ Novom Sadu Serbia, National Nuclear Research University Moscow, University of Dubna, yandex and corbina systems Russia, Royal Institute of Technology Stockholm, DEU Turkey, Meti Univ Turkey, UCW Aberystwyth, Brighton, Astronomy Cambridge, St Catherine's College Cambridge, Churchill College Cambridge, Chemical Engineering Cambridge, Unix central system Cambridge, Engineering Cambridge, UW Cardiff*, City College Sunderland, Daresbury Laboratory, Durham*, King's College London, Luton, Manchester, Newcastle upon Tyne, Queen's Univ Belfast*, UW Swansea, University College London, York, British Library, Royal Society, Richard Rose Foundation.

Rest of World : Universities, Institutes and Similar.

Melbourne, IMPA Brazil, UFC Brazil, Federal Univ Rio, Gov British Columbia Canada, Dalhousie, McMaster, Pontifical Univ Valparaiso Chile, Univ Valle Colombia, ITB Indonesia, UNDIP Indonesia, RCIL Gov India, Tata Institute India, Kyoto, Nitech, Osakafu Japan, New Pakistan Univ, Old Pakistan Univ., UAF Pakistan, Chula Thailand, NCTU, NCU, NTU Taiwan.

15 - 30 January

US Universities, Institutes and Similar

Boston Architectural College, Boston University, Buffalo, Caltech*, Claremont, Columbia, Case Western Reserve, Fairmont State, Florida Institute of Technology, Georgia Tech., ICCS Karachi (on edu), Indiana, Louisiana State, MIT*, Milwaukee School of Engineering, North Carolina State University, National Institute of Technology India (on edu), New York University, Ohio State, Pasadena, Princeton, Purdue*, Rowan Cabarrus Community College, Southern University and A and M, Syracuse, Texas A and M, UC Davis, UC Santa Barbara, UC Santa Cruz, UC San Diego, Univ Minnesota Twin Cities, UNC Chapel Hill, UNC Charlotte, Northern Iowa, New Mexico, Pennsylvania, Utah, Texas, NASA Jet Propulsion Technology, Pacific Northwest National Laboratory, 24th US Airforce Network Operations Centre Lackland Base, US Airforce Kirtland, US Army AMRDEC, US Navy NMCI, Lockheed Martin, Northrop Grumman, US Archives San Francisco (supported by NSF, Library of Congress and others), Champaign Schools, Sacramento Library, Grace Elementary School District, Upper Merion School District Pennsylvania.

Europe and Asia Minor: Universities, Institutes and Similar

Free Univ. Brussels, Catholic Univ. Leuven, Ghent, CERN, EPF Lausanne, ETH Zurich, German Synchrotron Facility, Red Cross Freiburg, KFA Juelich, RWTH Aachen, Siemens*, TU Chemnitz, TU Darmstadt, Bonn, Hannover*, Karlsruhe, Leipzig, Stuttgart, Autonomous Barcelona, Complutensian Madrid, Malaga, Hernandez de Elche, National Spanish Distance Education University, Madrid Polytechnic, Zaragoza, Valladolid, Spanish Xunta (Parliament), HUT Finland, JYU Finland, IFPEN France, French National Particle Laboratory, INSA Toulouse, Pierre et Marie Curie, Polytechnique de France, Poitiers, Rennes 1, HCMR Greece, NTUA Greece, OSSKI Hungary, PTR Ireland, Technion Israel, ICTP Trieste, INFN Padua, Turin Polytechnic, RP Engineering Italy, Florence, Siena, KTU Latvia, CSS Latvia, University of the Hague, TU Delft, TU

Eindhoven, UIO Norway, AGH Poland, Lodz, Monho Portugal, CEC Romania, UBBCLUJ Romania, UICCLUJ Romania, MSU Russia, RIKT Russia, yandex Russia, KACST Saudi Arabia, Laser Physics KTH Stockholm, Kiev*, Bristol*, Churchill College Cambridge, Applied maths and physics Cambridge, St John's College Cambridge, Newnham College Cambridge, Durham* Greenwich, Imperial*, Lancaster, Manchester*, Nottingham, Physics Oxford*, Wadham College Oxford, Rutherford Appleton, St Andrews, Swansea, Isle of Wight Council.

Rest of World: Universities, Institutes and Similar

UNL Argentina, Sydney Australia, FAJ Brazil, UERJ, UFPR, UFSC Brazil, McGill Canada, MUN Canada*, Perimeter Institute, Alberta, Calgary, Laval, Montreal, Quebec Trois Rivieres, Toronto, Waterloo*, Univ. Chile, UCN, UFRO Chile, Engineering Univ. Colombia, Espol Ecuador, RCLIL, RRCAT, IMSC, RRI India, IMS Japan, Titech, Fukui, Tokyo, ITESM Mexico, New Pakistan, Old Pakistan, Quest Pakistan, KACST Saudi Arabia, Chula Thailand, NCTU, Academica Sinica*, Taipei Taiwan, Limpopo South Africa, UFCTC South Africa.

Summary of February 2012

For www.aias.us there were 85,040 hits, 14,110 distinct visits, 41,352 files downloaded, 5.741 gigabytes downloaded, 2462 documents read from 94 countries led by USA, Mexico, Germany, Austria, France, Britain, Australia, India,...

For www.atomicprecision.com there were 34,423 hits from 5,871 visits, 11,351 pages downloaded and 4.71 gigabytes.

For www.upitec.org there were 5,993 hits from 714 visits, 1,092 pages downloaded and 1.01 gigabytes.

This gives a total of 125,456 hits from 20,695 distinct visits, 53,795 pages downloaded and 11.46 gigabytes.

All UFT papers studied, led by 25, 94, 4, 43, 166, 150-B, 140, 88, 140, 41, 107, 18, 54, 13, 206, 37, 159(Spanish), 12, 116, 166(Spanish), ...

All essay broadcasts heard, led by 3 (Light Deflection by Gravity), Nobody is Perfect, 27, 49, 48, 5, 43, 6, 9, 28, 8, 29, 44, 41, Nikola Who?, 4, 24, 25, 22, 33, 34, 36, 42, Universe 1, 35, 20, 23, Nobody's Perfect, 12, 32, 17, 10, 11, 13, 14, 15, 16, 19, 21, 30, 37, 38, 45, ...

All documents read led by Felker3 (Spanish), Autobiography, 2D by Doug Lindstrom, Felker4 (Spanish), Space Energy Devices (Spanish), Etherington Report, Johnson Magnets, CV. ..

US Universities, Institutes and Similar

Berkeley, Buffalo, Caltech, Carnegie Mellon, Colorado, Columbia, City Univ. New York, Case Western Reserve, Duke*, Fitchburg State, Harvard*, Illinois Purdue Indianapolis, Lehigh, Maine, Colorado School of Mines*, Missouri Univ. Science and Technology, North Carolina A and T, North Carolina State, North Dakota*, Northwestern, New York University, Oklahoma Christian, Ohio State, Penn State, Rochester, St Peter's College, Stanford, Texas A and M, Navarra Spain (on edu), Texas Tech, Central Florida, UCLA, Connecticut, UC Santa Barbara, Florida, Houston, Indiana Urbana Champaign, Louisiana Monroe, U Mass., Michigan, Missouri Kansas City, Minnesota, Chapel Hill, Utah*, Washington, Western New England, US Gov. BOP, US Navy NAVO, Naval Research Laboratory, Dupont, Ford, General Electric, Raytheon, US Archives* (Library of Congress, NSF, and others), Berkeley Library.

Europe and Asia Minor: Universities, Institutes and Similar.

KF Graz, Vienna, Antwerp, German Bund (Government), Red Cross Freiburg, Free University Berlin, KFA Juelich, TU Chemnitz, TU Darmstadt, Bielefeld, Erlangen, Frankfurt*, Karlsruhe, Leipzig, Marburg, Saarland, Ulm, Aarhus Denmark, Danish Technical University, Tartu Estonia, Gov. Canaries, Autonomous Barcelona, Autonomous Madrid, Barcelona, Complutensian Madrid, Granada, Leon, Polytechnic Catalonia, Polytechnic Valencia, Valencia, European Government, Helsinki, Oulu Finland, Turun Finland, French Atomic Energy Commission, University of Paris Central, French National Particle Laboratory, French Mining Solutions (Noopsis), French National Synchrotron Laboratory, Bordeaux 1, Strasbourg, Poitiers, Democritus Thrace Greece, Aristotle Thessaloniki, Crete Technical, Nansen Institute Norway, University of Oslo, Budapest Technical and Economics, International Centre for Theoretical Physics Trieste, INFN, Bari, INFN Milan, Insubria, Padua, Royal Society, Government of Poland, Warsaw, Minho Portugal, Lisbon Technical, Cluj Napoca Romania, Corbina system Russia, yandex system Russia, Gov. Irkutsk, Uppsala Sweden, Keiv Ukraine, Odessa Ukraine, Aberystwyth, Corpus Christi Cambridge, central unix system Cambridge, Jesus College Cambridge, Cardiff, Edinburgh, Glamorgan, Heriot Watt, Lancaster, Brasenose College Oxford, Christchurch College Oxford, Queen's Belfast, University College London, UMIST, Queen's College London Independent School, Essex Schools, National Library of Wales, North Yorkshire Schools,

Rest of World: Universities, Institutes and Similar.

ITBA, UNC, UNSA Argentina, Australian National, Griffith, Sydney, Gov. British Columbia Canada, Brock, McGill*, Memorial Newfoundland, Perimeter, Quebec Trois Rivieres, Toronto, Victoria, Waterloo*, UDEA Colombia, UNRI Indonesia, Raman Research Institute India, Gfu, Nagoya, NIFS Tokushima Japan, UAM, UNAM Mexico, Auckland New Zealadn, AU, LHR-NU, New Pakistan, Chula Thailand, AIMS, UL, UWC, VUT South Africa.

15-29 February

US Universities, Institutes and Similar.

Arizona, Baylor, Berkeley, Caltech, Colorado*, Colorado State, Cornell, CSU Long Beach, CSU Monterey Bay, Eastern Oregon, Georgia Tech, SUNY Geneseo, Harvard*, MIT, Illinois, Johns Hopkins, Minnesota State Mankato, Michigan State, National Institute of Technology Tiruchirappali India (on edu), Ohio State, Mississippi*, Oregon State, Oklahoma, Rowan, Stanford, UC Merced, UC San Diego, Houston, Idaho, Illinois Urbana Champaign, Maryland, Michigan*, Minnesota, new Hampshire, Pennsylvania, Washington St Louis, Yale*, US Govt. NOAA, US Department of Agriculture, US Navy NSWC, US Airforce Materiel Command, Dupont, Ford, General Electric, Intel, Northrop Grumman, Raytheon, US Archives*, Berkeley Public Library, Champaign Schools, Brigham Young, Iowa State, New Mexico Tech., Purdue, Central Florida, UCLA, US Army Fort Sam Houston, Microsoft.

Europe and Asia Minor: Universities, Institutes and Similar

ATT Austria, STH Austria, TU Vienna, Virtual Campus Graz, KU Leuven, ETH Zurich, Geneva, Cyprus, Czech Academy of Sciences, Charles Univ. Prague, Masaryk Univ., BESSY Germany, Red Cross Freiburg, Fraunhofer Institute, IFH, Jacobs, Albert Einstein Institute, Siemens, TU Berlin, TU Chemnitz, TU Darmstadt, Erlangen, Karlsruhe, Kiel, Leipzig*, Tuebingen, Heidelberg, Wuerzburg, Aarhus, Danish Technical University, Aragon, Andalucian Gov., Madrid Autonomous, Madrid Complutensian, Murcia, Valencia Polytech., Valencia, Univ. Basque Country, Helsinki, Oulu, Bordeaux, Paris Psud (IEF and LPS), Lyon 1, INSERM Toulouse, Poitiers, Aristotle Greece, IRB Croatia, ICTP Trieste, INFN Naples, Florence, Milan, Pisa, RAC Netherlands, Free Univ. Amsterdam, UMB Norway, IMA Poland, Torun, UM Warsaw, Technical Univ. Lisbon*, City of Novosibirsk, CN Russia, Corbina system, IHC Russia, Moscow State University, Krasnoyarsk, Troitsk, KAU Saudi Arabia, Umea Sweden, Uppsala, Jozef Stefan Slovenia, Iyte Turkey, Kiev, students Aberystwyth, Christ's Cambridge, unix system Cambridge,

Durham, Greenwich, Imperial, King's College London, Lancaster, Leicester, Manchester, Manchester Metropolitan, University College Oxford, Merton, St Edmund Hall Oxford, Somerfield, Wadahm Oxford, Queen's Univ. Belfast, Reading, Southampton, St Andrew's, Trinity St David, Central Lancashire, North Yorkshire, Royal Society of Chemistry, City of Bratislava..

Rest of World, Universities , Institutes and Similar.

Abu Dhabi Investment Authority, GBA Gov. Argentina, Australian National University, Sydney, DU Bangla Desh, UGJH Brazil, UFPB, Sao Paolo, AEI Canada, Simon Frazer, Calgary, Laval, Manitoba, McGill, New Brunswick, Quebec Trois Rivieres, Toronto, Waterloo*, UDLA Colombia, UNAL Colombia, BUH India, RVRJCCE India, IGCAR, RCIL India, Tokyo, Japanese Particle Lab. (KEK), KAIST, Postech, Sogang Korea, UNAM Mexico, UEM Mozambique, Auckland, PUCP Peru, LUMS Pakistan, New PU, PIEAS, QAU, IUB, UET Pakistan, National University of Singapore, NWU South Africa, Univ. of South Africa, UNP, UWC, Witwatersrand South Africa.

Summary of March 2012

For www.aias.us there were 102,598 hits from 18,817 distinct visits, 12.97 gigabytes downloaded, 68,253 page views, 2,503 documents read from 102 countries, led by US, Argentina, Britain, Australia, Germany, Canada, Mexico, ...

All UFT papers read, led by 25, 43, 4, 41, 166, 54, 159(Spanish), 16, 88, 166(Spanish), 177(Spanish), 209,........

All essays heard, led by: 3 (Deflection of Light by Gravity), Nobody is Perfect, 5 (Reform of the University of Wales), 46, 8, 37, 24, 9, 28,29, 32, 6, 14, 4, 35, 4, 48, Nobody is Perfect (RC), 16, 17, 20, 25, 7, 34, 50, 38, 47, 26, 64,40, 36, 45, 33, 10, 18, 61, 11, 62, 15, 27, ,63, 60, 21, 43, 59, 42, 19, 22, 23, 52, 30, 41, 58, 39, 53, 54, 44, ...

All Articles and books read, led by Felker3 (Spanish), Etherington Report, Space Energy Devices, Johnson Magnets, Autobiography, Spacetime Devices, Levitron, Space Energy, ...

For www.atomicprecision.com there were 37124 hits, 12,462 page views, 6,493 visits and 4.86 gigabytes.

For www.upitec.org there were 6.702 hits, 700 visits, 1,101 page views and 1.00 gigabytes downloaded.

This makes a total of 146,424 hits, 26,010 visits, 81,816 page views and 18.83 gigabytes downloaded.

US Universities, Institutes and Similar.

Arizona, Arizona State, Boston University, Caltech*, Catholic Healthcare West, Central Michigan, Colorado, Cornell, California State University Long Beach, Georgia Tech*, Harvard*, Hawaii, Harvey Mudd, Kansas State, King Saud (on edu)*, MIT*, New Mexico State, North Dakota University system, Northwestern*, Naval Postgraduate School, Oberlin, Oklahoma, Pacific Lutheran, Sinteg Leska, Ana Mendez, Southwest Research Institute, Chicago, UC Irvine, UC Santa Barbara, Florida*, Georgia, Ibadan (on edu), Illinois Chicago, Illinois Urbana Champaign, Incarnate Word, Maryland, Missouri Kansas City, Nevada Lincoln, Texas Arlington, West Florida, Vanderbilt, Virginia, Washington, Wisconsin, Western Washington, US Government NOAA, Clayton State, US Courts, West Virginia, US Airforce AFNOC, US Airforce Kirtland, Army Research Laboratory, US Army Mepcom, US Military Disa, US Navy NMCU*, Intel, Microsoft, US Archives*, Baldwin County Public Schools, Chicago Public Library, Institute of Electric and Electronic Engineers.

Europe and Asia Minor: Universities, Institutes and Similar.

JKU Austria, EPF Lausanne, VSPJ Czechia, German Aerospace Centre, Red Cross Freiburg, Albert Einstein Institute, TU Braunschweig, TU Darmstadt, TU Munich, Goettingen, Hannover, Karlsruhe, Leipzig, Paderborn, Rostock, Danish Technical University, College Net Denmark, Talin Estonia, Cartif Spain, INTA Spain, Gov. Andalucia, RENFE, Autonomous Barcelona, Complutensian Madrid*, Cordoba, Granada, Autonomous Madrid, Miguel Hernandez de Elche, Canaries, Polytechnic Madrid, Polytechnic Valencia, Salamanca, Valencia, Helsinki*, TPU, UTU Finland, CHU Caen(, CNRS Orleans, French Observatory, Lille 1, Poitiers, Rennes 1, Savoie, Bcenter Hungary, DIMA Hungary, Univ Transylvania Romania, Tel Aviv, SISSA Italy, Trieste, Milan, PFI Lithuania, MPE Latvia, Physics Leiden, TU Eindhoven, HIN Norway, Lodz Poland, Corbina Russia, Novosibirsk, yandex system, Uppsala Sweden, VRG Sweden, Anadolu Turkey, Iyte, Aberystwyth, Bristol, Brookes, Durham*, Edinburgh*, Glasgow, Imperial*, King's College London Students' Union, Liverpool John Moores, Keele, Lancaster, Leeds, Luton, Manchester, Newcastle, Nottingham*, Lady Margaret Hall Oxford, NSMS Oxford, Physics Oxford*, Queen's Belfast, Swansea, British Library, Orkney Council, NHS, Ysgol CCC, Surrey, Royal Society of Chemistry.

Rest of World: Universities, Institutes and Similar.

MDP Argentina, UNLP Argentina, ANU Australia, Sydney*, Defence Asutralia, McGill*, OCDSB, Gov. Quebec, Queen's, Alberta, British Columbia, Victoria*, Biobio Chile, UCN Chile, UDEA Colombia, UNAL Colombia, IITK India, IITM India, BARC India, TUT Iran, Tohoku Japan, Japanese National Particle Lab (KEK), KAIST* Korea, Korea, UGTO Mexico, Auckland New Zealand, Otago, UCT South Africa, UNP South Africa.

15 - 30 March

US Universities, Institutes and Similar.

Arizona*, Arizona State, Berkeley, Buffalo, Caltech, Carleton, Central Michigan*, Colorado*, Cornell, Fox Valley Technical College, Hampshire College, Iowa State, Kansas State, Karlsruhe Institute of Technology (on edu), Kansas State, Maine, MIT*, Minnesota State Colleges and Universities, Missouri University of Science and Technology, New Mexico Tech., Ohio State, Oregon State, Oklahoma, Penn State, Reed, Stonybrook, Texas A and M, Florida, Illinois Chicago,

Michigan*, Minnesota, Catalonia Polytechnic on edu, South Carolina, Wyoming, Washington, Wayne State, Wesleyan, Wheaton, Wisconsin*, Brookhaven, District of Columbia, Food and Drug Administration, NASA, National Institutes of Health*, NOAA, Dept. Of Agriculture, Virginia, Missile Defense Agency, General Electric, International Education Corporation, Lokheed Martin, Northrop Grumman*, Raytheon*, US Archives, Baltimore Police, IEEE, Methacton School District, Kentucky Schools, Peak Organization.

Europe and Asia Minor: Universities, Institutes and Similar

Austrian Institute of Technology, Linz, Graz, UACG Bulgaria, Sofia, CERN*, VSPJ Czechia, Red Cross Freiburg, Max Planck Institute for Physics Munich, Siemens, TU Darmstadt*, Hannover, Leipzig*, Muenster, Saarland, Niles Bohr Institute Denmark, Talin Estonia, Autonomous Madrid, Barcelona, Complutensian Madrid*, Cordoba*, Polytechnic Madrid, Juan Carlos, Santiago de Compostela, Valencia, Spanish Parliament, TPU Finland, ENS Lyon, IMAG, Jussieu, Caen, Limoges, Poitiers, Rennes 1, Tours, Toulouse, UPS Toulouse, UTC; JCT Israel, Weizmann, Iceland, INFN Pavia, rome1, Rome2, Milan Polytechnic, Modena, Siena, Turin, Trento, MPE Latvia, TU Eindhoven, Utrecht, Oslo, Lodz, Torun, PK, Gdansk, UVT Romania, Corbina Russia, Albanova Sweden, LTH, LTU, UU Sweden, NLB Slovenia, NTS Serbia, Anadolu Turkey, Iyte, Metu, Kiev, Odessa, Aberystwyth, Brooklands, Brunel, Cranfield, Edinburgh, Heriot Watt, Leicester, Nottingham*, Hertford

College Oxford, Oxford University, Queen Mary London, Sheffield*, Swansea, NHS, Royal Society, Royal Society of Chemistry.

Rest of World: Universities, Institutes and Similar

UNC Argentina, Gov. Argentina, Gov.Australia, UFAL, UFPB, UFPR, UFRN Brazil, Acadia, McGill, Queen's, Alberta, British Columbia*, Calgary, Trois Rivieres*, Saskatchewan, Sherbrooke, Waterloo, IIA Chile, Biobio, UCN, UCV Chile, XJTU China, UNAL, UDEA Colombia, IIAP India, IUT Iran, Kure-NCT Japan, Sony*, KAIST Korea, UTHM Malaysia, NCP Pakistan, New Pakistan, Chula Thailand, NCTU, NCU,NTHU Taiwan, SUN, UNP South Africa.

Summary of April 2012

For www.aias.us there were 91,842 hits from 16,052 distinct visits, 10.02 gigabytes downloaded, 59,937 page views, 2,491 documents read from 106 countries, led by US, Mexico, Italy, Germany, Britain, France, Australia, Japan ......

All 216 UFT papers read led by: 43, 25, 94, 4, 107, 88, 159(Sp), 54, 158, 166, 177(Sp), 177, 140 (Sp), 142, 175 (Sp), 170 (Sp), 142 (Sp), 169 (Sp), 140, 18, 39, 41, 116, 165 (Sp), 169, ...

All 65 essay broadcasts heard led by: Light Deflection by Gravitation, Nobody is Perfect, 7, 28, 44, 6, 9, 29, 32, 34, 47, 5, 48, 25, 45,46, 50, 35, 54, 8, 27, 49, 56, 42, 51, 58, 62, 64, 41, 53, 60, ...

Documents led by: OO461, OO594, OO595, OO472, OO446, OO593, Felker 3 (Sp), OO591, OO462, Space Energy (Sp), Human Soul Jackson, Autobiography, Nikola Who broadcast by Robert Cheshire, ...

For www.atomicprecision.com there were 38,381 hits, 6,180 visits, 11,661 page views, and 4.76 gigabytes downloaded.

For www.upitec.org there were 6,589 hits, 660 visits, 1,165 page views and 1.05 gigabytes downloaded.

This gives a total of 136,812 hits, 22,892 visits, 72,763 page views and 15.83 gigabytes downloaded.

1 - 15 April

US Universities, Institutes and Similar.

Andrews, Arizona State, Berkeley*, Bryn Mawr, Boston University*, Colorado*, Columbia, Cornell*, California State University Fresno, Depaul, Eastern New Mexico, Florida State University, Georgia Southern, Harvard*, Humboldt, Loyola, Maine, Miami, Colordo School of Mines, MIT*, Montana, Medical University of South Carolina, New York University, Ohio State, Oklahoma, Pittsburgh, Plymouth, New York

Polytechnic, Princeton, Penn State, Purdue, Redwoods*, Rose Hulman, Rutgers, Southeast Missouri State, Stonybrook, SUNY Stonybrook, Southwest Research Institute, Texas A and M*, Tufts, Arkansas, Central Florida*, Chicago, UCLA, Illinois Chicago, Iowa, Michigan*, Nevada Lincoln, South Dakota, Toledo, Virginia, Vermont, West Florida, Western Washington, Navy Spawar, NASA JPL, NASA NDC, US Gove Parcel Service, Virginia, General Electric, Lockheed Martin, US Archives*, Berkeley Public Library, Lee Academy, Peak, Peertech.

Europe and Asia Minor: Universities, Institutes and Similar.

Royal Society of Chemistry, EPF Lausanne, Zurich, SLU Czechia, JYU Finland, German Synchrotron Facility, Humboldt, IKB, Albert Einstein Institute, RWTH Aachen, Stuttgart, Wuppertal, AENA Spain, CSIC Spain, EHU Spain, Autonomous Barcelona, Autonomous Madrid, Barceloan, Complutensian Madrid, Iovi, Valencia, CHI Elbeuf Louviers, ESPCI France, INSA Lyon, Polytechnique, Nice, Poitiers, Tela Aviv, Technion, INFN Florence, Trieste, Rome 1, Turin, TU Eindhoven, UVT Romania, Amres Serbia, UNS Serbia, MTS-NN Russia, Corbina, yandex, HCK Slovakia, Association of Commonwealth Universities, Bristol, Cambridge*, Cardiff, Exeter, Imperial*, Luton, Newcastle, Oxford, Sheffield Hallam, Stranmillis, University College London, University of East Anglia, Warwick.

Rest of World: Universities, Institutes and Similar.

UNLP Argentina, UNT, Australian National, UNESP Brazil, McGill*, Perimeter Institute, Simon Frazer*, Saskatchewan, Waterloo, Quebec Trois Rivieres, TIE Chile, Univ. Chile, UCV, UTA Chile, UNAL Colombia, UGM Indonesia, Hokudai, Tohoku, Tokaha, Tokyo, Sony*, KAIST Korea, INAOEP Mexico, FJWU Pakistan, LUMS, QAU, National Univ. Rwanda, SQU Oman, National Taiwan Univ., AIMS South Africa.

15 - 29 April

US Universities, Institutes and Similar

Andrews, Arizona State, Bowdoin, Boston University, Central Michigan, Colorado, Cornell*, Case Western Reserve, Florida State, Georgia Tech., Harvard, Harvey Mudd, Johns Hopkins, National Aerospace Ukraine on edu (KHAI), Kansas State, Lousiana Community and Technical College System, Lehigh, Loyola New Orleans, Mercer, Miami, MIT, Michigan State*, North Carolina State, NIT Poducherry India (on edu), Northern Kentucky, New Mexico Tech., North Dakota System, Northwestern, New York University, Ohio State Medical Center, Oklahoma, Pittsburgh, New York Polytechnic, Penn State, Rutgers, St

Olaf, Stonybrook, Texas A and M, Tufts, Central Florida, Chicago, University of California San Diego, U Mass, Minnesota Twin Cities, Chapel Hill, UNC Charlotte, Nevada Las Vegas, Nebraska Lincoln, U Penn., Utah*, Texas, Virginia, Washington, Yale, Bureau of Prisons, National Research Council, Army Stuttgart, US Coast Guard, Northrop Grumman, Raytheon, US Archives*, Lee Academy.

Europe and Asia Minor: Universities, Institutes and Similar.

TU Graz, Vienna, Linz, Siemens Austria, KU Leuven, Morris Chapman Belgium, Ghent, CCUT Czechia, HU Berlin, Albert Einstein Institute, Ruhr Bochum, RWTH Aachen, Siemens*, STW Bonn, TU Darmstadt, Karlsruhe, Leipzig, Magdeburg, Muenster, Tuebingen, Ulm, CSIC Spain, Gov. Anlalucia*, Ministry of Defence Spain, Autonomous Barcelona, Autonomous Madrid, Complutensian Madrid, Iovi, Polytechnic Madrid, Valencia, Spanish Parliament, JYU Finland, ENS Lyon, INSA Lyon, LPTHE Jussieu, UJF Grenoble, Lille 1, Marseilles, Paris 13, Poitiers, South Britanny, AUTH Greece, UC Dublin, Hebrew Jerusalem, Tel Aviv, Csi Italy, ENI Italy, INFN Bari, Milan Polytech, Modena, Utrecht, Free Amsterdam, NTNU Norway, UIB Norway, UTL Russia, St Petersburgh Russia, Chalmers Sweden, Joseph Stefan Slovenia, Novosibirsk Avademy of Sciences, Iyte Turkey, Metu, Royal Society of Chemistry, European Trade Union, Abeyrystwyth , Birmingham Central, Cambridge*, Durham, Edinburgh*, Imperial*, Lancatse, Leicester, Luton*, Manchester*, Nottingham, Chemistry Oxford, Oriel College Oxford, VPN Oxford, Sheffield, Southampton, Stranmillis, Warwick, Emannuel CTC Tyneside.

Rest of World: Universities, Institutes and Similar

ITBA Argentina, Citefa Gov. Argentina, ITBA Argentina, Latrobe Australia, Monash*, RMIT Australia, UFRGS Brazil, Rio de Janeiro, Sao Paolo, Banff Centre Canada, Canadian Gov. Nuclear Safety, McGill, Laval, Quebec Trois Rivieres, Toronto, Waterloo, UCN Chile, UDEC, UFRO, UTA, UTFSM Chile, UDEA Colombia, UNAD, UNAL, Ustatunja, Jijel Algeria, Ambou Ethiopia, ULEAM Ecuador, IGCAR Gov. India, RCIL Gov. India, Keio Japan, Tohoku, Japanese Advanced Science Organisation (RIKEN), Korean Advanced Science Institute*, UGTO Mexico, UNAM*, Auckland New Zealand, Otago*, CIIT Lahore Pakistan, LUMS Pakistan, New PU, KYU Taiwan, NTU* Taiwan, Kwa Zulu Natal South Africa.

END OF EIGHT YEARS OF DAILY RECORDING

Summary of May 2012

For www.aias.us there were 92,084 hits, 14.07 gigabytes downloaded, 18,527 distinct visits, 58,718 page views, 2,512 documents read from 99 countries led by US, Germany, Ukraine, Brazil, Mexico, France, Britain, Canada, Australia .....

For www.atomicprecision.com there were 43,760 hits, 13,697 page views, 7,361 visits, 5.08 gigabytes.

For www.upitec.org there were 5,847 hits, 1,010 pages, 698 visits, 0.95 gigabytes.

This gives a total of 141,691 hits, 73,425 pages, 26,586 visits, 20.1 gigabytes downloaded.

All UFT papers studied, led by: 25, 94, 177(Sp), 166, 41, 88, 4, 142, 198, 216, 159(Sp), 177, 54, 170(Sp), 56, 107, 157(Sp), 38, 45, 30, 149, 217, 140, 43, 175(Sp), 9, 116, 150B,196, 2, 110, 174, 65, ...

All essays read and broadcasts heard, led by 3, 50, 32, Intro139, Nobody is Perfect, 48, 29, 46, 49, 24, 6, 38, 8, 45, 34, 44, 35, 36, 62, 7,37, 4, 40, 56, 5, 53, 9, 27, 28, 30, 10, 12, 20, 47, ...

All other material read, led by: Felker3 (Spanish), Spacetime Devices, Felker5 (Sp), Levitron, Autobiography volume one, LCR Resonant, CV, Space Energy, Galaxies (Sp), Triple Orbit Simulation, Marquis Who's Who 2007, ...

1 - 15 May

US Universities, Institutes and Similar

Arizona, Brandeis, Bryn Mawr, Boston, California Polytech., Caltech, Colorado*, Cornell, Drexel, Florida State, Geneseo, Indiana, Indiana Purdue Indianapolis, Kansas, Macon State, Colorado School of Mines*, MIT*, Northern Arizona, Northwestern, Naval Postgraduate School, New York, New York Polytechnic, Princeton*, Penn Statae, Purdue, Reed, Rider, Texas A and M, Thiagaraj Engineering India (on edu), Temple, Texas Tech., Univ Chicago*, UC Irvine, UC Riverside, UC Santa Barbara*, UC San Diego, Illinois Chicago, Illinois Urbana Champaign, U Mass., Maryland, North Carolina Charlotte, Nevada Las Vegas, Pompeu Fabra (on edu), Texas*, Washington*, Wisconsin, Xavier, Brookhaven, NOAA Alaska Weather Service, US Archives*, Grand Island Public Schools Nebraska, Mesquite Independent School District, General Electric, Lockheed Martin, Raytheon

Europe and Asia Minor: Universities, Institutes and Similar

Vienna, Virtual Campus Graz, EPF Lausanne, ETH Zurich*, Zurich, Brno, FH Duesseldorf, Max Planck Berlin History of Science, Ruhr Bochum, RWTH Aachen, Siemens, TU Darmstadt, Freiburg, Marburg*, Regensburg, Stuttgart, South Denmark, ENDESA Gov. Spain, Gov. Canaries, Autonomous Madrid, Complutensian Madrid*, Granada, National Distance Education University, Canaries, Polytechnic Madrid, Valencia*, French Atomic Energy Authority, French National Synchrotron Facility, French National Particle Laboratory, Paris Observatory Meudon, Poitiers*, National Technical University Athens, ATT Greece, Eotvos Lorand Hungary, Hebrew University Jerusalem*, INFN Milan, Milan Polytechnic, Trieste, Florence, Modena, Perugia, MII Lithiania, Leiden, Radboud, Groningen, TU Delft, Royal Society of Chemistry, Wroclaw, NIPNE Romania, NI Serbia, TLT Russia, Vologda, Chalmers Sweden, Karolinska Institute, Lund, Stockholm, Uppsala*, Istanbul Technical University, Metu*, Kiev, Lviv, Aberdeen, Aberystwyth, Bristol, Cambridge, Jesus College Cambridge, Cardiff, Durham*, Glasgow*, Heriot Watt, Imperial, Leeds, Luton, New College Oxford, Sommerfield College Oxford, Reading, Southampton, Strathclyde, Neath Port Talbot Council.

Rest of the World: Universities, Institutes and Similar.

MDP Argentina, Adelaide, Charles Sturt, Monash, South Australia, Melbourne*, Western Australia, Sydney, PSI Brazil, UFG, UFMG, UFPB, UNB, Unicamp Brazil, Dalhousie, EC Gov Quebec, Perimeter Institute, British Columbia, Quebec, Quebec Trois Rivieres Saskatchewan, Waterloo, Central Chile, Univ. Chile, UFRO, Mayor, UTA, UTFSM Chile, Tsinghua* China, UDEA Colombia, National Univ Colombia*, University of Technology Colombia, RCIL India*, IMSC, Physics Research Laboratory India, Shahed Iran, Keio Japan, Osaka, Tokyo, RIKEN Gov. Japan, National Advanced Institute Japan, Korean Advanced Institute of Science and Technology*, UGTO Mexico, Uruguay, Cape Town, Pretoria, Office of the President of South Africa.

15 - 30 May

US Universities, Institutes and Similar.

Bryn Mawr, Boston, Buffalo, Caltech, Carleton, Chapman, Colorado, Cornell, Dartmouth, Drexel, Duke, Florida Atlantic, Georgia Tech., Georgetown, Harvard, Hawaii, Middlebury*, Montana, Morgan, New Mexico State, Oberlin, Ohio State*, Ohio, Princeton, Renselaer, Santarosa, Stanford, SUNY Stonybrook, Texas Southern, Texas Tech., Chicago, UC Santa Cruz*, Houston, Illinois Urbana Champaign, Minnesota*, North

Carolina Charlotte, New Hampshire, Washington, Brookhaven, Lawrence Livermore, NASA, US 24th Airforce AFNOC, US Army Strategic Command, US Army HUA, US Navy NMCI, US Archives*, DCBOCES Poughkeepsie, Freemont Union High School, J Sterling Morton High School, Montgomery County Public Schools, University Medical Center of Princeton, Conoco Phillips, Dupont, Hewlett Packard Houston, Lockheed Martin, Northrop Grumman, Philips, Wolfram.

Europe and Asia Minor: Universities, Institutes and Similar.

CERN, ETH, Bern, Zurich, Charles University Prague, UPCE Czechia, Upol Czechia, Bavarian State Library Munich, German National Synchrotron Facility, Fraunhofer, KFA Juelich, MAN Engineering*, Max Planck Plasma Physics, Max Planck SB, Students Bonn, TU Dresden, Freiburg, Hamburg, Jena, Regensburg, Aarhus* Denmark, Aarhus Hospitals, Argon, Astrophysics Canaries, Autonomous Madrid, Barcelona*, Murcia*, Spanish Distance Education University, Polytechnic Valencia, Santiago de Compostela, Valencia*, Free University, Valladolid, Spanish Parliament, European Union Parliament, INSA Lyon, SNECAM France, FComte, Poitiers*, AUTH Greece, CYTA Greece, ICTP Trieste, INFN Naples, INRIM Italy, Univ Trieste, Rome 2, Udine, Ancona, Groningen, UIN Norway, Polish Academy of Sciences, Silesian University, Porto. PUB Romania, AMRES Serbia, UNS Serbia, JINR Dubna, LU Sweden, Uppsala, Istanbul Technological University, Metu*, Kocaeli, Aberystwyth, Cambridge*, Darwin College Cambridge, St John's College Cambridge, Wolfson College Cambridge, Cardiff, City University London, Durham*, Edinburgh, Exeter, Imperial*, Manchester*, Oxford, New College Oxford, University College Oxford, Queen's University Belfast, Reading, Sheffield, Southampton, North Yorkshire Schools.

Rest of World: Universities, Institutes and Similar.

UNLP Argentina, UTN Argentina, Monash*, Melbourne, Queensland, Sydney, RBA Gov Australia, Gov South Australia, UFPB Brazil*, UFPE Brazil, UFRJ, UFSM, Unicamp, Sao Paolo*, AEI Canada, Perimeter Institute, British Columbia*, Quebec Trois Rivieres, CECS Chile, Central, UCN, UFRO, URA, Tsinghua China*, Andes Colombia, Valle, Gov. Colombia, Esoch Ecuador, UGM Indonesia, IMSC India, TIFR India, Nagaoka, Tohoku, KAIST, UGTO Mexico*, USON, MMU Malaysia, Otago, PUCP Peru*, AIOU Pakistan, New University Pakistan, NUST, PIEAS, QAU, NCTU Taiwan, NCU, NFU, National Institute of Physics Uruguay, Gov Uruguay, ORT Uruguay, UC Venezuela, Rhodes Univ South Africa, Northwestern University South Africa.

Summary of June 2012

For www.aias.us there were 80,162 hits from 16,130 distinct visits, 52,419 page views, 8.625 gigabytes downloaded, 2,564 documents read from 101 countries, led by US, Britain, Ukraine,   Mexico, Canada, Russia, Argentina, ...
For www.atomicprecision.com there were 32.105 hits, 9,336 page views, 4,736 visits and 3.58 gigabytes downloaded.
For www.upitec.org there were 5,690 hits, 1,082 page views, 694 visits, 0.89 gigabytes.
This gives a total of 117,957 hits, 62,837 page views, 21,560 visits, 13.1 gigabytes.

All UFT papers read, led by 25, 88, 18, 41, 4, 107, 159(Sp), 43, 63, 157(Sp), 177, 140(Sp), 29, 175(Sp), 94, 155(Sp), 170, 165(Sp), 2, 140, 33, 166, 142(Sp), 146(Sp), 142, 15, 30, 37, 42, 149, 57, 7, ...
All Essays heard, led by: 3 (Light Reflection by Gravitation), 49 (Mythical Higgs Boson), 37, 45 (Future of General Relativity), 6, 38, Nikola Who?, 25, 35, 32, 50, 8, 28, 46, 48, 5, 47, 43, 7, 44, paper 139 introduction, 29, 24, 30, 53, 9, 51, 62, 41, 15, 16, 54, 56, 57, 58, 59, 13, 18, 26, 39, 4, 40, 42, 52, 44, 41(Sp), ...
All articles and books read, led by: Felker3 (Sp), LCR Resonant, Autobiography, Sapce Devices (Sp), ...

1-16 June

US Universities, Institutes and Similar.
Bowling Green State, Boston University, Cornell*, California State Bakersfield, Case Western Reserve, El Camino, Emory, Georgetown, Hawaii, Indiana, Morgan, Michigan Technological, NITT India (on edu), New Jersey Institute of Technology, New York University, Ohio State, Rose Hulman, San Diego Supercomputer Center, Trinity College Hartford, Chicago, UC Irvine, UCLA, Connecticut, UC Santa Barbara, UC Santa Cruz*, Florida, Iowa, Maryland, Michigan, Tennessee Knoxville, Vermont, Washington, William and Mary, U. S. Gov. Homeland Security, Library of Congress, NOAA, US Airforce AFNOC, US Archives*, Colonial Hills Christian School Georgia, BAE Systems, Dupont, Honeywell, Lockheed Martin.

Europe and Asia Minor: Universities, Institutes and Similar.
Innsbruck Medical Austria, Linz, Leoben, Leuven Belgium, CERN, EPF Lausanne, Belfort Switzerland, UPCE Czechia, ZCU; Fraunhofer Germany, Bielefeld*, Goettingen, Heidelberg, Jena, Karlsruhe, Oldenburg,

Stuttgart, Ulm, WH Stuttgart, Aarhus Denmark, Gov. Andalucia, Gov. Canaries, Autonomous Barcelona, Autonomous Madrid, Complutensian Madrid*, Granada*, Spanish Distance Learning University, Malaga, Cartagena Polytechnic, Valencia; IMAG France, French National Particle Laboratory, IRCAM, ISAE, ANN Pierre et Marie Curie LIRMM, SORHEA, Nice, Limoges, Poitiers, Zagreb, NUIM Ireland, Hebrew University Jerusalem, INFN Florence, INFN Ferrara, INFN Rome 2, Milan Polytechnic, Siemens Italy, Bologna*, Trieste, Oslo Norway, Jagiellonian Krakow, Warsaw University of Medicine, Technical University Lisbon Portugal, Serbian Academy, EPS Serbia, Russian Academy of Sciences Institute of Nuclear Research, Metu University Turkey*, Cambridge*, King's College Cambridge, Churchill College Cambridge, Carnegie College Fife, Cranfield, Eastleigh, Imperial*, Manchester, Magdalen College Oxford, Sheffield, British Library.

Rest of World : Universities, Institutes and Similar.

MDP Argentina, UNL, MECON Gov. Argentina, Australian National, Queensland, Victoria Gov., UFPE Brazil*, Rio de Janeiro, UFU*, Sao Paolo*, CST Bhutan, UCDSB Ontario Canada, SSHA, Trois Rivieres, Toronot*, CECS Chile, UDEC, UFRO* Chile, HA China, UNAD Colombia, UNAL* Colombia, ULEAM Ecuador, RCIL Gov. India, IITK Inida, IIT Karanpur, IMSC India, Kurume NCT Japan, Tohoku*, Toshiba Corporation, Torikyo, Korea Advanced Institute*, Postech South Korea, National Autonomous Mexico, INAOEP Mexico, Otago New Zealand, Auckland University of Technology, Auckland, UNP Peru, Oxfam Mexico, MAPUA Philippines, QAU Pakistan, PERN Pakistan, KKU Thailand, NTHU Taiwan, NTU, UC Venezuela, Capetown South Africa.

15 - 29 June

US Universities, Institutes and Similar

Arizona*, Armstrong, Caltech, Drexel, Florida State, Georgia Tech., Georgia State, Illinois State*, Indiana, Johns Hopkins, Macon State, Michigan State, North Carolina State, Notre Dame, NITT India (on edu), New Jersey Institute of Technology, New York University*, Ohio State*, new York Polytechnic, Southern Illinois, Stanford, Chicago, UC Irvine, Connecticut, UC Santa Barbara, UC San Diego, Florida, Medicine and Dentistry New Jersey, New Mexico, North Texas, Pennsylvania, South Carolina, Vermont, Washington*, Wisconsin, Wright, Brookhaven, US Homeland Security, Fermilab, Lawrence Livermore, NASA, Department of State, Navy Oceanographic, General Electric, Lockheed Martin, Northrop Grumman*, Siemens, US Archives, Marshall Medical.

Europe and Asia Minor: Universities, Institutes and Similar.

Education Group Linz Austria, Louvain, IMEC Belgium, KU Leuven, Bulgarian Academy of Sciences, CERN*, EPF Lausanne, ETH Zurich, Fraunhofer*, Humboldt Berlin, KFA Juelich, Student net Bonn, TU Berlin*, TU Darmstadt, TU Dresden, TI Ilmenau, Bielefeld, Bonn, Heidelberg*, Leipzig, Paderborn, Saarbrucken, Stuttgart*, Tuebingen, Aragon, Autonomous Barcelona*, Almeria, Autonomous Madrid, Barcelona, Polytechnic University of Catalonia, Santiago de Compostela, Valencia, Vigo, Spanish Parliament, KTL Finland, French National Particle Lab., Marseilles 3, Informatics Lab. Paris 6, Limoges, Broest, Poitiers, Trinity College Dublin, Hebrew University Jerusalem, WeizmannInstitute*, Ghislieri College Italy, Insubria, Padua, Perugia, Leiden, ONS Student net Netherlands, Radboud, Tabor College, TU Eindhoven, Free Amsterdam, Bergen, Oslo, Jagiellonian Krakow, Gdansk, Lublin, Warsaw, UTL Serbia, EPS Serbia, Gov. Serbia, NN Russia, Yaroslavl, Russian National Nuclear Institute Moscow, NS Russia, Stockholm, Uppsala, NLB Slovenia, Ljubljiana, Kiev, Bristol, King's College Cambridge, Edinburgh, Glasgow, Imperial*, Lancaster, Magdalen College Oxford, Reading.

Rest of World: Universities, Institutes and Similar.

MDP Argentina, UNLP, UNR, GBA, INTI, Mecon Gov., Argentina, Latrobe Australia, RMIT, Western Australia, UFMG Brazil, UFPA, UFRJ, USP* Brazil; AEI Canada, Memorial Newfoundland, Simon Fraser, British Columbia, Laval, Quebec Trois Rivieres, Toronto, Waterloo*, UCT Chile, UTFSM*, UDEA* Chile, UNAL Colombia, RRI India, Kobe, Tohoku*, Tokyo*, KAIST Korea, Postech Korea, UGTO Mexico, UNAM*, USON* Mexico, New Pakistan University, NUST, Upesh Pakistan, NTU Taiwan*, NCHC Taiwan, UCU Uganda, Physics Institute Uruguay, Capetown.

Summary of July 2012

For www.aias.us there were 78,948 hits, 16,877 distinct visits, 50,427 pages downloaded, 7.40 gigabytes, about 2,600 documents read from 99 countries, led by: USA, Italy, Ukraine, Germany, Canada, Mexico, Britain, Australia, ...

All UFT papers read, led by 218, 193, 183, 25, 197, 177, 223, 63, 19, 157(Sp), 166(Sp), 222, 159(Sp), 177(Sp), 43, 13, 88, 150B, 2, 41, 4, 140, 172, 208, 116, 144(Sp), 61, 28, ...

All essays heard or read, led by: 3, 49(Sp, pdf), 44, 42, 48, 6, 36, 9, 41, 49(pdf), 45, 46, 5, 37, 38, 4, 40, 43, 47, Nobody's Perfect, 18, 20, 50, Nicola Who?, 24, 39, 53, 57, 64, 29, 30, UME1, 28, 34, 52, 35, 49, 16, ...

All Articles and books read, led by: Felker3(Sp), UNCC4, LCR Resonant, GGLtr1, Autobiography, Felker5(Sp), Triple Orbit Simulation by Robert Cheshire, Space Energy, felker1 (Sp), Levitron, CV, ...

http://www.atomicprecision.comhttp://www.upitec.orgFrom this point in an estimated total is made by adding 50% to the www.aias.us total. On avergae this is about the contribution made by www.atomicprecision.com and www.upitec.org. So the estimated total for July 2012 is 118,422 his, 25315 distinct visits, 11.1 Gbytes.
1 - 15 July

US Universities, Institutes and Similar.
Berkeley*, Buffalo, Caltech, Columbia*, Duke, Illinois, Oberlin, Purdue, San Diego Supercomputer Center, Stanford, Southwest Research Institute, Thiagarajar (on edu), Arkansas, Universiy Collaboration on Atmospheric Research, UC Santa Barbara*, Houston, Maryland, Minnesota, North Texas, South Dakota, South florida, Utah*, Texas, Washington, Wisconsin, Ames Laboratory, Department of State, Gov. Virginia, Airforce Operations Center (AFNOC), Army Research Laboratory, US Navy, US Archives.

Europe and Asia Minor: Universities, Institutes and Similar.
FIEGL Graz, Bern, Charles University Prague, Fraunhofer, Helmholtz, Munich, Albert Einstein Institute, Max Planck Garching, Max Planck Heidelberg, RZ Berlin, Tuebingen, OWI Aachen, TU Berlin*, TU Clausthal, TU Darmstadt, TU Munich, Bonn, Frankfurt, Hannover*, Leipzig, AENA Spain, Aragon, IAC, Gov Andalucia, MCU, RENFE, Autonomous Barcelona, Barcelona, Complutensian Madrid, UMA, Iovi, UPCT, UVA, Vigo Spain, Aalto, Joensuu Finland, Ramboll, INTRA CEA France*, CNRS Grenoble, ENS LPT, Poitiers, UPS Toulouse, CYTA Greece, Crete, Tel Aviv, European Space Agency*, ICTP Trieste, INFN Naples, SNS, Rome 1, Turin, RU Groningen, TU Delft, UPC, Twente, Jagiellonian Wroclaw, Warsaw, UVT Romania, WLB Slovenia, RZ-HM Slovenia, Bristol, Greenwich, Imperial, King's College London, Luton, York, BBC*.

Rest of World: Universities, Institutes and Similar.
UNLP, UNLPAM, Gov. Argentina*. Latrobe, QUT Australia, IMPA Brazil, UFTTJ, UFU, UNESP, Sao Paolo Brazil, British Columbia, Gov. British Columbia, McGill, Perimeter, Queen's, Laval, Toronto Canada, Univ. Chile, UFRO, UFTSM, CHLA Colombia, Funandi, UDEA, Valle,

IIT Karagpur India, PRL India, Gov. Mexico, ITM Malaysia, UNAN Nicaragua, UAF Pakistan, CMU Thailand, Academica Sinica Taiwan, Tlabs South Africa.

15 - 30 July

US Universities, Institutes and Similar.
   Auburn, Berkeley, Boston, Colorado*, Cornell, Denver, East Carolina, Harvard, MIT, Michigan State, New York, New York Polytechnic*, Stevens, UC San Diego, Florida*, Michigan, Texas, Wisconsin Lutheran College; US Gov. Centers for Disease Control and Prevention, Lawrence Livermore, Oak Ridge, Aerospace Corporation, US Archives*, General Electric, Lockheed Martin*, Northrop Grumman.

Europe and Asia Minor: Universities, Institutes and Similar
   FWF Austria, UIBK Austria, Fiegl Graz Austria, Catholic Univ. Leuven, CERN, ETH, Humboldt Berlin, MHN, Max Planck KG, Siemens*, DIMA TU Berlin*, EECSIT TU Berlin*, Bonn, Frankfurt, Cologne*, Saarland, Siegen, Tuebingen*, Wuppertal, Aarhus, CSIC Spain, TSAI Spain, Autonomous Madrid, Zaragosa, Polytechnic Madrid*, Polytechnic Valencia, Autonomous Valencia, IAP France, INRIA, LISIF, Lyon1, Marseilles, Univmed, Bioacademy Greece, Crete, DBVK Hungary, Hebrew Jerusalem, Tel Aviv, INFN Italy LNF, INFN Turin, Pisa, Royal Society, Royal Society of Chemistry, Lodz, Lisbon Polytech., Ioffe Institute Russia, Utel system Ukraine (extensive downloads), Durham, King's College, Lancaster, Leeds Metropolitan, Manchester, Nottingham.

Rest of World: Universities, Institutes and Similar.
   Citefa Gov. Argentina, Defence Australia, Abu Dhabi Investment, Justice Ministry Brazil, Sao Paolo, UGDSB Ontario Canada, Waterloo, Western Ontario, PUC Chile, Biobio, UFRO, UTFSM Chile, LN China, UDEA Colombia*, UFPS Colombia, Andes Colombia, Espoch Ecuador, City University Hong Kong, SXCCE India, IITKGP India, TIFR India, Kyoto Japan, Osaka, KAIST South Korea, Postech South Korea, UNAM Mexico, CIIT Lahore Pakistan, LHR-NU, New PU Pakistan, KMITL Thailand, MUT Thailand, UC Venezuela, North Western South Africa, Capetown South Africa.

Summary of August 2012

For www.aias.us there were 74,512 hits, 7.40 gigabytes downloaded, 16,073 distinct visits, 47,781 page views, 2,637 documents read, from 98 countries, led by USA, Ukraine, Mexico, Germany, Canada, Australia, Argentina, Italy, ...

All UFT papers read, led by 25, 218, 177(Sp), 43, 166, 225, 224, 4, 140, 159(Sp), 41, 107, 157(Sp), 150B, 110, 140(Sp), 142(Sp),.........

All essays heard, led by Nobody's Perfect, 3, 5, 6, Nikola Who?, 48, 37, 50, 46, 44, 8, 7, 40, 49, 35, 38, 4, 45, 49(Sp), 35, Intro139, 42, 54, 20, 63, 9, 18, 41, 43, Nobody is Perfect (MWE), 24(Sp), 17, 16, 22, 55, 60, 10,12, 19, 23, 27, 30, 34, 53, 59, 62, 64, 29, 15,21, 25,39,47, 52, 56, ...

Other Documents led by: Felker3 (Sp), Spacetime Devices, Autobiography, Felker5(Sp), Felker4(Sp), 2D Douglas Lindstrom, ...

Estimated total 111,768 hits, 24,111 distinct visits, 11.1 gybytes.
http://www.atomicprecision.com

1 - 15 August

US Universities, Institutes and Similar.

Armstrong, Arizona State, Caltech, California State Northridge, City University of New York, Denver, Georgia Tech, Hawaii, Indiana, Michigan State, North Carolina A and T, New Mexico Tech, Oregon State, Penn State, Purdue, San Diego Supercomputer Center*, UC Santa Cruz, Illinois Urbana Champaign, Maryland, Michigan, Texas Arlington, Texas, U. S. Government Ultimate Network Information Center, Library of Congress, U. S. National Archives and Records Administration, U. S. Gov. National Oceanographic and Atmospheric Administration, Sandia National Laboratory, Fermilab, King County Library System, U. S. National Center for Genome Resources, BAE Systems, Ford, Lockheed Martin*, Siemens US, U. S. Army AMRDEC.

Europe and Asia Minor: Universities, Institutes and Similar.

GEG Austria, Free University of Brussels, EPF Lausanne, CERN, Univ. Basel, Vitkovice Czech Republic, HU Berlin, Max Planck Extraterrestrial Physics, Max Planck Dresden, TU Dresden, Hannover, Heidelberg, Karlsruhe, Oldenburg, Rostock, Complutensian Madrid, Polytechnic Catalonia, Georgian Technical University Tbilisi, AHRT Hungary, Tel Aviv, Italian Prime Minister's Office Palazzo Chigi, Free Univ. Amsterdam, Warsaw, UGAL Romania, Kubang Russia, SAO Russia, Uppsala, Ljubjliana, UTEL system Ukraine, physics Cambridge, Imperial, Manchester, Nottingham.

Rest of World: Universities, Institutes and Similar.

INPI Gov. Argentina, INTI Gov. Argentina, Australian National University, James Cook University, Monash, Queensland, Melbourne*, Wollongong*, CNEN Gov. Brazil, UEL Brazil, Federal Rio de Janeiro,

Dalhousie Canada, Gov. Quebec, NCF, Univ. Alberta, Univ. Saskatchewan, UCT Chile, UCV, UFRO, UTFSM Chile, HA China, UNAL Colombia*, Espol Ecuador*, SAHA India, NKN, IMSC, SCFBIO-IITD India, Osakafu Japan, Torikyo, GSI Gov. Japan, UASLP, UGTO*, UNAM*, USON Mexico, Auckland* New Zealand, New Pakistan Univ., Peshawar, NTHU Taiwan, Academica Sinica* Taiwan, National Physics Institute Uruguay, Limpopo South Africa, Witwatersrand South Africa*, Bindura University ZW.

15 - 31 August

U. S. Universities, Institutes and Similar.

Arizona, Caltech*, Central Michigan, Colorado, Columbia, City University of New York*, Duke*, Georgia Tech., Harding*, Hawaii, Iowa State*, Illinois Institute of Technology, Kent State, Lousiana State, North Carolina State, New Mexico State, Nova Fort Lauderdale, Ohio State, Pittsburgh, Penn State, Rhodes, Rice, Rochester, Stanford, Syracuse, Akron, Central Florida, UC San Diego, Iowa, Illinois Urbana Champaign*, North Carolina Charlotte, Nebraska Lincoln, Southern California, Texas, Brookhaven, U. S. Dept. of Energy Nevada Nuclear Test Facility, Alaska Regional HQ, US Weather Service, U. S. Patent and Trademark Office, U. S. Airforce Network Operations Center; U. S. Army Joint Staffs Hampton Roads, U. S. Naval Marine Command; BAE, Ford, Honeywell, Horthrop Grumman, Boston Museum of Science, Palacio School District Texas.

Europe and Asia Minor: Universities, Institutes and Similar.

TU Vienna*, KHK Belgium, KU Leuven, TEA Bulgaria, German Synchrotron Facility, KFA Juelich, Max Planck Extraterrestrial Physics, Siemens*, TU Darmstadt, TU Dresden, TU Ilmenau, Leipzig, Oldenburg, Tuebingen, Niels Bohr Institute Denmark, AENA Spain, GVA Spain, Ministry of Defence Spain, Social Security Spain, Complutensian Madrid, Jaen, Valencia, HUT Finland, IFPEN France, Education DSL Greece, Hebrew University Jerusalem, ITB CNR Italy, Polito Italy, RP Engineering, Milan, Rome 1, Free University Amsterdam, Atman Poland, Triburie Romania, Corbina, Yandex Russia, TLT Russia, BORAS Sweden, DFRI Sweden, NLB Slovenia, GTS Slovakia, ITL Ukraine, UTEL* , Christ's College Cambridge, Computing Cambridge, Nanoscience Cambridge, Durham, Imperial, Manchester, Strathclyde, Sussex, Swansea.

Rest of the World: Universities, Institutes and Similar.

IESE Argentina, UNCU, Deaking University Australia, Griffith, Latrobe, Monash, Concordia, Queensland, Western Sydney, INPE Brazil, UEL, UFMG, CST Buhtan, Defence Canada, Gov. Quebec, Simon Fraser,

Alberta*, Saskatchewan, Western Ontario, Univ. Chile, UTFSM Chile, Tsinghua China, UNAL Colombia, Valle, Colombia, ESPOL Ecuador*, Valle Gatemala, Guayaquil Ecuador, Indonesia, SCSITS India, IIT Kharagpur, Railnet, NKN India, Nagoya Japan, NIFS, KAIST South Korea, ETRI, UASLP Mexico, UFTO*, UNAM, MPU Taiwan, Academica Sinica*, Capetown South Africa.

Summary of September 2012

For www.aias.us there were 72,982 hits, 46,629 page views, 17,456 distinct visits, 9.70 Gbytes downloaded, 2,547 documents read from 105 countries, led by USA, Britain, Mexico, Germany, Ukraine, Italy, Canada, Russia, ...
    All UFT papers read, led by: 226, 43, 177(Sp), 107, 227, 25, 159(Sp), 166(Sp), 157(Sp), 140, 41, 166, 150B, 4, 177, 142(Sp), 170(Sp), 94, 171(Sp), 149, ...

All Essays Heard, led by Light Deflection, UME1, 9, 30, Nikola Who?, 47, 50, 6, 38, 8, 32, 5, Nobody's Perfect, 49, 24, 40, 53, 45, 52, 48, Nobody is Perfect, 7, 4, 36, 51, 58, 49, 59, 60, 63, 37, 29, 54, UFT139 Introduction, 64, 20, 24, 14, 21, 22, 17, 10, 12, 13, ...
    All documents and books read, led by Felker3 (Sp), Felker5(Spanish), LCR Resonant, UNCC4, Autobiography, Felker4(Sp), 2D, ...

Estimated total for September is 109,473 hits, 26,184 distinct visits, 14.55 gigabytes.

1 - 15 September

US Universities, Institutes and Similar
    Arizona, Boston College, Berkeley, Boston University, Caltech., Central New Mexico Community College, Columbia, Cornell*, CSU East Bay, Harvard*, Iowa State, Kansas, Lehigh, Missouri, MIT, North Dakota, Oberlin, Ohio, Oklahoma, Pittsburgh, New York Polytechnic*, Princeton, Penn State, Purdue, Rochester, San Jose State, Southern Maine Community College, Stonybrook, Texas A and M, Toronto (on edu), Central Florida, UCLA, Connecticut*, Illinois Chicago, Michigan*, North Carolina Charlotte*, New Mexico, Wisconsin Milwaukee, Vermont, Washington, Western Carolina, Wisconsin; U. S. Federal Aviation Authority, JPL NASA, U. S. Gov National Institute of Science and Technology (NIST), U. A. Gov. National Oceanographic and Space Administration (NOAA), U. S. Army tardecren, Honeywell, Northrop

Grumman, Philips, Raytheon*, U. S. Archives*, King County Library, Sioux City Central Library, Highland.

Europe and Asia Minor: Universities, Institutes and Similar.
Erwin Schroedinger Institute Univ. Vienna, Univ. Vienna, EPF Lausanne, ETH Zurich, Paul Scherrer Institute, Geneva, Fraunhofer Foundation, Max Planck Halle, Ruhr Univ Bochum, Siemens, Clausthal, Freiburg, Hamburg, Karlsruhe, Tuebingen*, Wuppertal, Andalucian Gov., TSAI Spain, Autonomous Barcelona, Granada, Malaga, Miguel Hernandez de Elche, Spanish National Distance Education University, Polytechnic Madrid, Valencia, Helsinki, Siauliai Lithuania, Latecoere Aircraft Company France, Paris Observatory, Polytechnique de France, Paris Psud, Angers, Poitiers, Hungarian Academy of Sciences Campi Budapest, Patient Treatment Register Ireland, Weizmann Institute, CNR Florence Physics Laboratory, TU Delft, TU Eindhoven, Utrecht, Lodz Poland, Lebedev Institute of the Russian Academy of Sciences, Moscow Physics Engineering University, Obninsk, Yaroslavl, Utech Russia, Chalmers Sweden, Fatburen Sweden, Linkoping, Joseph Stefan Institute Slovenia, Ankara, Bilkent Turkey, Aberystwyth, Birmingham Central, Physics Cambridge, Cardiff, Kent, mathematics Oxford, physics Oxford, British Library, National Library of Wales, Oxford University.

Rest of World: Universities, Institutes and Similar.
MDP Argentina, FLENI Argentina, Sydenham Catholic Australia, Monash, Queensland Tech., New South Wales, Western Australia, Queensland, FOA Brazil, Lactec Brazil, UFMG, UFRGS, UNESP Brazil, Gov. British Columbia Canada, McGill, Perimeter, British Columbia, UQ Trois Rivieres, York Univ. Toronto*, Providencia Chile, Unov. Chile, UDA, UFRO, UNAB, UTFSM* Chile, UDEA Colombia, UNAL*, Unitec Colombia, IIT Karagpur India, Indian Institute of Health, Gov. India IGCAR, Gov. India RCIL, TIFRH India, Tokushima Japan, Riken research System Japan, Korean Advanced Institute (KAIST), UNAM Mexico, USON Mexico, Waikato New Zealand, Wellington Tech., Aula Virtual of Sedes Sapientiae University Peru, UPAO Peru, Gov. Peru, IIU Pakistan, New Pakistan Univ., National Univ. Singapore, Gov. El Slavador, SAFA Uruguay, Capetown South Africa, Johannesburg, Western Cape South Africa.

15 - 29 September

US Universities, Institutes and Similar.
Arizona State, Berkeley*, Brown, Caltech, Rhode Island Community College, Colorado*, Columbia, Cornell*, California State East Bay, Case

Western Reserve, Duke, Florida International, Gustavus Adolphus, State Univ. New York Geneseo*, Harvard, Hawaii, Johns Hopkins, Kansas State, Louisville, Louisiana State*, Notre Dame, NIT Tiruchirapalli (on edu); Ohio State, Ohio State, Pittsburgh, Penn State, Purdue*, Reed*, Rice, Rochester, San Diego Supercomputer Center, Southern Illinois Edwardsville, Stevens Tech*, UC Santa Cruz*, Idaho, Illinois Urbana Champaign, U Mass., Maryland*, Michigan*, Minnesota, North Carolina Charlotte*, Nevada Las Vegas, New Mexico, Pompeu Fabra (on edu), Utah, Texas Dallas, Texas, Tennessee Knoxville, Vanderbilt, Virginia, NASA Goddard Spaceflight Center, NASA Jet Propulsion, NIST, US Airforce AFNOC, US Army AMRDEC, US Army Tank Construction (TARDECDREN), US Marines Command Norfolk, US Naval Research Laboratory, Chevron, Intel, Lockheed Martin, Raytheon, US Archives*, King's County Library system, Mohawk Regional Information Center, Novant Health, Seattle Public Library.

Europe and Asia Minor: Universities, Institutes and Similar.

TU Vienna, Innsbruck, Mons Hainaut Belgium, ETH Zurich*, Bruker, FH Potsdam, Max Planck Institute for the Science of Light, Max Planck Institute for Colloid Science, Siemens, TU Munich, Leipzig, Regensburg, Egmont College Denmark, Aarhus University, Jaelland Hospitals, Aarhus Hospitals, Univ Roskilde Denmark, Tartu Estonia, Basque University, Andalucian Government, Autonomous Madrid*, Juan Carlos III, Complutensian Madrid, Valencia Polytechnic, Salamanca, Santiago de Compostela, Government Canary Islands, Helsinki, INSERM Lyon France, ISAE, Paris Observatory, French Aerospace Lab ONERA, Angers, Poitiers*, ATT Schools Greece, UVG Gibraltar, Hebrew Univ. Jerusalem*, Iceland, INFN Parma, INFN Turin, Milan Polytechnic, SISSA, Scuola Normale Superiore Pisa, Free University Amsterdam, Adam Mickiewicz Poznan Poland, Wroclaw, Minho, Analytical Center of the Russian Federation Government, Russian National Research Nuclear University, Nizhny Novgorod Region, Karolinska Institute Sweden, Stockholm, Uppsala, Kiev, Birmingham Central, Bristol, Cardiff, Manchester, Nottingham, Swansea, Western Scotland, National Health Service, Hertfordshire Learning Grid, Parliament, British Library.

Rest of World: Universities, Institutes and Similar.

UNSA Argentina, INIT Gov. Argentina, Adelaide, Western Australia, UEL Brazil, Unicamp*, Sao Paolo, Laurentian Canada, McGill, Memorial Newfoundland, Perimeter, British Columbia*, New Brunswick, Guelph, Quebec Trois Rivieres, Waterloo*, Central Univ, Chile, Engineering Univ. Colombia, UDEA, Andes, Norte, Tecnologia Colombia, UO Cuba, INF Cuba, Agriculture Gov. Ecuador, BHU India, IIT Hyderabad, IIT

Karakhpur, Kyoto Japan, NIFS, KAIST Korea, SNU Korea, SAT Gov Mexico, UGTO, UNAM* Mexico, PUCP* Peru, UNAC Peru, QUA Pakistan, NU Singapore, KMITL Thailand, NCTU, NCU, NTHU, NTU Taiwan, Institute of Physics Montevideo Uruguay, Capetown South Africa, Limpopo, Pretoria, Witwatersrand.

Summary of October 2012

For www.aias.us there were 80,701 hits from 20,315 distinct visits each of n real visits, 7.60 Gbyte downloaded, 51,023 page views, about 2,500 documents read from 102 countries, led by US, Mexico, Germany, Ukraine, Canada, Britain, France, ...
All UFT papers read, led by 177(Sp), 25, 166(Sp), 159(Sp), 157(Sp), 140(Sp), 142(Sp), 150B, 169, 177, 175(Sp), 227, 54, 88, 4, 43, 94, 170(Sp), 229, 140, 41, 11, 228, 50, 142, 57, 165(Sp), 175, ...
All essay broadcasts read and essays read, led by: Light Deflection by Gravitation, 7, 6, 5, 38, 49, 35, 27, IntroUFT139, 34, 58, Nobody's Perfect, 46, 8, 48, 36, 39, Univ MWE Part 1, 28, 4, 60, 66(Sp), 52, 64, 44, 45, 24, 25, 32, 54, 59, 33(Sp), 49(Sp), 41(Sp), 9, 63, 62, 40, 61, 29, ...
Leading articles were: Felker3(Sp), F5(Sp), LCR Resonant, Autobiography Volume One, F4(Sp), Nikola Who? (broadcast by Robert Cheshire), Space Energy (Sp), My CV, ...

Estimated total 121,052 hits, 30,473 distinct visits, 11.4 gigabytes.

1- 15 October

US Universities, Institutes and Similar.
Auburn, Berkeley, Boston, Buffalo*, Caltech, Chattanooga State, Clemson, Colorado, Columbia*, Cornell*, CSU Pomona, Case Western Reserve, Dartmouth, Georgia Tech*, Harvard, Holy Family, Illinois Institute of Technology, Illinois State, Johns Hopkins, Lehigh, Macon State, Maine, Missouri, MIT*, Mississippi State, Tiruchirappalli India (on edu), Northwestern*, New York, Portland State, Pittsburgh, Princeton, Penn State, Purdue*, Renselaer, San Diego Supercomputer Center*, SUNY Stonybrook*, Texas A and M*, Tufts, Alabama Birmingham, Arkansas Little Rock, Central Florida, Chicago, UCLA, UC Riverside*, UC Santa Barbara, Houston, Illinois Urbana Champaign, Michigan*, Chapel Hill, UNCC, Nevada Las Vegas, U Penn, Utah State, Utah, Texas*, William and Mary, Vermont, Western Michigan, Yale*, Brookhaven, Fermilab, NIST, Sandia*, Sedgwick County*, US Patent and Trademark Office, 24th Airforce Operations Center; Space and Naval Warfare Command; General Electric, Hewlett Packard, Intel, Raytheon,

Rockwell Collins, US Archives, Ludwig von Mises Institute, New York State Department of Environmental Conservation.

Europe and Asia Minor: Universities, Institutes and Similar.
Ascus Art and Science Cooperation Austria, Salzburg Research, Belarus Academy of Sciences, EPF Lausanne, Basel, Geneva, German Synchrotron, German Aerospace, FBH Berlin, Free Univ. Berlin, Siemens*, TU Munich, Heidelberg, Karlsruhe, Leizpig*, Trier, Tuebingen*, Aragon, Spanish Electricity Endesa, IFAE Spain, Ministry of Defence, Autonomous Madrid, Barcelona, Catalonia, Distance Education University, Canaries, Polytech Catalonia, Polytech Madrid*, Polytech Valencia, Vigo, Xunta, French Particle Lab., Latecoere, Onera, Nantes, Joseph Fourier Grenoble, Lyon 1, Poitiers*, Renens 1, Vladan, Merom Golam Isreal, Turin Polytech., Genoa*, Free Univ. Amsterdam, NTNU Norway, Oslo, Coimbra, Minho, Bolyai Romania, CN Russia, DSI Russia, MTS-NN Russia, NN, ORN, Yaroslavl, Uppsala, Josef Stephan, Mariboru Slovenia, Gazi Turkey, NKU Turkey, Kiev, Aberystwyth, Berking Dagenham, Cambridge Computing, Pembroke College Cambridge, Cardiff, Durham, Edinburgh, EPSRC, Exeter, Imperial, Lancaster, Lincoln, Manchester, St. John's College Oxford, Queen's University Belfast, Southampton, Strathclyde, Susses, UWIC, Warwick, York, CGP Books, Essex Schools, Oxford University Registry, Audenshaw School Tameside.

Rest of World: Universities, Institutes and Similar.
Patagonia, Adelaide, Monash, Melbourne, Sydney, UFAM Brazil, ABTLUS Brazil, Sao Paolo, Agriculture Canada, RCMP-GRC Canada, Simon Frazer, Quebec Trois Rivieres, Saskatchewan, Toronto, Biobio Chile, Chile, UFRO, UNAB, UTFSM Chile*, Tsinghua China, Engineering University Colombia, Andes, Americas Ecuador, UTPL Ecuador, UV Guatemala*, Bhabha Atomic RC India, Railnet, Rrcat India, Japanese Instituet for Advanced Research, Waseda, MESH, KAIST Korea, Samsung Korea, UNAM Mexico, USON Mexico, Victoria Wellington New Zealand, Unisza Malaysia, Unimas Malaysia, National Univ. Peru, LUMS Pakistan, PERN, National Univ. Rwanda, National Univ. Singapore*, KMITL*, SU Thailand, CCU Taiwan, NCTU*, NTHU Taiwan, National Institute of Physics Uruguay, Johannesburg South Africa, Vaal Univ. of Technology.

15 - 30 October

US Universities, Institutes and Similar

Arizona*, Arizona State, Buffalo, Carnegie Mellon, Columbia*, Cornell*, CSU Long Beach, CS Northridge, Georgia Tech.*, Georgia Perimeter College, Illinois, Ithaca College, Johns Hopkins Medical, Johns Hopkins, Jacksonville State, Miami, MIT*, Metropolitan State Denver, Michigan State, North Dakota, Ohio State, Oklahoma State, New York Polytechnic*, Princeton, Purdue, Rochester, San Diego Supercomputer Centre, Sarah Lawrence College, Texas A and M, Tufts, Arkansas, Chicago, IC Irvine, UCLA*, Connecticut, Florida, Massachusetts Dartmouth, Maryland, Michigan, Chapel Hill, Southern California, Texas*, Virginia, Yale, NASA Langley, NIST, NOAA ARC, Sandia*, US Airforce AFNOC, US Navy NAVO, American Institute of Physics, US Archives*, Oak Hammock Florida, American Anthropological Association, San Francisco Public NTU, Library, Hewlett Packard Houston, Intel, Northrop Grumman*, Raytheon.

Europe and Asia Minor: Universities, Institutes and Similar

Innsbruck*, Ascus Arts and Science Salzburg, Leuven, HMTI Belorussia, CERN*, EPFL*, Basel, FZU Czechia, DLR Germany, FU Berlin, GU Berlin, Max Planck Mathematics Leipzig, Max Planck Colloids, Siemens*, Mainz, Regensburg, Students Bonn*, TU Clausthal, TU Freiburg, Bielefeld, Heidelberg*, Mainz, Regensburg, Wuerzburg, Madrid Arts and Science, Complutensian Madrid*, Murcia, Canaries, Polytechnic Cartagena, Rovira and Virgili, Seville, Vigo, Helsinki*, Jyvaskula, Vaasa, ECP France, IMCE, French National Nuclear Particle Lab., Jussieu ADMP6, French Atomic Energy Commission, Polytechnique*, French National Synchrotron Laboratory, Paris 10, Caen, Lemans, Poitiers, UPS-TLSE, Ioannina Greece, Hebrew Univ. Jerusalem, Weizmann Institute, Naples Observatory, INFN LNL, Italian National Computer Centre Salerno, Salento, Polish Academy, Transylvania, Met Office Serbia, Debryansk, Ioffe Institute Russia, Stockholm, Joseph Stefan, Kiev, Maths Cambridge, Cambridge*, Girton, Magdalene, Newnham Cambridge, Cardiff, Durham*, Edinburgh, Glasgow, Imperial*, Lancaster, Luton, Manchester*, Oxford*, Plymouth, Queen's Belfast, Reading, Sussex, University College London*, Central Lancashire, Ministry of Defence*, Essex Schools, Barking Dagenham College.

Rest of World: Universities, Institutes and Similar.

UNGS Argentina, UNP Argentina, GBA Gov. Argentina, Monash* Australia, Fairholme, Melbourne*, PUCRS Brazil, UFAL, UFPB, UFRRJ, YNB, UNESP, Sao Paolo* Brazil, Gov. British Columbia, Concordia, Dalhousie*, McGill, McMaster, Perimeter, Univ. British Columbia*,

Montreal, Quebec Trois Rivieres, Toronto, INACAP Chile, PUC*, Biobio, Central, UTFSM, Tsinghua China, National Engineering University Colombia, UNAL*, Andes*, Valle*, UTPL Ecuador, Infom Guatemala, University of Science and Technology Hong Kong, IITH India, CAT Gov. India, NKN, CCMB, RRI India, IUT Iran, Shahed Iran, Keio Japan, Metropolitan Tokyo, Titech, Waseda, Chiba, KAIST South Korea, Postech, UGTO Mexico, UIA, UNAM*, USON, Unisza Mexico, Univ. of Panama, Piura Univ Peru, Peruvian Navy, UNAC Peru, New Pakistan Univ., NUST Pakistan, CCU Taiwan, NDHU*, NTHU, NTUST Taiwan, VPC Venezuela, SUN South Africa, Capetown.

Summary of November 2012

For www.aias.us there were 82,043 hits, 18,770 distinct visits, 10.902 gigabytes downloaded, 53,504 page views, 2,566 documents read from 98 countries, led by US, Mexico, Germany, Canada, Argentina, Ukraine, Australia, Britain, France, China, ...

All 232 UFT papers read, led by: 25, 43, 166(Sp),157(Sp), 177, 169, 140(Sp), 4, 150-B, 107, 140, 170(Sp), 88, 94, 166, 18, 175(Sp), 142, 41, 159(Sp), 33, 85, 150, 11, 175, 38, 13, 81, ...   All essays heard or read, led by: 3, Nobody's Perfect, Nobody is Perfect, 50, 46, 45, 48, 35, 57, 52, 9, 49, 6, 39, 56, 54, 29, 53, 7, 59, 47, 60, 41, 61, 44, 58, 34, 42, 24, 30, 64, 4, 12, 16, 20, 43, 51, ...

All other items read, led by F3(Sp), F5(Sp), Spacetime Devices, Spacetime Devices (Spanish), HPExp2, Triple Orbit Simulation, ECE overview, Autobiography, ...

The total estimated returns from www.aias.us, http://www.atomicprecision.com , www.atomicprecision.com and www.upitec.org are calculated by adding 50% to the above figures, giving 123,063 hits, 28,155 distinct visits, 16.38 Gbytes downloaded and 80,256 page views.

1 - 15 November

US Universities, Institutes and Similar.

Bloom, Buffalo, Brigham Young, Caltech*, Fullerton, Georgia Tech*, George Mason, Hawaii, Illinois Institute of Technology, Miami, Missouri*, MIT, NITT India (on edu), Northwestern, New York, Old Dominion, Ohio State*, Penn State*, Purdue, Quinnipiac, Rochester, Rennselaer, San Diego Supercomputer Center*, Stanford*, SUNY Stonybrook, Texas A and M*, Alabama, UC Davis*, Chicago, UCLA, UC Santa Cruz*, Georgia, Iowa, North Texas, South Dakota, Texas,

Wisconsin Milwaukee, Wyoming, Virginia, Wake Forest, Wisconsin, William and Mary*, Washington State Pullman, Yale, US Dept. of Energy Hanford, Los Alamos*, Lawrence Livermore, Gov. Virginia, Intel, Motorola, US Archives*, Evergreen Health Care, Highline Schools, Ludwig von Mises Institute, Prince George's County Public Schools Maryland.

Europe and Asia Minor: Universities, Institutes and Similar.

Antwerp, CERN, EPF Lausanne, ETH Zurich, Charles Univ. Prague*, Ruetlingen, FU Berlin, Max Planck Muelheim*, RWTH Aachen, Siemens*, Siemens ERLM, TU Chemnitz, TU Clausthal, TU Darmstadt, Bielefeld, Bonn, Erlangen, Frankfurt, Cologne, Saarland*, Wuerzburg, Gov. Andalucia, Spanish Ministry of Defence*, Autonomous Barcelona, Barcelona*, Cadiz, Iovi, Polytechnic Valencia*, Seville, Valencia*, Spanish Parliament (Xunta), CNRS Grenoble*, Curie Institute Paris, Ecole Normale Superieure Lyon, French National Particle Lab., INPL Nancy, Fremch Synchrotron Lab, Strasbourg, Nice, Angers, Paris Diderot, Poitiers8, Rennes 1, UPS Toulouse*, ATT Schools Greece, Patras, NUI Galway, Trinity College Dublin, Bar Ilan, CNR Institute of Science and Technology Pisa, ICTP Trieste, INFN Florence, SISSA, International School for Advanced Studies, Modena, Padua, Rome 2, Salerno, Groningen, Free Univ. Amsterdam*, Serbian Academy, Kaluga Russia, Sclipetsk Russia, Linkoping*, Uppsala, Iyte Turkey, Metu, Aberdeen, Bath, Bristol, Cambridge*, Durham, Glasgow, Hull*, Imperial*, Kingston, Lancaster, Leeds*, Manchester, Nottingham, Oxford*, reading, Rutherford Appleton, University College London*.

Rest of World: Universities, Institutes and Similar.

UNC, UNJU, UNLP, UNSL Argentina. Adelaide, Monash, CBPF Brazil, UFPB, UFSC, Unicamp, USP Brazil, Carleton, Concordia, Perimeter, Queens, Canadian Particle Lab., British Columbia, Laval, Quebec Trois Rivieres*, Toronto, Waterloo*, PUC, UFRO, Mayor, UTFSM Chile, Engineering School, UDEA*, UNAL Technology, Valle Colombia, PUCE Ecuador, IITH, IITKGP, IMSC, IRDE India, Gakushuin, Hiroshima, KYU Tech., Orakafu, Ritsumei Japan, KAIST, Korean Government, Postech* Korea, INAOPE, IPN, UGTO*, UNAM*, UNIMAS Mexico, Panama, UNP, IPEN, Gov. Peru, National Univ. Singapore, Chula, SU Thailand, FJU, NCTU Taiwan, UC, UPEL, Gov., USB Venezuela.

15-29 November

US Universities, Institutes and Similar

Boston College, Boston University, Caltech*, Caribbean, Columbia, Cornell, CUNY*, Dallas Community College, Georgia Tech., Georgia State, Harvard, Indiana Purdue Indianapolis, Kennesaw, Karlsruhe (on edu), Louisiana State, Maricopa, MIT, Michigan State, Miami University Ohio, Northwestern, New York, Ohio State*, Pittsburgh*, San Diego Supercomputer Center*, Alabama, U. S. Airforce Academy, UC Davis*, Chicago, UCLA, UC Riverside, UC Santa Barbara, Delaware, Girona (on edu), Florida, Illinois Urbana Champaign, Michigan, Missouri Kansas City, Minnesota, Chapel Hill, New Hampshire, North Texas, Puerto Rico, Southern California, U. S. Naval Academy, Texas, Vanderbilt, Wake Forest, Wisconsin, Yale, Los Alamos, U. S. Gov. Regional Weather Center Alaska, U. S. Army Detrick, U. S. Archives*, Otto von Mises Institute, Montefiore Clinic Albert Einstein School of Medicine New York City, Prince George's County School District Maryland, World Wide Web Organization, Honeywell, Intel, Goodrich, Northrop Grumman.

Europe and Asia Minor: Universities, Institutes and Similar

Telering Austria, Vienna, EPF Lausanne, ETH Zurich, Lausanne, Technical University Prague, Bosch, German Synchrotron Facility, FU Berlin, LRZ, Siemens, ERLM Siemens, Darmstadt, Bielefeld, Druisburg, Erlangen, Frankfurt, Karlsruhe, Koeln, Marburg, Paderborn, Regensburg, Saarland, Wuerzburg, BNAA Denmark, National Library of Estonia, Ministry of Social Security Spain, Santa Pau Hospital, SEPSA Spain, Barcelona*, Granada*, Polytechnic Catalonia, Polytechnic Valencia, Seville, Valencia, Spanish Parliament*, European Space Agency, Jyvaskyla Finland*, Tontut, French Atomic and Alternative Energy Commission, CROUS Grenoble, Cergy, Strasbourg, Nice, Nantes, Paris Diderot, Poitiers, Rennes 1, Patras Greece, Hungarian Development Agency, Trinity College Dublin*, University College Dublin*, Hebrew University Jerusalem, Tel Aviv, INFN Aquita, INFN Florence, INFN Naples, INFN Rome 1, Milan Polytechnic, International School of Advanced Studies (SISSA), Pisa, KIS Lithuania, Groningen, Eindhoven, NTNU Norway, Wroclaw, Aveiro Portugal, Lisbon, Icelandic Meteorology Office, Serbian Academy of Sciences, Belgrade, Joint Institutes for Nuclear Research Dubna Russia, Kaluga, Sclipetsk, Vologda, MIS Russia, Stockholm, Uppsala, Ljubljana*, Ankara, Iyte, Selcuk, Donetsk, DP Ukraine, IN Ukraine, Kharkov, Poltava, TPK Ukraine, Bath, Bristol, Oxford Brookes, Cambridge*, Pure Mathematics Cambridge, Durham, Glasgow, Glyndw^r, Imperial, Kingston, Lancaster*, Leeds, Luton, Nottingham, Christchurch Oxford, Mathematics Oxford, St John's Oxford, Queen Mary University of London, Reading, Sheffield, Southampton, St Andrew's, Strathclyde*, Forestry Commission, Concord

College* Shrewsbury, Connection St Martin's in the Fields, Essex Schools Chelmsford.

Rest of World: Universities, Institutes and Similar.

UNLP Argentina, UNLPAM, UTN, Curtin University Australia, CTA Brazil, IMPA, Unicamp, Sao Paolo, SCE Bhutan, Government British Columbia Canada, Langara, ORL British Columbia, Carleton, Dalhousie, McGill, McMaster, Perimeter, Queen's, Quebec Trois Rivieres*, Waterloo, UNIL Chile, Collahuasi, UFRO*, UTFSM, Tsinghua China, UDEA Colombia, UNAL*, Andes, Valle, Espol Ecuador, Mansoura Egypt, UST Hong Kong, UNS Indonesia, IITH India, ISICAL, MNIT, Bhabha, IMSC, Raman Institute, TIFRH, Chuo University Japan, Ehime, Kyoto, Rikkyo, APIG Japan, Korea, Postech Korea, UASLP Mexico, UGTO, UIA, USON, Unimas Malaysia, AQP Peru, CIIT Lahore Pakistan, IIU, New Pakistan, PERN, New University Singapore, NCHU Taiwan, NCTU*, NTU*, Norte Uruguay, UPEL Venezuela, SUN South Africa, Limpopo*.

Summary of December 2012

For www.aias.us there were 99,081 hits from 18,326 distinct visits, 12.74 gigabytes downloaded, 66,754 page views, 2,568 documents read from 102 countries, led by US, Ukraine, Iran, Britain, Germany, Mexico, Canada, ...

These data are incremented by an estimated 50% from www.atomicprecision.com and www.upitec.org to give totals of 148,621 hits, 19.11 gigabytes downloaded, 27,489 distinct visits, 100,131 page views, 2,568 documents read from 102 countries.

All UFT papers read, led by 229, 140, 177, 226, 25, 159, 231,166, 54, 230, 142, 43, 4, 169, 177(Sp), 226, 25, 159, 231, 166, 54, 230, 142, 43, 4, 169, 177(Sp), 88, 107, 227, 89, 203, 175, 232, 58, 63, ...

All Essays heard or read, led by 3(Light Deflection, 8, 7, 49, Nobody's Perfect Nobody is Perfect, 32,58, 4, 50, 6, Intro139, 36, 5, 26, Nikola Who, 30, 48, 52, 53, 9, 9, 27, 44, 60, 20, 24, 37, 40, 61, 63, 64, 42, 54, 55, 56, 28, 17, 11, 10, 12, ...

All articles read, led by F3(Sp), LCR Resonant, UNCC4, CEFE Leaflet, F5(Sp), Human Soul by Jackson, Autobiography, Universe of ME Part 1, CEFE Preprint, My CV, Simulation Circuit (Spanish), Triple Orbit Simulation by Robert Cheshire, ...

1 - 15 December

US Universities, Institutes and Similar

Berkeley, Boston University, Buffalo, Caltech, Claremont, Carnegie Mellon, Colby, Colorado, Columbia*, Cornell, Duke, Florida State, Gettysburg, George Mason*, Harvard, Indiana, Lipscomb, MIT, Montana, North Carolina State, New Mexico State, New Mexico Tech., Northwestern, New York, Ohio State*, Pacific Lutheran, Rochester, San Diego Supercomputer Center*, Sonoma, St Petersburg, Stanford, Stritch, St Thomas, Texas A and M, Tufts, Alabama, University Corporation for Atmospheric Research, UC Davis, Chicago, UCLA, UC Santa Barbara, UC Santa Cruz, Florida, Illinois Chicago, Illinois Urbana Champaign, Missouri Kansas City, New Hampshire, Nebraska Lincoln, U Penn*, University of Pittsburgh Medical Center, Southern California, South Florida, Utah, Tennessee Chattanooga, Texas Dallas, Texas, New Mexico, Vanderbilt, Washington*, Brookhaven, U. S. Equal Employment Opportunity Commission, Lawrence Berkeley, NASA Langley Research Center, Oak Ridge, Pacific Northwest, U. S. Navy Advanced Traceability and Control; Naval Research Laboratory, U. S. National Archives*, Contra Costa County Library California, Ludwig von Mises Institute, Good Neighbors Now, Trendnet, Ford, IBM, Lockheed Martin*, Microsoft.

Europe and Asia Minor: Universities, Institutes and Similar

Innsbruck, Vienna, KABSI Austria, OMA Belgium, Bern, Freiburg, Bergdorf, Fehrer Czechia, Bosch, German Synchrotron Facility, FH Bielefeld*, FH Giessen, FU Berlin, HU Berlin, Ruhr Bochum, Siemens*, TU Berlin, TI Harburg, Bonn, Leipzig, Mainz, Wuerzburg, AAU Denmark, KBHAMT Denmark, Niels Bohr Institute, Campusnet Estonia, EHU Spain, Spanish Ministry of Defence, Barcelona, Cadiz, Complutensian Madrid, Polytechnic Catalonia, Polytechnic Madrid*, Valencia*, Vigo, Helsinki, Jyvaskyla*, Tampere Finland, AC Rouen*, CNRS Grenoble, ENS Lyon, Strasbourg*, Nice, BP Clermont, Marseilles, Poitiers*, Rennes 1, Aristotle Thessalonika Greece, ESRI Hungary, Hebrew Jerusalem, Tel Aviv, Simnet Iceland, INFN Padua, Netherlands National Particle Lab., Utrecht, Free Univ. Amsterdam*, Wroclaw, Szczecin, Serbian Academy, Moscow State, Chalmers, Royal Institute of Technology Stockholm, IYTE Turkey, Middle East technical University Turkey, Bath, Birmingham, Clare College Cambridge, Cambridge, Imperial, Keele, Royal Holloway, Rutherford Appleton, Staffordshire, Sussex, Swansea, University College London*, National Library of Wales / British Library*, Dimensions Organization.

Rest of World: Universities, Institutes and Similar.

Monash Australia, UFRJ Brazil, British Columbia Canada, Montreal, Quebec at Montreal, Quebec Trois Rivieres, Saskatchewan, Collahuasi

Chile, IIA Chile, UDEC, UTA, UTFSM Chile, EIA Colombia, EDEA*, National Univ. Colombia, Valle, UGM Indonesia, Indira Ghandi India, KNTU Iran, Kagoshima Japan, Kyoto, Osakafu, Tokyo*, CNC, Jaxa Japan; KAIST South Korea, Korea, CENAM Mexico, UGTO, UNAM; Unimas Malaysia, Nicaragua, Pontifical Peru, UPLB Philippines, CIIT Lahore, IIU, New Pakistan, NCTU Taiwan, NTU*, USB Venezuela, UNP South Africa, Government of South Africa.

15 - 30 December

US Universities, Institutes and Similar

Arizona, Cornell, East Carolina, George Mason, Johns Hopkins, Missouri, Maine Maritime Academy, UC San Diego Supercomputer Center / US Archives, San Francisco State, Stonybrook, Texas A and M, UC Davis, UCLA, UC San Diego, Maryland, Pennsylvania*, Toledo, Vermont, Washington State, US Gov. Department of Transport, U. S. Gov Social Security System, United States Antarctic Program; US Department of Agriculture, US Navy Marine Command Internet Jacksonville Florida*, Genreal Electric, U. S. Archives, Ludwig von Mises Institute, Kentucky Schools.

Europe and Asia Minor: Universities, Institutes and Similar.

TU Vienna, UDL Catalonia, Xtec Catalonia CERN, EPF Lausanne, German Synchrotron Facility*, FH Friedburg, Siemens, Erlangen, Heidelberg, Leipzig, Stuttgart, Gov. Andalucia, RENFE Spain, Autonomous Madrid, Barcelona, Juan Carlos III Madrid, Polytechnic Valencia, Salamanca, Spanish Parliament*, INPL Nancy (Univ. Lorraine), Limoges, Marseilles*, Paris Diderot, Poitiers*, AMIS Croatia, FSB Croatia, BME Hungary, Hebew Univ. Jerusalem, Tel Aviv*, Weizmann Institute, INFN Milan, INFB Pisa, MII Lithuania, PAVB Latvia, Jagiellonian Krakow, Gdansk, Wroclaw, Technical Univ. Lisbon*, Joint Institutes of Nuclear Research Dubna, Lipetsk, MTS Russia, Nizhnyi Novgorod, Samara, Saransk, KTH Stockholm, City of Stockholm, Uppsala*, SDU Turkey, Gov. Ukraine, Kiev*, Cambridge, London School of Economics, Balliol College Oxford, British Library.

Rest of World: Universities, Institutes and Similar.

UNC Argentina, Gov. Argentina, Campinas Brazil, Toronto, Beijing, Gov. Dominican Republic, Iwate Japan, Tohoku, KAIST South Korea, IIU Pakistan*, Lahore Univ Management Science, Chula Thailand, KKU Thailand, NCTU Taiwan, National Taiwan.

Summary of January 2013

There were 82,267 hits from 15,801 distinct visits, 8.40 gigabytes downloaded, 54,499 page views, 2602 documents read from 107 countries, led by USA, Germany, Canada, Australia, Britain, Mexico, Japan, Switzerland, ...

All UFT papers read, led by 43, 25, 177, 177(Sp), 169, 166, 107, 140(Sp), 175, 41, 166(Sp), 88, 4, 142, 157(Sp), 140, 18, 142(Sp), 150B, 232, 225, 159(Sp), 33, 50, ...

All Essays heard and read, led by: 3 (Light Deflection by Gravity), Nobody's Perfect, 5, 6, 32, IntroUFT139, 49, 35, 26, 46, 50, 48, 57, 39, 7, Nobody is Perfect, 36, 40, 51, 63, 19, 29, 41, 44, 45, 9, 37, 42, 60, 27, 15, 14, 38, 17, 34, 4, 12, 16, 10, 13, Universe of MWE Part 1, 11, 20, 25, Nikola Who?, 28, 24(Sp), 18, 22, ...

All Articles read, led by F3(Sp), Johnson Magnets, F5(Sp), Triple Orbit, Human Soul, LC resonant, Autobiography, CV, Space Devices (Sp), ...

To take account of www.atomicprecision.com and www.upitec.org the returns for www.aias.us are rounded up by an estimated 50% giving 123,400 hits from 23,702 distinct visits, 12.60 gigabytes downloaded and 81,749 page views.

1- 16 January

US Universities, Institutes and Similar

Airforce Graduate University AFIT, Augusta State, Bard, Brigham Young, Caltech*, Carleton, Cornell*, Harvard, Lousiana State, Macon State, Mary Baldwin, MIT, Michigan State, Notre Dame, Northwestern, Old Dominion, Pittsburgh, Rochester Institute of Technology, San Diego Supercomputer Center, Stevens, Chicago, Girona (on edu), Maryland, UNCC*, Nebraska Lincoln, Federal Aviation Authority, Nuclear Regulatory Commission, Sandia, Patent and Trademark Office, Airforce AFNOC, Army Redstone Arsenal, Honeywell*, Nothrop Grumman, US Archives*.

Europe and Asia Minor: Universities, Institutes and Similar.

Mons, Ghent, Lleida Catalonia, EPF Lausanne, ETH Zurich, Univ. Zurich*, Charles Univ. Prague, German Synchrotron Facility (DESY), Bielefeld University of Technology, Max Planck Plasma Physics, Max Planck Halle, Max Planck Heidelberg, Otto von Guericke, Siemens*, Goettingen, Bonn, Erlangen, Jena*, Cologne, Saarland, CIT Denmark, Spanish Airports (AENA), ESA Spain, Gov. Andalucia, Spanish Railways, Social Security Ministry, Barcelona, Granada, Oviedo, Valencia*, Spanish

Government, INPL Nancy, Polytechnique Paris, Starsbourg, Paris Diderot, Poitiers, Rennes 1*, EduDSL Greece, Technion Israel, INFN Florence, Italian Institute of Statistics, Pontifical Gregorian Rome, Turin, Gorzow Poland, Olsztyn, BP Portugal*, Porto, UTCLUJ Romania, Siberian Academy of Sciences, Boras Sweden, Selcuk Tukey, Aberystwyth, Bristol, Cambridge, Edinburgh, Imperial*, Lancaster, Leicester, Manchester, Robert Gordon Aberdeen, Sheffield, Southampton, University College London, Ceredigion, Headlands School Yorkshire, Royal Society.

Rest of World: Universities, Institutes and Similar.
Melbourne, Sydney, Ministry of Defence Australia, UFS Brazil, NCF Canada, Simon Fraser, Alberta, Quebec Trois Rivieres*, Waterloo, TJ China*, UDEA Colombia, ITS Indonesia, MNIT India, Kyoto, Ritsumei, Tokyo Metropolitan, Waseda, KAIST Korea, UFTO Mexico, National Univ. Mexico, UNP Peru, UPAO Peru, New Pakistan, NUST, QAU, UE Pakistan, Gov. San Salvador, KMITL Thailand, Academica Sinica Taiwan, UC Venezuela, UCAB Venezuela, AIMS South Africa.

15 - 30 January

US Universities, Institutes and Similar
Georgia Regents, Berkeley, Caltech, Colorado*, Colorado State, Columbia*, Cornell*, Iowa State, Indian River, Johns Hopkins, Lehigh, Lousiana State, Colorado School of Mines, MIT*, Michigan State, North Dakota*, Old Dominion, Pasadena, Purdue, Rochester*, San Diego Supercomputer Center for U. S. Archives, Southern Illinois, Stanford, Texas A and M, Trinity College Connecticut, Chicago, UCLA*, UC Santa Barbara, UC San Diego, Girona (on edu), Maryland, Twin Cities Minnesota, North Carolina Charlotte*, North Florida, Utah, Texas Dallas, Texas, Wisconsin, U. S. Gov NOAA, Oak Ridge, US Gov Patent and Trademark Office, US Airforce Network Operations Center, US Defense Intelligence Agency, US Naval Marine Command Internet, Ford, General Electric, Honeywell, IBM, Intel, Northrop Grumman*, US National Archives*, Northwest Pennsylvania Tri County School District, Ludwig von Mises Institute Alabama, SENSE Indianapolis School District.

Europe and Asia Minor: Universities, Institutes and Similar.
Vienna, Xtec Catalonia, UDL Catalonia, EPF Lausanne*, Geneva, DLR Germany, KFA Juelich, MH Hannover, Albert Einstein Institute, Max Planck Halle, Siemens*, Goettingen, TU Berlin, TU Clausthal, TU Dresden, Bielefeld, Bonn, Jena, Leipzig, Regensburg, Rostock, CRT Denmark, KMD Denmark, Dinamic Orea Spain, EHU, ESA, Andalucian Parliament, Social Security, Autonomous Madrid, Complutensian Madrid,

Granada, Polytechnic Madrid, Seville*, Valencia, Valladolid, Spanish Parliament, Jyvaskyla Finland, Turku, Culture France, IHES, Engineering Institute Rouen, Nice Observatory, Poitiers, Rennes 1, Cyta Greece, Greek Orthodox Yazou, Trinity College Dublin Ireland, Hebrew University Jerusalem, Weizmann Institute, FS net Iceland, EMFCSC Sicily (INFN), INFN Florence, INFN Naples, Pontifical Gregorian University Rome, Wroclaw Poland, Lisbon Technical University, UTCLUJ Romania, Serbian Academy, Lebedev Institute Russian Academy, KM Russia, Kaluga, Nizhny Novgorod, DFRI Sweden, UMB Slovakia, Anadolu Turkey, Computing Cambridge, Maths Cambridge, Trinity Hall Cambridge, Hugh Baird Liverpool, Imperial*, Lancaster, Magdalen Oxford, Mansfield Oxford, Worcester Oxford, Peter Symonds College, Southampton*, York*.

Rest of World: Universities, Institutes and Similar.

INTI Gov. Argentina, Western Australia, BA Gov. Brazil, IMPA, PSI, UFC, UFS, Sap Paolo Brazil, GC Canada, Guardian Chemistry, McMaster, Simon Fraser, Manitoba, Ottawa, Quebec Trois Rivieres*, Toronto*, York University Toronto, UTFSM Chile, Beijing China, HA China, Engineering University Colombia, UDEA, UFPS Chile, ICRT Cuba, UEW Ghana, City University Hong Kong, IIT Kharagpur India, Bhabha Atomic Research Centre, India Railways, NKN, IMSC India, Kagoshima Japan, Tokyo, Japanese Particle Accelerator (KEK). UASLP Mexico, UGTO Mexico, Massey New Zealand, National Aids Council Secretariat Papua New Guinea, GIKI Pakistan, QAU, UOM Pakistan, Zeop Reunion Island, National University of Rwanda, NCTU Taiwan*, NTU Taiwan, Gov. Venezuela, AIMS South Africa, NMRL Zimbabwe.

Summary of February 2013

The overall returns for www.aias.us are given below in the Feb. 15 - 27 section. It is estimated that www.atomicprecision.com and www.upitec.org at 50% so the total returns for February 2013 are115,493 hits, 22,274 distinct visits each of an estimated order of ten readings; 77,690 page views, 12.03 gigabytes downloaded, and 2,618 documents read from 96 countries.

1-15 February

US Universities, Institutes and Similar.

Berkeley*, Buffalo, Caltech*, Claremont, Colorado, Columbia*, Cornell*, CSU Long Beach, Derxel*,Duke, Geneseo, George Mason, Gerogia Perimeter, Illinois, Macon State, Colorado School of Mines,

MIT*, Northwestern*, Oregon State*, Purdue North Central, Rutgers, San Diego Supercomputer Center, Anna Mendez System, Thiagarajar (on edu), UC Davis, Chicago, UC Merced, UC Santa Barbara*, Houston, Iowa, Illinois Urbana Champaign*, Michigan*, Minnesota Twin Cities*, Nebraska Lincoln, Pennsylvania*, Pompeu Fabra (on edu), University of Pittsburgh Medical Center, Texas*, Washington, Wisconsin*, Washington State, Fermilab, US Oceanographic and Atmospheric Administration, US Nuclear Regulatory Commission*, US Department of Agriculture, Government of Virginia, US Army Tobyhanna, US Archives*, Calhoun Intermediate School District Michigan, Ludwig von Mises Institute Alabama, Western Suffolk Regional Education Board, Cray Inc., Honeywell, Lockheed Martin, Microsoft, Philips.

Europe and Asia Minor: Universities, Institutes and Similar.

Vienna, Linz, EDIS Austria, Xtex Catalonia, CERN, Zurich, Neuchatel, ETH, UJEP Czechia, FH Bielefeld, HU Berlin, Jacobs, Ruhr Bochum, Siemens*, Freiburg, Oldenburg, Aarhus Denmark, Campusnet Estonia, CSIC Spain, TSAI, Barcelona*, Complutensian Madrid, Granada*, Valencia Polytechnic, Valencia, Valladolid*, AC Nantes, CH Rochefort, ENSI Caen, ENST Bretagne, INP Grenoble*, Polytechnique, UJF Grenoble, Lyon 1, Marseilles, Poitiers, Rennes 1, Savoie, Forthnet Greece, BME Hungary, UCD Ireland, Hebrew Jerusalem, Technion, INFN Padua, Rome 1, PRTC Lithuania, TU Eindhoven, UPC Netherlands, Free Univ. Amsterdam, UMB Norway, Warsaw, Mofnet Gov. Poland, Polish Army, Szczecin, Ministry of Culture Portugal, TUIASI Romania, MTS Russia, LTH Sweden, UMB Slovakia, MU Turkey, Tubitak Turkish Government, Kiev, Aberystwyth, Cambridge, Engineering Cambridge, Peterhouse Cambridge, Pembroke Cambridge, Durham*, Esher, Exeter*, Hills Road Sixth Form College Cambridge, Imperial*, Lancaster, National Army Museum, Lady Margaret Hall Oxford, Physics Oxford*, VPN System Oxford, Oxford University system, Southampton, Strathclyde, Penfro Educational System, SOTA Educational system.

Rest of the World: Universities, Institutes and Similar.

Argentine Nuclear Energy Department; UFPB Brazil, UFRJ, Unicamp, RDCRD Alberta, Banff Centre, Dalhousie, McGill, Perimeter, British Columbia, Calgary, Manitoba, Quebec Trois Rivieres, Toronto, Waterloo*, UDEC Chile, HB China, UNAL Colombia, Andes, IGA Cuba, City Hong Kong*, IIT Kharagpur India, IMSC India, Kagoshima Japan, Tokyo, TUA Technology Japan, Waseda, AMA Net, Japanese Gov. Kishou, MRT Sri Lanka, UAEH Mexico, INAOEP, UNAM, NU Singapore, KSU Taiwan, NCTU, Univ. South Africa, Pretoria, Witwatersrand.

15 - 27 February

For www.aias.us there were 76,995 hits, 14,849 distinct visits, 51,793 page views, 8.017 Gbytes downloaded, 2,618 documents read from 96 countries led by USA, Germany, Mexico, Canada, Britain, Australia, Greece, ...

All UFT papers read led by 25, 43, 88, 142, 177(Sp), 140, 140(Sp), 166, 177, 166(Sp), 157(Sp), 38, 107, 228, 54, 84, ...

All Essays read led by: 3, NP, 50, 35,6,9, 5,7, 64, 61, 51, 8, UME1, 32, 41(Sp), 48, 49, 29, 34, 55, 30, 40, 37, 36, 38, 47, 52, 26, 42, 57, 46, 25, 27, 24, 12, 62, 63, 56, 24(Sp), 20, 16, 60, 19, 39, 41, 53, 54, NIP, 11, 15, 44, 58, 10, 13, 14, ...

All documents and books read, led by: F3(Sp), Spacetime Devices, Spacetime Devices, F5(Sp), Autobiography, Spacetime Devices (Spanish) Johnson Magnets, HSD2 (Bangor), HSJ, F1(Sp), CV, ... ,

U. S. Universities, Institutes and Similar

Arizona, Berkeley, Caltech*, Cornell*, CSU Northridge, Duke, Fort Lewis, Georgia Tech., Harvard*, Johns Hopkins, Kansas State, Lehigh, Macon State, Miami, MIT*, New York, North Dakota State, Oklahoma State, Princeton, Penn State, Rochester, San Diego Supercomputer Center for US Archives, Southern Adventist, Stonybrook, Syracuse*, Chicago, UC Riverside, Delaware, Florida, Houston, Illinois Urbana Champaign, Massachusetts, Nebraska Medical Center, Texas, Texas Pan American, Western Illinois, U. S. Dept. Homeland Security, Lawrence Berkeley, Gov. Montana, Sandia National Lab., U. S. Army Medical Department, U. S. Archives, Ludwig von Mises Institute Alabama, General Electric Corporation Research, Lockheed Martin, Northrop Grumman, School Districts in Missouri, Oregon and Texas.

Europe and Asia Minor: Universities, Institutes and Similar.

Ghent, EPF Lausanne, ETH Zurich, Charles Univ. Prague, Helmholtz Geavy Ion Institute, Max Planck Physics Munich, Siemens Company, Goettingen, Regensburg, Rostock, Tuebingen, AAU Denmark, Campus Net Estonia, Tartu, Reykjavik Iceland, Spanish Gov. Presidencia, Spanish Ministry of Culture, Spanish Research Council, Barcelona, Juan Carlos III Madrid, Malaga, Polytechnic Catalonia, Seville, Helsinki, CNRS CRPP Bordeaux, Ecole Normale Superieure, Aviation Civile French Gov., Paris Institute of Astronomy, INSA Lyon, Maths Polytechnique, Marseilles 3, Angers, Nantes, Poitiers*, Rennes 1, XMCO France, HRBI Croatia, BME Hungary, Trinity College Dublin Ireland, University College Cork, International Centre for Theoretical Physics Trieste, Pisa, TU Delft, Institute of Physics Polish Academy of Sciences, Silesia at Katowice, Gdansk, Kiev, Christ's College Cambridge, Computing Cambridge,

Glamorgan, Keele, Middlesex, Nottingham, Hertford College Oxford, Jesus College Oxford, Oriel College Oxford, St. Edmund Hall Oxford, Plymouth, Sussex, Penfro* distance learning (Pembroke County Council).

Rest of World: Universities, Institutes and Similar.

MDP Argentina, DPF Gov. Brazil, UFPR Brazil, BCCRC Canada, McGill, Alberta*, UQTR*, Waterloo, York, Andes Colombia, Norte, IDU Gov. Colombia, Indian Statistical Institute Calcutta, Tezu, IMSC India, Kyoto Japan, Oita, Tokyo, UASLP Mexico, National Autonomous Univ Mexico, New Pakistan Univ., NCU Taiwan, TMU Taiwan, Pretoria South Africa.

Summary of March 2013

For www.aias.us there were 88,786 hits from 15,770 distinct visits, 10.805 gigabytes downloaded, 52,117 page views, 2,611 documents read from 93 countries, led by US, Ukraine, Britain, Russia, Mexico, Germany, Canada, and Colombia, ...

It is estimated that www.atomicprecision.com and www.upitec.org will increase thi sinterst by 50%, giving a total of 134,733 hits; 16.21 gigabytes downloaded, 23,655 distinct visits (of the order 240,000 readings) and 78,175 page views.

All UFT papers read, led by 25, 43, 107, 177, 177(Sp), 166(Sp), 88, 159(Sp), 26, 169, 140(Sp), 142, 54, 157(Sp), 140, 155, 41, 171(Sp), 175(Sp), 238, ...

All essays heard, led by 3(Light Deflection by Gravity), 9, Nobody's Perfect, 35, 139 Intro, 37, 36, 10, 62, 7, Univ. Myron Evans 1, 8, 5, Nicola Who?, Nobody is Perfect, 47, 57, 38, 19, 6, 30, 53, 54, 4, 42, 49, 25, 40, 50, 51, 48, 46, 29, 24(Sp), 11, 14, 18, 58, 61, 13, 20, 21, 12, 15, 16, 22, 26, 27, ...

All articles and books read, led by F3(Sp), Autobiography, F5(Sp), ...

1- 15 March

U. S. Universities, Institutes and Similar

Berkeley, SUNY Binghampton, Caltech, Central Michigan, Columbia, Cornell, Case Western Reserve, Drexel, Harvard, Los Rios, Colorado School of Mines*, MIT*, North Dakota, Northwestern, Princeton*, Rochester, San Diego Supercomputer Center, Stanford, Truman, Arkansas Little Rock. Chicago*, UC Irvine*, UCLA, UC Riverside, Delaware, Florida, Iowa, North Carolina Charlotte, North Texas, South Carolina, Washington, Wisconsin, West Virginia, Yale, US Department of Agriculture, US Airforce AFNOC, US Airforce SOAR, US Naval

Research Laboratory, Ford Corporation*, General Electric Corporation, Northrop Grumman Corporation*, U. S. National Archives*, US Certification Board of Nuclear Cardiology, Dalton School NYC, Milton Hershey School PA, Ludwig von Mises Institute Alabama, Trinity Health.

Europe and Asia Minor: Universities, Institutes and Similar.
   Medical University Vienna, TU Vienna, Xtec Catalonia, EPF Lausanne, FH Trier, Max Planck Heidelberg, Ruhr Univ. Bochum, Siemens Company*, TLS Tautenberg, TU Darmstadt, Frankfurt, Cologne, Oldenburg, Tuebingen, Aarhus, Spanish Air Ministry (AENA), Deusto, Andalucian Parliament, Ministry of Defence, Barcelona, Cadiz, Castile La Mancha, Complutensian Madrid, Oviedo, Polytechnic Madrid, Rivira i Virgili, Valencia, Helsinki, French Atomic and Alternative Energy Commission, Pierre et Marie Curie, Paris Observatory, Paris Psud, Angers, Lille 1, Poitiers*, Weizmann Institute Israel*, European Space Agency, Aquila, Pisa, Groningen, TU Delft, Oslo*, Univ. Life Sciences Norway, Wroclaw, Institute of Physics Belgrade, Moscow Engineering Physics Institute, Institute for High Energy Physics Moscow, Uppsala, Lund, Metu*, Kherson Ukraine, Odessa, Zaporizhe, Abeystwyth*, Bath, Cambridge*, Durham, Imperial, Nottingham*, Ravensbourne College, York*, British Library, Windsor and Maidenhead, Buckingham Schools.

Rest of World: Universities, Institutes and Similar
   Gov. Argentina, New South Wales, Defence Gov. Australia, UFV Brazil, Sao Paolo, Gov. Canada, Lakehead, Alberta, UQTR*, Toronto, Victoria, Waterloo*, Adventist Chile, Antioquia Colombia, Univ of the Americas Ecuador, Dokkyo Medical Japan, Tokyo, Nikon Co. Japan, Gov. Korea, Kyunghee Korea, National Autonomous University of Mexico, New Pakistan University, National Chiao Tung University Taiwan, National Taiwan University, Academica Sinica, African Institute of Mathematics Capetown South Africa, Stellenbosch University.

15 - 30 March

U. S. Universities, Institutes and Similar
   Berkeley, Binghampton, Caltech*, CSU Long Beach, Duke, Fullerton, Mayo, North Carolina State University*, Northeastern, Oklahoma State, Olin, Princeton, Purdue, Toronto, Chicago*, UC Irvine, UC San Diego, Illinois Chicago, Illinois Urbana Champaign, Massachusetts, Michigan*, Nebraska Medical Center, Wake Forest, Wisconsin, Western Michigan, U. S. Govt. NOAA, U. S. Govt. Oak Ridge Institute for Science and Education, U. S. Sandia National Laboratory, U. S. National Archives*,

Mormons, Ludwig von Mises Institute, Goodrich Corporation, Lockheed Martin, Northrop Grumman.

Europe and Asia Minor: Universities, Institutes and Similar.

TU Graz, Leoben Austria, TU Sofia, EPF Lausanne, ETH*, FHNW Switzerland, Zurich, FH Trier, FU Berlin, Siemens, TU Clausthal*, TU Ilmenau, Dortmund, Frankfurt, Tuebingen, Junta de Analucia*, Spanish Junta or Parliament, Spanish Ministry of Defence, Autonomous Barcelona, Castile La Mancha, Complutense Madrid*, TU Cartagena, TU Madrid, TU Valencia, Seville, Salamanca, Valencia, French Atomic Energy Commission, Institut Optique, Paris Observatory, UJF Grenoble*, Angers, Marseilles, Poitiers, Rennes 1, ATT Schools Greece, KBCCSM Croatia, Hungarian Academy Campi Budapest, Engineering Pisa, UASLP Mexico*, Autonomous National University Mexico, National Technical University of Norway Trondheim, UMB Norway, Jagiellonian Krakow Poland, Szczecin, Corbina Russia*, Royal Institute of Technology Stockholm, University of Stockholm High Energy Institute Moscow, Sumdu Ukraine, Kherson, Kiev, Odessa, Poltava, Aberystwyth*, Bristol, Mathematics Cambridge, Manchester, Nottingham, Strathclyde, British Library.

Rest of World: Universities, Institutes and Similar.

Latrobe, UFSC Brazil, UNESP Brazil, Campinas, Dalhousie Canada, Greater Victoria Public Library, Memorial Newfoundland, MB, Sheridan County, Quebec Government, Queen's, Quebec Trois Rivieres*, Toronto*, Victoria, Waterloo*, Western Ontario, HA China, UNAL Colombia, Valle Colombia, Science and Technology Hong Kong, CMC Vellore India, IITIDR India, National Knowledge Network India, IMSC India, BB Niigata Japan, UASLP Mexico*, UNAM Mexico, PERN Pakistan, AIMS South Africa, SUN South Africa.

Summary of April 2013

For www.aias.us there were 89,493 hits (files downloaded) from 16,851 distinct visits (order of 160,000 readings), 62,562 page views, 7.77 gigabytes downloaded, 2,636 documents read from 96 countries led by USA, Germany, Mexico, Canada, Austria, Italy, Britain, ...

All UFT papers read, led by 169, 43, 88, 177, 177(Sp), 166(Sp), 107, 166, 41, 157(Sp), 201, 94, 25, 142, 159(Sp), 140(Sp), 228, 175, 85, 237, 239, 63, 116, 150(Sp), 149, 150-B, 114, 146, 238, 2, 140, 152, 12, 168, 8, ...

All essays heard or read, led by: 3, 48, 42, 50, 40, NP, 56, TUO1, 46, 43, 19, 44, 41, 47, 9, 7, 26, 5, 6,30, 51, 14, 34, 38, 45, 53, 8, 57, I139, 55, TU1c, 24, 41(Sp), 11, 18, 29, ...

All articles and books read, led by: Fe3(Sp), Spacetime Devices (Sp), 2D, CV, Autobiography Part 1, ...

The total from www.aias.us, www.upitec.org and www.atomicprecision.com is estimated by adding 50% to the results from www.aias.us, giving 134,240 hits, 25,277 distinct visits, 11.66 gigabytes downloaded and 93,843 page views.

1 - 15 April

U. S. Universities, Institutes and Similar.

Berkeley*, Baldwin Wallace, California Polytechnic, California Institute of Technology*, Claremont, CSU Northridge, Dartmouth, Fullerton, Harvard, Indiana Purdue Indianapolis, Johns Hopkins, Los Rios, MIT, North Carolina State, North Dakota*, New York University, Ohio State, Olin, Princeton, Penn State, Purdue, Rhode Island School of Design, Rose State, Rutgers, Roger Williams, St. Olaf, Central Florida, Chicago*, UC Santa Cruz*, Illinois Urbana Champaign, North Carolina Charlotte, Southern California, Texas*, Tennessee Knoxville, Wisconsin Milwaukee, Worcester Polytechnic Institute, Brookhaven, Alaska Regional Weather Center, Oak Ridge, Naval Marine Corps, U. S. National Archives, Poudre School District, World Wide Web, Northrop Grumman.

Europe and Asia Minor: Universities, Institutes and Similar.

OEAN Austria, TU Vienna*, Linz*, JCT Austria, EPF Lausanne, Bern, Zurich, Charles Univ. Prague, CVUT Czechia, TUL Czechia, Max Planck Heidelberg, Siemens, TU Berlin, TU Darmstadt, Bielefeld, Heidelberg*, Leipzig, Oldenburg, Xtec Catalonia, Astrophysics Institute Canaries, Junta Andalucia, Spanish Social Security, Autonomus Barcelona, Barcelona*, Murcia, Spanish Distance Learning University, Technical University of Madrid, Seville*, Salamanca, Jyvaskyla Finland, Strasbourg, UHP Nancy*, Angers, Marseilles, Poitiers, NTU Athens, Trinity College Dublin, SIVIT Israel, INFN Ferrara*, Padua, Perugia, Groningen, Free Univ. Amsterdam, Joint Institutes for Nuclear Research Dubna Russia, NLB Slovenia, STUBA Slovakia, Middle East Technical University Turkey, Aberystwyth, Bristol, Brunel, Castlereagh College, Imperial College, Manchester, Oxford*, Eduroam Oxford, Sheffield, Southampton, Swansea, University College London.

Rest of World: Universities, Institutes and Similar.

UNR Argentina, RMIT Australia, Defence Australia, UFRJ Brazil, UNESP, Campinas, Sao Paolo*, British Columbia Canada, PWGSC Gov. Canada, TPSCG Gov. Canada, McGill*, Perimeter, Quebec Trois Rivieres*, Saskatchewan, Toronto*, Waterloo*, Atento Chile, Tsinghua China, Distance Learning Colombia, National Univ. Colombia, Norte Colombia, IIT Kharagpur India, the Institute of Mathematical Sciences India, Shahed Iran, Osaka Japan, Pohang South Korea, Autonomous Mexico, Auckland, Otago, UPA Peru, European Space Organization Southern Hemisphere, Presidency Paraguay, Lahore University of Management Sciences, National Taiwan*, Limpopo South Africa*.

15 - 29 April

U. S. Universities, Institutes and Similar.

Berkeley*, SUNY Buffalo, Caribbean, Claremont, Carnegie Mellon, Columbia*, Cornell*, Columbus State CC, CSU Chico, Dartmouth, Magnet Laboratory Florida State, Georgia Tech., Iowa State*, Lehigh, Louisiana State, Mississippi State, Michigan State, North Carolian State*, North Dakota, Mississippi, Pittsburgh, Princeton*, Rochester, Rutgers, Southeast Missouri State*, Texas A and M, Tufts, UC Davis, Central Florida, Chicago, UC Santa Barbara*, Florida, Missouri St Louis, Nebraska Lincoln, Utah, Texas Dallas, Tennessee Knoxville, Vermont, Wisconsin, William and Mary, Western Michigan, West Virginia, Los Alamos, NIST, Alaska Weather Center of NOAA, Navy Marine Command, U. S. National Archives*, Ludwig von Mises Institute Alabama, Radnor Township School District Pennsylvania, Northrop Grumman, Raytheon.

Europe and Asia Minor: Universities, Institutes and Similar.

Linz, EPF Lausanne, ETH Zurich, Charles Univ. Prague, TUL Czechia, Bosch, Free Univ. Berlin, RWTH Aachen, Augsburg, Konstanz, Oldenburg, SDU Denmark, European Trade Union, SMIT Estonia, IAC Spain, Spanish National Aerospace Institute, Ministry of Defence, Ministry of Railways, Almeria, Madrid Complutense, Madrid Polytechnic, Salamanca, Valencia*, Vigo, Polytechnique de France, Sergy, Marseilles, Poitiers*, Rennes 1, EDUDSL Greece, Universit College Dublin, INFN Florence, Modena, Pisa, Leiden, Free Amsterdam, PW Poland, UT Lisbon, Corbina Russia, High Energy Institute Moscow*, Aberystwyth, Bristol, Cambridge*, Sidney Sussex Cambridge, Castlereagh, Deeside, Durham, Goodenough, Imperial, Luton, Manchester, Newcastle upon Tyne*, Hertford College Oxford, Sheffield Hallam, Ministry of Defence*,

Llyfrgell Genedlaethol Cymru (National Librray of Wales), Oxford University Eduroam.

Rest of World: Universities, Institutes and Similar.

MDP Argentine Government, Library of the National Argentine Congress, Ministry of Economics, Ministry of Planning, New South Wales Australia, Queensland, UF Santa Catarina Brazil, Sao Paolo, Gov. British Columbia Canada, McGill*, McMaster, Quebec Trois Riviers*, regina, Toronto, Chile*, UCT Chile, Tsinghua China, Ceva Colombia, Distance Learning, National Univ. Colombia, Valle, UCLV Cuba, Science and Technology Hong Kong, Banaras Hindu India, Indian Statistical Institute Calcutta, Indian Railway, RRCAT India, IIAP, IMSC India, tokyo Medical Japan, Tohoku, Tokyo, Waseda*, Aerospace Exploration Japan, Korean Advanced Institute of Sceince and Technology, Kynghee, Postech, SKKU Korea, UAEH Mexico, ININ Gov. Mexico, IPN, UIA, National Autonomous* Mexico, Auckland New Zealand*, QAU Pakistan, National Tawan*, Academica Sinica, Limpopo South Africa, Natal Pietermaritzburg, Pretoria, St. John's College Johannesburg, National Physical Institute Uruguay.

END OF NINE YEARS OF RECORDING

Summary of May 2013

For www.aias.us there were 92,806 hits from 17,351 distinct visits, 12.54 gigabytes downloaded, 64,324 page views, 2,691 URL's from 102 countries, led by US, Germany, Mexico, Canada, Russia, Argentina, China, France, Spain, Brazil, India, Colombia, Britain, Japan, ...

All UFT papers read, led by 25, 177(Sp), 88, 94, 166, 39, 142, 43, 228, 157(Sp), 169, 140, 159(Sp), 142(Sp), 110, 175(Sp), 114, 149, ...

All essays heard, led by: 3(LDG), 14, 16, NP, 11, 10, 9, 52, 12, 20, 18, 53, 54, 55, 24, 25, 56, I139, 58, 62, NW, LME1, 24, 22, 34, 38, 4, 40, 15, 21, 23, 32, 35, 37, 39, 41, 42, 13, 27, 30, 36, 17, 19, 26, 44, 5, 45, ...

All Articles etc. read, led by: F3(Sp), F5(Sp), Potential Waves, Autobiography Volume One (Auto1), Galaxies (Sp), ...

The total from www.aias.us, www.atomicprecision.com and www.upitec.org is estimated to be 50% higher than www.aias.us giving 139,209 hits, 26,027 distinct visits, 18.81 gigabytes.

U. S. Universities, Institutes and Similar
Baylor, Berkeley, Boston University, Caltech, Carleton, City Colleges of Chicago, Colby, Colorado, Georgia, Southern, Hawaii*, Hobart and William Smith Colleges, Karlsruhe Institute of Technology (on edu), Louisiana Tech., Maine, McDaniel, Miami, Missouri, MIT, Minnesota State Mankato, Montana, Tiruchirappalli India (on edu), Northern Kentucky, Northwestern*, New York University, Pittsburgh, Regis, Santa Clara, Southern Methodist, Stanford, St Cloud State, Tufts, Alabama, UCLA, UC Santa Barbara*, UC Santa Cruz, Univ. Iowa*, U Mass, Michigan, Minnesota Twin Cities, North Carolina Chapel Hill*, Nevada Las Vegas, University System of Maryland, Texas*, Tennessee Knoxville, Wisconsin Parkside, Virginia, Vermont, Washington*, Wisconsin*, Yeshiva; U. S. Govt. FAA, NASA (Glen), NOAA (Alaska Regional), U. S. Airforce AFNOC, U. S. Army AMRDEC, U. S. Navy NSWC, U. S. National Archives*, Army and Navy Academy, Christiana Healthcare Delaware, Houston PC Network Texas, Los Angeles Public Library, Queens Medical Center Hawaii, General Electric.

Europe and Asia Minor: Universities, Institutes and Similar
Johannes Kepler Austria, Edis Austria, JKU Austria, Neuchatel Switzerland, Bosch Company Germany, Max Planck Leipzig for Mathematics in Science, Siemens Company, Students Bonn (STW), TNG, Bonn, Heidelberg, Magdeburg, Muenster, Aarhus Denmark, Niels Bohr (Copenhagen), Endesa Spain, Junta Andalucia, STL Spain, TSAI Spain, Navarra, Extremadura, Pablo de Olavide Seville, Valencia Tech, Seville, Valencia*, VIAS Spain, Jyvaskyla Finland, Oulu Finland, EC Nantes, ENPC France, Bordeaux 1, Poitiers, Trinity College Dublin Ireland, Univeristy College Dublin, Aruba Italy, ENEA, INFN Bologna, INFN Padua, Milan Tech., Turin, IST Lithuania, Free Univ Amsterdam, University of Oslo, Jagellonian Krakow Poland, UNS Serbia, Corbina network Russia, Chalmers Sweden, IBE Slovenia, Matej Bel Slovakia, Middle East Tech Turkey, Aberystwyth, Bristol, Churchill College Cambridge, Cambridge*, St John's Cambridge, Manchester, Trinity College Oxford, Oxford*, Sanger Institute, Sheffield Hallam, Southampton, Strathclyde, Sussex, Swansea, University College London*, East Anglia, Llyfrgell Genedlaethol (National Library of Wales).

Rest of World: Universities, Institutes and Similar.
Argentine Ministry of Economics, Adelaide, Monash, Sidney*, Melbourne*, Sao Paolo*, Dalhousie Canada, Memorial Newfoundland, Perimeter, Montreal Tech., British Columbia, Winnipeg, Pontifical

Catholic Chile, Pontifical Catholic Valparaiso, University of the Frontier Chile, Federico Santa Maria Chile, Tsinghua China, Distance Learning Colombia, Antioquia Colombia, Andes Colombia, UCF Cuba, Tezu India, Indian Railways*, Ritsumei Japan, Titech , Tokoshima, Waseda, Osaka Institute, Mexican National Polyetchnic Institute, Iberoamericana Mexico, National Autonomous* Mexico, Otago New Zealand, Pontifical Catholic Peru, Antenor Orrego Peru, National Chiao Tung Taiwan, National Taiwan*, Academica Sinica Taiwan.

15 - 30 May

U. S. Universities, Institutes and Similar.

Central Connecticut State, Lehigh, MIT, Michigan State, Texas A and M Commercial, Univ. California (UC) Merced, UC San Diego, Maryland, Virginia, NASA Langley, Alaska regional Weather Center of NOAA, US Army Fort Monroe, US National Archives*, Ludwig von Mises Institute Alabama, Bosch (US), Hewlett Packard, Lockheed Martin, Northrop Grumman, Scarsdale Schools New York City.

Europe and Asia Minor: Universities, Institutes and Similar.

Johannes Kepler Linz*, Xtec Catalonia, Geneva, CHMI Czechia, Audi Germany, Max Planck Dresden, MWN*, Siemens*, TU Chemnitz, TU Darmstadt, TU Dortmund, Bonn, Bremen, Heidelberg*, Karlsruhe, Leipzig, Mainz, Wuerzburg, AAU Denmark, Junta de Andalucia*, Autonomous Barcelona*, Autonomous Madrid, Barcelona, Complutense Madrid*, Malaga, Canaries, Catalonia Tech., Pablo de Olavide Seville,Valencia*, Spanish Parliament or Xunta, Brukenthal Museum Romania, Jyvaskyla Finland, UTA Finland, Nantes*, ENPC France, INPG, UHP Nancy, Rochelle, Paris Diderot, Poitiers*, Rennes 1, Aristotle Thessaloniki Greece, Meliusz Hungary, Hebrew Univ. Jerusalem, ICTP Trieste, INFN Rome 1, Pavia*, TU Eindhoven, Twente, Jagiellonian Krakow Poland, Lisbon Tech Portugal, ECRO Romania, INFIM Romania, STSISP Romania, UAIC Romania, Corbina System Russia, Glonass-IAC Russia, SWEB Russia, VPSNow Russia, HIAST Syria, NLB Slovenia, HUN Turkey, Middle East Tech*, Aberystwyth*, Birmingham Central, Bristol, Churchill College Cambridge, Cambridge University, St. John's College Cambridge, Edinburgh, Imperial*, Lancaster, Manchester, St Edmund Hall Oxford, Oxford University (Eduroam), Warwick, York, Ministry of Defence*, Secret Intelligence Service (MI6).

Rest of World: Universities, Institutes and Similar

UNLP Argentina, Adelaide, Latrobe, Monash*, Sydney, Melbourne, Univ. Sydney, Dakhar Bangladesh, UFPE Brazil, YRBE Ontario,

Perimeter Institute, Saskatchewan, INACAP Chile, UFRO Chile, UTFSM Chile, ZJXU China, UNAL Columbia*, Tecnologica, Valle*, RIMED Cuba, UH Cuba, ESPOL Ecuador, UGM Indonesia, IUAC India, Kyoto Japan, Tsukuba, Tokyo, Waseda, KCT Japan, Korea Advanced Institute of Science and Technology, University of Korea, SNU Korea, MRT Sri Lanka, CICESE Mexico, SESLP Gov. Mexico, UAEM Mexico, National Autonomous* Mexico, Auckland Tech New Zealand, UPAO Peru, UTP Peru, Panama, QCU Pakistan, IBA_SUK Pakistan, NTU Taiwan, Academica Sinica Taiwan.

Summary of June 2013

For www.aias.us there were 79,172 hits from 16,156 distinct visits, 11.76 Gbytes downloaded, 55,349 page views, 2721 documents read from 96 countries, led by USA, Britain, Mexico, Germany, Canada, China, France, Spain, ...

All UFT papers read, led by: 177(sp), 166(Sp), 88, 43, 140, 25, 177, 107, 150B, 166, 142(Sp), 142, 170(Sp), 149, 94, 140(Sp), 169, 159(Sp), 26, 175, 2, 175(Sp), 228, 42, 39, 171(Sp), 57, 169(Sp), 38, 65, 18, 41, 118, ...

All Essays heard, led by 3, Nobody's Perfect, 18, 14, 111, 51, 61, 9, LME1, 29, 54, 26,12, 20, 44, 24, 25, 27, 30, 7, 45, 42, 60, 48, 8, 139(Intro), 40, 46, 6, LME, 58, 10, 13, 15, 22, 23, 35, 36, 38, 19, 32, 34, 37, 41, 21, 4, 47, ...

All Articles read, led by F3(Sp), Autobiography Volume One, Potential Waves, Johnson Magnets, HS Jackson, Triple Orbit Simulation, LCR Resonant, F5(Sp), Homopolar, Levitron, F1(Sp), CEFE, LCR Resonant Part 2, Space Engineering (Sp), Galaxies (Sp), 2D, CV, Autobiography Volume Two, ...

The total from www.aias.us, www.atomicprecision.com and www.upitec.org is estimated to be 50% higher than www.aias.us, giving 117,888 hits, 24,234 distinct visits, 17.64 gigbytes, 83,023 page views during the month.

1 - 15 June

US Universities, Institutes and Similar

California Polytechnic, Chemeketa, Columbia, Illinois, Lipscomb, Macon State, MIT, Montgomery College, Northwestern, Old Dominion, Ohio, Rochester, Chicago, UC Los Angeles, UC Santa Barbara, UC San Diego, Massachusetts, Massachusetts Boston, Maryland, Michigan, Utah, Washington St Louis, NASA, US Gov. Nuclear Regulatory Commission, US Naval Marine Command, Naval Research Laboratory, US National

Archives*, Mormon Organization, Ludwig von Mises Institute Alabama, Chevron, General Electric*, Lockheed Martin, Northrop Grumman, Norton, Raytheon, Rockwell Collins.

Europe and Asia Minor: Universities, Institutes and Similar.
Linz*, Vienna, EDIS Austria, Johannes Kepler, Catholic Univ. Leuven, Xtec Catalonia, CERN, Swiss Federal Institute Lausanne, ETH Zurich, Norwest Swiss Tech (FHNW*), Bern, CHMI Czechia, KFA Juelich, Max Planck Science of Light, Siemens*, Dortmund, Frankfurt, Heidelberg*, Muenster, Regensburg, Tartu Estonia, Spanish Research Council, Valencian Autonomous Community, Andalucian Parliament (Junta), Autonomous Barcelona*, Autonomous Madrid, Complutense Madrid*, Spanish National Distance Education, Oviedo, Polytechnic Catalonia, Seville, Valencia, Spanish Parliament (Xunta)*, ESPCI France, French National Particle Laboratory, LPTMS Paris Psud, Marseilles, Poitiers, Eszszk Hungary, Meliusz Hungary, Hebrew University Jerusalem, Vedur Iceland (Metereological Office), Astronomy Bologna, Itagile Italy, Rome 1, RISTENANational Education and Research Luxembourg, TU Eindhoven*, Groningen, Lodz Poland, Poznan, UTCLUJ Romania, UNS Serbia, MPEI Moscow, Pereslavl, Vologda, Volkswagen Sweden, Bath, Bristol, Cambridge*, Heriot Watt, Imperial*, Luton, Manchester, Somerville College Oxford, Eduroam Oxford, Southampton, University College London, University of Wales Institute Cardiff, British Library site download (Safari and WCT), BBC* (cwwtf*, mh*, telhc*, thls), Middle East Technical University, Turkey.

Rest of World: Universities, Institutes and Similar.
Australian National*, MG Gov. Brazil, UERN, UFRGS, UNESP, Sao Paolo*, Edmonton, YRBE Ontario, Montreal Tech., Simon Fraser*, Guelph, Toronto*, Waterloo, Pontifical Catholic Valparaiso Chile, Federico Santa Maria Tech Chile, Chinese National Supercomputer Center (tj.cn), National Ditance Learning Colombia, Indian Railways, Aoyama Japan, Hokudai, Kyoto, National Institute for Fusion Science, Osaka, Tokyo, Waseda*, Gov. Mexico, INAOEP Mexico, IPN, UIA, National Autonomous University Mexico*, Auckland University of Technology New Zealand, UNP Peru, UPC Peru, UOM Pakistan, PERN Pakistan, ISEAS Singapore, National Taiwan, Capetown South Africa.

15 - 29 June

US Universities, Institutes and Similar.
Arizona, Bard, Carnegie Mellon, Colorado, Columbia*, Cornell, City University New York, Florida State, Georgia Tech., Illinois State,

Lousiana State, MIT, Michigan State, Notre Dame, NorthWestern Health Sciences University, Ohio, Stanford*, UCLA, North Carolina Charlotte, Wisconsin Lacrosse, South Florida, Texas Arlington, Wisconsin, Lousiana Government, NIST, Social Security Administration, US Army Dugway Proving Ground, Naval Marine Command, Ludwig von Mises Institute, Nothrop Grumman*, Norton, Raytheon.

Europe and Asia Minor: Universities, Institutes and Similar

Free Univ. Brussels, EPF Lausanne*, Northwest Swiss Tech., Zurich, German Synchrotron, FH Duesseldorf, TU Berlin, TU Ilmenau, Heidelberg*, Jena, Leipzig, Mainz, Regensburg, Spanish Research Council IATA, Spanish Research Council IFF, Barcelona, Complutense Madrid, Granada, Polytechnic Catalonia, Polyetchnic Madrid, FDN France, Paris Observatory, French Aviation Authority Osiris, Poitiers8, Sivit Israel, UPS Toulouse, INFN Milan, Otalige, NGI, Milan Tech., Bologna, Milan Bicocca, Groningen, Eindhoven, Oslo, Warsaw, Joint Institutes of Nuclear Research Dubna Russia*, Russian Particle Laboratory,Vympelstroy, Uppsala Sweden, Middle East Technical University*, Aberystwyth, Durham, Glamorgan, Nottingham, British Library, BBC*, Oxfam.

Rest of World: Universities, Institutes and Similar

Australian Gov. Defence, RN Gov. Brazil, UFJF, UFSC Brazil, ODY Canada, Victoria, Waterloo, UNACH Chile, HDU China, Tienjin China*, UCF Cuba, UGM Indonesia, SGSITS India, IIAP India, Chiba Japan, Hiroshima, Osakam Univ. Electrocommunications, Waseda, MESH Corporation* Japan, New Pakistan, PERN, UST Hong Kong.

Summary of July 2013

For www.aias.us there were 77,313 hits, 15,314 distinct visits, 52,301 page views, 8.87 gigabytes downloaded and 2,733 documents read from 105 countries, led by USA, Germany, Mexico, Canada, Britain, Colombia, Australia, ...

The total from www.aias.us, www.atomicprecision.com and www.upitec.org is estimated to be about 50% higher, giving 115,970 hits, 13.31 gigabytes, 22,971 distinct visits and 78,452 page views.

All UFT papers read, led by 43, 54, 243, 88, 177(Sp), 166, 177, 166(Sp), 85, 25, ...

All essay broadcasts heard, led by: 3, 32, 12, 26, 27, NP, 35, 34, 23, 24, 59, UOMEd, UOMEa, 25, 52, 64, 20, 56, 21, 47, 55, 37, 62, 19, 33, 46, I139, UOMEb, 18, 29, 30, 63, 28, 36, 40, 60, 10, 4, 50, 7, 49, 54, 58, 25, 38, 41, 42, 44, 45, 57, 9, 11, 14, 15, 39, 5, 51, 53, 6, 61,UOME,22, 8, ...

All articles read, led by F3(Sp), Auto1, Potential Waves, Droca Dailogue, Johnson Magnets, F5(Sp), CEFE, LCR Resonant, Leviron, Double Slit, MWE CV, Human Soul Jackson, Hompolar, Triple Orbit Simulation, Auto2, ...

1 - 15 July

US Universities, Institutes and Similar

Boston College, Biola, Chicago, Columbia*, Harvard, Karslruhe Inst. Technology (on edu), Missouri, Michigan State, New Mexico State, Penn State, St. John's College, Texas A and M, UC Santa Cruz*, Maryland*, Nevada Las Vegas, Texas Dallas, Wisconsin, Alaska Regional Center NOAA, Gov. Phoenix, Virginia, US Navy NMCI Jacksonville, US Navy SPAWAR, US Special Operations Command, Argonne National Lab*, U. S. National Archives, Ludwig von Mises Institute, Kentucky Education Department.

Europe and Asia Minor: Universities, Institutes and Similar

Innsbruck*, CERN, ETH, Siemens, Bonn, Frankfurt, Karlsruhe, TU Kaiserslautern, Stuttgart, Tartu Estonia, Basque Country, Junta de Andalucia, Spanish Ministry of Defence, Barcelona, Seville, French Atomic and Alternative Energy Commission, OVH France, Marseilles 3, Strasbourg, Poitiers, Technion Israel, Arcetri Observatory Italy, ICTP Trieste, INFN Naples, NCS Group Italy, Eurocentre Pavia, Radboud Netherlands, Groningen, Royal Society, Warsaw, Lisbon Tech., Kubang Russia, Metu Turkey, Selcuk, Aberystwyth, University College London, Oxford.

Rest of World: Universities, Institutes and Similar.

Monash, University of Western Australia*, British Columbia Canada, Waterloo, CCU Chile, HA China, Tienjing China*, Engineering School Colombia, UCLV Cuba, UO Cuba, UGM Indonesia, Tezu India, Indian Rail, Tata Institute Bangalore, Aizu Japan, Waseda*, Otago New Zealand, National Univ. Singapore, VNU Taiwan, Capetown South Africa, Zululand.

15 - 30 July

US Universities, Institutes and Similar.

Berkeley, Buffalo, Cornell*, Fort Hays Kansas, MIT, Mississippi State, Northern Arizona, Northwestern, Oregon State, Princeton, Penn State, St John's College California, SUNY Stonybrook, Arkansas, UC Santa Cruz, Michigan*, UNCC*, UNC Greensboro, Oregon, Utah*, Texas, Wisconsin

Whitewater, Vermont, Wisconsin, Yale, Brookhaven, Alaska Regional NOAA, Navy NMCI*, Navy USMC, US National Archives*, Buffalo Library, Ludwig von Mises Institute Alabama, World Wide Web Consortium, Intel Corporation, Lockheed Martin, Raytheon, Sunnyvale California, Kentucky Schools.

Europe and Asia Minor: Universities, Institutes and Similar.
   Vienna, ETH Zurich, Max Planck Nuclear Physics Heidelberg, RWTH Aachen, Siemens*, Students STW Bonn, TU Berlin, TU Darmstadt, TU Dresden, Bayreuth, Essen, Hannover Medical Centre, Hannover, Jena, Karlsruhe, Oldenburg, Bioacademy Greece, Madrid Autonomous, Madrid Complutense, Madrid Polytechnic, Juan Carlos University, Valencia, ENS Lyon, French Gov. Ecology etc Aviation Department*, French National Particle Laboratory, Marseilles 3, Nantes Brittany, Poitiers*, Rennes, LCAR Toulouse, Patient Treatment Register Ireland, Hebrew Univ Jerusalem, INFN Milan, Calabria, Pisa, ONS Student Net Eindhoven, Silesian Data Centre* Poland, Wroclaw, NIPNF Romania, Serbian Academy, Samara Russia, Kubang, Vympelstroy*, Kiev Ukraine, Bristol, Cambridge*, Keele, Univ. Wales Lampeter, Lancaster, Univ Wales Swansea,

Rest of World: Universities, Institutes and Similar.
   Dakhar Bangla Desh, UFPEL Brazil, UFPA, UFBP Brazil, Memorial Newfoundland Canada, National Capital Freenet, Queen's, Simon Frazer, British Columbia, Waterloo, CCU Chile, Tienjing China*, Engineering University Colombia, Pontifical Bolivarian University Colombia, IITK India, IITKGP India, Tokyo Tech., Tohoku, Aizu, Tokyo, Wasedu, Yokohama National Japan, UPD Philippines, ITS Uruguay.

Summary of August 2013

There was a server crash on 28 / 8 / 13, so the www.aias.us feedback is known from 1 - 27 August during which there were 56,845 hits from 9,775 distinct visits. This extrapolates to 65,266 hits from 11,223 distinct visits for 31 days. The total from the three sites is as usual estimated to be 50% greater giving 97,839 hits from 16,834 distinct visits. Other types of feedback data were lost due to the server crash (i.e. UFT paper, essays and articles and books). The three ECE sites are as usual www.aias.us, www.atomicprecision.com and www.upitec.org.

1 - 15 August

US Universities, Institutes and Similar.

Amrita (edu), Bowie State, Cornell*, Dartmouth, Emory, Harvard, Information Sciences Institute, Karlsruhe Institute of Technology (on edu), Princeton, Rochester, San Diego Supercomputer Center, SUNY Stonybrook, Texas A and M, Temple, Illinois Urbana Champaign, Tennessee Knoxville, Washington*, Yale, US Census Bureau, NASA, NIST, NSF, Argonne Guest House, US Archives*, IMC Net, Ludwig von Mises Institute, AEP, Honeywell, Hewlett Packard, Northrop Grumman, Wolfram.

Europe and Asia Minor: Universities, Institutes and Similar.

Swiss Federal Institute Lausanne, IBM Zurich, Bosch, Siemens, SPECS Berlin, TU Clausthal, TU Erlangen, Muenster, Tartu, Complutense Madrid, Technical University of Cartagena, French Government Aviation Civile, UPS Toulouse, Tel Aviv, Technion, Italian INAF, Florence, TU Eindhoven, Silesian Data Center, Wroclaw, Technical Univ. Lisbon, SGA Slovenia, EFX Romania, INFIM Romania, EMS Serbia, Avant Guard Russia, Bashtel, Corbina system, Kubang, Orel, PTL system, Samara, Vympelstroy, Chalmers Univ. Sweden, Bristol, Cambridge, Oxford, Swansea, Penfro Distance learning, SOTA system, National Library of Wales.

Rest of World: Universities, Institutes and Similar.

Deakin Australia*, Melbourne, Ministry of Defence Australia, Arcadia Univ. Canada, McGill, NCF Canada, Simon Fraser, Alberta, Tianjin China*, UDEA Colombia, Inca Univ. Colombia, UPB Colombia, RRCat (Railway System) India, Tsukuba, Tokyo, Waseda, Samsung South Korea, CICESE Mexico, IPN Mexico, UNAM Mexico, UITM Malaysia, National Institute of Technology Peru, National University Singapore, KMITL Thailand, NDHU Taiwan, NWU South Africa, Capetown.

15 - 27 August

U. S. Universities, Institutes and Similar

Alashka County Public Schools, Minnesota Supercomputer Institute, Cornell*, Georgia Tech., Yale, Harvard, Monroe Community College Rochester*, U. S. Archives*, Illinois Urbana Champaign, UC Irvine, Texas, Fermilab, Raytheon, Indianapolis Public Library, Loyala Health Center, State University of New York Stonybrook, Los Alamos, Minnesota, NIST, Maths, Illinois Urbana Champaign.

Europe and Asia Minor: Universities, Institutes and Similar

Pierre et Marie Curie, Lisbon Tech., Duesseldorf, Duisburg Essen, Esen, Greek Research and Technology Network, UW Cardiff Graduate

Centre*, Poitiers, Silesian Data Center Poland, Geocentrum Czechia, Manchester, TU Dresden, Jyvaskyla Finland, Marie de Liverdun, Engineering Cambridge, CU Leuven, Zurich, Frankfurt, Wroclaw, German Aerospace Centre, Siemens, Aerospace Institute Paris, Tel Aviv, Agder Norway, Durham, Ysgol CCC, Humboldt Berlin, Bar Ilan, Maths Oxford, Maths ETH Zurich, Istella Italy, Gov. Baden Wurttenberg, Royal Society of Chemistry, Chemistry Oxford, IBM Zurich, SPECS Berlin.

Rest of World: Universities, Institutes and Similar.

McGill, Ecuador Ministry of Agriculture, Tezpur India, Mexican National Polytechnic Institute*, National Univ. Mar del Plata Argentina, Harish Chandra Institute India, Johannesburg, Pretoria South Africa, ANU Mexico, Witwatersrand South Africa, Enaex Mining Chile*, Antioquia Colombia, IM Sceinces Chennai India, Inst, Materials Research NAU Mexico, IMS Sao Paolo Brazil, Tianjing China, IMS Chennai India, Science Museum NAU Mexico, Catolica del Norte Chile, Japanese Atomic Energy Authority, NU Peru Piura, Presidency of Uruguay Montevideo, NU Singapore, Australian National University, Gov. British Columbia Canada, TU Chile, NTU Peru, KMITL Thailand.

Summary of Sept. 2013

There was an outage from 28th August to 6th September due to a server crash so the feedback starts from 7th. September as follows.

For www.aias.us there were 52998 hits, 12332 distinct visits, 34145 page views and 9.585 gigabytes downloaded from 7th to 30th Sept., which scales up to 69109 hits, 12.50 gigabytes downloaded, 16080 distinct visits and 44525 page views from 1 - 30 Sept. Visits from 7 - 30 Sept. Studied 2578 documents from 97 countries, led by USA, Britain, Germany, Mexico, Canada, China, France, ...

All UFT papers read, led by 25, 228, 166(Sp), 166, 43, 177(Sp), 142, 177, 243, 237, 169, 88, 57, 244, 142(Sp), 159(Sp), 175, 157(Sp), 140, 94,41,149, 150-B, ...

All essays heard and read, led by: 4, 40, 39, 38, 34, 36, 62, 3, 45, 42, 8, 32, 29, 30, 35,25, 59, 44, 53, 6, 41, UMEa, 56, 28, 12, 61, 64, 60, 46, 46,49, 51, 54, 63, 9, 22, 24, 47, 50, 52, 55, NW, NP, NIP, 24(Sp), 27, 37, 57, 7, 23,26, 48,5, UMEd, ..

All articles read, led by F3, F5, Autobiography volume one, Space Energy (Spanish), LCR Resonant, ...

For the three ECE sites www.aias.us, www.atomicprecision.com and www.upitec.org these totals increase by an estimated 50% to 24,120 distinct visits, 103,663 hits, 18.75 gigabytes downloaded, and 66,788 page views

7 - 15 September

U. S. Universities, Institutes and Similar.
    Colorado, Colorado State, Columbia, Fullerton, Georgia Tech, Illinois, Kentucky Community and Technical College System, Mineral Area College Missouri, Monroe Community College Rochester, Ohio State, Princeton, Rice, University of California San Diego Supercomputer Center for US National Archives*, Stanford, Stonybrook, Texas A and M*, Central Florida, UCLA, Georgia, Michigan, Minnesota Twin Cities, Utah*, Texas*, Virginia, Wheaton, Wisconsin, Washington State, Alaska Regional Weather Center, US Naval Marine Command, Philips Research, US National Archives*, Eden Prairie Schools Minnesota, Department of Regulatory Agencies Gov. Colorado, Western Oklahoma State College.

Europe and Asia Minor: Universities, Institutes and Similar.
    Department of Justice Belgian Government, EPF Lausanne, Helmholtz Centre Berlin, Max Planck Plasma Physics, Max Planck Nuclear Physics Heidelberg, Ruhr Bochum, RWTH Aachen, TU Clausthal, Duisburg Essen, Karlsruhe, Mainz, Tartu Estonia, Valencia Spain, French Atomic and Renewable Energy Authority, EC Lyon, Poitiers*, Milan, ISP Fabriek Netherlands, Silesian Data Center Poland, Sverdlovsk Orthopaedic Enterprise Russian Federation, Blekinge College of Technology Sweden, Human Relief Foundation, Lviv National Sceintific Library Ukraine; Imperial College London, British Library, Birmingham City Council, Financial Times.

Rest of World: Universities, Institutes and Similar
    Australian National, Sydney, UFRGS Brazil, UNESP Brazil, Quebec Trois Rivieres, Engineering and Informatics Association of Chile, HA China, Tianjing China, UDEA Colombia, NISER India, IIT Kharagpur, Harish-Chandra, Osaka, Waseda, PDN Sri Lanka, CUDI Mexico, ITESM Mexico, ADDU Philippines, University edu system Pakistan, Chainat Thailand, NTHU Taiwan, TFN Taiwan.

15 - 30 September

US Universities, Institutes and Similar
    Boston University, Buffalo, Central New Mexico Community College, CSU Fresno, Catholic University of America, De Paul, ECPI, Georgia Tech., Harvard, Hawaii, Illinois, Kansas State, MIT*, Monroe Community College, Michigan State, North Carolina State*, New York, Mississippi, Princeton, Penn State, Rice, Rutgers, South Carolina, UC San Diego

Sueprcomputer Center, SUNY Stonybrook*, Texas A and M*, Temple, Cincinnati, Ibadan (on edu), Illinois Chicago, Massachusetts, Minnesota Twin Cities, North Carolina Asheville, North Carolina Charlotte, Nebraska Lincoln, Nevada Las Vegas, Utah, Texas Dallas, Texas, Texas Pan American, Vermont, Wheaton, Wisconsin*, Lawrence Livermore, NASA Goddard, Alaska Weather Center, Aok Ridge, Naval Marine Command, US Archives*, Marquette General Hospital, City of Sunnyvale, City of South Bend.

Europe and Asia Minor: Universities, Institutes and Similar.
   IIASA Vienna, TU Vienna, Free Univ. Brussels, CERN, Bern, Zurich, Zurich Applied Sciences, Jacobs International Bremen, RWTH Aachen, TU Berlin, Heidelberg, Saarland, Royal Library Denmark, Niels Bohr Institute, Tartu Estonia, Dinamicarea Spain, Basque Country, Madrid Autonomous, Navarra, Barcelona Tech., Jyvaskyla Finland, Ecole Normale Superieure, French Particle Physics Lab., LPTHE Jussieu, LPGP Paris Psud, Strasbourg, LOAFA Paris Diderot, Perpignan, Poitiers*, Compiegne University of Technology, Ioannina Greece, INFN Rome 1, European University Institute Florence, Radboud Netherlands, Bergen Norway, Polish Academy Krakow, Wroclaw, Research and Education Network Vladivostok Russia, Saratov, Voronezh, Samara, Vologda, Chalmers Sweden, National Library of Sweden, Institute for Low Temperature Physics Kharkov Ukraine, Kiev, Edinburgh, Glasgow, Leicester, Chemistry Oxford, Mathematics Oxford, British Government Science and Technology Facilities Council, Susses, Swansea, University College London, British Library, Ceredigion, Penfro, Essex Schools, Ombudsman Wales, Royal Society.

Rest of World: Universities, Institutes and Similar.
   ARBA Argentina, CITEFA Argentina, UFMG Brazil, UFRJ, UNESP, USP Brazil, Industry Canada, McGill, Memorial Newfoundland*, National Capital Freenet Ottawa, Sim oe County District School Board, College Huntsic Quebec, Quebec Trois Rivieres, Quebec, Windsor, Western Ontario, York University Toronto, Biobio Chile, Chile*, La Frontera Chile, Tsinghua China, Tianjing China, edu system Colombia, EDEA Colombia, National University Colombia, Andes Colombia, NISER India, Indian Railways, PRL India, MPIO Chihuahua Mexico, UNAM Mexico, IPB Namibia, Osaka Japan, CIMS Japan, Mitsuba, Japan Space Exploration, ITN Peru, ADDU Philippines, edu system Pakistan, QAU Pakistan, NU Singapore, KMITL Thailand, NTU Taiwan, Pretoria South Africa.

For www.aias.us there were 76,319 hits, 16,379 distinct visits, 12.48 gigabytes downloaded, 51,162 page views, 2752 documents read from 89 countries led by US, Canada, Germany, Mexico, Russia, Romania, Britain, Poland, ...

All UFT papers read, led by 142, 43, 177, 228, 177(Sp), 88, 11, 150B, 107, 175, 149, 213, 157(Sp), 166, 12, 159(Sp), 42, ...

All essays heard, led by: 3, 4, 44, 40, UMEc, 45, 5, 37, 9, 46, 47, 49, 6, 24, 26, 25, 38, 35,24(Sp), 28,49(Sp),41, 63,50,59,30, 60, 61, 20, 51, 52, 53, 55,64, 57, 62,NW,29, NIP,22, 23, 26, 27,7,....

All articles read, led by: F3(Sp),F13(Sp), ECE Article, F5(Sp), Auto1,Space Energy (Sp), Potential Waves, Human Soul Jackson, Experimental Refutation Heisenberg, LCR Resonant, Spacetime Devices, Parametric Simulation(Sp), Space Energy, Auto2, ...

For the three ECE sites: www.aias.us, www.atomicprecision.com and www.upitec.org the www.aias.us total increases by an estimated 50% giving 114,479 hits, 24,569 distinct visits, 18.72 gigabytes, 76,743 page views.

1 - 15 October

U. S. Universities, Institutes and Similar

Aims Community College Colorado, Arizona State, Barnard, SUNY Buffalo, Caltech, Carnegie Mellon, Columbia*, Cornell, Drexel, ECPI, Florida State, Georgia Tech., Illinois, Illinois Mathematics and Science Academy, Maine, Colorado School of Mines, Missouri, MIT, Montgomery College, Missouri University of Science and Technology, NITT Bangalore (on edu), Northwestern, Princeton, Penn State*, Purdue, Rice, Rochester Institute of Technology, San Francisco State, SUNY Stonybrook, Texas A and M*, Toronto, Central Florida, UC Irvine, UCLA, UC Santa Barbara, Delaware, Illinois Chicago, New Hampshire, Washington St. Louis, Yale, Lawrence Livermore, Alaska Weather Headquarters, US Archives, Ludwig von Mises Institute Alabama, Princeton University Organisation, City of Sunnyvale California, Cit yof South bend Indiana, City of Danville PA, Huron Valley.

Europe and Asia Minor: Universities, Institutes and Similar.

Leoben Austria, Vienna, SALK Austria, CHMI Czechia, TMCL Czechia, TU Munich, Bonn, Erlangen, Regensburg, Saarland, Spanish Airports (AENA), Basque Country, Andalusian Junta, Tecnun, Autonomus Barcelona, Barcelona, Juan Carlos 3rd, La Laguna, Barcelona Tech., Madrid Tech., Seville, Santiago de Compostela, Valencia*, Zertia Spain,

TUT Finland, YOK Finland, INRIA France, Polytechnique, Poitiers*, Pierre et Marie Curie, HI Iceland, Milan, INFN Rome 1, Erasmus Rotterdam, TU Delft, TU Eindhoven, Utrecht, FUN Poland, Gdansk, Silesian Data Center, Mari El Republic Russia, Volgograd, Voronezh, Moscow Nuclear Research University (MEPHI), Uppsala Sweden*, Institute of Physics Ukranian Academy Kiev, IPP Portugal, NTS Soviet Union, Odessa, Bristol, Edinburgh, Exeter, Imperial*, Keele, Luton, Manchester, Nottingham, Physics Oxford, University College Oxford, Queen Mary London, St Andrews, Warwick, British Library, BBC, Penfro (Pembroke), Llyfrgell Genedlaethol Cymru (National Library of Wales), Ashby School, Durham Schools.

Rest of World: Universities, Institutes and Similar.

National University of La Plata Argentina, Adelaide, Australian National, Deakin, Queensland University of Technology, Justice Department Brazil, Abtlus Organization Brazil, FU Rio Grande do Sul, Sao Paolo, Brandon Canada, Canadian Department of Defence, Canadian Government, McGill*, British Columbia, Quebec Trois Rivieres*, Waterloo*, Chile, TJ Chile, UNAL Colombia, UPB Colombia, ICRT Cuba, Ambo Ethiopia, UNS Indonesia, Indira Ghandi Centre for Atomic Research India, Indian Railways, UNAM Mexico*, UEM Mozambique, QAU Pakistan, NU Singapore, NCTU, NTU Taiwan, CES, FING, UMM Uruguay, Pentech South Africa, UFS South Africa.

15 - 30 October

U. S. Universities, Institutes and Similar

Barnard, California Polytech San Luis Obispo, Caltech, Columbia, Drexel, Fullerton, George Mason, Harvard, Hawaii, Indiana Purdue Indianapolis, Karlsruhe Institute of Technology (on edu), Kansas State, Louisville, Colorado School of Mines, Montgomery College, Michigan Tech, North Carolina State*, New Jersey Tech, New Mexico Tech, Ohio, Mississippi*, Portland State, Princeton, Penn State*, Puerto Rico Tech, Reed, Rochester, Rutgers, Stetson, SUNY Stonybrook, Toronto (on edu), Trinity College, Tusculum, Alabama, Central Florida, UCLA*, UC Santa Barbara, UC Santa Cruz, Houston, Maryland, Michigan, Nebraska Lincoln, Barcelona Tech (on edu), Washington, Wisconsin, Worcester Polytech, Lawrence Livermore, NOAA Alaska Regional Weather Center, US Navy Command Support Center, US Navy Marine Command, US National Archives, Mathematical Sciences Research Institute Berkeley, Nieman Lab Harvard, Ford Corporation.

Europe and Asia Minor: Universities, Institutes and Similar

CERN, ETH, Bern, Fribourg, Xtec Catalonia, FU Berlin, Helmholtz Centre, RWTH Aachen, TU Ilmenau, Heidelberg, Karlsruhe, Wuerzburg, Dinamicarea Spain, Basque Country, Andalucian Junta, Autonomous Barcelona, Autonomous Madrid, Barcelona, Pontifical Comillas Madrid, Seville, Santiago de Compostela, Valencia, RTGI Europe, Helsinki, LPN CNRS France, ENSTA, INPG, LIAFA Paris Diderot, Poitiers, Rennes 1*, LKB Pierre et Marie Curie, Ioannina Greece, Technion Israel, INFN Rome 1, Turin Tech, Free Amsterdam*, NTNU Norway, Troms County Norway, PL Poland, Silesian Data Center, Coimbra Portugal, UFP Portugal, BH Romania, Akado Urals, Samara, Volgograd, NIC Russia, Chalmers Sweden, Miev. Odessa, Poltava, Aberystwyth*, Birmingham City, NAT Cambridge, Girton College Cambridge, Edinburgh, Imperial*, Lancaster, New College Oxford, Somerville College Oxford, Sussex, Swansea, UCL, BBC Cardiff, Penfro, NHS, Eduroam Oxford, St. Peter's School York.

Rest of World: Universities, Institutes and Similar

UNT Argentina, Adelaide*, ANU, Queensland Tech, Sydney, INPE Brazil, UE Rio, UFPE, UFSC, Unicamp, Sao Paolo, Brock Canada, Dalhousie, Queen's, UBC*, Quebec Trois Rivieres, Waterloo*, York Toronto*, Chile, UCN, UTFSM, UDEA Colombia, Andes, Norte, Bhaba ARC India, RIKEN Japan, Postech South Korea, Samsung Corporation South Korea, IPN Mexico*, ITESM, UNAM Mexico, International Peace Bureau Namibia, Wellington Girls' College New Zealand, National University Singapore, NCTU Taiwan, CES Uruguay, UFH South Africa, Limpopo South Africa.

Summary of November 2013

For www.aias.us there were 73,286 hits, 9.838 Gbytes downloaded, 17,024 distinct visits, 50,346 page views, 2,690 documents read from an estimated 60 countries, led by US, Canada, Russia, Mexico, Czechia, France, Germany, Colombia, Britain, ...

All UFT papers read, led by: 166(Sp), 88, 25, 43, 236(Sp), 240(Sp), 238(Sp), 177, 232(Sp), 237(Sp), 235(Sp), 142(Sp), 166, 168, 198,41, 228, 145(Sp), 206(Sp), 157(Sp), 159(Sp), 140(Sp), 213(Sp), 230(Sp), 160(Sp), 170(Sp), 202(Sp), 215(Sp), 169, 20, 182(Sp), 212(Sp), 226(Sp), 227(Sp), 175, 85, ...

All essays heard ro read, led by: 36, UMV1, 44, 45, 46, 24, 6, NIP, 37, Nicola Who?, 8, 3, 5, 50, 55, 53, 61, 54, 49, NIP (2), 20, 27, I139, 49, 33, 40,35, 4, 10, 39, 59, 28, NIP(Sp), ...

All articles read, led by: F3(Sp), Auto1, Evans Equations, Parametric Simulation 2 (Sp), Triple Orbit Simulation, Levitron, LCR Resonant,

Johnson Magnets, Simulation Circuit Spanish, Space Energy (Sp), F9(Sp), Auto2, ...

For the three ECE sites www.aias.us, www.atomicprecision.com and www.upitec.org the totals increase by an estimated 50% giving 25,536 distinct visits, 109,929 hits, 14.76 Gbytes, 75,519 page views.

1 - 15 November

U. S. Universities, Institutes and Similar

Albany, Berkeley, Baldwin Wallace, Brigham Young, Caltech*, Cerritos, Columbia, Cornell, Drexel, Delaware Technical CC, Duke, Hamilton, James Madison, Michigan Tech., Oklahoma State*, Oneanta, Ohio Northern, Ohio, Princeton, Reed, Rochester Tech., Rochester, Rose-Hulman, Samford, South Florida State, San Jose State, Sarah Lawrence, Stanford*, Stevens Tech., Texas A and M, Tylor, Trinity College, Central Florida*, Chicago, UCLA, UC Santa Barbara, Iowa, Illinois Urbana Champaign, Michigan*, North Carolina Charlotte, Nevada Las Vegas, U Penn, Ursinus, South Carolina Upstate, Utah*, Washington, Worcester Tech., US Naval Marine Command, US Archives, Bank of America, Ford, Marquis, Motorola.

Europe and Asia Minor: Universities, Institutes and Similar.

SRV Catalonia, Vaud Tech Switzerland, Zurich, CHMI Czechia, Fraunhofer Radar, Albert Einstein Institute, Max Planck Computing Centre Garching; RWTH Aachen, TU Ilmenau, Hannover, Heidelberg, Cologne, Mainz, Aarhus*, Niels Bohr Institute, Granada Spain, Murcia, Madrid Tech., Salamanca, Santiago de Compostela, Vigo, CNRS LPN France, ENS Lyon, Paris South, Poitiers, St. Etienne, ENEA Frascati Italy, INFN Lecce, SISSA, Milan, Rotterdam Netherlands, Leiden, Radboud, Free Univ Amsterdam, Signo Foundation Norway, Silesian Data Center Poland, Warsaw Tech., ASE Romania, IGCTI Romania, CTTC Ryazan Russia, Metronv, SAN, Stockholm Sweden, Uppsala, Metu Turkey*, Kiev, Odessa, Aberystwyth, Birmingham, Churchill College Cambridge, Computing Cambridge, Trinity Hall Cambridge, Cardiff, Exeter, Imperial*, Leeds*, London School of Economics, Brasenose College Oxford, Keble College Oxford, St. Catherine's College Oxford, Edoroam system Oxford University, Strathclyde, Sussex, Swansea, University College London, Llyfrgell Genedlaethol Cymru (National Library of Wales).

Rest of World: Universities, Institutes and Similar.

MDP Argentina, UNR Argentina, Australian National, Melbourne, Victoria, Dhaka Bangladesh, CBPF Brazil, RJ Brazil, UFF, UFPB,

UNESP, Sao Paolo Brazil, BCGS Canada, Dalhousie, Ahuntsic, Simon Fraser, British Columbia, Quebec Trois Rivieres, Victoria, Waterloo, York, CFM Chile, UTFSM Chile, Shanghai China, Tianjing, UDEA Colombia, Andes, UPB Colombia, Tezu India, Indian Rail, IMSC, Raman Research Institute, Tohoku Japan, MESH Japan, Korean Advanced Tech (KAIST), IPN Cicata Mexico, UNAM Mexico*, UNAN Leon Nicaragua, ITN Peru, QAU Pakistan, Edu System Pakistan, National Univ. Singapore, NCTU Taiwan, NTU Taiwan, Northwestern University South Africa.

15-29 November

U. S. Universities, Institutes and Similar.

Caltech, Columbia, Cornell, Drexel, Florida Tech., Harvard, Illinois, Karlsruhe Tech (on edu), Los Rios, MIT*, Missouri Tech., Miami Univ Ohio, Notre Dame, New York, Oberlin, Ohio State, Oklahoma State, Portland State, New York Polytechnic, Princeton, Purdue, Rochester*, Samford, San Francisco State, Sarah Lawrence, Stanford*, Southern A and M, Trinity College Hertford, UC Davis, Central Florida, Houston, Illinois Urbana-Champaign, Michigan Flint, Michigan*, New Hampshire, Oregon, Texas, Wisconsin Milwaukee, Walla Walla, West Kentucky, Brookhaven, US NOAA, US Missile Defense Agency, US Naval Marine Command, AMAS Charter School, US National Archives, Cass Health, Hallmark Health, Ford Corporation, Intel Corporation.

Europe and Asia Minor: Universities, Institutes and Similar.

Vienna Tech, Johannes Kepler Linz, KU Leuven, SRV Catalonia, XTec Catalonia, CERN, Vaud Tech, Paul Scherrer Institute, Tianjin China, CHMI Czechia, TMCZ Czechia, Munich, German Synchrotron (DESY), Fraunhofer*, FU Berlin, Jacobs, RWTH Aachen, Jena, Karlsruhe, Kiel, Cologne*, Mainz, Regensburg, MNT Estonia, Spanish Rail (RENFE), Autonomous Barcelona, Barcelona, Complutense Madrid, Granada, Seville, Salamanca, Valencia, Helsinki, IGUA Finland, ENIB France, Pasteur Institute, Paris Psud, Strasbourg, Limoges, Bordeaux, Poitiers, Meliusz Library Hungary, GN Gov. Ireland, Trinity College Dublin, University Colleeg Dublin, Israli Military, Tel Aviv, Arab College Israel, INFN Milan, INFN Trieste, Genoa, Insubria, Rome 3, Radboud Netherlands, Utrecht, Silesian Data Center Poland, Coimbra Portugal, Corbina system Russia, Kirov, Tomsk, DFRI Sweden, KTH Stokholm, Metu Turkey, Aberystwyth, Brunel, Computing Cambridge, Physics Cambridge, Selwyn College Cambridge, Cardiff, Durham, Eastleigh, Edinburgh, Exeter, Imperial*, King's College London, Kent, Luton, University College Oxford, Wadham College Oxford, Eduraom system

Oxford, Queen's University Belfast, Sheffield, Southampton, Swansea, Warwick, Essex Schools, Peterborough Schools.

Rest of World: Universities, Institutes and Similar.

UNC Argentina, CNEA Argentina, Adelaide, UNESP Brazil, Carleton Univ. Canada, Dalhousie, McGill, NSHealth, Manitoba, Qubec Trois Rivieres, Regina, Toronto, Waterloo, UFRO Chile, UTFSM Chile, Tianjin China, UDEA Colombia, UNAL Colombia, Andes Colombia, ESPE Ecuador, ITB Indonesia, IIT Kharagpur India, Tezu, Indira Ghandi Atomic Research Centre, Indian Rail (RCIL), Raman Research Institute, NIMS Japan, RIKEN Japan, SNU South Korea, IPN Mexico*, ITESM, UNAM, UTM Mexico, ITN Peru, Panama, NU Singapore, NCUT Taiwan, NTU Taiwan.

Summary of December 2013

For www.aias.us there were 84,273 hits, 16,366 distinct visits, 18.56 Gbytes downloaded, 58,831 page views, 2,635 documents read from 91 countries, led by USA, Britain, Germany, Colombia, Mexico, France, ...

For the three ECE sites www.aias.us, www.atomicprecision.com and www.upitec.org the totals are increased by an estimated 50% giving 126,410 hits, 27.84 Gbytes downloaded, 24,549 distinct visits, 88,247 page views

All UFT papers read led by 177, 88, 25, 43, 85, 170(Sp), 150(Sp), 94, 150B, 228, 175, 166, 183, 140(Sp), 107, 54, 140, 63, 169, 42, ...

All Essay Broadcasts heard and Essays Read, led by: 38, 8, 50, 7, 52, UMEa, 57, 47, 29, 51, 12, 6, 61, 44, 45, 59, 10, 64, 3, 23, 55, 37, NP, 49, 5, 54, NW, UMEb, 9, 27, 46, 53, 60, 11, 25, 62, 24, 49(Sp), ...

All Articles Read, led by: F3(Sp), Overview of ECE, Auto1, Spacetime Devices (Sp), levitron, Double Slit, F5(Sp), Triple Orbit Simulation, HSJ, LCR Resonant, CV, Space Energy, Evans Equations, CEFE, Auto2, Book of Scientometrics, ...

1 - 15 December

U. S. Universities, Institutes and Similar

Arizona State, Berkeley, Boston, Caltech, Colby, Cooper, Duke, Florida State, Gustavus Adolphus, Georgia Tech., Harvard, Hawaii, Johnson and Wales, Louisville, Maine, Memphis, Colorado School of Mines, MIT, Michigan State, Olin, Ohio State, Princeton, Rochester, Santa Clara, Trinity College, Central Florida, Chicago, UC Merced, Michigan, North Carolina Charlotte, Oregon, Southern California, South Alabama,

Utah, Texas, William and Mary, Vermont, Williams, Worcester Polytechnic Institute, Western Washington, Yale, NSA Johnson Space Center, Alaska Regional HQ of NOAA, Naval Marine Command, Naval Research Laboratory, U.S. Missile Command (Naval Marine Command), Ford Corporation, Lockheed Martin, Northeast Georgia Health, United States Archives.

Europe and Asia Minor: Universities, Institutes and Similar.

UIBK Austria, Vienna, CESCA Catalonia, SRV Catalonia, Catholic Univ. Liege, CERN, Zurich, Mendel Univ. Brno, Masaryk University Czechia, Bulgarian Academy of Sciences, Fraunhofer, RWTH Aachen, SAP-AG, TU Darmstadt, Frankfurt, Hamburg, Kiel, Munich, Regensburg, Aarhus Denmark, Copenhagen, Dinamicarea Spain, Iberbanda, Andalucian Junta, Autonomous Barcelona, Granada, Salamanca, Valencia, Polyechnique France, Strasbourg, Joseph Fourier Grenoble, Orleans, Poitiers, Troyes Univ Technology, Croatian National Institute of Physics, Budapest Univ. Technology and Economics, Weizmann Institute Israel, International Centre for Theoretical Physics Trieste, Italian Institute of Technology, International School for Advanced Studies Trieste, Genoa, Leiden, Groningen, Eindhoven, City of Wroclaw, City of Krakow, Silesian Data Centre, Hadara Palestine, Timisoara Polytechnic University Romania, Serbian Academy of Sciences, Moscow State University, Royal Swedish Institute of Technology Stockholm, Gazi Turkey, Computing Cambridge, Magdalene College Cambridge, Exeter, Manchester*, Nottingham, Bodleian Libraries Oxford, Southampton, Swansea, British Library, Llyfrgell Genedlaethol Cymru (National Library of Wales).

Rest of World: Universities, Institutes and Similar.

UNC Argentina, BCN Gov. Argentina, Cerider Argentina, Monash Australia, HNE Health New South Wales, Pontagrossa Gov. Brazil, City of Niteroi Brazil, Ibrati, UFF, Unicamp, DPCDSB Ontario Canada, Canadian Ministry of Defence, McGill, Simon Fraser, Quebec Trois Rivieres, Waterloo, PUC Chile, Talca Chile, IUT Iran, Toyama Japan, KAIST South Korea, UAFM Mexico, Unimas Malaysia , Supreme Court Nicaragua, Panama, Piura Peru, IIU Pakistan, PIEAS Pakisatn, National Univ Singapore, Overseas Family School Singapore, NCU Taiwan, NTHU, NTU, African Institute of Mathematics Capetown South Africa.

15 - 30 December 2013

US Universities, Institutes and Similar.

Arizona State, Atlanta Public Schools, Prince George's County Maryland Public Schools, Columbia, Illinois, Iowa, Karlsruhe Tech (on edu), New York University, Princeton*, Rochester, Samford*, Texas A and M Corpus Christi, UC Irvine, Worcester Polytechnic Institute, Ford Corporation, United States National Archives.

Europe and Asia Minor: Universities, Institutes and Similar.

Swiss Federal Institute Lausanne, Bruker Company, RWTH Aachen, Freiburg, Hannover, Jena, Barcelona, Vigo, AC Lyon, French Atomic and Alternative Energy Commission, CNRS Grenoble, INP Grenoble, French National Particle Laboratory (in2p3), Jussieu, Polytechnique de France, Poitiers, Leiden, Delft, Lodz, Silesian Data Center, CTTC Ryazan, Kursk, Volograd, Lviv, Edinburgh, Imperial College London, Nottingham, Oxford.

Rest of World: Universities, Institutes and Similar.

AP Gov. Brazil, Western University Cuba, PUC Chile, UNS Indonesia, Harish Chandra Institute India, Tata Research Institute India, IMS Japan, Tohoku, Tokushima, UEC Japan, IPN Mexico*, Osiptel Government Peru, Edu system Pakistan, PIEAS, PERN, UOG Pakistan, African Institute of Mathematical Sciences South Africa.

END OF 2013.

Summary of January 2014

For www.aias.us there were 68,037 hits, 12.49 gigabytes downloaded, 13,438 distinct visits, 45,052 page views from 95 countries, led by USA, Latvia, Germany, Italy, Mexico, Czechia, Canada, Britain, Turkey, Japan, ... At least 6% of the returns were lost due to timer problems at the beginning and end of the month so the total is estimated to be 1.06 multiplied by the above, which gives 72,117 hits, 13.24 gigabytes, 14,244 distinct visits, 47,755 page views. As usual this total is increased by an estimated 60% by the other two ECE sites, www.atomicprecision.com and www.upitec.org, giving 108,179 hits, 19.86 gigabytes, 21,366 distinct visits, 71633 page views.

All UFT papers read, led by 88, 177, 25, 43, 166(Sp), 169, 228, 41, 140, 255, 239, 58, 166, 63, 140(Sp), 158, 175, 214-1b, 170(Sp), 177(Sp), 26, 157(Sp), 146, 155, 201, 42, 107, 170, 38, 143, ...

All Essays heard and read, led by 46, 23, 5, 12, UME1, 28, 45, 24, 8, 53, 57, 47, 44, 6, 9, 37, 50, 59, 64, 29, 3, 27, 25, 52, Nicola Who?, 35, 4, Intro139, 41, 61, 60, 62, Nobody is Perfect, Nobody's Perfect, ...

All books and articles read, led by F3(Sp), Potential Waves, Autobiography Volume One, Overview of ECE, Space Energy Devices (Sp), Autobiography Volume Two, Human Soul Jackson, LC Resonant, Book of Scientometrics, My CV, Evans Equations, CEFE, ...

1 - 15 January

U. S. Universities, Institutes and Similar

Blair College, Johnson and Wales, MIT, New York, Pittsburgh, Princeton, Samford, University of California Irvine, UC Santa Barbara, UC San Diego, U Mass, Univ. Arizona Medical Center Tucson, Missouri Kansas City, Minnesota Twin Cities, Naval Research Laboratory, U. S. Special Operations Command, NASA JPL, NIST, St. Andrew's School Savannah, US Archives.

Europe and Asia Minor: Universities, Institutes and Similar.

AAU Austria, Innsbruck, Vienna, ETH Zurich, JBI Gymnasium Czechia, German Synchrotron Facility (DESY), RWTH Aachen, TU Dortmund, TU Munich, Freiburg, Karlsruhe, Oldenburg, Bavarian Academy Computing Centre Munich, Copenhagen, Tartu Estonia, UT Estonia, CSIC Spain, European Space Agency Newton Project Spanish Facility, Eberri, Barcelona Autonomous, Madrid Complutense, Polytechnic Valencia, French Atomic Energy Agency, French Particle Laboratory*, French Graduate School of Engineering, Nantes, Perpignan, Poitiers, Troyes, Trinity College Dublin, University College Cork, Secom Israel, Univ Iceland, Italian ENEA Frascati, International School of Advanced Studies Trieste, Genoa, Twente Netherlands, Univ. Norway Trondheim, Jagiellonian Krakow Poland, Silesian, Wroclaw, UTCLUJ Romania, Ates Medica Russia, CTTC Ryazan, Russian Ministry ofEnergy Moscow, HIAST Aleppo Syria, Ankara Turkey, Gazi, Cambridge, Leeds, Luton, Oxford Univ Libraries, Somerville Oxford, Oxford Eduroam, Sheffield, Sussex, National Health Service, Eduroam Oxford, Gov Canary Islands, IT Portugal.

Rest of World : Universities, Institutes and Similar.

Southern Cross Australia, LNCC Brazil, UFPE Brazil, Unicamp, Gov. British Columbia Canada, Concordia, Humber, Perimeter, Quebec Trois Rivieres, Victoria, Waterloo, UDEA Colombia, ESPE Ecuador, Espoch Ecuador, Espol Ecuador, Einstein Science Union Ecuador, UNS Indonesia, IITD India, IITIDR India, CAT Gov. India, RCIL Gov. India, IUAC India,

Chiba Univ Japan, IPN Mexico*, UNP Peru, BIMCS Pakistan, PIEAS Pakistan, NCTU Taiwan, NSYSU Taiwan.

15 - 29 January

U. S. Universities, Institutes and Similar.
   Blair, Clemson, Catholic University of America, Illinois, Michigan State, New Jersey Institute of Technology, New Mexico State, Pittsburgh, Princeton, Rochester, SUNY Stonybrook, Texas A and M, UCLA, UC Santa Cruz*, UC San Diego, Illinois Chicago, Minnesota Twin Cities, Utah, Tennessee Chattanooga, Wisconsin, Oak Ridge, US Social Security Administration, Naval Marine Command, Chevron, CISCO, United States National Archives.

Europe and Asia Minor: Universities, Institutes and Similar.
   Mons Belgium, EPF Lausanne, ETH Zurich, Bruker Company, Leibnitz Centre Munich, TU Berlin, Bonn, Freiburg, Jena, Oldenburg, Aarhus Denmark, Copenhagen, Valencian Government, Autonomous Madrid, Madrid Complutense, Granada, Polytechnic Valencia, Valencia, Galician Government, Jyvaskyla Finland, French Atomic and Alternative Energy Commission, INP Grenoble, French National Particle Lab., Franche Comte, Versailles St Quentin en Yvelines, Poitiers, MTA Csilleberc Hungary, Bar Ilan Israel, Weizmann Institute, Iceland Gardur, Iceland RHI, INAF Italy, SISSA Italy, Genoa, Pavia, TU Delft, Twente, Silesian Poland, Romanian National Institue of Material Physics, Brasov, Timisoara, Institute of Physics Belgrade, Moscow Power Engineering Institute, Aviaren Russia, Kazan National Research Technological University, Kristianstad Sweden, Landmateriet Sweden, Slovenian Environmental Agency, Aberystwyth*, Cambridge computer centre, Cambridge Department of Applied Mathematics and Theoretical Physics, King's College Cambridge, Edinburgh, Leeds, Nottingham, Oxford Libraries, University College Oxford, Eduroam Oxford, Queen Mary University of London, Birmingham City Council, Hampshire County Council, Ampleworth College.

Rest of World: Universities, Institutes and Similar.
   Southern Cross Australia, Brazilian National Scientific Computational Laboratory, Hospital Regional, Sao Paulo*, CCRA-ADRC Gov. Canada, Quebec Trois Rivieres*, UNAL Colombia, Gov. Ecuador, Einstein Scientific Ecuador, IIT Delhi India, Indian Rail, Harish Chandra, Indian Institute of Plasma Research, Indian Inter University Accelerator Centre, Kumamoto Japan, Tokushima, Hyogo, Postech South Korea, Gov. Mexico

IPN, UNAM* Mexico, FIJSUNAC Peru, UPD Philippines, edu system Pakistan, National University Singapore, KU Thailand.

Summary of February 2014

For www.aias.us up to 27th February there were 60,807 hits, 8.86 Gbytes downloaded, 13,239 distinct visits, 41,280 page views, an estimated 2,900 documents read from 93 countries, led by USA, Germany, Mexico, Russia, Czechia, Canada, Britain, Australia, ... It is estimated that about 6% of these totals were lost due to timer resetting and that about 50% must be added from the other ECE sites, www.atomicprecision.com and www.upitec.org, giving final totals of 94,859 hits, 13.82 Gbytes, 20,652 distinct visits, 64,397 page views.

All UFT papers read, led by 85, 25, 88, 43, 140, 175, 42, 26, 152, 38, 239, 177, 177(Sp), 144, 149, 150-B, 166, 107,168, 7, 214-1b, 155, 228, 39, 81, 160(Sp), 146, 169, 201, 54,....

All essays heard or read, led by: 3, Nobody's Perfect, 29, 44, 45, 9, Nikola Who?, 38, 50, 59, 47, 6, 4, The Universe of Myron Evans Part 1, 5, 52, 58, 8, 34, 53,49, 37, Intro to UFT139, 60, 62, 61, 64, 39, 41, 24(Sp), 41(Sp), ...

All articles and books read, led by: F3(Sp), Engineering Model, Autobiography Volume One, HSJ, F1(Sp), F5(Sp), JM, Triple Orbit, Animation Archive, Space Energy (Sp), Space Energy, Autobiography Volume Two, Book of Scientometrics, Spacetime Devices, CV, Potential Waves, Overview of ECE, Homopolar Generator, levitron, "Llais", Numerical Article 3, ...

1-15 February

U. S. Universities, Institutes and Similar

ASU, Baylor, Berkeley*, Caltech, Calvin, Claremont, Cleveland State Univ. Ohio, City Univ New York, Illinois, Karlsruhe Institute of Technology (on edu system), North Dakota State, Ohio State, Temple, UC Irvine, UC Santa Cruz, UC San Diego, Houston, Massachusetts, Michigan, North Carolina Charlotte, New Hampshire, New Mexico, Texas Arlington, Utah, Wisconsin Milwaukee, NASA Johnson Space Center, U. S. Naval Marine Command, U. S. Archives, Internet Archives, Measured Progress Organization.

Europe and Asia Minor: Universities, Institutes and Similar.

SRV Catalonia, Xtec Catalonia, ETH Zurich, Mazaryk Univ. Brno, Bosch Company, German National Synchrotron Facility, Ruhr Univ Bochum, TU Braunschweig Computer Centre, TU Cottbus, TU Munich,

Erlangen, Max Planck Hannover, Karlsruhe, Oldenburg, Paderborn, Armed Forces Munich, Aarhus Denmark*, Dinamicarea Spain, Autonomous Madrid*, Murcia, Polytechnic Madrid, Polytechnic Valencia, UTA Finland, French Atomic and Alternative Energy Authority, Paris 10, Nantes, Poitiers*, Paul Sabatier Toulouse, IRB Croatia, ENVI Italy, International Centre for Theoretical Physics Trieste, Bari, Genoa, IST Latvia, Silesian Data Centre Poland, Sprint Data Center Poland, SBB Serbia, Joint Institutes for Nuclear Research Dubna Russia, Urals Region, High Energy and Particle Physics Institute Moscow, DLL County Council Sweden, Royal Institute of Technology Stockholm, Lulea University of Technology Sweden, HCK Slovakia, Computing Cambridge, Pembroke College Cambridge, Durham, Lancaster, Jesus College Oxford, Strathclyde, Eduroam system Oxford University, Cynghor Penfro County Council, Hertfordshire Grid for Learning.

Rest of World: Universities, Institutes and Similar.
   Government of Corrientes Argentina, McGill Univ. Canada, Mcmaster, Alberta, Quebec Trois Rivieres*, Toronto, ITB Indonesia, Kanagawa Univ Japan, Osaka, Tohoku, ITESM Mexico, UGTO Mexico, National Autonomous Univ. Mexico*, Edu system Pakistan, ULA Venezuela, North West university South Africa, TUT South Africa, UL South Africa.

15 - 26 February

US Universities, Institutes and Similar.
   Brockport, Caltech, Calvin, Case Western Reserve, Harvard, Illinois Urbana-Champaign, Lehigh, Missouri, MIT, Notre Dame*, NITT India (on edu), Oklahoma State, Penn State, Purdue, Rochester Institute of Technology, St Mary's Notre Dame, Texas A and M*, Temple, Toronto, Truman, Texas Tech University, UC Santa Barbara, UC Davis, Delaware, Georgia, Illinois Chicago, Iowa, Utah, Toledo, Washington, Fermilab, Naval Marine Command, United States National Archives, GVO Data Center, FNF Archive, Atlanta Public Schools.

Europe and Asia Minor: Universities, Institutes and Similar.
   Vienna, ETH Zurich, DIFU Germany, KFA Juelich, TU Berlin, TU Munich Computer Centre, Albert Einstein Institute, TU Munich, Heidelberg, Mainz, Muenster, Univ Basque Country, Barcelona*, Complutense Madrid, Granada, Cantabria, Madrid Polytechnic, Bruno Kessler Foundation, Lisbon Tech., Jyvaskyla Finland, Tampere Finland, CNRS Grenoble, OVH France, Lorraine Nancy, Anges, Marseilles, Nantes, Poitiers*, Paris Diderot, DEBIS Hungary, Tel Aviv Israel, Technion, Florence, Genoa, INFN Padua, Italian Television, Free Univ.

Amsterdam, NTNU Norway, Institute of Mathematics Polish Academy of Sciences, Wroclaw, BH Romania, Serbian Academy of Sciences, Novosibirsk, Lund, Stockholm, ILT Kharkov Ukraine, Cambridge computing, Department of Applied Mathematics and Theoretical Physics Cambridge, Magdalene College Cambridge, Pembroke College Cambridge, Trinity Hall Cambridge, Imperial College London, Mathematics Imperial, Physics Imperial, Physics King's College London, Liverpool, University College Oxford, Eduroam system Oxford, VPN System Oxford, Wadham College Oxford, Plymouth, Wolverhampton, York.

Rest of World: Universities, Institutes and Similar.

Australian National, MacQuarrie, Dhaka Bangladesh, State Univ Rio de Janeiro Brazil, UFPR, State University Sao Paolo Brazil, Gov. Quebec, Memorial Newfoundland, PPOE Canada, british Columbia, Quebec Trois Rivieres, Toronto, Victoria, Waterloo*, Wightman, UDEA Colombia, Engineering University Bogota, National Univ Colombia, City University Hong Kong*, AC Higher Education System India, NITS India, Indira Ghandi Atomic Research Centre, Tezu University, KCT Japan, ETRI South Korea, UTJ Mexico, ITESM Mexico, Autonomous National Univ Mexico, IIU Pakistan, PERN Pakistan, NCTU Taiwan, Capetown South Africa.

Summary of March 2014

For www.aias.us there were 80,357 hits, 16.52 Gbytes downloaded, 15,628 distinct visits, 50,850 page views, with an estimated two days lost due to timer problems, so an estimated total of 85,540 hits, 17.59 Gbytes downloaded, 16,636 distinct visits, 54,130 page views from 99 countries, led by USA, Britain, Germany, Canada, Ukraine, Mexico, Italy Poland, ...

The other two sites www.atomicprecision.com and www.upitec.org are estimated to increase these totals by 50% giving 128,310 hits, 26.39 gigabytes, 24,954 distinct visits, 81,195 page views

All UFT papers read, led by 177, 238b, 88, 239, 166, 43, 41, 61, 155, 169, 25, 177(Sp), 168, 175(Sp), 26, 140(Sp), 157(Sp), 2, 158, 140, 150B, 149(Sp), 85, 141, 146, 175, 258, 67, 19, ...

All Essays heard and read, led by: 60, 62, 6, 55, 61, 59, 58, 24, 9, 57, Nobody's Perfect, 54, 46, 4, 44, 8, 49, 50, 64, 24, ...

All Articles read, led by: F3(Sp), Auto1, Space Energy Devices (Sp), F5(Sp), Engineering Model, Space Energy, Johnson Magnets, Numerical Article 3, PW, F9(Sp), Levitron, CV, Book of Scientiometrics, F1(Sp), Numerical 2D, Auto2, Spacetime Devices, CEFE, Found. Phys. Leaflet,

LCR Resonant, Sapcetime Devices (Sp), A Basic Error in the EFE, HSJ, Galaxies, ECE Article, Hompolar, LCR Resonant 2f, Evans Equations, Device Development, Overview of ECE, ECE (Sp), Infinite Solenoid, ...

1-14 March

U.S. Universities, Institutes and Similar

Arizona, Berkeley, Caltech, Carleton, Columbia, Cornell, Earlham, Georgia Tech., Harvard, Indiana Univ. Pennsylvania, Kettle Moraine School District, MIT, Notre Dame, Northwestern, Princeton, Penn State, Rutgers, Sonoma, University of California, Nevada Las Vegas, U Penn, South Carolina, Washington*, Wisconsin, Worcester Polytechnic, United States National Archives, Seattle Public Library, Fermilab, Los Alamos, U. S. Army Picatinny Arsenal, U. S. Navy Information Dominance Command, Public Schools of the Tarrytowns.

Europe and Asia Minor: Universities, Institutes and Similar.

OEAW Austria, TU Graz, TU Vienna, Innsbruck, Linz, Vienna*, City Government of Vienna, UPC Austria, Free University of Brussels, CERN Geneva, Xtec Catalonia, EPFL Lausanne, CXU Czechia, Fraunhofer Foundation, HFT Stuttgart, Max Planck Heidelberg, TU Darmstadt, Jena, Marburg, Munich, Oldenburg, Wuerzburg, GJK Denmark, Niels Bohr Institute, Caja Madrid, EHU Spain, Gov. Valencia, Autonomous Madrid, Complutense Madrid, Granada, Polytechnic Madrid, Valencia, Helsinki, HUT Finland, French Particle Laboratory, Polytechnique, Paris 5, Poitiers, Tours, Gov. Ireland, Weizmann Israel, SISSA Italy, NGI Italy, Genoa, ICTP Trieste, INFN Naples, INFN Padua, Leiden, Utrecht, Birmingham City Council, CAL Poland, Jagiellonian University Krakow, ISTS Poland, Silesian Data Center, Wroclaw, Elie Wiesel Institute Romania, Serbian Academy, HC Russia, Omsk, LTH Sweden, Metu Univ Turkey, Aberystwyth, Anglia, Brunel, Cambridge, Cardiff, Durham, Edinburgh, Heriot Watt, King's College London, Lancaster, Manchester, Oxford, University College London, UMIST, Univ Wales Registry, Wolverhampton, British Library.

Rest of World: Universities, Institutes and Similar.

UNSA Argentina, UTN Argentina, Australian National, Univ. Sunshine CoastAustralia, UFJF Brazil, UFPB Brazil, Sao Paolo, Gov. Canada, McGill*, National Canadian Research Council, PPPOE Canada, Alberta*, New Brunswick, Quebec Trois Rivieres, Waterloo*, Western Ontario*, Windsor, Chile, edu system Colombia, National Colombian University, China extensive, NITS India, Bhabha, Indira Ghandi, IGCAR, TIFR India, Sony Corporation Japan, IPQ Japan, IPN Mexico, UGTO Mexico,

University of the Sea, National Autonomous Mexico, Otago New Zealand, UNAC Peru, NCU Taiwan, NDHU Taiwan, City Hong Kong, Univ. Science and Technology Hong Kong, FQ Uruguay, AIMS South Africa, Pretoria.

14-30 March

US Universities, Institutes and Similar

Los Alamos*, Northwestern, US Naval Academy, Bard, Illinois*, Maine, Ohio State, Berkeley, Georgia State, New York, Stonybrook, Stanford, Princeton, Old Dominion, Naval Marine Command, Cornell, Ohio, Stanford Linear Accelerator, Tarrytowns. Western Washington, Maine, MITRE Organization, Texas A and M Commercial, University of North Carolina Charlotte, U. S. National Archives.

Europe and Asia Minor: Universities, Institutes and Similar.

UTU Finland, Romanian Materials Science Institute, Jena, Munich, Caja Madrid, EHU Spain, Polytechnique*, Poitiers*, NGI Italy, MTS Russia, Brunel, MORS Belorussia, GJK Denmark, French Particle Laboratory, Inet Finland, Andalucian Junta, KYLA Finland, genoa, Durham, Computer Centre TU Munich*, Navarre Spain, Cardiff, Physics Oxford, Foerster Group, Autonomous Barcelona, Dauphine Paris, St. Etienne, Computing Cambridge, Newcastle, Nottingham, Inerco Spain, Polytechnic Valencia, French Atomic Energy Commission, Paris 10, International Graduate School Trieste, Bruker, Hannover, Computing Karlsruhe, Hull, Bath, TU Berlin, AABH Denmark, UPCT Spain, USC Spain, IGUA Finland, Grenoble, Ecole Normale Superieure Architecture, Engineering Gdansk Poland, ASTRAL Romania, CERN, Bern, Autonomous Madrid, TU Darmstadt, National College of Ireland, University College London, Barney School, SRV Catalonia, ALU Metal Switzerland, UDC Spain, Polytechnic Milan, Silesian Data Center, Cambridge, Jacobs International University, MWN Germany, TU Dresden, Hamburg*, Niels Bohr Institute, Galician Parliament, Helsinki, Bashtel, Saratov, Corbina, Tomsk, Vologda, British Library, JEN Czechia, Isle of Wight Council Community Learning, Muenster, INFN Bologna, University of Warsaw.

Rest of World: Universities, Institutes and Similar.

Alberta, Sony Corporation Japan, Mexican National Autonomous, Otago New Zealand, UST Hong Kong*, McGill, Perimeter, Melbourne, Quebec Trois Rivieres*, Korean Advanced Institute of Science and Technology, National Peru, National Central University Taiwan, Memorial University Newfoundland, Andes Colombia, Kyoto, Ryukyu*, MDP

Argentina, McMaster*, Stanford, INEGI Gov. Mexico, UPD Philippines, INACAP Chile, UAD Colombia, Chiba Japan, PDN Sri Lanka, Sao Paolo Brazil, British Columbia, Chile, Indian Statistical Institute, Education Secretariat Government of Mexico, Pakistan Research and Education Network, Chennai Institute of Mathematics, National Taiwan University, Seoul University

Summary of April 2014

For www.aias.us Up to 30th April there were 14,222 hits, 87,111 distinct visits, 9.51 Gbytes downloaded, 41,364 page views and 2,696 documents read from 95 countries, led by USA, China, Ukraine, Germany, Japan, Mexico, Russia, Italy, ...

All UFT papers read, led by 85, 88, 43, 25, 177, 175, 177(Sp), 169, 140(Sp), 239, 159(Sp), 11, 259, 166, 233, 157(Sp), 228, 63, 107, 41, 42, 142, 178, 255, 26, 85(Sp), 175(Sp), 157, 65, 4, 97, 152, 8, ...

All Essays heard, led by 7, 63, 45, 24, 8, 40, 3, 64, 50, 59, 5, 52, 58,53, 60, 6, 44, 49, ...

All Other Documents read, led by: F3(Sp), Auto1, Potential Waves, LCR Resonant, Space Energy (Sp), F5, F1, Auto2, CEFE, "Llais", ECE Engineering Model, Evans Equations, Book of Scientometrics, ...

For the three sites www.aias.us, www.atomicprecision.com and www.upitec.org the above totals increase by an estimated 50% to 21,333 visits, 130,667 hits, 14.27 gigabytes, 62,046 page views.

1- 15 April

U.S. Universities, Institutes and Similar.

Arizona State, Berkeley, Caltech, Colorado, California Polytechnic State San Luis Obispo, City University New York, Hamilton, Illinois, Indiana, Maine, Michigan State*, New York, Ohio State, Princeton, Rochester, San Jose State, UCLA*, UC Santa Cruz, Illinois Urbana-Champaign, Maryland, Utah, Vermont, Washington St. Louis, NASA Goddard, Oak Ridge, Sandia, Gov. Washington, U. S. Army PICA, Naval Research Laboratory, U. S. National Archives.

Europe and Asia Minor: Universities, Institutes and Similar.

IBM EMEA, Royal Society of Chemistry, TU Vienna, Brunauer Centre Salzburg, Ghent, Vitebsk Belorussia, SRV Catalonia, ALU Metal Switzerland, EPF Lausanne, Zurich, General Electric Russia, German National Aerospace Center, Jacobs International University, TU Cottbus, TU Dortmund, Hamburg, Marburg, RENFE Spain, Complutense Madrid, Malaga, Polytechnic Valencia, Rovira i Virgilli, French Atomic and

Alternative Energy Commission, Limoges, Lille 1, Metz, Nantes, Perpignan, Poitiers*, Greek Army, IT Sligo Ireland, Agemont Italy, International Centre for Theoretical Physics Trieste, INFN Legnaro National Laboratories, INFN Pavia, Genoa, Schola Europaea Luxembourg, Free Univ. Amsterdam, Silesian Data Center, IPB Serbia, Joint Institutes for Nuclear Research Dubna Russia, Moscow State University, Lund University Sweden, Hacettepe University Turkey, Selcuk Turkey, Kiev, Aberystwyth, Birmingham Central, Bristol, Engineering Cambridge, Durham, Edinburgh, Imperial, Kent, Manchester*, Newcastle, Magdalen College Oxford, Robert Gordon Aberdeen, West Thames, British Library, Eduroam Oxford University.

Rest of World: Universities, Institutes and Similar.

UNCU Argentina, UNLP Argentina, Citefa Gov. Argentina, JCU Australia, Monash, New South Wales, Gov. Western Australia, Australian Ministry of Defence, UFJF Brazil, UFPE Brazil, UNESP, Sao Paolo Brazil, Brandon Canada, Gov. Canada, McGill, British Columbia, Trois Rivieres*, Toronto, York, Chile, Extensive visits from China, Colombia, Antioquia Colombia, Gov. Dominican Republic, City University Hong Kong, Bhabha Atomic Research Centre India, Kyoto SE, Kyoto, UIA Mexico, UNAM Mexico*, UPAO Peru, UPC Peru, NTU Taiwan*.

15 - 28 April 2014

US Universities, Institutes and Similar

Boston University, Arizona State University, Caltech, City University of New York, Harvard, Illinois, Kansas State, Massachusetts College of Pharmacy and Health Sciences, Morgan State, Notre Dame, New York University, Ohio State*, Pennsylvania State University, Rochester, Southern Illinois Edwardsville, San Jose State, University of California Los Angeles, University of California San Diego, University of Nevada Las Vegas, University of Pennsylvania, Wisconsin, Wright, Yale, Fermilab, Government of Virginia*, Naval Research Laboratory, United States National Archives, Mercy Ships, Poudre School District, Arizona, Knox, PSU*, UC Santa Cruz, Southern California, Yale.

Europe and Asia Minor: Universities, Institutes and Similar.

TU Vienna, SRV Catalonia, ETH Zuerich, TMCZ Czechia, WIA, ZCU Czechia, Benteler, Fraunhofer, Max Planck Heidelberg, TU Berlin, Heidelberg, Karlsruhe, Carlos III Madrid, Valencia, Jyvaskyla, Ecole National Superieure, Institute for Advanced Studies, Limoges, Poitiers*, Rennes 1, IE System, University College Dublin, TAL Iceland, Aruba Italy, ICTP Trieste, Genoa, Padua, Pisa, Groningen, Free University

Amsterdam, GD Poland, Polish Ministry of Finance, Silesian Data Center*, Bashtel Russia, JINR Dubna, MTW Russia, ISP Russia, Metu, Aberdeen, Cambridge, Durham, Kent, Lancaster, Leeds, Manchester, Newcastle, Merton College Oxford, Oxford VPN System, Portsmouth, Hull, British Library, Oxford Eduroam, Free University of Brussels, TU Braunschweig, Frankfurt, Durham, Luton, Max Planck Astrophysics, RWTH Aachen, Samara Aerospace University, York, Fujitsu Spain.

Rest of World: Universities, Institutes and Similar.

Melbourne*, UFPL Brazil, UERJ, UFPB Brazil, Sao Paolo, McGill, McMaster, Memorial University Newfoundland, National Capital Freenet Ottawa, Waterloo, York, Univ. Chile, UTFSM Chile, Tezu, IMSC India, PRL India, JAIST Japan, Kyushu, Lolipop Japan, KAIST Korea, Mexican Ministry of Health, IPN Mexico, Science Museum National Autonomous University of Mexico, UEM Mozambique, Gov. Dominican Republic, Auckland*, NU Singapore, National Taiwan, Bandes Venezuela, UFU Brazil, EAD University of Auckland, UEM Mozambique, UFSC Brazil, State University of Sao Paolo, Windsor University Canada, Tezu India, Raman Research Institute.

END OF TEN YEARS OF RECORDING

Summary of May 2014

For www.aias.us up to 30/5/14 there were 991,130 hits from 14,024 distinct visits, 37,303 page views, 10.354 gigabytes downloaded and 2,148 documents read from 102 countries. Led by USA, Germany, France, Mexico, Russia, Italy, Canada, Poland, Australia, Japan, Czechia, Brazil, Argentina, Ukraine, Britain, Netherlands, ... However a large number of hits were due to a site from China repeatedly downloading UFT107. Also, returns were not available for 31/5/14. Using the average ratio of hits to page views for the last six months the genuine number of hits is estimated to be about 60,000.

The total for the three sites www.aias.us, www.upitec.org and www.atomicprecision.com is estimated as usual to be about 50% higher, and allowing for the missing stats on 31/5/14 the totals are 22,523 distinct visits, about 93,000 hits, 16.85 gigabytes downloaded, 60,087 page views, 2,154 documents read from 102 countries, led by U. S., Germany, France, Mexico, Russia, Italy, Canada, Australia, ...

All UFT papers read, led by 25, 88, 177, 177(Sp), 159(Sp), 43, 166, 169, 142, 41, 85, 175, 215, 38, 157(Sp), 140, 94, 42, 116, 228, 2, 64, 107,

150B, 90, 161(Sp), 175, 178, 214-1b, 170(Sp), 201, 239, 7, 158, 102, 155, 149, 15, 237, 261, 26, 53, 89, 247, 63, ...

All essays heard, led by 24, 58, 62, 63, 4, UME1,.....

All books and articles read, led by: F3(Sp), Auto1, Book of Scientometrics, CV,   HSJ, LCR Resonant 2f, Potential Waves, F5(Sp), Auto2, Spacetime Devices (Spanish), CEFE, Evans Equations, Spacetime Devices, "Llais", F1(Sp), F9(Sp), Galaxies (Sp), Engineering Model, Space Energy, 2D paper, F4(Sp), AIAS Fellows, Cosmology in Crisis by Stephen Crothers, ...

1 - 15 May

U. S. Universities, Institutes and Similar.

Berkeley, Boise State, Caltech, Carleton, Columbia*, Cornell, Drexel, Emory and Henry, Harvard, Knox, Mines, Michigan State, Michigan Tech., New Mexico Tech., New York, Oakton, Oklahoma, Penn State*, Rochester, Stanford, Iowa, Minnesota Twin Cities, North Carolina Greenboro, Southern California, Utah, Tennesse Knoxville, Toledo, Western Florida, Wisconsin, Gov. Virginia, U. S. Army Signals, U. S. Naval Marine Command, U. S. National Archives, CUE Art Foundation New York City, Equal Access Foundation, Kern High School District California, Lake Worth Independent School District, Seattle Public Library; est Suffolk Regional Education Agency.

Europe and Asia Minor: Universities, Institutes and Similar.

Medical Institute University of Vienna, CERN, Paul Scherrer Institute Zurich, Humboldt Berlin, KFA Juelich, Max Planck Nuclear Physics Heidelberg, Max Planck Astrophysics Garching, RWTH Aachen, Frankfurt*, Greifswald, Tartu Estonia, IST Lithuania, Granada, Malaga, Spanish Distance Learning University, Cantabria, Cartagena, Valencia*, ENS Lyon, ENSA Rouen, Hebrew University Jerusalem, Vedur (Weather Centre) Iceland, INFN Rome 1, Genoa, Pisa, Rome 2 Sapienza, WUM Poland, Krakow, Lublin, Torun, Silesian Data Center*, Ural, Institute of Laser and Information Technologies Russia, Mari El Republic Russian Federation, SUTTK Russia, Space Plasma Physics Uppsala Sweden, Unila Slovakia, Aberdeen, Aberystwyth, Bristol, Cambridge, Cardiff, Durham, Edinburgh, Imperial, King's College London, Leeds, Manchester, Corpus Christi Oxford, Lady Margaret Hall Oxford, Lincoln Oxford, East Anglia, York, National Library of Wales.

Rest of World: Universities, Institutes and Similar.

MDP Argentina, Adelaide, Autralian National, Monash, Queensland Tech., Melbourne, INPE Brazil, UFSC, Unicamp Brazil, NCF Canada,

Perimeter Institute, Montreal, Quebec Trois Rivieres, Toronto, Talca Chile, UTFSM Chile, Large interst from China, UDEA Colombia, UNAL, Andes, Valle, UPB Colombia, EPN Ecuador, City Hong Kong, UGM Indonesia, UNS Indonesia, NITS India, Tezu, Raman Research Institute, Canon Corporation Japan, UIA Mexico, UNAM* Mexico, Unimas Malaysia, Uspot Malaysia, IPN Nepal, QAU Pakistan, PERN Pakistan, National Taiwan, Academica Sinica Taiwan, North Western South Africa, University of South Africa.

15 - 30 May

U. S. Universities, Institutes and Similar
    Amherst, Boston University, Caltech, Carleton, Columbia*, Cornell, Duke, Georgia Tech., George Mason, Illinois, North Dakota, Stanford, Texas A and M, Chicago, University of California Irvine*, University of California Riverside, University of California Santa Barbara*, Michigan, University of Nevada Las Vegas, United States Naval Academy, Utah*, Toledo, Wisconsin, Oak Ridge National Laboratory, U. S. Army AMRDEC, U. S. Navy NMCI, United States National Archives, Seattle Public Library, LARC NASA, U. S. Army NRO, .

Europe and Asia Minor: Universities, Institutes and Similar.
    Technical University Vienna, EPF Lausanne, Geneva, Charles University Prague, Bosch, Bochum, Electron Synchrotron Facility (IFH), Albert Einstein Institute, Technical University Cottbus, TU Ilmenau, Augsburg, Bielefeld, Erlangen, Kiel, Mainz, Munich, Regensburg, Stuttgart, Tuebingen, Cantabria, Valencia, Spanish Virtual Campus, Vigo, Complutense Madrid, ENS Cachan France, Angers, Poitiers*, University College Dublin, Tel Aviv Israel, CNR Bologna Italy, Genoa, Salento, Amsterdam, IFE Norway, Gazeta Poland, Extensive Visits Russian Federation, Aberdeen, Aberystwyth, Cambridge computing, Cambridge physics, Cardiff, Durham, Imperial College London library, Imperial Theoretical Physics, All Souls Oxford, Sceince Facilities Council British Government, St. Andrews, Swansea, High Energy Physics University College London, Melrose Press, Oxford Eduroam.

Rest of World: Universities, Institutes and Similar.
    MDP Argentina, UNCU Argentina, Australian National, Monash, Queensland Tech., New South Wales, IMPA Brazil, UFRGS Brazil, IFSC, Sao Paolo*, Government British Columbia, Physics and Astronomy British Columbia, Arcadia, Brandon, Perimeter, Quebec Trois Rivieres*, INACAP Chile, UTFSM Chile, Extensive visits from China, City Hong Kong, Tezu India, Vaves India, Kyoto, Osaka, Ritsumeikan, Tohoku,

Tokyo, Canon Corporation, SKKU South Korea, UAM Mexico, UGTO Mexico, Auckland New Zealand, NCU Taiwan, NTU Taiwan.

Summary of June 2014

For www.aias.us up to 30/6/14 there were 71,812 hits from 12,379 distinct visits, 11.435 Gbytes downloaded, 50,562 page views, 2,793 documents read from 88 countries, led by USA, Ukraine, Britain, Germany, Russia, China, Czechia, Australia, Mexico, Brazil, Canada, ...

It is estimated that one day approximately of recording was lost because of the way the timer goes from one month to the next, so these results are multiplied by 1.033 to give 74,182 hits, 12,788 distinct visits, 11.812 gigabytes downloaded, 52,231 page views.

The complete returns from the three ECE sites, www.aias.us, www.atomicprecision.com and www.upitec.org are estimated to be 50% higher, giving: 111,273 hits, 19,182 distinct visits, 17.718 gigabytes downloaded, 78,347 page views.

All UFT papers read, led by: 166(Sp), 25, 88, 43, 177, 94, 166, 107, 150(Sp), 169(Sp), 26, 46, 177(Sp), 175, 41, 90, 157(Sp), 140, 111, 2, 227, 243, 244, 38, 64, 239, 85, 89, 150B, 155, 54, 63, 68, 262, 42, ...

All Essays heard, led by: 25, 49, 6, 61, 22, 23, 26, 40, 45, 30, 32, 37, 38, 46, 3, 34, 36, ...

All Articles etc. read, led by: Autobiography Volume One, F3(Sp), F5(Sp), Llais, Universe of MWE Part Three, CV, F1(Sp), HSD31, Autobiography Volume Two, Space Energy (Sp), 2D, Potential Waves, Book of Scientometrics, Crisis in Cosmology, ...

1 - 15 June

U. S. Universities, Institutes and Similar.

Colorado, Columbia, Cornell, Maine, Mines, Michigan Tech., North Carolina State, North Dakota, Ohio State, IC Irvine, Oregon, Pompeu Fabra (Catalonia) on edu, Utah*, U. S. Airforce AFNOC, U. S. Naval Marine Command, U. S. Archives, South Brunswick Public Schools.

Europe and Asia Minor: Universities, Institutes and Similar.

Sofia, Xtec Catalonia, CERN, ETH, MWN Germany, Ruhr Bochum, Hannover, Heidelberg, Mainz, Jaume 1st Spain, Murcia, Cantabria, Seville, Aalto Finland, Polytechnique, Paris Psud, Strasbourg, LA Rochelle, Poitiers*, Versailles St. Quentin en Yvelines, Eotvos Lorand Hungary, Tel Aviv, Weizmann, INFN Padua, SIGQU Italy, Catania, Genoa, Groningen, Academ Russia, AGH Poland, Jagiellonian Krakow, Silesian Data Center, Lisbon Tech, National Institute of Physics Romania, Ministry of Social

Security Romania, Moscow University, Joint Institutes of Nuclear Research Dubna Russia, Russian Academy of Sciences Siberian Branch, KTH Stockholm Sweden, Uppsala, HCK Slovakia, Kiev Ukraine, Aberystwyth, Computation Cambridge, Fitzwilliam College Cambridge, Cardiff, Imperial, British Library, Llyfrgell Genedlaethol Cymru, Hereford County Council.

Rest of World: Universities, Institutes and Similar.

MDP Argentina, UCH, UNC, UNR, UNS Argentina, Aeronautics Research Institute (ITA) Brazil, Pontifical Catholic Rio de Janeiro, Sao Paolo*, Brandon Canada, NCF Canada, Polytech Montreal, Quebec Trois Rivieres*, Victoria, Chile*, China extensive, Distance Learning Colombia, National Colombia*, ICTDCA Cuba, ICRT Cuba, Raman Research Institute India, Vaves India, Shinshu, Tokyo Tech, Tokyo, Waseda, Mesh Corporation, Lolipop Japan, UAEM Mexico, National Autonomous Mexico, Edu System Pakistan, National Taiwan, Academica Sinica Taiwan, Gov. Taiwan., Gov. South Africa.

15 - 30 June

U. S. Universities, Institutes and Similar.

Iowa State, Northern Michigan, Ohio State, UC Santa Cruz, South Florida, Utah, Tennessee Knoxville, Fermilab, NASA Goddard Space Flight Center, United States National Archives*, United States Army Forces Command, United States Defense Department Missile Defense Agency, Global Virtual Opportunities Data Center, Lockheed Martin, Lightspeed Systems, Los Angeles Library, Pomona Unified School District California.

Europe and Asia Minor: Universities, Institutes and Similar.

TU Vienna, ETH Zurich*, Paul Scherrer Institute, TMCZ Transport Czechia, TU Brno Czechia, Humboldt Berlin, Max Planck Heidelberg, Ruhr Bochum, Stadtwerke Karlsruhe, Erlangen, Frankfurt, Jena, Ulm, Hamburg, Mannheim*, Xtrinsic, Aalborg Denmark, ETK Company Denmark, Aristotle University Thessaloniki Greece, Spanish Research Council Astrophysics Institute, Autonomous Barcelona, Vigo, Polytechnic Valencia, IPGP France, Supelec France, Strasbourg, Henri Poincare Nancy, Poitiers*, Zagreb Croatia, Weizmann, ICTP Trieste, INFN Turin, SIGQU Italy, Genoa, Silesian Data Center Poland, Institute of Experimental Physics Wroclaw, AMRES Serbia, Chelny, Samara, Russian Academy of Sciences Institute of Laser and Information Technologies, Solomono Archives Russia, Izhevsk, St. Petersburg Research University, CN Ukraine, Cardiff, Imperial, Nottingham, Salford, Staffordshire,

Swansea, University of Wales Registry, British Library (19/6/14 2.868 Gbyte download), BBC CWWTF, BBC TELHC, Nebosh Ltd.

Rest of World: Universities, Institutes and Similar.

MRSE Company Argentina, IMPA Brazil, National Institute of Physics Brazil, Sao Paolo, British Columbia Canada, Quebec Trois Rivieres*, TU Federico Santa Maria Chile, China extensive, National University Colombia, Valle Colombia, Pontifical Catholic Ecuador, ITB Indonesia, Government of India Official Portal, Tata Institute of Fundamental Research India, Hokudai Japan, Tokyo Metropolitan, Tokyo*, Canon Company Japan, Keio, Pohang South Korea (Postech), National Autonomous Mexico, International Peace Bureau Namibia, WBS South Africa.

Summary of July 2014

There were an estimated six days lost from www.aias.us in July 2014 due to an intermittent timer bug, giving 64,349 hits, 15,389 distinct visits, 9.844 Gigabytes downloaded, 43, 341 page views and 2,544 documents read from 85 countries, led by USA, Germany, Russia, Turkey, Italy, Brazil, Czechia, ...

For the three sites www.aias.us, www.atomicprecision.com, and www.upitec.org there an estimated 96,524 hits, 14.77 gigabytes downloaded, 23,084 distinct visits, 65,012 distinct visits.

All UFT papers read, led by 88, 177(Sp), 177, 25, 43, 166(Sp), 94, 142, 85, 26, 140, 18, 41, 175, 157(Sp), 239, 107, 166, 264, 121, 89, 140(Sp), 58, 150(Sp), 224, 2, 167, 42, 138(Sp), 159(Sp), 146, 150-B, 169, 236, 254, 265, ...

All Essays head, led by: 25, 23, 24, 22, 33, 49, 26, 35, 36, 37, 38, 44, 45, 46, 47, 50, 53, 54, 55, 61, 63, 9, 27, ...

All articles and books read, led by Auto1, F3(Sp), Auto2, CV, Potential Waves, Crisis in Cosmology, HSSD31, Llais, ...

1- 15 July

U. S. Universities, Institutes and Similar.

Arizona State, Central Michigan, Indiana, Johns Hopkins Medicine Baltimore, Michigan State, North Dakota, Princeton, Michigan, Utah, Tennessee Knoxville, U. S. House of Representatives, NASA, U. S. Army Medical Division, U. S. National Archives, Los Angeles Free Net Medical Organization, Iparadigms Corporation.

Europe and Asia Minor: Universities, Institutes and Similar

IRPHE Armenia, Johannes Kepler Austria, Ackle Computing Switzerland, EPF Lausanne, ETH Zurich, German Synchrotron Facility, Max Planck Plasma Physics Garching and Augsburg, Frankfurt, Mannheim, ETK Denmark, Spanish Research Council Institute for Agriculture (IATA), Generality of Valencia, Spanish National Institute for Aerospace Technology (INTA), Polytechnic Catalonia, Polytechnic Madrid, Valencia, French Atomic Energy Authority Cadarache Research Centre, Paris Diderot, Poitiers, Ghislieri College Pavia Italy, Genoa, ISTS Poland, Wroclaw (City), Silesian Data Centre, Lisbon Tech Portugal, Murmansk (City), St Petersburg National Research University, Kubang (City), Uppsala Sweden, Okan Turkey, Applied Mathematics and Physics University of Cambridge, Imperial College, British Library, Ysgol CCC.

Rest of World: Universities, Institutes and Similar.

MRSE Argentina, UNC Argentina, Australian National, IMPA Brazil, UFPE Brazil, Qubec Trois Rivieres* Canada, Chile, Federico Santa Maria Chile, China extensive, Jiao Tong University China, UNS Indonesia, IIT Delhi India, Indira Gandhi Atomic Research Centre, Tata Institute for Fundamental Research India, Keio Japan, Ritsume, Canon Corporation, Okinawa Institute for Science and Technology, Korean Institute for Advanced Research, Pohang University of Science and Technology, Mexican National Polytechnic Institute, Edu system Pakistan, National Central University Taiwan, National Dong Hwa University Taiwan, Faculty of Chemistry National University of Uruguay.

15 - 28 July

U. S. Universities, Institutes and Similar

Columbia, Cornell*, Denver*, Harvard, Indiana, New Mexico State, Purdue, U. S. Department of Energy Nevada, U. S. Naval Marine Command, U. S. National Archives*, Georgia Baptist Organization, Los Angeles Medical Free-Net, Cummins Aerospace, Data Foundry, Intel Inc., Sprint Data Center.

Europe and Asia Minor: Universities, Institutes and Similar.

CERN, FU Berlin, Landes Zeitung, Max Planck Plasma Physics Augsburg, Max Planck Heidelberg, Students Bonn, Frankfurt, Jena, Tartu Estonia, Agricultural University Athens Greece, CSIC IATA Spain, Ministry of Defence, Complutense Madrid*, Jyvaskyla Finland, Tampere Finland, Helsinki, Poitiers*, NGI Italy, Milan Polytech., Bari, Genoa, Bashtel Russia, Umea Sweden, Kocaeli Turkey, Odessa, Birmingham City, Department of Chemical Engineering Cambridge, Department of Applied

Mathematics and Physics Cambridge, Cardiff, Durham, Imperial, Nottingham, Balliol College Oxford, Luton, Quad Consult.

Rest of World: Universities, Institutes and Similar
     Adelaide, Australian National, INT Gov. Brazil, IMPA, TJRN Court Brazil, UFPE Brazil, NCF Canada, Alberta, UQTR*, Waterloo, Toronto, SEC Chile, Extensive China, IITD India, Kyoto, Osaka, Canon Inc., Osaka Institute of Science and Technology, University of Tokyo, Law Society of Kenya, UDG Mexico, Chainat Thailand.

Summary of August 2014

There were 64,053 hits from 13,329 distinct visits, 13.33 gigabytes downloaded, 42,696 page views, 2,274 documents read from 97 countries, led by : U.S.A., Germany, Russian Federation, Mexico, Australia, Brazil, Britain, Czechia, Canada, ...

These figures are increased by 50% for combined returns from www.aias.us, www.atomicprecision.com and www.upitec.org.   Plus an estimated outage of two days To give 20,860 individual visits, 100,243 hits, 20.87 Gigabytes.

All UFT papers read, led by: 25, 166(Sp), 177(Sp), 177, 107, 88, 39, 153,43, 166, 41, 94, 63, 158, 169, 138(Sp), 150(Sp), 158(Sp), 214, 90, 169(Sp), 155, 142(Sp), 11, 161(Sp), 2, 175, 136(Sp), 134, 168, 175, 201, 227, 30, 42, 140, 152, 239, 4, 140(Sp), 59(Sp), 33, 146, 167, 228, 67, 99, ...

All essays read or heard, led by: 25 (broadcast), 24(Sp), 7, 32, 11, 25 (pdf), 43 (pdf), 43(Sp), 35(pdf), ...

     All articles and books read, led by: F3(Sp), Auto1, Book of Scientometrics, Auto2, LCR Resonant, ...

1 - 15 August

U. S. Universities, Institutes and Similar.
     Caltech, Denver*, Indian Purdue Indianapolis, MIT, Ohio State, Pittsburgh, Princeton, University of California Santa Cruz. Michigan, Texas, NASA Goddard, Sandia, U. S. National Archives, Hospice of Queen Anne's, Integromedia B Medical Data Bank UC San Diego Supercomputer Center, Maine Health, Mitre Organization, U. S. Naval Marine Command.

Europe and Asia Minor: Universities, Institutes and Similar

Free University Brussels, ETH Zurich, TU Clausthal, Augsburg, Karlsruhe, Mainz*, Munich, Bornholm Denmark, State Library Aarhus, Seville, Paris Observatory, Patras Greece, European Space Agency Centre Darmstadt, Genoa, Naples Parthenope, Movement for the Protection of Rural Wales, Free University Amsterdam, Wieniawski Lublin, IMF Iasi Romania, Major Domo Russia, Cambridge, Cardiff, MRU Unit Oxford, British Library.

Rest of World: Universities, Institutes and Similar

INTI Argentina, QUT Australia, Ibrati Brazil, UFU Brazil, Sao Paolo, Waterloo, PUC Chile, UNAC Colombia, UPB Colombia, Municipal Ministry Colombia (Envigado), China extensive, Tezu University India, Gov. Mexico INEGI, UNAM Mexico.

15 - 29 August

U.S. Universities, Institutes and Similar.

Berkeley, Columbia, Denver*, Georgia Tech., Iowa State, Illinois, Colorado School of Mines, MIT, Oregon State, Purdue, UC San Diego, Montana, Vermont, City of Arlington Texas, NASA, Airforce Network Operations Command, Naval Marine Command, U. S. National Archives, Iparadigms California., Indianapolis Public Library

Europe and Asia Minor: Universities, Institutes and Similar.

Rolls-Royce, Silesian Data Center, ETH Zurich*, Homolka Czechia, Mazaryk Univ. Brno, Beuth University Germany, Humboldt Berlin, Mosaik Berlin, Albert Einstein Institute, Bonn, Erlangen, Karlsruhe, Kiel, Koeln (Cologne), Mainz, Munich, Munich Institute of Philosophy, Spanish Research Council IATA, Andalucian Gov., Spanish Ministry of Finance, Seville, Spanish Trade Unions, Helsinki. Oulu, Orleans, Paris Diderot, LCAR Paul Sabatier Toulouse, Genoa, Milan, Rome 1 Sapienza, Wroclaw Tech., extensive Russian Federation, Kiev, Odessa, Bristol, Nottingham, Neath Port Talbot Colleges, Salford, British Library, Ombudsman Wales.

Rest of World: Universities, Institutes and Similar.

Gov. Corrientes Argentina, INTI Argentina, Monash Australia, Melbourne, National Library of Australia, Sao Paolo Brazil*, Geotech Canada, National Research Council Canada, Dipreca Hospital Chile, INACAP Chile, Bio Bio Chile, Extensive China, UDEA Colombia, Envigado Colombia, Tezu India, Institute for Plasma Research India, Osaka Japan, Canon Corporation Japan, Gov. Mexico, INEGI, UNAM,

Academia Sinica Taiwan, Limpopo South Africa, Faculty of Medicine Univ. Uruguay.

Summary of September 2014

For www.aias.us to 28/9/14 there were 76,550 hits, 13.78 gigabytes downloaded, 14,053 distinct visits, 51,650 page views, 2,786 documents read from 92 countries, led by USA, Mexico, Romania, Slovakia, Britain, Germany, China, Australia, ...

These figures are increased by an estimated factor of 1.533 for the three websites www.aias.us, www.atomicprecision.com and www.upitec.org. To allow for an estimated 50% increase from the two other ECE sites and a day lost due to timer adjsutments. This gives 117,451 hits, 21.12 gigabytes downloaded, 21,543 distinct visits, and 79, 180 page views.

All UFT papers read, led by: 25, 166, 177, 43, 88, 177(Sp), 42, 54, 41, 175, 2, 94, 172, 169, 168, 239, 107, 39, 15, 64, 67, 33, 150B, 18, 157, 11, 29, 139, 13, 16, 57, 146, 158, 118, 170, 12, 269, 89, 141, 167, 161(Sp), 142, 22, 147, 8, 113, 158(Sp), 155, 60, ..

All Essays Heard or Read, led by: 24(Sp), 40(Sp), 24, 7, 25, 32(pdf), 11, 83, 41, 44(Sp), ...

All Articles and Books read, led by: F3(Sp), Auto1, Book of Scientometrics, Evans Equations, Simulation Circuit (Sp), Engineering Model, CV, Potential Waves, Auto2, CEFE, Space Energy, Llais, Spacetime Devices, Latest Family History, AIAS Fellows, ...

1 - 15 September

U. S. Universities, Institutes and Similar.

Boston, Colorado, Cornell, Denver, Florida International, Illinois, Kansas State, Maine, Mississippi State, Maharishi Univ of Management (on edu), National Optical Astronomy Observatory, Ohio State, Ohio University, Palermo (on edu), Purdue, San Francisco State, Southern Illinois, Stanford, University of California, Chicago, Minnesota Twin Cities, North Texas, Pennsylvania, Southern California, NASA Kennedy Space Center, United States National Archives, Indianapolis Public Library.

Europe and Asia Minor: Universities, Institutes and Similar.

Technical University Vienna, UTA Austria, Mons Belgium, ETH Zurich, CHMI Czechia, Mazaryk Brno, VSB Czechia, Max Planck Institute for Biophysical Chemistry Goettingen, KFA Juelich, Augsburg, Aalborg Denmark, Copenhagen, Spanish Ministry of Defence, Oviedo, Polytechnic Valencia*, Student Village Network Turku Finland, French

Atomic and Alternative Energy Commission, Nice, Marseilles, Poitiers, Schenenyi Istvan Univeristy Hungary, Weizmann Institute Israel, Italian CNR Institute of Microelectronics and Microsystems, DISI Genoa, Institute for Energy Technology Kjeller Norway, Zgora Poland, UAIC Romania, Bashtel Russia, KIS Russia, Solomono Educational System Russia, Lund Sweden, Uppsala Sweden, Anadolu Turkey, Cambridge, Durham, Imperial, Newcastle, Southampton, British Library, Hampshire County Council, Llyfrgell Genedlaethol Cymru / National Library of Wales, Ombudsman Wales.

Rest of World: Universities, Institutes and Similar.

UNSA Argentina, INTI Gov. Argentina, Australian National, Swinburne, Western Australia, UFMG Brazil, UFPR, Sao Paulo*, Perimeter Institute Canada, China extensive, Andes Colombia, MTOP Gov. Ecuador, Bose Institute India, DU Iran, IUT Iran, Yamagata Japan, IPN Mexico*, UASLP, UGTO, UIA Mexico, USMP Peru, National Taiwan.

15 - 30 September

U.S. Universities, Institutes and Similar

Arizona, Arizona State, Carnegie Mellon, Colorado, Cornell, Denver, Georgia Tech., Harvard, Illinois State*, Indiana, Kansas State, Mercyhurst, Mount St. Mary's, Notre Dame, New Mexico Tech., Ohio, Princeton, Penn State, Chicago, U Mass., Minnesota Twin Cities, New Hampshire, U Penn., Rafael Belloso Chacin (on edu), Southern California, Tennessee Knoxville, Texas A and M, Washington, Williams, United States Naval Marine Command, Wisconsin, Dept. Of Transport California, Lawrence Livermore, Oak Ridge, US National Archives, World Wide Web, General Electric (Russia).

Europe and Asia Minor: Universities, Institutes and Similar

CERN, Centrio Czechia, Charles Univ. Prague, TMCZ Heavy Transport Czechia, Bayer Pharmaceuticals, Beuth University Berlin, German Synchrotron Facility (DESY), Max Planck Heidelberg, Fraunhofer, TU Berlin, Darmstadt, Bonn*, Hamburg, Karlsruhe, TTU Estonia, Autonomous Barcelona, Valencia, YOK Finland, Civil Aviation France, Poitiers, TU Troyes, Kapodistrian Univ. Athens; Ioannina Univ Greece, UC Cork Ireland, UC Dublin, ICTP Trieste, INFN Genoa Italy, Genoa, Radboud Netherlands, IFE Kjeller Norway, NTNU Norway, Bashtel Russian Federation, Tomsk State Univ. of Control Systems and Radio Electronics, Royal Institute of Technology Stockholm, Bristol,

Heriot Watt, Newcastle, Oxford Maths, Oxford Physics, Oxford general, Cambridge, Salford, Southampton.

Rest of World: Universities, Institutes and Similar.
UNGS Argentina, Macquarie Australia, Swinburne, Queensland, Hospital Regional Brazil, UFMG Brazil, UNICAMP, Sao Paulo, DND Canadian Armed Forces, McGill, McMaster, Perimeter, Toronto, Waterloo, Windsor, UCT Chile, China extensive, UDEA Colombia, Valle, ICRT Cuba, Gov. Ecuador, Tezpur India, Indian Rail, DU Iran, Kanagawa Univ. Japan, Tokyo Univ., Canon Corporation Japan, CICESE Mexico, Gov. Mexico, UASLP Mexico, UNAM Mexico*, NDHU Taiwan, NTHU Taiwan, Gov. Venezuela, Witwatersrand South Africa.

Summary of October 2014,

For www.aias.us there were 87,101 hits, 16,333 distinct visits or reading sessions, a new twelve year record high of 19.97 gigabytes downloaded, 54,470 page views, 2,819 documents read from 97 countries led by U. S. A., Germany, Romania, China, France, Mexico, Britain, Czech Republic, ...
These figures are increased by an estimated factor of 1.53 for www.aias.us, www.atomicprecision.com and www.upitec.org to give 133,264 hits, 24,989 distinct visits, a twelve year record high of 30.55 gigabytes downloaded, 83,339 page views.
All UFT papers read, led by: 145, 166(Sp), 25, 177, 42, 63, 7, 88, 97, 214, 273, 243, 175, 169, 166, 107, 18, 43, 271, 170(Sp), 94, 158, 39, 239, 17, 41, 155, 46, 15, 55, 57, 159(Sp), 11, 4, 5, 150(Sp), 12, 16, 1, 10, 13, 157(Sp), 2, 140, 153, 157, 171, 14, 21, 3, 76, 116, 19, 81, 152, 29, ...
All essays heard or read, led by 32, 25, 38, Finite Photon Mass, 26, 37, 69, UFT139 Intro, 66, ECE Versus Standard Model, 19, 31, 44, ...
All books and articles read, led by F3(Sp), Auto1, Book of Scientometrics, Spacetime Devices, Auto2, Engineering Model, Englynion (Book of Poetry in Welsh and English), Llais, CEFE, ...

1 - 15 October

U. S. Universities, Institutes and Similar
Amherst, Arizona State, Boston, Buffalo*, Caltech, Carleton College, Colorado*, Denver, Duke, Florida State, Northwestern, New York*, Ohio State*, Princeton, Penn State, Purdue*, Rice, Rutgers, Stanford, Stonybrook, Texas A and M, Maryland, New Hampshire, Nebraska Lincoln, Ursinus, U. S. Naval Academy, Texas, Tennessee Knoxville,

Vermont, Washington, Wisconsin, Washington St Louis, Yale, Intel Corporation, U. S. Archives, World Wide Web, Rolling Meadows Library.

Europe and Asia Minor: Universities Institutes and Similar.

EDIS Austria, Paul Scherrer Institute Switzerland, Suedo STS Switzerland, CVUT Czechia, Jemnice, VUTBR Germany, Beuth University Berlin, Max Planck Garching, TU Berlin*, Karlsruhe, Atlas Electronik, University of the Basque Country, Government Valencia, Barcelona, Malaga, Granada, Spanish National University of Distance Education, Valencia*, BOB Finland, Helsinki, ENS EEIHT Toulouse France, French National Particle Laboratory, Sorbonne, SCSE, St Euverte St Croix School Orleans France, Poitiers, European Space Agency ESOC and ESTEC, Eotvos Lorand Hungary, Szeged Hungary, IFS Croatia, PTR Ireland, Genoa, Leiden, Utrecht, Eniro Norway, AGH Poland, Lisbon, Tomsk Region, Regional Joint Computer Network Education Science and Culture St. Petersburg, Saratov State University, Kiev, Buckingham University, Cambridge*, Edinburgh, Lancaster, Leeds, Exeter College Oxford, Physics Oxford, Sheffield, Southampton, York College, Westminster School.

Rest of World: Universities, Institutes and Similar.

Citefa Gov Argentina, Adelaide* Australia, Griffith, Monash, New South Wales, National Library of Australia, CBPF Brazil, UFU, UNESP, Sao Paulo, SCE Bhutan, Carleton Canada, McGill, Toronto, Waterloo, Chile, China extensive, UDEA Colombia, Valle Colombia, ICRT Cuba, City University Hong Kong, Tezu India, Indian Railways, Indian Institute of Mathematical Sciences, Tokyo Tech, University of Tokyo, Canon Corporation Japan, IPN Mexico, IIU Pakistan, NCTU Taiwan, NSYSU Taiwan, IVIC Venezuela, Witwatersrand South Africa, Zimbabwe Education Network.

15 - 30 October

U. S. Universities, Institutes and Similar

Berkeley*, Caltech, Columbia, CSU Northridge, Case Western Reserve, Denver*, SUNY Geneseo, Harvard*, Iowa State, Illinois, North Dakota State, New York, Portland State, Penn State, Purdue*, Stanford, Texas A and M*, UCLA, UC Santa Cruz, Delaware, Maryland, North Florida, Utah State, Toledo, Vermont, Washington, Washington St Louis, U.S. Department of Energy Nevada, U. S. Dept. Of Justice, Aerospace Corporation, Intel, Brookhaven*, U. S. National Archives, NASA Advanced Supercomputing Facility, Los Angeles Arboretum,

Europe and Asia Minor: Universities, Institutes and Similar

Liege, Ghent, Free Univ. Brussels, Xtec Catalonia, CERN, EPF Lausanne, Paul Scherrer Institute, Bern, Beuth University Berlin, German Synchrotron Facility, Aachen, Karlsruhe, Aarhus Denmark, Gironia Spain, Basque Country, Granada, Rioja, Polytechnic Cartagena, Valencia, Bruno Kessler Institute Italy, Turku Finland, French National Particle and High Energy Laboratory (in2p3), SUPELEC, Bordeaux, Poitiers, Technical Univ. Troyes, Eotvos Lorand Hungary, Haifa Israel, Genoa Italy, Pavia, Lebanese American Univ., Bergen Norway, Norwegian National Health Institute (STAMI), Polish Academy Krakow, Silesian Data Center, St Petersburg Distance Learning Russia, Bristol, Cambridge, Edinburgh, Imperial, Lancaster, Leeds, Oxford, Sussex, University College London, Engineering Cambridge.

Rest of World: Universities, Institutes and Similar

MRSE Argentina, UNLP Argntina, Adelaide Australia, Monash, CBPF Brazil, SCE Bhutan, McMaster Canada, Toronto, Unibe Chile, Pontifical Catholic Chile, Univ. Chile, China extensive, UDEA Colombia, City Hong Kong, IITD India, Gov. Jamaica, Kanazawa Japan, Nagoaka Tech., Titech, Tokyo Univ. Science, IPN Mexico, UAEM, UGTO Mexico, CIICAP Mexico, Auckland New Zealand, Massey, National Univ Singapore, Singapore Bible College, Pohang South Korea, NCTU Taiwan, TWU Taiwan, Witwatersrand South Africa.

Summary of November 2014

For www.aias.us there were 73,141 hits, 13,858 distinct visits, 18.625 gigabytes downloaded, 50,620 page views, 2,848 documents read from 96 countries, led by USA, China, Germany, Russia, Mexico, Italy, Australia, Britain, Canada, ...

All UFT papers read, led by 145, 25, 243, 43, 169, 168, 166, 88, 170(Sp), 214, 269, 94, 177(Sp), 26, 42, 157(Sp), 170, 13, 140, 152, 39, 177, 149, 150B, 228, 142, 139, 159(Sp), 33, 84, 175, 141, 11, 206, 158, 10, 15, 2, 155, 140(Sp), 17, 46, 47, 54, 59, 6, 75, 134, 159, 38, 5, ...

All Essays read and heard, led by 27, 15, 19, 43, 25, 29, 37, 53, 22, 69, 70, 26, 38, ...

All books and other items read, led by Auto1, F3(Sp), LCR Resonant 2f, CV, Auto2, Book of Scientometrics, Potential Waves, Spacetime Devices, Englynion, ...

These figures are increased by an estimated factor of 1.53 when www.atomicprecision.com and www.upitec.org are taken into account, and allowance made for an estimated one day lost as the timer changes over from month to month. So the total is 21,203 visits, 111,906 hits, 28.50 gigabytes downloaded, 77,449 page views.

1 - 15 November

US Universities, Institutes and Similar.
   Amherst, Berkeley, Boston University, Caltech, College of DuPage, Cornell, Dartmouth, Illinois Urbana Champaign, Manhattanville, New York University, Princeton, Penn State, Rutgers, Scripps Institute, Texas A and M, Alabama Huntsville, California system, UC Davis, UC Santa Cruz*, North Carolina Chapel Hill, New Hampshire, U. S. Naval Academy, Wisconsin, Brookhaven National Laboratory, US Food and Drug Administration, Lawrence Livermore, U. S. Archives San Francisco, Tri-Rivers Educational Computer Association of school districts Ohio, Chevron Corporation.

Europe and Asia Minor: Universities, Institutes and Similar
   Innsbruck, Free University of Brussels, Catholic University Leuven, CERN, ETH Zurich, German Synchrotron Facility (DESY), Leibniz Research Centre Munich, Max Planck Augsburg, TU Darmstadt, Kaiserslautern, Paderborn, Tuebingen, Aarhus Denmark, Institute of Catalysis and Petrochemicals Spain, Andalucian Government, Complutense Madrid, Cordoba, La Laguna, Malaga, Polytechnic Valencia, Turku, Lyon 1 France, Poitiers, Haifa, Israel, Galliera Hospital Genoa, Projects International Netherlands, NTNU Norway, Warsaw Polytechnic, Lisbon Tech., Moscow State University, Siberian Academy Institute of Nuclear Physics, Swedish University of Agricultural Sciences, Uppsala, Ege Turkey, Gazientep Turkey, City of Poltava Ukraine, Social Services Romanian Government, Cambridge, DAMTP Cambridge, Edinburgh, Imperial, Leeds, Nottingham, Statistics Oxford, St Hilda's Oxford, Sussex, Swansea, College of Law, Gordonstoun School.

Rest of World: Universities, Institutes and Similar.
   Wollongong Australia, UF Pernambuco Brazil, UF Santa Catarina, UF Uberlandia, McGill Canada, Perimeter Institute, British Columbia, Quebec Trois Rivieres, Sherbrooke, Toronto*, Windsor, China extensive, Colombo Britanico Colombia, National University Colombia, Andes, TIE Chile, UFRO Chile, ESPOL Ecuador, City Hong Kong, Indian Institute of Technology, National Institute of Technology Silchar, Indian Railways, DU Iran, UFTO Mexico, National Autonomous Mexico, Diliman Philippines, Edu system Pakistan.

15 - 27 November

US Universities, Institutes and Similar

Caltech, Dartmouth, Denver, Gustavus Adolphus, Indiana, Kansas State, Louisiana Tech, Manhattanville, Oberlin, Purdue, Texas A and M, UC Davis, UCLA, UC Santa Cruz, Florida, Massachusetts Boston, North Carolina Greensboro, U Penn, Texas Dallas, Wisconsin, Washington, California Department of Transport, Los Alamos, Naval Research Laboratory, United States National Archives*, World Wide Web.

Europe and Asia Minor: Universities, Institutes and Similar

TU Vienna, ETH Zurich, Ostroj Mining Machinery Czechia, TMCZ Heavy Transport Czechia, Jan Evangelista Purkine, Beuth University Berlin, IPHT Jena, Leibniz Research Center Munich, Students Bonn, TU Darmstadt, Bonn, Hamburg, Jena, Cologne, Copenhagen, Institute of Astrophysics Tenerife Spain, Andalucian Government, Cadiz, Malaga, Polytechnic Catalonia, Seville, Helsinki Finland, Turku, French Atomic and Alternative Energy Commission, Pierre et Marie Curie, ENSEEIHT, Institute of Space Astrophysics Orsay, Lille 1, Poitiers, St. Etienne, Italian CNR Institute of Informatics and Telematics, INFN Gran Sasso Science Institute Italy, Genoa, ISP Fabriek Netherlands, Amsterdam, Free Univ Amsterdam, Eniro Norwegian Univ. Science and Technology, Poznan Poland, TU Warsaw, Lisbon Tech Portugal, Landau Institute Russian Academy of Sciences, Siberian Branch Russian Academy of Sciences, Bashtel region, Middle East Technical University Ankara Turkey, Cities of Yalta, Odessa and Poltava Ukraine, Cambridge Computing, Jesus College Cambridge, Robinson College Cambridge, St Edmund's College Cambridge, Cardiff, Durham, Kent, Nottingham, St. Anne's College Oxford, University College Oxford, University of Oxford, British Library.

Rest of World: Universities, Institutes and Similar.

Library of Congress of Argentina, UNC Argentina, INTI Argentina, Wollongong Australia, Office of the Brazilian President Planalto, National Capital Freenet Canada, Carleton Canada, Alberta, British Columbia, Waterloo, Chile, China extensive, National Univ. Colombia, Andes, City Hong Kong, ITB Indonesia, NITS India, Tezpur, IMSC India, Kyoto Japan, Osaka*, Riken Research Organization, American Univ. Beirut Lebanon, Cicata Altamira Mexico, UITM Malaysia, ADMU Philippines, NUST Pakistan, QAU Pakistan, Physics National Univ Uruguay, Cape Town South Africa.

Summary of December 2014

For www.aias.us there were 82,984 hits, 16.87 gigabytes downloaded, 12,072 distinct visits, 57,542 page views, 2,862 documents read from 90 countries, led by USA, Germany, Britain, Czechia, China, Austria, Italy, Russia, ... and for the year 2014 there were 1,797,040 hits.

For www.aias.us, www.atomicprecision.com and www.upitec.org These figures are increased by a factor of 1.55 to allow for an estimated one day lost by timer readjustment and for a 50% increase in traffic as usual from the other two sites. Sothis gives 128,625 hits, 26.15 gigabytes downloaded, 18,712 distinct visits and 89,190 page views for December 2014.

All UFT papers read (survey of late December on blog), led by: 177, 25, 88, 243, 169, 175, 166, 140, 150B, 39, 43, 18, 107, 155, 42, 94, 201, 41, 152, 56, 142, 67, 13, 228, 8, 158, 214, 229, 27, 157, 4, 61, 19, 44, 65, 85, 159(Sp), 170, 279, 47, 57, 159, 10, 58, 161, 168, 234, 239, 3, 144, 160, 224, 36, 116, 150(Sp), 146, 35, 64, 153, 174, 12, 15, ...

All Books and articles read, led by: Autobiography Volume One, F3(Sp), UNCC Saga1, CV, Autobiography Volume Two, LCR Resonant 2f, Space Energy, Numerical Article, Book of Scientometrics, Potential Waves, HPexp2, Definitive Proof One, Universe of Myron Evans, Space Energy Devices (Sp), CEFE, Spacetime Devices 2, ..

All Essays read and heard, led by: 24(pdf), ECE Theory of Hydrogen Bonding, Implications of Finite Photon Mass, 26(pdf), 43(pdf), 7(pdf), 11(pf), 24Sp(pdf), 44(pdf), ...

1 - 15 December

U .S. Universities, Institutes and Similar.

Amherst, SUNY Buffalo, Caltech, Columbia, Cornell. CSU Chico, Cleveland State, City University New York, Case Western Reserve, Florida Institute of Technology, Iowa State, Illinois*, Johns Hopkins, New York University, Purdue, San Francisco State, Syracuse, Chicago, UCLA, UC Santa Cruz, Maryland, U Pennsylvania, Washington*, Washington State, U. S. Archives, Valley View School District; Chevron Corporation.

Europe and Asia Minor: Universities, Institutes and Similar.

Philips Campus Eindhoven Netherlands, Charles University Prague, Mazaryk University Brno, Kplan Engineering Germany, TU Berlin*, Bonn, Heidelberg, Potsdam, Tuebingen, Ulm, Spanish Research Council Institute for Catalysis, Basque Country, Seville, Salamanca, Enterare Europe, Lacompany France, Le Floch, Poitiers, Ioaninna Greece, Genoa Italy, Greenwich, Imperial, British Library*, Llyfrgell Genedlaethol Cymru (National Library of Wales), Radboud University Nijmegen Netherlands, Government Canary Islands, Bauman Moscow State Technical University, Leicester, Bristol, Nottingham, Durham, Firzwilliam Colleeg Cambridge, Swansea, Southampton.

Rest of World: Universities, Institutes and Similar.

INTI Argentina, Adelaide, GEGEP Limoilou Quebec, Waterloo, China extensive, Valle Columbia, ITS Indonesia, NITS India, Indian Railways, Tokyo, RIKEN Japan, Hitachi, IPN Mexico, UASLP, UDG, UNAM Mexico.

15 - 31 December

U. S. Universities, Institutes and Similar.

Caltech, Carnegie Mellon, Hawaii, NewYork, Michigan, Oak Ridge, U. S. Archives, Chicago Public Library, Cisco, Yonkers Public Schools New York City.

Europe and Asia Minor: Universities, Institutes and Similar.

Andorra, OEAN Educational Agency Austria, TU Vienna, EPFL Switzerland, Institute of Physics, Czech Academy of Sciences, SXG Software Czechia, TU Berlin*, TU Darmstadt, Erlangen, Heidelberg*, Karlsruhe, Kiel, Koeln, Konstanz, Tuebingen, Salamanca Spain, CEA France, Ioaninna Greece, NDCO Iran, Informatics Institute Italian CNR, Genoa, LAU Lebanon, Utrecht, Amsterdam, Academ Russia, Max Planck Software Systems. Serbian Academy, Lipetsk Region Russia, City of Siguna Sweden, Siberian Academy of Sciences, Anadolu Turkey, Bilkent Turkey, Odessa Ukraine, Cambridge portal, King's College London, Keele, Southampton, Swansea.

Rest of World: Universities, Institutes and Similar.

UNGS Argentina, INACAP Chile, Sernapesca (Ministry of Fisheries) Chile, UFRO Chile, China extensive, Envigado Gov. Colombia, ITB Indonesia, UNPAD Indonesia, Bose India, IMSC India, Tokyo, Nikon Company, Korea, KMITL Thailand, NTTU Taiwan.

END OF 2014

Summary of January 2015 : New Record High

For www.aias.us There were 14,203 distinct visit, 101,598 hits, 77,244 page views, 16.71 gigabytes downloaded, 2,903 documents read from 93 countries led by USA, Germany, Latvia, Italy, China, Czech Republic, Canada, Britain, Russian Federation, ...

The combined total for the three sites www.aias.us, www.atomicprecision.com and www.upitec.org is estimated to be 60%

higher from the 2014 total of hits, and one day is estimated to be lost as the timer changes over. Therefore the adjustment factor for January is 1.632. This gives a total of 23,051 distinct visits, 165,807 hits, 27.12 Gigabytes downloaded and 125,367 page views in January 2015.

All UFT papers read, led by: 93, 89, 25, 54, 88, 155, 43, 169, 42, 47, 177(Sp), 39, 300, 177, 166(Sp), 41, 107, 166, 140, 239, 94, 85, 152, 158, 191, 170, 175, 228, 243, 46, 159(Sp), 170(Sp), 157(Sp), 67, 142, 146, 86, 38, 2, 61, 149, 116, 83, 63, 168, 15, 290, 8, ...

All Books and articles read, led by: Autobiography volumes 1 and 2, Principles of ECE Theory, Book of Scientometrics, F3(Sp), Engineering Model, Second Book of Poetry, ...

All essays heard or read, led by: 24, 24(Sp), 37, 9(Sp), 11, 86, 43, 49(Sp), 9, 31(Sp), 32(Sp), ...

1 - 15 January

U. S. Universities, Institutes and Similar
California State Polytechnic Pomona, Clemson, Denver*, Georgia Tech., Georgetown, Iowa State, Illinois, Michigan State, Northwestern, New York, Rutgers, UC Irvine, UC Santa Barbara, UC Santa Cruz, Univ. Georgia, U Penn, U. S National Archives, Los Angeles Public Library.

Europe and Asia Minor: Universities, Institutes and Similar,
EPF Lausanne Switzerland, Valais Canton, Wagner and Grimm Engineering, Charles Univ. Prague, TU Dresden, Augsburg, Heidelberg, Karlsruhe, Kassel, Munich, Tuebingen, Spanish Research Council Institute of Science and Materials Madrid, Granada, Complutense Madrid, Catalonia Polytech., Cartagena Polytech., Seville, JG Engineering Spain, Turku Finland, Patras Greece, Technion Israel, Science and Tech University Krakow Poland, City of Gdansk, Physics Moscow State University, City of Belgorod, Chalmers Sweden, Karlstad Sweden, Cities of Kiev, Odessa and Rovno Ukraine, Brunel, Physics and Applied Maths Cambridge, Cardiff, City, Imperial, Leicester*, Middlesex, Sussex, Swansea, York.

Rest of World: Universities, Institutes and Similar.
Deakin Australia, Monash, Perimeter Institute Canada, China extensive, International Inst. Information Tech. Bangalore India, Bhabha Atomic Research Institute, Indian Rail (RCIL), Azarbaijan University Iran, Kyushu Japan, Nagoya, Osakafu, Sophia, Hyogo, KEK High Energy Research Institute, Hitachi Corporation, Kongju University South Korea, Pohang Univ. Sceince and Technology, UDG Mexico, Mexican National

Polytechnic Institute, Gujrat Pakistan, National Chiao Tung Taiwan, National Tsing Hua Taiwan, National Taiwan.

15 - 29 January

U. S. Universities, Institutes and Similar.
Arizona State University, Caltech, Clemson, Case Western Reserve, Denver*, Georgia Tech*, Iowa State, Illinois, Merrimack, MIT, Northwestern, Princeton, University Corporation for Atmospheric Research, UC Irvine*, UC Santa Barbara, UC San Diego, U Mass, U Penn, Texas, Washington*, Wisconsin, U. S. Navy Marine Command, Los Angeles Public Library.

Europe and Asia Minor: Universities, Institutes and Similar.
Ghent, CERN, ETH, Zurich, Institute of Physics Czech Academy, German Aerospace Corporation (DLR), IDS Manheim, Munich, Oldenburg, Spanish Research Council Institutes*, Autonomous Barcelona, Granada, Murcia, Cartagena Tech, Seville, Helsinki, Tampere Tech Finland, Henri Poincare Nancy, Paris Diderot, Poitiers*, Croatian Geophysics Institute, Eotvos Lorand Hungary, University College Dublin, Iceland, Astrophysical Observatory Catania, Insubria, School of International Studies Trieste, HVIV Internet Freedom Netherlands, Radboud Nijmegen, Norwegian National Tech., Warsaw University Foundation, Academ Russia, Kurgan region, City of St. Petersburg, Swedish Society for Digital Freedom, Karlstad University Sweden, Royal Institute of Technology Stockholm, Cities of Kiev and Odessa Ukraine, Bristol, Cambridge*, Cardiff*, Durham, Edinburgh*, Lancaster, Leicester, Newcastle, Swansea, Royal Society of Chemistry, BBC, News Corporation International ("The Times"), Ysgol CCC.

Rest of World: Universities, Institutes and Similar.
Monash Australia, Sao Paulo State Brazil, Campinas, Simon Fraser Canada, Montreal, Quebec Trois Rivieres, Toronto*, University of Chile, China extensive, UCLV Cuba, UEES Ecuador, Tezpur India, Asahikawa NCT Japan, Sophia, Tokyo, Waseda, Hitachi Corporation, KEK High Energy and Nuclear Physics Organization, Mexican National Tech, Monterrey Tech., Guadalajara, National Autonomous University of Mexico*, National University of Singapore, National Taiwan Ocean University.

Summary of February 2015

For www.aias.us to 26/2/15 there were 88,167 hits, 12.62 Gigabytes downloaded, 12,142 distinct visits, 68,926 page views, 2,960 documents read from 95 countries, led by USA, Germany, Mexico, Australia, Russia, Ukraine, Czech Republic, Britain, France, Japan, ...

The combined total for the three sites www.aias.us, www.atomicprecision.com and www.upitec.org is estimated by increasing the above totals by 60% and allowing for one day lost as the timer changes over from month to month. This means that the above totals are multiplied by 1.636. This gives 144,241 hits, 19,864 distinct visits, 112,763 page views, 20.65 gigabytes downloaded

All UFT papers read, led by 93, 89, 88, 43, 25, 166(S), 177(S), 94, 171(S), 107, 39, 243, 157(S), 175, 42, 170(S), 155, 85, 228, 239, 46, 41, 169, 26, 152, 304, 140, 149, 58, 116, 63, 99, 159(S), 142, 255, 2, 161(S), 159, 156, 33, 4, 61, 151, 15, 75, 110, ...

All Essays heard or read led by 36(S), 106(S), 105(S), 24(S), 24, ...

All Books and articles read led by: F3(Sp), Auto1, CV, Book of Scientometrics, Universe of Myron Evans (Sp), Auto2, Evans Equations, Engineering Model, Englynion, CEFE, ...

1 - 15 February

U. S. Universities, Institutes and Similar.

Barnard, Berkeley*, Boston University, Columbia*, Cornell, Denver, Embry-Riddle Aeronautical University, Iowa State, MIT*, Northwestern, Reed, Florida, Illinois Urbana Champaign, U Mass, US Naval Academy, Texas Pan American, Washington*, Yale, U. S. Airforce Network Operations Command, U. S. Naval Marines Command, U. S. National Archives, Association of Missouri Trial Attorneys, Oak Ridge National Laboratory, Gov. Virginia.

Europe and Asia Minor: Universities, Institutes and Similar.

Liege, Leuven, CERN, EPF Lausanne, ETH, Geneva, Fraunhofer ILT, Koenigsbruhn, Albert Einstein Institute of the Max Planck Foundation, Duesseldorf, Heidelberg*, Jena, Ulm, Danish Tech., Spanish Ministry of Defence, Spanish Ministry of Finance, Procisa Corporation Spain, Autonomous Barcelona, Seville, Valladolid, Jyvaskyla Finland, Turku, Sorbonne Universities Paris, Paris Psud, Nantes, Poitiers*, Rennes 1, Toulouse, Athens, INFN Naples, Scuola Normale Superiore Pisa, Turin, Heart Internet Freedom Netherlands, European Space Agency HQ Netherlands, TU Delft, Royal Society of Chemistry, Institute of Mathematics Polish Academy of Sciences Warsaw, Institute of Molecular

Physics Polish Academy of Sciences Poznan, Polish Space Research Centre Warsaw, City of Lipetsk, Moscow State University, Solomono Russia, STK Construction St. Petersburg, City of Kiev Ukraine, Cities of Lviv, Poltava and Sidorovshina, Bristol, Cambridge computer centre, Girton College Cambridge, Durham, Imperial, Lancaster*, Central Lancashire, Lincoln College Oxford, Oxford eduroam, Welsh Industrial Networking.

Rest of World: Universities, Institutes and Similar.

UNLP Argentina, Adelaide, UFMG Brazil, Sao Paolo, McMaster Canada, Quebec Trois Rivieres, Toronto, Victoria, Waterloo, China extensive, Antioquia Colombia, National Univ. Colombia, ESPE Ecuador, Univ Science and Technology Hong Kong, Indian Railways, Osaka, National Autonomous Mexico*, UPAO Peru, Edu system Pakistan, National University of Singapore, Limpopo South Africa.

15 - 26 February

U. S. Universities, Institutes and Similar

Albany, Berkeley, Caltech, Claremont, Florida Institute of Technology, Georgia Tech., Johns Hopkins, McDaniel, Ohio State, Ohio State University, Renselaer Polytechnic Institute, Stonybrook, University of Texas Southwestern Medical Center, Texas A and M University, UC Davis, Chicago, UC San Diego, Georgia, Wisconsin Milwaukee, Virginia, Vermont, Wisconsin Madison, Washington State, Yale, Library of Congress, Oak Ridge, United States Naval Marine Command, South Coast Repertory California, United States National Archives, World Wide Web Consortium.

Europe and Asia Minor: Universities, Institutes and Similar

Graz, EPF Lausanne, Charles University Prague, Czech Technical Univ. Prague, Bosch Corporation, German Aerospace Corporation, Leibniz Research Centre, TU Berlin, Kaiserslautern, Ulm, Copenhagen, Spanish research Council Institute for Fundamental Physics, Eberri Company Spain, Madrid Complutense, Oviedo, Salamanca, Vigo, Poitiers*, University of Nancy Hospital Centre, ENSEEIHT Toulouse, Pierre et Marie Curie Paris, Metz, Paris 1, Greek Schools, Trinity College Dublin, University of Iceland Gardur, INFN Pisa, INFN Rome 1, Heart Internet Freedom Netherlands, Radboud, Eindhoven, Chalmers, City of Poltava, City of Sebastapol, Aberdeen, Bath, Girton College Cambridge, Durham, Imperial, King's College London, Balliol College Oxford, Physics Oxford*, Eduroam Oxford, Swansea, Welsh Networking, Xtec Catalonia.

Rest of World: Universities, Institutes and Similar.

UNC Argentina, MPD Gov. Argentina, Adelaide, Monash, Wollongong, Toronto*, China, Antioquia Colombia, National Univ. Colombia, Cuenca Ecuador, Tata Institute India, NITT India, Nagoya, Osaka, MESH Corporation Japan, Hitachi, Mexican Government Institute of Statistics, IPN Mexico, ITESM, UNAM Mexico, Otago, NCU Taiwan, PSU Thailand, Witwatersrand South Africa.

Summary of March 2015 : New Record High of Gigabytes Downloaded.

For www.aias.us up to 30th March there were 14,033 distinct visits, 78,480 hits, 22.23 Gbytes downloaded, 55,082 page views, 2,919 documents read from 104 countries, led by USA, Germany, Ukraine, Britain, Russia, Luxembourg, Japan, Czech Republic, ...

All UFT papers read, led by 25, 177, 88, 169, 42, 177(Sp), 175, 43, 170(Sp), 107, 166(Sp), 39, 41, 46, 166, 94, 157(Sp), 140(Sp), 61, 8, 214, 243, 155, 85, 150(Sp), 142, 235, 159(Sp), 150B, 99, 214-1b, 255, 26, 2, 65, 7, 57, 10, 89, 56, 15, 116, 140(Sp), 159, 33, 110, 171, 239, 54, 64, 93, 170, 176, 203, 218, 228, 137(Sp), 161, ...

All Essays read or heard, led by Light Deflection by Gravitation, 24, 79, 35, 92, 94, 11, 41, 26, 73, 9, 43, 107, 32, ...

All articles read, led by: F3(Sp), Auto1, Auto2, CV, LCR Resonant, Potential waves, Space energy devices, Spacetime devices, Engineering Model, F1(Sp), CEFE, ...

The combined total for www.aias.us, www.atomicprecision.com, and www.upitec.org is estimated by multiplying by 1.632, estimated by a 60% increase over www.aias.us by feedback data from the other two sites, plus allowance for an estimated on day lost by timer characteristics. This gives 22,902 distinct visits, 133,452 files downloaded, 37.78 gigabytes downloaded, 93,685 page views, 2,919 documents read from 104 countries.

1 - 15 March

U. S. Universities, Institutes and Similar

Arizona State, Case Western Reserve, Georgia Tech., Iowa State, Kennesaw, Karlsruhe Tech (on edu), New Jersey Institute of Technology, Portland State*, Princeton, Penn State, Texas Christian, Maryland, Michigan, Vermont, Washington, Wisconsin, Yale, United States Department of Energy Nevada, Oak Ridge, Naval Marine Command, U. S. National Archives, Calyx Institute.

Europe and Asia Minor: Universities, Institutes and Similar.

TU Vienna, Free University of Brussels, CERN, ETH Zurich, Geneva, Bern, Mazaryk Univ Brno, Max Planck Institute for Colloid Physics, TU Berlin, Augsburg, Erlangen, Jena, Karlsruhe, Mainz, Osnabrueck, Tuebingen, Hamburg, Granada, La Laguna, Murcia, Student Village Turku, Lycee Dautet, Paris Psud, Marseilles, Poitiers, ICTP Trieste, INFN Naples, Genoa, Milan, Pisa, Amsterdam*, UMB Norway, Silesian Data Center Poland, Torun, NIPNE Romania, Timisoara Tech Romania, Royal Institute of technology (KTH) Stockholm, University of Stockholm, Uppsala, Iyte Turkey, City of Kiev, City of Lviv, Birmingham, Cambridge, Cardiff, Durham, Lancaster, Leeds, Manchester, Nottingham*, Nottingham Malaysia, Oxford, Sheffield, York, British Library*, Llyfrgell Genedlaethol Cymru (National Library of Wales).

Rest of World: Universities, Institutes and Similar.

Melbourne, Wollongong, TO Gov. Brazil, Sao Paolo, Calgarry, Toronto, China, Chile, Federico Santa Maria Tech Chile, Edu system Colombia, City Univ. Hong Kong, UNRI Indonesia, Amrita India, Indian Rail (RCIL), National Aerospace Technologies India, Tottori Japan, National Autonomous Mexico, UEM Mozambique, Otago, Quaid i Assam Pakistan, National Sun Yat-Sen Taiwan, National Taiwan.

15 - 30 March

U. S. Universities, Institutes and Similar.

Amherst*, Berkeley, Cornell, CSU Long Beach, George Mason, Iowa State, Illinois, Kansas State, Lycoming, Maryland, MIT*, Oregon State, Pittsburgh, Princeton, Stanford, Texas A and M, Alabama Birmingham, Colorado Colorado Springs, UCLA, UC Santa Barbara, UC San Diego, Iowa, Michigan*, Nevada Las Vegas, New Mexico, U Penn, Vermont, Western Illinois, William and Mary, NASA Johnson Space Center, NOAA Pacific Laboratory, Oak Ridge, US Navy SPAWAR, US National Archives*.

Europe and Asia Minor: Universities, Institutes and Similar.

TU Vienna, SRV Catalonia, Geneva, Charles Univ. Prague, Kassel, Talin Estonia, Murcia Spain, Navarra, Santiago de Compostela, Helsinki, Jyvaskyla, Turku, Turku Student Village Finland, Autonomous Madrid, Valencia, Ecole Normal Superieure Paris, Institute for Advanced Scientific Studies Bures sur Yvettes, ISAE Toulouse, Limoges, Poitiers*, Aristotle Thessaloniki Greece, SISSA Trieste, Milan, Groningen, Utrecht, Amsterdam*, Oslo, TU Wroclaw, Joint Institutes for Nuclear Research Dubna Russia, City of St. Petersburg, Sidorovshina Ukraine, Bristol,

Cambridge*, Durham, Hertfordshire, Imperial*, King's College London, Luton, Physics Oxford, Swansea, Met Office, Neath Port Talbot, Llyfrgell Genedlaethol Cymru (National Library of Wales).

Rest of World: Universities, Institutes and Similar.

Monash, Western Australia, UF Paraiso Brazil*, UF Paraiba Brazil, Unicamp, Sao Paulo, NPCRS British Columbia Canada, McGill. UQTR Canada, Chile, Antioquia Colombia, National Univ. Colombia, UNRI Indonesia, City Hong Kong*, Amrita India, Ritsumei Japan, Waseda, Korea Advanced Institute of Science and Technology (KAIST), Postech Korea, National Autonomous Univ. Mexico*, Mexican National Polytechnic, Nottingham Malaysia, UTHM Malaysia, Auckland*, Edu Pakistan, ICSI Romania, National Taiwan*, National Tsing Hua Taiwan. Edu Uruguay, Witwatersrand South Africa.

Part 3: Monthly Feedback Statistics

(May 2002 to Date)

History for aias.us

Month	Hits	Giga bytes	Visits	Page Views
May-02	859	0.02	155	592
Oct-03	13104	1.245	3098	10780
Dec-03	12577	1.589	3049	10432
Jan-04	19792	2.216	3990	15435
Feb-04	31987	3.071	6952	27895
Mar-04	19425	2.486	5052	15801
Apr-04	22179	3.246	4460	18000
May-04	23481	3.426	3749	19507
Jun-04	23447	3.054	4316	19460
Jul-04	22178	2.501	4534	18890
Aug-04	27097	3.909	5745	22867
Sep-04	36153	5.286	6248	31592
Oct-04	49821	5.247	7569	43362
Nov-04	53097	5.047	6807	38177
Dec-04	41177	4.329	7633	28136
Jan-05	67965	5.559	11823	44406
Feb-05	56280	4.298	11736	34252
Mar-05	63748	5.001	11466	37572

Oct-05	101000	6.4	12000	45300
Nov-05	114939	4.7	15435	66062
Jan-06	97488	4.9	14489	46619
Feb-06	85296	4.722	12376	41007
Mar-06	122868	5.961	25172	78403
Apr-06	105833	5.161	19883	44325
May-06	78083	3.75	11733	42429
Jun-06	79369	3.263	11314	47264
Jul-06	141767	4.923	12885	58554
Aug-06	84311	4.991	11706	46948
Sep-06	171160	6.958	17761	88413
Oct-06	289040	6.949	20297	152500
Nov-06	342042	8.078	15366	253550
Dec-06	78749	3.942	9291	31173
Jan-07	84660	4.808	8558	37955
Feb-07	76453	6.826	7972	34705
Mar-07	72118	6.947	8709	29065
Apr-07	65921	3.871	8344	29211
May-07	52678	3.058	7664	22489
Jun-07	59182	5.325	8757	27298
Jul-07	59417	8.069	8152	29872
Aug-07	52367	3.978	7249	26133
Sep-07	59396	6.192	6981	30865

Oct-07	56148	3.703	6821	25269
Nov-07	56597	6.055	7308	28006
Dec-07	52508	11.452	7255	28021
Jan-08	69803	12.882	7497	38137
Feb-08	47682	4.178	7011	22074
Mar-08	47515	2.211	8403	22035
Apr-08	55285	4.964	10281	29066
May-08	60941	4.38	11271	33685
Jun-08	61357	8.976	9636	34169
Jul-08	66872	6.433	9267	34470
Aug-08	51527	4.014	7887	26598
Sep-08	63878	6.534	7598	38709
Oct-08	60781	5.419	7772	34084
Nov-08	66895	5.258	8449	33069
Dec-08	85241	9.646	9898	54196
Jan-09	71215	6.611	8604	42008
Feb-09	59770	4.86	10149	31563
Mar-09	85479	7.330	11878	46622
Apr-09	66354	8.656	9414	36391
May-09	68655	8.102	10881	36724
Jun-09	78866	6.793	10288	46229
Jul-09	71928	6.504	10786	41111
Aug-09	75,464	5.023	11,145	43,287

Sep-09	72,364	6.336	12,725	39,828
Oct-09	84,816	9.043	13,786	50,571
Nov-09	89,459	8.792	13,510	54,837
Dec-09	75,628	8.248	10,910	43,951
Jan-10	112,688	10.060	14,465	71,331
Feb-10	75,536	5.586	11,381	46,309
Mar-10	91,428	7.370	12,631	59,165
Apr-10	84,648	8.564	11,619	53,274
May-10	74,481	7.432	12,678	44,338
Jun-10	100,336	8.664	15,896	65,194
Jul-10	90,480	6.581	11,468	55,636
Aug-10	85,902	10.069	12,632	51,281
Sep-10	74,184	9.243	12,704	42,000
Oct-10	114,756	11.233	15,706	79,592
Nov-10	112,586	6.511	16,357	75,334
Dec-10	104,401	6.766	12,524	68,648
Jan-11	119,315	7.738	15,350	80,350
Feb-11	104,860	8.948	12,596	74,230
Mar-11	105,227	7.917	14,579	69,795
Apr-11	93,005	7.826	13,688	58,354
May-11	109,303	13.530	16,301	73,150
Jun-11	95,620	8.752	13,670	58,361
Jul-11	77,626	6.799	13,313	46,198

Aug-11	84,636	13.962	13,942	50,677
Sep-11	78,226	10.026	11,858	40,721
Oct-11	90,011	7.9810	14,326	52,805
Nov-11	72,983	5.7940	13,679	40,554
Dec-11	95,927	14.1820	15,334	60,938
Jan-12	93,176	7.7150	15,501	51,581
Feb-12	85,040	5.7410	14,110	41,352
Mar-12	102,598	12.996	18,817	68,253
Apr-12	91,407	10.019	16,052	59,937
Total to date	7,620,769	613.0543	1,054,532	4,425,496

Combined total hits with atomicprecision.com 20 June 2008 to April 2009

Month	aias	atomic precision	Total Hits
Jun-08	61718	208073	269791
Jul-08	67154	185488	252642
Aug-08	51978	222637	274615
Sep-08	63878	205660	269538
Oct-08	60781	180752	241533
Nov-08	66895	109007*	175902*
Dec-08	87033	95652*	182685*

Jan-09	71781	178550	250331
Feb-09	60314	134982	195296
Mar-09	86126	160708	246834
Apr-09	63395	196865	260260

* partial because of feedback site outage.

In these 11 months there was a total of 802,292 hits for www.aias.us and 1,919,015 hits for www.atomicprecision.com, (also known as www.unifiedfieldtheory.info) making a combined total of 2,721,307 hits in 11 months, or about three million hits a year taking into account the outage in Nov and Dec. 08 caused by a computer failure.

Combined Feedback for aias.us and atomicprecision.com

Month	Visitors	Hits	Documents	Giga bytes		Countries
Jun-09	25007	144161	1627	6.79	(aias)	85
Jul-09	22309	121410	1650	6.50	(aias)	92
Aug-09	27527	137610	1673	4.76	(aias)	96
Sep-09	28265	136924	1668	9.15	(total)	93
Oct-09	29135	145990	1629	11.67	(total)	87
Nov-09	27235	149462	1651	11.71	(total)	94
Dec-09	22344	135488	1687	11.02	(total)	88
Jan-10	26230	172466	1687	13.23	(total)	86
Feb-10	21657	128523	1686	8.94	(total)	94
Mar-10	23873	144693	1687	12.52	(total)	94
Apr-10	22700	131520	1692	11.53	(total)	85

May-10	23239	119482	1712	7.43	(aias)	100
Jun-10	24436	446339	1758	8.66	(aias)	87
Jul-10	19741	134485	1892	6.4	(aias)	87
Aug-10	20520	124766	1973	10.07	(aias)	95
Sep-10	19567	111846	1911	9.24	(aias)	97
Oct-10	22242	152008	1868	11.23	(aias)	102
Nov-10	23516	151756	1919	6.51	(aias)	103
Dec-10	18946	143424	2042	6.77	(aias)	98
Jan-11	21333	162804	1947	8.874	(total)	93
Feb-11	16838	136656	1963	10.191	(total)	88
Mar-11	20433	143704	2006	13.91	(total)	96
Apr-11	18496	129191	2066	14.87	(total)	94
May-11	21153	143688	2117	16.57	(total)	98
Jun-11	18846	127251	2154	11.43	(total)	92
Jul-11	18999	109629	2036	9.37	(total)	89
Aug-11	19602	118636	2218	14.64	(total)	99
Sep-11	18424	115113	2234	10.51	(total)	97
Oct-11	20431	127580	2246	8.54	(total)	102
Nov-11	20604	113652	2200	6.27	(total)	102
Dec-11	21884	137755	2372	14.79	(total)	106
Jan-12	22592	140256	2315	14.77	(total)	96
Feb-12	20695	125456	2462	11.46	(total)	94
Mar-12	26010	146424	2503	18.83	(total)	102
Apr-12	22892	13681	2491	15.83	(total)	106
Totals	777720	5,040,792				

Average	22,220	144,023				95
Avg / Yr	266640	1,728,276				
May-12	26,586	141,691	2512	20.1	(total)	99
Jun-12	21,560	117,957	2564	13.1	(total)	101
July-12	25,315	118,422	2600 (est)	11.1	(total)	99
Aug-12	24,111	111,768	2673	11.1	(total)	98
Sept.-12	26,184	109,473	2547	14.55	(total)	105
Oct.-12	30,473	121,052	2500 (est)	11.4	(total)	102
Nov.-12	28,155	123,064	2566	16.38	(total)	98
Dec.-12	27,489	148,621	2568	19.11	(total)	102
Totals	987,593	6,024,319				
Av. 2012	22,967	140,100				98
Total 2012	275,604	1,681,200				
Jan 13	23,702	123,400	2602	12.60		107
Feb 13	22,274	115,493	2618	12.03		96
Mar 13	23,655	134,733	2611	16.21		93
Apr 13	25,277	134,240	2636	11.66		96
May 13	26,027	139,209	2691	18.81		102
June 13	24,234	117,888	2721	17.64		96
July 13	22,971	115,970	2733	13.31		105

Aug 2013	16,834	97,839	server crash		
Sept. 2013	24,120	103,663	2578	18.75	97
Oct. 2013	24,569	114,479	2752	18.72	89
Nov. 2013	25,536	109,929	2690	14.76	60
Dec. 2013	24,549	126,410	2635	27.84	91
Total 2013	283,748	1,306,969			
Av.	23,646	108,914			
Jan 14	21,366	108,179	~ 3000	19.86	95
Feb 14	20,652	94,859	2900	13.82	93
Mar 14	29,954	128,310	2815	26.39	99
Apr 14	21,233	130,667	2696	14.27	95
May 14	22,523	93,000	2148	16.85	102
June 14	19,182	111,273	2793	17.72	88
July 14	23,084	96,524	2544	14.77	85
Aug 14	20,860	100,243	2274	13.33	97
Sept 14	21,543	117,451	2786	21.12	92
Oct 14	24,989	133,264	2819	30.55	97
Nov 14	21,203	111,906	2848	28.50	96
Dec 14	18,712	128,625	862	26.15	90
Total 14	265,301	1,238,538			
Average	22,108	103,212			

Jan 15	23,051	165,807	2903	27.12	93
Feb 15	19,864	125,619	2960	20.65	95
Mar 15	22,902	133,452	2919	37.78	104

* For www.aias.us only.

Year of 2008 : 731,994 hits from 104,970 visitors, average of 60,999.5 hits per month 8,747.5 visitors a month.

Year of 2008 : Combined total of 2,792,358 hits, average of 232,696 hits per month.

Year of 2007: 747,445 hits from 93,770 visitors, average of 20 62,287 hits per month from 7,814 visitors a month.

Footnotes

Over a span of about a decade it is seen that the interest in the ECE theory is unprecedented and constant on a "high plateau". The main site www.aias.us has been archived in the National Library of Wales and British Library on www.webarchive.org.uk, science and technology section.

PART 4: A Survey of Papers Read in December 2014.

This survey is by files downloaded for the first two hundred and eighty unified field theory papers in this twenty five day time interval. Each paper is one file because each paper is a continuous document in pdf. So the number of files downloaded is also the number of times the paper has been downloaded for study. The survey is typical of the intense interest in ECE theory that has been monitored for almost twelve years in the attached Book of Scientometrics and monthly fee dback back to May 2002. The Survey shows that all the UFT papers in the English language are studied intensively and continuously in essentially all the best places in the world, universities, institutes, large corporations, government departments and so on, and by a vast private sector. This has been going on for nearly twelve years. I will send out a separate survey for the Spanish language UFT papers. The survey is in order of UFT paper numbers with the number of times downloaded from 1 - 1 5 / 1 2 / 1 4 .

1 (44), 2 (53), 3 (49), 4 (49), 5 (33), 6 (39), 7 (50), 8 (52), 9 (29), 10 (46), 11 (37), 12 (46), 13 (54), 14 (22), 15 (42), 16 (41), 17 (39), 18 (53), 19 (51), 20 (42), 21 (37), 22 (34), 23 (31), 24 (33), 25 (151), 26 (38), 27 (36), 28 (23), 29 (38), 30 (27), 31 (31), 32 (18), 33 (18), 34 (24), 35 (45), 36 (46), 37 (34), 38 (40), 39 (71), 40 (30), 41 (60), 42 (62), 43 (74), 44 (50), 45 (31), 46 (43), 47 (46), 48 (35), 49 (42), 50 (41), 51 (37), 52 (44), 53 (40), 54 (30), 55 (43), 56 (56), 57 (46), 58 (44), 59 (38), 60 (38), 61 (49), 62 (27), 63 (40), 64 (45), 65 (47), 66 (36), 67 (53), 68 (35), 69 (30), 70 (25), 71 (32), 72 (30), 73 (22), 74 (41), 75 (33), 76 (39), 77 (38), 78 (30), 79 (26), 80 (31), 81 (34), 82 (31), 83 (47), 84 (54), 85 (50), 86 (37), 87 (54), 88 (115), 89 (101), 90 (45), 91 (47), 92 (48), 93 (27), 94 (66), 95 (55), 96 (26), 97 (30), 98 (31), 99 (36), 100 (33), 101 (25), 102 (35), 103 (27), 104 (28), 105 (27), 106 (20), 107 (68), 108 (37), 109 (33), 110 (41), 111 (32), 112 (27), 113 (32), 114 (31), 114 (33), 116 (47), 117 (28), 118 (31), 119 (27),
120 (36), 121 (36), 122 (23), 123 (27), 124 (24), 125 (28), 126 (40), 127 (28), 128 (29), 129 (30), 130 (22), 131 (29), 132 (18), 133 (52), 134 (38), 135 (34), 136 (25), 137 (27), 138 (21), 139 (39), 140 (89),
141 (28), 142 (55), 143 (38), 144 (42), 145 (38), 146 (43), 147 (31), 148 (36) *1*149 (62), 150 (91), 151 (28), 152 (57), 153 (45), 154 (56), 155 (62), 156 (38), 157 (48), 158 (51), 159 (45), 160 (48), 161 (45),
162 (45), 163 (22), 164 (26), 165 (28), 166 (107), 167 (43), 168

(46), 169-(85), 170 (48), 171 (37), 172(29), 173 (22), 174 (42), 175 (89), 176 (17), 177 (237), 178 (23), 179 (21), 180 (33), 181 (27), 182 (33), 183 (22), 184 (29), 185 (22), 186 (24), 187 (11), 188 (29), 189 (29), 190 (33), 191 (39), 192 (21), 193 (24), 194 (19), 195 (12), 196 (29), 197 (36), 198 (32), 199 (37), 200 (21), 201 (67), 202 (38), 203 (28), 204 (21), 205 (24), 206 (28), 207 (25), 208 (11), 209 (18), 210 (25), 211 (21), 212 (31), 213 (42), 215 (41), 216 (36), 217 (42), 218 (18), 219 (30), 220 (22) 221 (23), 223 (33), 224 (48), 225 (27), 226 (24), 227 (28), 229 (54), 230 (34), 231 (26), 232 (19), 233 (31), 234 (45), 235 (28), 236 (20), 237 (33), 238

249 (28), 250 (25), 251 (13), 252 (17), 253 (36), 254 (.24255 (31), 256 (25), 257 (33), 258 (22), 259 (26), 260 (21), 261 (16), 262 (28), 263 (23), 264 (23), 265 (38), 266 (37), 267 (33), 268 (30), 269 (23),
270 (16), 271 (25), 272 (17), 273 (24), 274 (14), 275 (19), 276 (24), 277 (27), 278 (39), 279 (48), 280 (34).

This gives a total of 9,217 paper downloads for 25 day:,, an average of 32.92 for each of the 280 papers. Over the 31 days of December 2014 this extrapolates to 11,429 paper downloads at an average of 40.82 per paper. The other two ECE sites increase traffic by an estimated 50%, giving 17,144 paper downloads at an average of 61.23 per paper. The other two sites are w\vw.upitec.org (http:l/v>cr.N\v.upitec.org) and
W
\
'

A
\/.atomicprecision.com (http:l!w…u.-.\/.atomicprecision.com). They also contain a list of the UFT papers.

In order of how many times downloaded the papers are as follows (some papers are on two different types of pdf names and so some numberings appear twice in the following raw data. The total is correctly added in the above survey):

177, 25, 166, 243, 88, 175, 150, 169, 166, 43, 39, 107, 2 0 t 94, 155, 42, 41, 152, 56, 142, 13, 229, 18,
228, 2, 67, 8, 158, 19, 44, 7, 85, 239, 3, 4, 61, 157, 160, 170, 224, 279, 150, 65, 116, 168, 10, 12, 36, 47,
57, 84, 153, 159, 1 6 t 234, 35, 61, t 52, 58, 146, 167, 240, 46, 55, 144, 149, 174, 15, 20, 213, 217, 49,
16, 50, 110, 126, 215, 38, 53, 63, 87, 139, 17, 191, 278, 6, 89, 134, 143, 145, 156, 202, 265, 26, 29, 33,
59, 60, 162, 171, 11, 199, 21, 266, 51, 108, 83, 86, 148, 154, 197, 214, 216, 253, 27, 66, 120, 121,.99,
48, 68, 102, 135, 22, 230, 280, 37, 180, 182, 190, 222, 223,227,244, 24, 257,267, 5, 76, 100, 109, 115,

133, 198, 71, 77, 111, 113, 147,212, 233,23, 255, 31, 45, 93, 114, 118, 92, 98, 129,219, 241, 268, 40,
54, 69, 95, 97, 128, 131, 172, 184, 188, 196, 74, 9, 238b, 125, 127, 141, 1 S t 165, 203, 206, 227, 235,
249, 262, 81, 104, 117, 123, 137, 181, 235, 238, 242, 277, 30, 62, 103, 105, 112, 119, 93, 164, 231, 245,
246, 248, 259, 80, 85, 91, 96, 136, 207, 210, 250, 256, 271, 70, 95, 101, 124, 193, 205, 226, 254, 273,
276, 34, 82, 90, 178, 186, 221, 263, 264, 269, 28, 122, 238bde, 130, 163, 173, 14, 183, 185, 220, 247,
258, 72, 75, 89, 138, 179, 200, 192, 204, 211, 260, 91, 133, 149, 154, 236, 79, 88, 106, 189, 194, 232,
275, 90, 132, 209, 218, 32, 78, 176, 252, 272, 92, 261, 270, 73, 189-4, 274, 87, 251, 195, 74, 78, 187,
208, 75, 83, 84, 72, 82, 73, 76, 77, 79, 81, 150, 80, 186, 215.

bookofscientometrics.docx
(https://drmyronevans.files.\
;ordpress.com/2014/12/bookofscientometricsl.docx)

Monthly Feedback Statistics for AIAS.docx
(https://drmyronevans.files.wordpress.com/2014/12/monthly
feedback statistics for aiasl.docx)

This entry was posted on December 28, 2014 at 7:38am and is filed under asott2. You can follow any responses to this entry through the RSS 2.0 feed. Both comments and pings are currently closed.

The KubrickTheme. Create a free website to blog at

VVordPress.com.

Entries (RSS) and Comments (RSS).